Enabling Strategies and Policies toward a Sustainable Environment

Enabling Strategies and Policies toward a Sustainable Environment

Topic Editors

Abdul Majeed
Judit Oláh

MDPI

Basel • Beijing • Wuhan • Barcelona • Belgrade • Novi Sad • Cluj • Manchester

Topic Editors
Abdul Majeed
School of Insurance and Economics
University of International Business and Economics
Beijing
China

Judit Oláh
Doctoral School of Management and Business Administration
John von Neumann University
Kecskemét
Hungary

Editorial Office
MDPI AG
Grosspeteranlage 5
4052 Basel, Switzerland

This is a reprint of the Topic, published open access by the journals *Economies* (ISSN 2227-7099), *Energies* (ISSN 1996-1073), *International Journal of Environmental Research and Public Health* (ISSN 1660-4601), *Resources* (ISSN 2079-9276), *Water* (ISSN 2073-4441) and *Sustainability* (ISSN 2071-1050), freely accessible at: https://www.mdpi.com/topics/SP39P65726.

For citation purposes, cite each article independently as indicated on the article page online and as indicated below:

Lastname, Firstname, Firstname Lastname, and Firstname Lastname. Article Title. *Journal Name* **Year**, *Volume Number*, Page Range.

ISBN 978-3-7258-5477-6 (Hbk)
ISBN 978-3-7258-5478-3 (PDF)
https://doi.org/10.3390/books978-3-7258-5478-3

Contents

About the Editors

Abdul Majeed

Abdul Majeed is an Academic Researcher at the School of Insurance and Economics, University of International Business and Economics (UIBE), Beijing, China. With a Ph.D. in Finance/Economics from UIBE, his research focuses on sustainable development, energy economics, financial innovation, and the role of emerging technologies in environmental policy. His work employs advanced econometric techniques to analyze the interplay between economic globalization, natural resources, and ecological sustainability. He has published extensively in high-impact journals such as *International Review of Economics and Finance*, *Sustainable Development*, and *Energies*, and serves as an Academic Editor for *Humanities* and *Social Sciences Communications (SSCI Q1)* and *PLOS ONE (SCI Q2)*. He has received multiple awards, including the China Scholarship Council Ph.D. scholarship and research excellence awards from the UIBE. Currently, he is investigating the environmental implications of digital transformation and policy pathways for green innovation in emerging economies.

Judit Oláh

Judit Oláh is a Professor at the Faculty of Economics and Business, University of Debrecen, Hungary, and a researcher at John von Neumann University. Her expertise spans sustainable supply chains, the circular economy, and digital transformation in business and environmental governance. With a strong interdisciplinary approach, she integrates econometric modeling and stakeholder theory to address sustainability challenges. She has co-edited several Special Issues in *Sustainability* and *Energies* and collaborated on international projects examining the socioeconomic dimensions of energy transitions. Her recent work explores the nexus between industrial policy, climate resilience, and equitable technology diffusion. Recognized for her academic contributions, she actively promotes transdisciplinary research to bridge policy gaps in sustainability.

Preface

The transition toward a sustainable environment demands innovative strategies, interdisciplinary collaboration, and evidence-based policymaking. This reprint, Enabling Strategies and Policies Toward a Sustainable Environment, brings together diverse research to address pressing challenges in sustainability through four key pillars: circular economy innovation, green financial systems, decarbonization pathways, and adaptive governance models. The 21 studies featured here employ advanced methodologies, from spatial econometrics to behavioral analysis, to uncover the systemic barriers and opportunities for sustainable development. Motivated by the urgent need to bridge theory and practice, this reprint highlights context-specific solutions, technological integration, and the role of stakeholder engagement in driving equitable transitions. It is designed for researchers, policymakers, and practitioners seeking actionable insights into sustainability challenges across industries, regions, and governance levels. We extend our deepest gratitude to the contributing authors for their rigorous scholarship, the reviewers for their constructive feedback, and the MDPI team for their support in realizing this reprint. Together, these efforts aim to inspire transformative action toward a resilient and inclusive sustainable future.

Abdul Majeed and Judit Oláh
Topic Editors

energies

MDPI

Editorial

Enabling Strategies and Policies Toward a Sustainable Environment

Abdul Majeed [1,*] and Judit Oláh [2,3]

1 School of Insurance and Economics, University of International Business and Economics, Beijing 100029, China
2 Doctoral School of Management and Business Administration, John von Neumann University, 6000 Kecskemét, Hungary; olah.judit@econ.unideb.hu
3 Faculty of Economics and Business, University of Debrecen, 4032 Debrecen, Hungary
* Correspondence: abdulmajeed@uibe.edu.cn

This Topic, Enabling Strategies and Policies toward a Sustainable Environment, addresses the gaps in the literature by synthesizing pioneering research on the mechanisms driving sustainable transitions. The contributions span four thematic pillars: (1) circular economy innovations, (2) green financial architectures, (3) decarbonization pathways in energy systems, and (4) polycentric governance models. Methodologically, this Topic breaks new ground by applying computational econometrics, agent-based modeling, and participatory action research, offering granular insights into structural inertia and leverage points within socio-technical systems. By framing sustainability as a "wicked problem" [1] characterized by competing priorities and irreducible uncertainties, this collection advances a critical epistemology that rejects technocratic reductionism in favor of adaptive, justice-centered solutions.

A salient cross-cutting theme is the dialectic between globalization and hyper-localized sustainability challenges. For instance, some studies dissect how urban agglomeration economies exacerbate resource depletion in the Global South [2], while others interrogate the gender-disaggregated impacts of climate policies in agrarian communities [3]. Such analyses reveal the inadequacy of universalist frameworks, underscoring the need for spatially and culturally attuned governance. However, critical asymmetries persist. While energy transitions dominate scholarly discourse, under-researched intersections, such as the health–environment–poverty triad, remain marginalized [4]. Similarly, circular economy paradigms often falter in resource-constrained regions because of institutional voids, necessitating context-specific innovation ecosystems [5].

Future research should prioritize two topics: technological integration and contextual scalability. Emerging technologies, such as blockchain-enabled supply chain traceability and AI-driven climate modeling, promise to optimize resource flows and enhance governance transparency [6]. Concurrently, comparative studies across the Global South—examining, for example, solar microgrid adoption in sub-Saharan Africa versus Southeast Asia—could illuminate scalable models for equitable technology diffusion [7]. Such endeavors must be coupled with epistemic pluralism by integrating Indigenous knowledge systems and grassroot innovations into the mainstream sustainability discourse. By convening this Topic, we aim to galvanize a transdisciplinary praxis that bridges the theory–policy–action chasm. The existential stakes of ecological collapse demand nothing less than a radical reimagining of collaboration that transcends disciplinary boundaries, centers marginalized voices, and prioritizes intergenerational equity.

The first study (Paper 1) in this Topic examines Slovakia's municipal solid waste (MSW) management challenges, critiques its landfill dependency, and advocates for circular economy transitions. This study identifies socio-technical barriers (e.g., public awareness gaps) and stresses the importance of stakeholder engagement through behavioral nudges (e.g., pay-as-you-throw schemes). The second study (Paper 2) analyzes the industrialization–environment nexus in the South African Customs Union (SACU) using quantile regression (2007–2021), revealing an Environmental Kuznets Curve (EKC): industrial expansion initially worsens ecological degradation (e.g., air pollution, biodiversity loss) but reverses after economic thresholds. In alignment with SDGs 9 and 13, this study advocates for decoupling industrial growth from resource depletion by redefining economic complexity to integrate ecological resilience. Another study (Paper 3) investigates the COVID-19 pandemic's disruption of Brazil's poultry supply chain, a critical node in global agri-food systems. The authors advocate for structural resilience through supply chain diversification, enhanced biosecurity protocols, and adaptive production models to mitigate future shocks.

The next study (Paper 4) investigates the heterogeneous relationship between green patenting activity and financial performance across firm-size cohorts in the US and EU (2010–2022). This study contributes to the ecological modernization theory by quantifying how firm-specific capacities mediate green innovation–performance linkages and advocates granular policy frameworks to accelerate low-carbon transitions without compromising competitiveness. The following study (Paper 5) investigates Polish households' adaptive strategies to rising energy costs under economic and climate policy shifts by employing a multivariate regression analysis of survey data (2020–2023). This study underscores the necessity of coupling market-based incentives with equity-oriented governance to ensure inclusive climate resilience. The systematic review of this Topic (Paper 6) synthesizes 138 studies to evaluate conceptual and methodological advancements in energy literacy research, defined as individuals' capacity to comprehend energy systems, environmental externalities, and sustainable decision-making. This synthesis advances strategies for democratizing energy knowledge and accelerating equitable low-carbon transitions by aligning energy literacy initiatives with transition principles.

The following study (Paper 7) quantified the ecological and economic value of Wuhan's urban lake wetlands by employing multidisciplinary valuation frameworks, including contingent valuation surveys, shadow pricing, and hedonic analysis, to assess ecosystem services (ES). This study advocates for a holistic valuation framework that integrates the Total Economic Value (TEV) model into municipal land-use policies, emphasizing adaptive governance to balance urbanization with socio-ecological resilience. Another study (Paper 8) investigates the spatial agglomeration of pollution-intensive industries (PIIs) in Hebei Province, China, using spatial econometric analysis (Moran's I and geographically weighted regression) to assess industrial clustering patterns and environmental externalities. The authors advocate embedding socio-ecological equity into regional planning, ensuring that industrial redistribution aligns with the "Dual Carbon" targets. The next study (Paper 9) evaluates Thailand's transport decarbonization strategies by analyzing the efficacy of electric vehicle (EV) incentives and public transit expansion in reducing sectoral emissions. Utilizing a hybrid methodology—integrating traffic simulation (VISUM), life cycle emissions modeling, and policy gap analysis—the research projects a 28% reduction in CO_2-equivalent emissions by 2030 under full policy implementation.

The following study (Paper 10) investigates the strategic interdependencies between power producers and policymakers in biomass-coal energy transitions by employing evolutionary game theory and system dynamics simulations. This study advocates for hybrid policy frameworks that combine performance-based subsidies, real-time carbon markets, and regulatory sandboxes to accommodate sectoral heterogeneities. The next study (Paper 11) investigates the innovation dynamics in the wind energy sector through the lens of competitor-weighted centrality. This network metric quantifies firms' strategic influence based on their position within competitive ecosystems. By embedding competitive centrality into industrial policy, stakeholders can harness network dynamics to meet the global renewable energy targets of the Paris Agreement, fostering resilience in decarbonization pathways. Another study (Paper 12) explored the relationship between corporate environmental responsibility and firms' involvement in green innovation networks. The authors suggest that firms actively participate in green innovation networks to enhance their environmental performance while fostering a corporate culture that values sustainability and ethical practice.

The next study (Paper 13) investigates the effects of water resource taxation on conservation behaviors in Hubei Province, China, employing a dynamic stochastic general equilibrium (DSGE) model to evaluate the long-term policy impacts. The analysis underscores the necessity of water-pricing frameworks that internalize environmental and social externalities and align economic signals with sustainability objectives. The following study (Paper 14) examines the socio-psychological determinants of sustainable water consumption among Generation Z tourists in Quintana Roo, Mexico, employing a quantitative survey methodology to assess attitudes, subjective norms, and perceived behavioral control. This research contributes to the discourse on youth-centric sustainability strategies in high-traffic destinations, emphasizing actionable frameworks for mitigating resource depletion amid escalating tourism-driven environmental pressures. Another study (Paper 15) explored the transformative role of digital technologies in national audits and their impacts on regional sustainable development. The findings reveal that digital transformation enhances the government's capacity to monitor environmental governance and corporate green innovation, leading to better alignment with sustainable development goals.

The next study (Paper 16) examines divergent marine spatial planning (MSP) paradigms—"soft" sustainability, prioritizing adaptive governance and stakeholder collaboration, and "hard" sustainability, emphasizing regulatory stringency and conservation targets—through a cross-national comparative analysis of seven early adopting nations. The findings reveal systemic challenges, including institutional fragmentation, uneven stakeholder inclusion, and tensions between ecological preservation and maritime economic development. The following paper (Paper 17) examines the health benefits of "blue spaces" such as rivers, lakes, and wetlands, emphasizing their potential to improve mental and physical well-being. The findings show that blue spaces offer unique benefits, such as stress reduction, improved mood, and enhanced physical activity, particularly for vulnerable populations, such as older people and children. The next study (Paper 18) constructs a composite Urban Energy Transition Index (UETI) to evaluate low-carbon progress in Turin, Italy, by integrating environmental, economic, and social metrics. The results reveal marked advancements in environmental performance, notably renewable energy adoption, and efficiency gains, but lagging socioeconomic equity outcomes, underscoring disparities in transition inclusivity.

Another study (Paper 19) addressed the growing number of doctor–patient disputes in China, which are exacerbated by information asymmetry and unequal access to medical resources. This study proposes a multistakeholder governance framework involving doctors, patients, government agencies, and third-party mediators to resolve disputes more effectively. The following study (Paper 20) assesses the efficacy of China's innovative city pilot policies in driving regional innovation through a quasi-experimental difference-in-differences (DID) analysis of the 2003–2016 panel data. This research advances the theoretical and practical discourse on spatial equity in national innovation strategies by elucidating the interplay between policy design, regional heterogeneity, and innovation ecosystems. The last study (Paper 21) explored innovative methods for reducing Cd contamination in soils using thermally activated nanomaterials combined with potassium dihydrogen phosphate (KH_2PO_3). These findings contribute to the growing knowledge of sustainable soil management and offer practical solutions for mitigating the environmental and health impacts of heavy metal pollution.

This Topic, Enabling Strategies and Policies for a Sustainable Environment, has brought together a diverse collection of studies that collectively address some of the most pressing challenges in sustainability. These 21 papers span a range of topics from waste management and industrialization to green innovation and water conservation, offering novel insights and practical recommendations for policymakers, businesses, and researchers. Together, these studies highlight the interconnectedness of environmental, economic, and social systems and the complexity of transitioning to a more sustainable future for the planet. This Topic underscores the critical need for a systemic approach to achieve sustainability. For instance, the emphasis on transitioning from landfills to a circular economy (Paper 1) reflects the broader challenge of reducing waste and resource inefficiencies. Similarly, the nuanced findings on industrialization in SACU countries (Paper 2) remind us that economic growth can only be sustainable if green technologies and policies accompany it. Other studies, such as those on green patents (Paper 4) and innovation networks (Paper 12), have demonstrated the power of technological advancement and collaboration to drive sustainability.

These studies also highlight the importance of integrating environmental considerations into everyday decision-making at various levels of governance and society. The behavioral adaptations of Polish households to rising energy costs (Paper 5), health benefits of blue spaces (Paper 17), and sustainable water consumption practices among Generation Z tourists (Paper 14) illustrate how individual and collective behaviors can shape environmental outcomes. Simultaneously, the role of governance and institutions is apparent in studies on water resource taxes (Paper 13), digital audits (Paper 15), and marine spatial planning (Paper 16), which show how effective policies and frameworks can incentivize sustainable practices and improve resource management. While these studies provide valuable insights, they highlight persistent gaps and challenges in achieving sustainability. Many studies have called for stronger collaboration between stakeholders, whether in managing doctor–patient disputes (Paper 19), reducing motor vehicle emissions in Thailand (Paper 9), or implementing innovative city pilot policies in China (Paper 20). Additionally, the findings emphasize the need for context-specific solutions that consider regional disparities, as seen in the uneven effects of industrialization in SACU countries and the varied progress of urban energy transitions (Paper 18).

This Topic highlights the exciting opportunities for future research. Emerging technologies like artificial intelligence, blockchain, and advanced materials like nano-serpentine (Paper 21) have immense potential for addressing long-standing environmental challenges. However, realizing this potential requires interdisciplinary research that bridges the technical, economic, and social perspectives. Moreover, as papers such as those on Wuhan's wetlands (Paper 7) demonstrate, the valuation and preservation of natural resources must remain central to sustainability efforts, especially in urban areas, where ecological services are often undervalued. In conclusion, this Topic provides a comprehensive and multidimensional view of the strategies and policies required to achieve a sustainable environment. These findings provide a foundation for future research, emphasizing the importance of innovation, collaboration, and tailored approaches to address environmental challenges. By engaging with these studies, we hope readers will gain new perspectives and be inspired to contribute to the ongoing global efforts to build a sustainable, equitable, and resilient future.

Author Contributions: Conceptualization, A.M.; Project administration, J.O.; Supervision, J.O.; Validation, A.M.; Writing—original draft, A.M.; Writing—review and editing, J.O. All authors have read and agreed to the published version of the manuscript.

Conflicts of Interest: The authors declare no conflicts of interest.

List of Contributions:

1. Tokarčíková, E.; Ďurišová, M.; Trojáková, T. Circular Economy: Municipal Solid Waste and Landfilling Analyses in Slovakia. *Economies* **2024**, *12*, 289. https://doi.org/10.3390/economies12110289.
2. Mosikari, T. Heterogenous Effect of Industrialisation on Environmental Degradation in Southern African Customs Union (SACU) Countries: Quantile Analysis. *Economies* **2024**, *12*, 71. https://doi.org/10.3390/economies12030071.
3. Belarmino, L.C.; Pabsdorf, M.N.; Padula, A.D. Impacts of the COVID-19 Pandemic on the Production Costs and Competitiveness of the Brazilian Chicken Meat Chain. *Economies* **2023**, *11*, 238. https://doi.org/10.3390/economies11090238.
4. Semenova, A.; Semenov, K.; Storchevoy, M. One, Two, Three: How Many Green Patents Start Bringing Financial Benefits for Small, Medium and Large Firms? *Economies* **2023**, *11*, 137. https://doi.org/10.3390/economies11050137.
5. Stańczyk, E.; Szalonka, K.; Niklewicz-Pijaczyńska, M.; Nowak, W.; Stańczyk, P.; Witczyńska, K.; Ziobrowska-Sztuczka, J. Rationalization of Energy Expenditure: Household Behavior in Poland. *Energies* **2024**, *17*, 5329. https://doi.org/10.3390/en17215329.
6. Santillán, O.S.; Cedano, K.G. Energy Literacy: A Systematic Review of the Scientific Literature. *Energies* **2023**, *16*, 7235. https://doi.org/10.3390/en16217235.
7. Du, J.; Shrestha, R.P.; Nitivattananon, V.; Nguyen, T.P.L.; Razzaq, A. Unveiling the Value of Nature: A Comprehensive Analysis of the Ecosystem Services and Ecological Compensation in Wuhan City's Urban Lake Wetlands. *Water* **2023**, *15*, 2257. https://doi.org/10.3390/w15122257.
8. Zhou, J.; Tian, J.; Wang, X.; Bai, X. A Study on the Spatial Distribution and Gradient Transfer of Atmospheric Pollution Intensive Industries Such as the Thermal Power Industry in Hebei Province. *Energies* **2023**, *16*, 5114. https://doi.org/10.3390/en16135114.
9. Thanatrakolsri, P.; Sirithian, D. Assessing the Additional Benefits of Thailand's Approaches to Reduce Motor Vehicle Emissions. *Energies* **2024**, *17*, 2336. https://doi.org/10.3390/en17102336.
10. Yu, D.; Zhang, C.; Wang, S.; Zhang, L. Evolutionary Game and Simulation Analysis of Power Plant and Government Behavior Strategies in the Coupled Power Generation Industry of Agricultural and Forestry Biomass and Coal. *Energies* **2023**, *16*, 1553. https://doi.org/10.3390/en16031553.
11. Zhao, R.; Zhang, H.; Zhang, M.Y.; Qu, F.; Xu, Y. Competitor-Weighted Centrality and Small-World Clusters in Competition Networks on Firms' Innovation Ambidexterity: Evidence from the Wind Energy Industry. *Int. J. Environ. Res. Public Health* **2023**, *20*, 3339. https://doi.org/10.3390/ijerph20043339.
12. Wang, J.; Lv, W. Research on the Impact of Green Innovation Network Embeddedness on Corporate Environmental Responsibility. *Int. J. Environ. Res. Public Health* **2023**, *20*, 3433. https://doi.org/10.3390/ijerph20043433.
13. Wang, L.; Muniba, L.; Lakner, Z.; Popp, J. The Impact of Water Resources Tax Policy on Water Saving Behavior. *Water* **2023**, *15*, 916. https://doi.org/10.3390/w15050916.
14. Lima-Vargas, A.E.; Martínez-González, O.; Geronimo-Cruz, J.; Lima-Vargas, S. Sustainable Behavior of Generation Z Tourists' Water Consumption. *Sustainability* **2024**, *16*, 9651. https://doi.org/10.3390/su16229651.
15. Zhang, Y.; Zhang, Y.; Wang, Z. Digital Transformation of National Audits and Regional Sustainable Development: Quasi-Natural Experiment on the Establishment of National Audit Digital Departments. *Sustainability* **2024**, *16*, 10830. https://doi.org/10.3390/SU162410830.
16. Rezaei, F.; Contestabile, P.; Vicinanza, D.; Azzellino, A.; Weiss, C.V.C.; Juanes, J. Soft vs. Hard Sustainability Approach in Marine Spatial Planning: Challenges and Solutions. *Water* **2024**, *16*, 1382. https://doi.org/10.3390/w16101382.
17. Grace, M.J.; Dickie, J.; Bartie, P.; Brown, C.; Oliver, D.M. Understanding Health Outcomes from Exposure to Blue Space Resources: Towards a Mixed Methods Framework for Analysis. *Resources* **2023**, *12*, 135. https://doi.org/10.3390/resources12110135.

18. Desogus, E.; Bompard, E.; Grosso, D. A Composite Index for Tracking the Evolution towards Energy Transition at Urban Scale: The Turin Case Study. *Energies* **2024**, *17*, 1281. https://doi.org/10.3390/en17061281.
19. Jiang, M.M.; Wu, Z.Y.; Tu, A.X. Research on the Cooperative Governance Path of Multiple Stakeholders in Doctor–Patient Disputes under the Environment of Information Asymmetry. *Int. J. Environ. Res. Public Health* **2023**, *20*, 1597. https://doi.org/10.3390/ijerph20021597.
20. Muniba; Yu, B. Does Innovative City Pilot Policy Stimulate the Chinese Regional Innovation: An Application of DID Model. *Int. J. Environ. Res. Public Health* **2023**, *20*, 1245. https://doi.org/10.3390/ijerph20021245.
21. Wang, X.; Liu, H.; Zou, H. Effect of Potassium Dihydrogen Phosphate Combined with Thermally Activated Nano-serpentine and Thermally Activated Nano Zeolite on Cadmium in Soil. *Water* **2023**, *15*, 538. https://doi.org/10.3390/w15030538.

References

1. Rittel, H.W.J.; Webber, M.M. Dilemmas in a General Theory of Planning. *Policy Sci.* **1973**, *4*, 155–169. [CrossRef]
2. Song, Y.; Yeung, G.; Zhu, D.; Xu, Y.; Zhang, L. Efficiency of Urban Land Use in China's Resource-Based Cities, 2000–2018. *Land Use Policy* **2022**, *115*, 106009. [CrossRef]
3. Assefa, E.; Gebrehiwot, G. Gender Dimensions of Climate Change Adaptation in Tigray, Ethiopia. *Glob. Environ. Change* **2023**, *82*, 102737. [CrossRef]
4. Alda-Vidal, C.; Khalid, R.; Foulds, C.; Royston, S.; Greene, M. Gender Imaginaries in Energy Transitions: How Professionals Construct and Envision Gender Equity in Energy Access in the Global South. *World Dev.* **2023**, *168*, 106258. [CrossRef]
5. Majeed, A.; Wang, L.; Zhang, X.; Muniba; Kirikkaleli, D. Modeling the Dynamic Links among Natural Resources, Economic Globalization, Disaggregated Energy Consumption, and Environmental Quality: Fresh Evidence from GCC Economies. *Resour. Policy* **2021**, *73*, 102204. [CrossRef]
6. Anukiruthika, T.; Jayas, D.S. AI-Driven Grain Storage Solutions: Exploring Current Technologies, Applications, and Future Trends. *J. Stored Prod. Res.* **2025**, *111*, 102588. [CrossRef]
7. Ukoba, K.; Yoro, K.O.; Eterigho-Ikelegbe, O.; Ibegbulam, C.; Jen, T.C. Adaptation of Solar Energy in the Global South: Prospects, Challenges and Opportunities. *Heliyon* **2024**, *10*, e28009. [CrossRef] [PubMed]

sustainability

MDPI

Article

Digital Transformation of National Audits and Regional Sustainable Development: Quasi-Natural Experiment on the Establishment of National Audit Digital Departments

Yuhan Zhang [1], Yuankang Zhang [2] and Zichen Wang [1,*]

[1] School of Management, Ocean University of China, Qingdao 266100, China; zhangyuhan9763@stu.ouc.edu.cn
[2] College of Environment and Safety Engineering, Qingdao University of Science and Technology, Qingdao 266100, China; 03844@qust.edu.cn
* Correspondence: wangzichen0205@foxmail.com

Abstract: Promoting the realization of sustainable development goals is an important aspect of the national audit's work. In practice, the digital transformation of the national audit, which emphasizes the integrated development of digital technology and traditional auditing methods, has emerged as a significant tool for advancing regional sustainable development. Based on the quasi-natural experiment of the establishment of a national audit digital department, the impact of national audit digital transformation on the effectiveness of regional sustainable development and the mechanisms through which it operates is empirically tested in this study. The findings reveal that national audit digital transformation significantly promotes regional sustainable development. A mechanism analysis indicates that national audit digital transformation promotes regional sustainable development by improving the performance of the government's environmental governance and corporate green innovation. A heterogeneity analysis further demonstrates that the positive effects of national audit digital transformation on regional sustainable development are more pronounced in regions characterized by a high reliance on secondary industries, intense government competition, and elevated public concern for environmental issues.

Keywords: digital transformation; sustainable development; national audit; difference-in-differences approach

1. Introduction

In September 2015, the "2030 Agenda for Sustainable Development" was jointly adopted at the United Nations Sustainable Development Summit, establishing a system of Sustainable Development Goals (SDGs) comprising 17 distinct objectives [1]. The SDGs represent a binding target indicator system developed by the United Nations, aimed at comprehensively addressing social, economic, and environmental development challenges across three dimensions. Among these, SDG11 to SDG15 specifically focus on environmental issues, emphasizing the co-ordination of economic development with the intrinsic relationship between environmental protection. As the world's largest developing country, China faces both unique opportunities and significant challenges in the pursuit of these sustainable development goals. While China's economy has experienced rapid growth in recent years, the traditional economic growth model characterized by high consumption, high pollution, and high emissions has resulted in increasingly severe ecological damage and environmental degradation. Thus, effectively balancing economic development with environmental protection has emerged as a critical issue in advancing the modernization of China's governance.

SCOOP (2016) defines digitalization as "the use of digital technology and data to create revenue/transform business processes". Digital transformation is a broader term; it involves a change to the whole organization, not just a particular process or project.

Early research on digital transformation in auditing primarily concentrated on technology adoption and utilization, often neglecting to consider the broader concept of digital transformation in its entirety [2–5].

From the current stage, the digital transformation of audits refers to the process of using digital technology to modernize audits, with the ultimate vision of improving audit quality and efficiency. Audit digital transformation is not merely another technical iteration; it may also signify a fundamental shift in the organization of how public sector auditing is organized and how decisions are made, implemented, and enforced [6]. There are three main characteristics of audit digital transformation: First, it is the use of modern technology to achieve the integration of technology and audit activities. Digital technologies can conduct audits more promptly and efficiently by enhancing the processes of collecting and analyzing data in both structured and unstructured formats and replacing a large amount of manual auditing with technology [7,8]; leveraging technologies such as big data enhances the accuracy of forecasts related to estimates, going concern assessments, fraud detection, and other issues pertinent to auditors [9]. Secondly, the digital transformation of audits has produced a new audit management model. Lean, flat [10] organizational structures may be more suitable for the audit management model under digital transformation; China's Supreme Audit Institution is working hard to promote an audit management model characterized by "comprehensive analysis, identification of suspicious items, decentralized verification, and systematic research". This digitally empowered audit organization has significantly enhanced audit efficiency in practice. Finally, audit digital transformation improves the formation, storage, indexing, and application of audit knowledge. The use of digital technology can assist users by offering problem-specific decision support, thereby enhancing decision-making. This support is grounded in the knowledge and expertise accumulated from previous users and experts [11]. Digital technology facilitates the formation, storage, and indexing of audit knowledge, enabling audit organizations to swiftly retrieve relevant audit models applicable to similar situations [12].

In practice, auditors have used technologies such as SQL, big data analysis, and visualization to closely monitor the usage of relevant environmental protection funds. Additionally, they utilize technologies such as drones and geographical information systems to assess the environmental quality of key areas, investigate the remediation of environmental pollution, and evaluate the status and safety of related natural resources. For example, in 2023, the Audit Institution of Xining City used big data analysis alongside site investigations to conduct an audit of ecological and environmental protection funds. This approach enabled an objective evaluation of the performance associated with the use of these funds, resulting in favorable audit outcomes. Similarly, the Audit Institution of Xiamen City used geographical information software to perform cross-departmental and multi-year comparative analyses of vector maps and images. By leveraging emerging technologies for environmental audits, the Audit Institution has encouraged local leaders to actively fulfill their environmental responsibilities. In recent years, China's Supreme Audit Institution has developed a digital audit approach characterized by "comprehensive analysis, spotting suspicious items, decentralized verification, and systemic research". This approach, grounded in data analysis, facilitates off-site audits, enhances the integration of relevant data, and continuously improves process control and decision support, thereby enhancing work efficiency and audit quality.

Research on the economic consequences of the digital transformation of national audits has been conducted by relevant scholars, focusing on aspects such as audit quality, audit fees, and regional environmental governance [12–14]. However, a significant practical challenge remains: how to scientifically measure the level of digitization in national auditing. Existing studies have actively explored measurement methods for national audit digitization, primarily focusing on investments in digital technology, quasi-natural experiments related to China's Golden Auditing Project, and textual analysis to assess the level of audit digitization at the provincial level. Nonetheless, several challenges persist, including the endogeneity problem associated with the measurement indicators, a lack of continuity and transparency in data disclosure, and measurement methods that are restricted to provincial-level data, which complicates the refinement of research. Additionally, while digital technologies have been widely adopted in audit work related to environmental and sustainable development, empirical research assessing the effectiveness of national audit digital transformation in fostering regional sustainable development remains absent.

In light of this, this study innovatively employs the establishment of a prefecture-level national audit digital department as a quasi-natural experiment to systematically investigate the impact of national audit digital transformation on regional sustainability. The findings in this article aim to shed light on the effectiveness of audit digital transformation and offer relevant suggestions for China to further advance its sustainable development goals. More importantly, a thorough evaluation of this policy's effectiveness can also serve as a reference for other countries seeking to implement similar initiatives.

This study found that national audit digital transformation can effectively foster regional sustainable development. Mechanism tests indicated that the construction of audit digital transformation positively impacts regional sustainable development by enhancing government environmental governance and promoting corporate green innovation. Further heterogeneity analyses revealed that the effect of national audit digital transformation on regional sustainable development is more pronounced in regions characterized by high industrial dependence, intense government competition, and high public environmental concerns.

2. Literature Review

The relevant literature related to this study can be broadly divided into two main strands: one focusing on drivers of sustainable development and the other on audit digital transformation.

2.1. The Drivers of Sustainable Development

The driver of sustainable development has attracted extensive attention and made great progress in recent years; relevant scholars have conducted a large number of studies and pointed out that environmental regulation, green innovation, the digitization process, and transportation factors are critical factors affecting sustainable development.

Environmental regulation is an effective tool to improve the sustainable development of the region, and with the introduction of the Porter hypothesis [15], scholars have recognized the profound impact of environmental regulations on sustainable economic growth. The government is promoting regional sustainable development through legislation, taxation, and implementing other orders and control policies [16]. Although energy saving and emission reduction policies may be ineffective in the short term, they are likely to be effective in the long term [17]. Liao et al. pointed out that the implementation of carbon trading promotes the sustainable development of the region [18]. Additionally, Wu and Jia found that while public voluntary and market-incentive environmental regulations may inhibit sustainable growth in the short term, they ultimately contribute positively to sustainable development in the long run [19].

Some scholars pointed out that sustainable development is closely related to the digitalization process. For instance, digital finance plays a key role in promoting sustainable development and, to some extent, in bridging the sustainable growth gap within the region [20]. Zou et al. conducted empirical tests and found that the digital economy has played a significant role in promoting the sustainable development of the region by realizing green technological progress [21], but some scholars have found that there is a U-shaped relationship between the two [22]. Aleksy et al. also concluded that digitalization is conducive to environmental sustainability [23].

At the same time, some scholars have argued that green innovation is an important approach to achieving sustainable development. Green technology innovation has significantly contributed to the formation of science and technology clusters, which play an important role in mitigating environmental pollution [24,25]. Both government and enterprises promote sustainable development by increasing R&D expenditure on green technology innovations [15].

In addition, regional transportation factors represent a crucial perspective in the study of regional sustainable development. Peng and Wang [26] found that the opening of high-speed rail promotes the realization of regional sustainable development by alleviating the distortion of resource allocation. The research of Wang et al. also further verified that the high-speed rail opening has had obvious spillover effects in promoting regional sustainable development [27]. Additionally, some scholars have noted a positive association between sharing economy users and sharing economy values with sustainable development [28]. The hosting of mega sporting events can significantly enhance the sustainable development of the host city by fostering green innovation [29]. Furthermore, research indicates that increased government transparency positively influences regional sustainable development [30].

2.2. Research on the Digital Transformation of Audits

The relevant research on the consequences of the audit digitalization process mainly focuses on the impact of audit digital transformation on the audit activity itself as well as its external implications. Judging from the existing literature, research on the impact of audit digitalization on audit activity mainly focuses on audit methods, models, and the effects of audit implementation.

The digital transformation of auditing can integrate electronic working papers, optimize decision-making aids, and establish a comprehensive knowledge database [31–33], thereby enhancing the quality of financial analysis [34]. It effectively assists auditors in detecting fraud and in better understanding and quantifying audit risks [5]. Furthermore, it improves the quality of audit data analysis, enabling more accurate identification of potential risks [35]. By employing advanced technologies, auditors can significantly reduce the time spent on data collection, allowing them to concentrate more on analysis and eliminate tedious data-management tasks [36]. Some studies indicate that big data analysis profoundly impacts auditors' judgment and decision-making behavior [5] and provides support for automatic data correction [37]. Additionally, Zhang et al. noted that big data analysis offers auditors an effective means of processing data, which can more effectively address issues such as data consistency, integrity, identification, aggregation, security, and confidentiality [38].

Scholars have highlighted the impact of digitalization on the effectiveness of audit implementation, noting that the integration of technology into the audit process can reduce labor intensity [39] and enhance production efficiency [40]. Chan et al. discovered that continuous auditing methods that leverage technology and automation improve both the efficiency and effectiveness of the audit process, enabling real-time assurance of audit work [41]. Lombardi et al. suggested that utilizing digital technology for continuous auditing facilitates more frequent and real-time auditing activities, thereby significantly enhancing the overall efficiency and effectiveness of audits [42]. Krahel and Titera reached comparable conclusions [36]. Furthermore, Cunningham and Stein emphasized that digital technology can assist auditors in more effectively detecting anomalies [43]. Additionally, as digital technology is increasingly applied, auditors are relieved from mechanical and repetitive tasks, which greatly enhances the overall quality of audits [42].

Based on exploratory measurements of audit digitalization levels, relevant scholars have conducted empirical tests on the external effects of audit digital transformation. By assessing the digital transformation of audits through investments in digital technology and IT training, Zheng et al. found that such transformation can significantly enhance the quality of government audits [13]. Utilizing a quasi-natural experiment involving China's Golden Auditing Project, Guo et al. noted that the development of national audit digitalization has considerably reduced audit fees for state-owned enterprises [14]. Furthermore, Fang et al. analyzed panel data from China's provincial audit institutions, employing textual analysis to gauge the extent of digitalization, and discovered that digitalization has contributed to the enhancement of audit functions, significantly improving the level of regional environmental governance [12].

In summary, concerning the economic consequences of the digital transformation of national auditing, scholars have investigated the effects of national audit digitalization on audit quality, audit fees, and regional environmental governance, among others. However, due to limitations in the measurement methods for assessing the level of digitalization, related research remains scarce. Existing studies on factors influencing regional sustainable development have begun to acknowledge the impact of the digitalization process; nevertheless, few have examined the relationship between digital transformation and regional sustainable development from the perspective of national auditing. This article aims to systematically explore the effects of national audit digital transformation on sustainable development and its underlying mechanisms, thereby expanding the research on factors affecting sustainable development. Additionally, it seeks to enrich the understanding of the economic consequences of national audit digital transformation while innovating the measurement methods for assessing audit digitalization levels. The findings of this research provide valuable references and empirical support for leveraging digital technology in national auditing to promote sustainable development.

3. Theoretical Analysis and Hypothesis Development

The realization of local sustainable development is closely tied to government environmental governance. Resources and the environment exhibit clear external characteristics. Since the market mechanism cannot entirely address these external effects, the government plays a crucial role as a "visible hand" in the process of achieving regional sustainable development. The government fosters environmental governance by formulating relevant laws and policies, conducting environmental supervision, and thus facilitating the achievement of sustainable development. Consequently, the extent to which the government fulfills its environmental governance responsibilities is a critical factor in attaining local sustainable development. If these responsibilities are not fully met, it may negatively impact regional sustainable development. In the context of competition among local governments in China, there is a tendency to prioritize projects with lower investment scales and quicker returns in order to maximize economic performance within a limited term; however, such projects are often associated with high pollution and energy consumption, potentially undermining the region's sustainable development [44]. Conversely, if local governments allocate resources towards technological innovation and effective environmental governance and direct superior resources to the region's advantageous industries based on actual development conditions and needs, this can enhance overall city productivity. This approach can effectively introduce new driving forces for sustainable development. Moreover, strict and appropriate environmental regulations can stimulate technological innovation among market entities [45–47], contribute to improving the level of environmentally friendly technology [48], reduce pollution emissions, and increase productivity, thereby achieving a "win-win" scenario between improved environmental quality and enhanced productivity.

The digital transformation of national auditing enhances the government's ability to fulfill its environmental governance responsibilities, positively impacting regional sustainable development. Firstly, this transformation has broadened the scope of national auditing activities [42,49], allowing for a more comprehensive integration of environmental protection funds, projects, and related policies into the auditing process, even in the context of limited resources. This approach facilitates the monitoring of local governments' adherence to their environmental governance responsibilities, enabling objective evaluation and verification that motivates local officials to take the initiative in fulfilling these responsibilities. Secondly, the application of big data processing, modeling analysis, and other advanced methods effectively addresses auditors' cognitive limitations, professional barriers, and reasoning challenges [2]. These techniques assist auditors in efficiently analyzing audit data and swiftly extracting valuable information from complex datasets [49]. This capability allows for the supervision of potential issues such as waste, diversion, and arbitrage in the utilization of environmental protection funds. Consequently, it helps mitigate problems associated with waste and inefficiency in environmental funding. Moreover, by leveraging digital technology, auditors can assess the scientific basis of relevant environmental project planning and ensure compliance and effectiveness in project implementation. This comprehensive auditing and supervision ultimately enhances the effectiveness of local government environmental governance in promoting sustainable development.

Market entities represented by enterprises serve as the primary driving force behind economic development, consuming substantial resources throughout this process. Consequently, they exert a direct influence on regional sustainable development. The escalating issues of environmental pollution underscore the shortcomings and risks associated with the traditional extensive growth model that companies have relied upon. Green innovation can enhance the efficiency of energy resource utilization while simultaneously reducing production costs [50]. However, corporate green technology is characterized by long development cycles, significant investments, and uncertain outcomes [51], which can lead to a lack of clarity regarding corporate motivations for pursuing green innovation.

The digital transformation of national audits can leverage technologies such as big data analysis and visualization to enhance compliance, legality, and efficiency in the utilization of relevant environmental protection funds. This approach supports the implementation of incentive environmental policies, thereby alleviating the resource constraints faced by corporate green innovation and reducing innovation costs [52], as well as diminishing enterprises' concerns regarding innovation uncertainty [53]. Furthermore, the digital transformation of auditing has intensified environmental administrative penalties and law enforcement by local governments, compelling enterprises to adhere strictly to environmental policies under legal pressure. This has led to increased investment in and research of environmental technologies and equipment, ultimately enhancing the green innovation levels of enterprises [54]. Finally, the digital transformation of auditing can utilize relevant digital technologies to effectively identify potential environmentally friendly formalistic behaviors within enterprises. This promotes decision-making characterized by long-term transformation, drives corporate transformation and upgrading, and facilitates regional sustainable development through the achievement of green innovation.

In summary, we propose the following research hypotheses:

H1. *National audit digital transformation can promote regional sustainable development.*

H1a. *National audit digital transformation promotes regional sustainable development by improving the performance of government environmental governance.*

H1b. *National audit digital transformation promotes regional sustainable development by improving corporate green innovation.*

4. Research Design

4.1. Model Specification

For organizations, permanent specialized institutions play an important role in facilitating digital transformation [55]. The establishment of a dedicated digital department can effectively integrate organizational and business process characteristics into the application and execution of digital technology, serving as a vital mechanism for unlocking the potential of such technologies. In practice, the specialized digital audit departments in each city act as a driving force for comprehensive planning, system implementation, and personnel training in local audit digitalization. The creation of these departments represents a significant opportunity for advancing the digital transformation of regional audit and marks a pivotal milestone in the progression of digital infrastructure. To investigate this further, we manually collected data on the establishment dates of national audit digitalization departments and employed this information as a quasi-natural experiment to explore the impact and implications of national audit digital transformation on regional sustainable development. By 2020, a total of 294 administrative areas (including municipalities, prefecture-level cities, regions, autonomous prefectures, and leagues) in China had established national audit digitalization departments. Although the names of these departments vary slightly among regions, their functions remain largely consistent. The distribution of establishment dates for specialized departments across different regions is illustrated in Figure 1, with the distinct batch characteristics and gradual progression of the establishment of specialized audit digital departments.

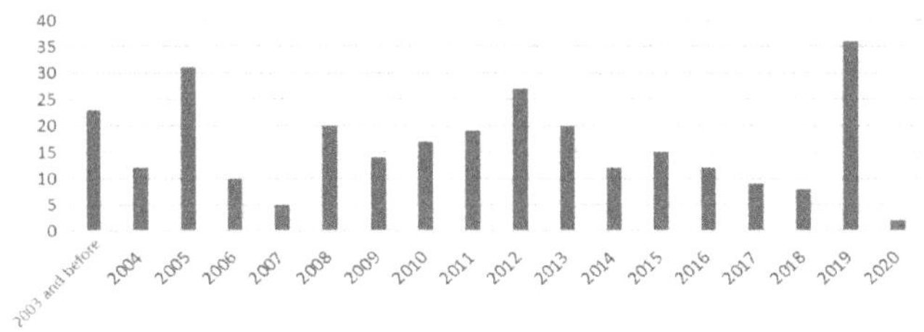

Figure 1. Time distribution for the establishment of audit digital departments.

The establishment of departments for national audit digitalization represents a quasi-natural experiment, offering a valuable opportunity to investigate the impact of national audit digital transformation on regional sustainable development. Specifically, we categorize cities that have established national audit digital departments as the experimental group, while cities that have not established such departments serve as the control group, using the difference-in-differences approach to empirically test the effects of national audit digital transformation on regional sustainable development.

To examine the relationship between the digital transformation of national auditing and the level of local sustainable development, this study designed the following model:

$$GTFP_{it} = \alpha_0 + \alpha_1 DID_{it} + \beta \sum Controls_{it} + \lambda_i + \eta_t + \varepsilon_{it} \quad (1)$$

DID_{it} represents the explanatory variable indicating whether city i has established a specialized department for audit digitization in period t. The dependent variable $GTFP_{it}$ denotes the regional sustainable development level of city i in period t. $Controls_{it}$ denotes the set of control variables, α_0 is the constant term, λ_i represents region fixed effects, η_t represents time fixed effects, and ε_{it} is the error term. The DID method can eliminate the influence of differences in system, geography, and culture in different cities and accurately assess the impact of national audit digital transformation on regional sustainable development. The analysis focuses on the coefficient α_1 of *DID*; if it is positive and statistically significant, then we deem that national audit digital transformation can improve regional sustainable development.

4.2. Variable Description

4.2.1. Explanatory Variables: *DID*

The explanatory variable, *DID*, is a dummy variable that indicates whether a city has established a national audit digitalization department. If a city has established such a department in a given year, the value is set to 1; otherwise, it is set to 0. Cities with established national audit digitalization departments serve as the experimental group, while cities without specialized departments act as the control group. The effect of national audit digital transformation on regional sustainable development is identified through a comparison between the experimental and control groups. The establishment date of the national audit digitalization department was manually gathered and compiled using the "China Audit Yearbook (2008–2021)".

4.2.2. Dependent Variable: *GTFP*

This study references the method developed by Li [56] to calculate green total factor productivity using the SBM-DDF model, thereby assessing the level of regional sustainable development. The input indicators for calculating green total factor productivity encompass labor input, capital input, and energy input. The output indicators for calculating green total factor productivity encompass desirable output indicators, and undesirable output indicators. The selection of input and output indicators is detailed in Table 1, drawing on the methodologies of Peng and Wang [26]. The relevant data were sourced from the "China Urban Statistical Yearbook" and the website of the National Bureau of Statistics of China.

Table 1. Input and output indicators.

Variable Type	Variable Name	Variable Definition
Input indicators	Labor	Number of employees in each city at the end of the year.
	Capital	The perpetual inventory method is used to measure the capital stock.
	Energy	Annual electricity consumption.
Output indicators	Desirable output indicators	Deflating GDP for each year, with 2000 as the base period.
	Undesirable output indicators	Industrial wastewater emissions, industrial carbon dioxide emissions, and industrial smoke and dust emissions measure undesirable output, and the entropy method is used to fit the above indicators into comprehensive environmental pollution indicators.

4.2.3. Control Variables

This study controls for potential factors that simultaneously affect both national audit digital transformation and sustainable development, such as the firm industrial rationalization degree (*INRA*); regional financial development level (*LNFIND*); urban education level (*EDU*); industrial upgrading degree (*INUP*); foreign direct investment (*FDI*); fiscal expenditure level (*EXPEND*); fiscal revenue level (*REVEN*); regional information development level (*IN*); economic development level (*GDP*); urban infrastructure construction level (*INFRA*). The definitions of each variable are provided in Table 2.

Table 2. Definitions of variables.

Variable Type	Variable Name	Variable Definition
Dependent variable	*GTFP*	Green total factor productivity of prefecture-level cities, calculated using the SBM-DDF method. The input and output indicators used to calculate GTFP are detailed in Table 1.
Explanatory variables	*DID*	If a city has established such a department in a given year, the value is set to 1; otherwise, it is 0. Cities with established national audit digitalization departments serve as the experimental group, while cities without departments act as the control group.
Control variables	*INRA*	The tertiary industry accounts for the proportion of GDP.
	EDU	Number of students enrolled in higher education institutions per 10,000 people in the city.
	INUP	The ratio of tertiary industry to secondary industry.
	FDI	The actual amount of foreign capital utilized by the city.
	INCOME	Local fiscal budget revenue.

Table 2. *Cont.*

EXPEND	Total government budget expenditure.
IN	Number of international internet users.
GDP	Per capita GDP.
INFRA	Highway passenger traffic.
LNFIND	Logarithm of year-end deposit balance of financial institutions.

4.3. Sample and Data

This study uses the balanced panel data of 279 prefecture-level cities in China (due to the limited availability of data on regions, autonomous prefectures, and leagues, this study focuses on prefecture-level cities as the research sample), covering the period from 2008 to 2020, to empirically analyze the impact of national audit digital transformation on regional sustainable development. The data on national audit digital transformation are sourced from the "China Audit Yearbook", while relevant data for calculating the level of urban sustainable development are obtained from the "China Urban Statistical Yearbook" and the website of the National Bureau of Statistics of China. Additionally, data on control variables, including fiscal level, industrial structure, and economic development level, are derived from the CSMAR database. To minimize the influence of extreme values on the estimation results, key continuous variables were winsorized at the 1% and 99% levels.

4.4. Descriptive Statistics

Table 3 presents the descriptive statistics of the variables. The maximum value of *GTFP* is 1.31, the minimum value is 0.754, and the standard deviation is 0.0244. Additionally, the other control variables are generally consistent with those reported in the existing literature, with no significant differences observed.

Table 3. Descriptive statistics.

	Sample Size	Mean	Standard Deviation	Minimum	Median	Maximum
GTFP	3627	1.000	0.0244	0.754	0.999	1.310
DID	3627	0.633	0.482	0	1	1
FDI	3627	42.12	76.85	0.0374	12.21	450
LNFIND	3627	16.52	1.060	14.34	16.43	19.37
INFRA	3627	0.724	0.898	0.0315	0.456	6.358
INRA	3627	39.12	9.263	18.70	38.20	66.40
INUP	3627	0.903	0.443	0.258	0.804	2.788
EDU	3627	0.0488	0.0421	0.00172	0.0361	0.196
INCOME	3627	149.6	208.9	7.039	80.01	1276
EXPEND	3627	275.8	252.0	28.59	203.6	1533
IN	3627	72.73	83.04	3.733	43	468
GDP	3627	4.378	3.025	0.698	3.535	15.94

5. Empirical Results and Analysis

5.1. Baseline Regression Results

To examine the impact of national audit digital transformation on sustainable development, this study performed a regression analysis using model (1). The results are summarized in Table 4. Column (1) presents the regression results without the inclusion of control variables and does not account for time- and region-fixed effects. Column (2) shows the baseline regression results that incorporate various control variables without time- and region-fixed effects. Column (3) is based on the findings from the preceding regression analysis, which includes control variables as well as time- and region-fixed effects.

Table 4. Baseline regression results.

	(1)	(2)	(3)
	GTFP	*GTFP*	*GTFP*
DID	0.0028 ***	0.0010 *	0.0016 ***
	(3.2905)	(1.8014)	(2.9263)
FDI		−0.0000	−0.0000
		(−1.3629)	(−1.3136)
LNFINDEP		−0.0000	0.0008
		(−0.0456)	(1.1276)
INFRA		−0.0011 **	−0.0011 **
		(−2.3835)	(−2.1048)
INRA		0.0002 ***	0.0004 ***
		(2.8769)	(4.1347)
INUP		0.0001	−0.0057 ***
		(0.1040)	(−3.8585)
EDU		−0.0071	−0.0128 *
		(−0.9233)	(−1.6843)
INCOME		0.0000	0.0000 **
		(0.4648)	(2.1313)
EXPEND		0.0000	−0.0000
		(0.5167)	(−0.8040)
IN		0.0000	0.0000
		(0.4677)	(0.2119)
GDP		0.0004 **	0.0002
		(2.1659)	(0.9006)
Year	NO	NO	YES
Region	NO	NO	YES
Constant	0.9986 ***	0.9905 ***	0.9827 ***
	(1.5×10^{-3})	(107.8774)	(96.4923)
N	3627	3627	3627
Adj. R2	0.0030	0.0957	0.3930

Note: ***, **, and * indicate significance at the 1%, 5%, and 10% levels, respectively.

As shown in column (1), when other control variables and fixed effects are not controlled, the coefficient of *DID* is 0.0028 and is significant at the 1% level. It can be seen from column (2) that after further adding control variables, the relationship between the digital transformation of national audit and the level of regional sustainable development is still significantly positive and passes the significance test at the 10% level. The results of adding control variables and further controlling time, region-fixed effects, are shown in column (3). It can be seen that the coefficient of the digital transformation of the national audit (*DID*) is 0.0016 and is significant at the 1% level.

5.2. Robustness Tests

5.2.1. Replacing the Dependent Variable

The dependent variable in this study is regional sustainable development. In order to test the robustness of the baseline regression results further, we replace the measurement method of the dependent variable. In the baseline regression, we use the SBM-DDF model to calculate the regional sustainable development level. In the robustness test part, we refer to the method of Quan [57], who used the SBM-GML (Global Malmquist–Luenberger) index model to recalculate the region's green total factor productivity, and the input indicators, expected output, and undesired output indicators are consistent with Table 1. We substitute the green total factor productivity (*GTFP1*) obtained by replacing the model with model (1) for re-regression. The results are shown in column (1) of Table 5. It can be seen that the coefficient of national audit digital transformation is 0.0037, meeting a 5% significance level; the regression results are basically consistent with the previous baseline regression.

Table 5. Robustness test results.

	(1)	(2)	(3)
	GTFP 1	*TOBIT*	*PSM*
DID	0.0037 **	0.0022 *	0.0043 **
	(2.5557)	(1.6801)	(2.1367)
FDI	−0.0000	−0.0000	0.0000
	(−1.4429)	(−1.5230)	(0.9554)
LNFINDEP	0.0006	0.0023	0.0038
	(0.3400)	(0.5338)	(0.4061)
INFRA	−0.0025 **	−0.0014 ***	−0.0008
	(−2.2294)	(−3.3466)	(−0.5219)
INRA	0.0009 ***	0.0005 **	0.0005
	(3.7050)	(2.4609)	(1.4059)
INUP	−0.0125 ***	−0.0093 ***	−0.0013
	(−3.2661)	(−2.8318)	(−0.2084)
EDU	−0.0269	−0.0381	−0.1145 **
	(−1.1802)	(−1.3960)	(−2.1533)
INCOME	0.0000 *	0.0000	0.0001 *
	(1.9340)	(1.6441)	(1.7445)
EXPEND	−0.0000	−0.0000	−0.0001 *
	(−0.6470)	(−1.0416)	(−1.7982)
IN	0.0000	−0.0000	−0.0000
	(0.6472)	(−0.7253)	(−1.5155)
GDP	0.0007	0.0006	−0.0009 *
	(1.4274)	(1.4421)	(−1.6917)
Year	YES	YES	YES
Region	YES	YES	YES
Constant	0.9739 ***	0.9602 ***	0.9375 ***
	(34.0954)	(12.7876)	(6.6213)
N	3627	3627	1578
Adj. R2	0.3093	-	0.1175

Note: ***, **, and * indicate significance at the 1%, 5%, and 10% levels, respectively.

5.2.2. Regression Analyses Using Tobit Model

In the baseline regression, this study uses the SBM-DDF model to measure the level of regional sustainable development. Given that the calculated indicators exhibit typical characteristics of censored data, parameter estimates may be biased when employing the least squares method for regression. Therefore, this study uses the Tobit model to regress model (1). The regression results are shown in column (2) of Table 5. It can be seen that the coefficient between national audit digital transformation and regional sustainable development is still significantly positive, indicating that national audit digital transformation can promote the realization of regional sustainable development; the test results are consistent with the baseline regression results.

5.2.3. Propensity Score Matching (PSM) Test

The condition for using the difference-in-differences (DID) method is that there is a common trend in each city before the establishment of a national audit digital department. However, various factors, such as economic foundations and political influences, often hinder the fulfillment of this condition, resulting in discrepancies between the experimental group and the control group. The lack of strict control over the initial conditions before policy intervention may lead to selection bias. Therefore, we use the PSM method to alleviate the endogenous problem caused by selectivity bias. Since characteristics such as urban economic development level, industrial structure, and regional informatization level will affect the degree of audit digitization in the region, we select the industrial structure, industrial advanced level, economic development level, and regional education level as covariates for matching. The nearest neighbor matching method was used to perform 1:1 matching to obtain the control group. Then, the successfully matched samples were used to examine the level of regional sustainable development before and after the establishment of the national audit digitalization departments. The regression results are presented in column (3) of Table 5, revealing that the regression coefficient for *DID* is

0.0043, which remains significantly positive at the 5% level. This indicates that the impact of national audit digital transformation on enhancing regional sustainable development levels is robust.

5.2.4. Counterfactual Test

Considering that there may be other policy events or economic and social trends that have led to pseudo-regression, we artificially advance the establishment time of the department and set a virtual time point for the establishment of the national audit digitalization department. If external factors interfere, altering the establishment date of the department should still yield a significantly positive coefficient for *DID*. Conversely, if the estimated coefficient for *DID* is no longer significantly positive, it would indicate that the observed improvement in sustainable development within the experimental group is attributable to the national audit, specifically due to the policy impact of digital transformation. Specifically, we advance the establishment time of national audit digitalization departments in prefecture-level cities by one, two, and three years and reconstruct the national audit digital transformation (*DID*) variable. The regression results are shown in columns (1) to (3) in Table 6. The coefficients of *DID* are −0.0009, 0.0013, and −0.0027, respectively, and they are no longer significant, supporting the robustness of the results.

Table 6. Counterfactual test results.

	(1)	(2)	(3)
	GTFP	*GTFP*	*GTFP*
DID	−0.0009	0.0013	−0.0027
	(−0.4874)	(0.5643)	(−1.2900)
FDI	−0.0000	−0.0000	−0.0000
	(−1.1326)	(−1.1297)	(−1.1289)
LNFINDEP	−0.0011	−0.0011	−0.0010
	(−0.2944)	(−0.2851)	(−0.2675)
INFRA	−0.0014 **	−0.0014 **	−0.0014 **
	(−2.2867)	(−2.2908)	(−2.3033)
INRA	0.0004 **	0.0004 **	0.0004 **
	(2.0536)	(2.0539)	(2.0710)
INUP	−0.0028	−0.0028	−0.0028
	(−1.0245)	(−1.0211)	(−1.0263)
EDU	−0.0203	−0.0205	−0.0207
	(−0.7068)	(−0.7098)	(−0.7196)
INCOME	0.0000	0.0000	0.0000
	(1.5970)	(1.5950)	(1.6009)
EXPEND	−0.0000	−0.0000	−0.0000
	(−1.4245)	(−1.4237)	(−1.4209)
IN	−0.0000	−0.0000	−0.0000
	(−0.8354)	(−0.8348)	(−0.8544)
GDP	0.0007 *	0.0007 *	0.0008 **
	(1.9632)	(1.9648)	(1.9793)
Year	YES	YES	YES
Region	YES	YES	YES
Constant	1.0107 ***	1.0102 ***	1.0091 ***
	(16.7939)	(16.6857)	(16.7236)
N	3627	3627	3627
Adj. R2	0.1242	0.1243	0.1245

Note: ***, **, and * indicate significance at the 1%, 5%, and 10% levels, respectively.

5.2.5. Placebo Test

Although this study involved a quasi-natural experiment, factors such as omitted variables and measurement errors may still interfere with the empirical results due to endogeneity issues. Based on this, we conducted a placebo test. Specifically, we randomly generated new national audit digital transformation variables 500 times and regressed model (1) to obtain a new estimated coefficient of national audit digital transformation. If there is no interference from omitted variables or other random factors, the distribution of this estimated coefficient will show an approximately normal distribution with a mean of 0. The kernel density distribution of the estimated coefficients of the placebo test is shown in Figure 2. It can be seen that the estimated coefficient of the digital transformation of national audit (*DID*) in the placebo test is distributed around the value of 0. It can also be seen that the positive impact of national audit digital transformation on regional sustainable development levels is not caused by other unobservable factors. It passed the placebo test, indicating that the results of this study are robust.

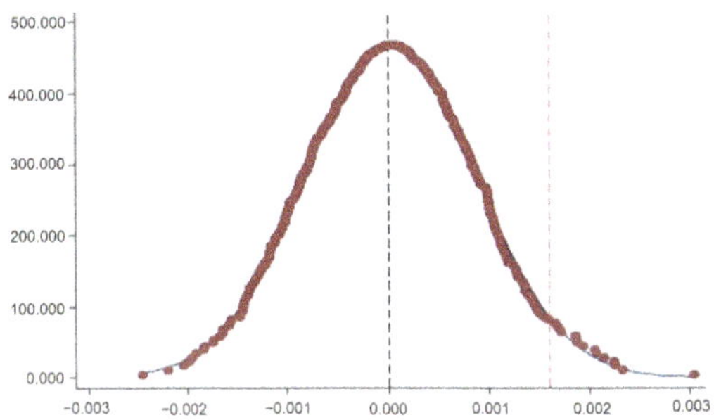

Figure 2. Placebo test results.

6. Further Analysis

6.1. Mechanism Test

The previous sections analyzed and tested the impact of national audit digital transformation on sustainable development. This section follows the theoretical framework proposed to investigate how national audit digital transformation influences regional sustainable development through two mechanisms: enhancing government environmental governance and promoting green innovation.

In order to verify the mechanism that national audit digital transformation promotes regional sustainable development by promoting the improvement of government environmental governance levels, this study uses the ratio of investment in pollution control to industrial output value in each region, which serves as a positive indicator of local government environmental governance (*EN*) [58]. The relevant data were sourced from the "China Industrial Economic Statistical Yearbook" and the "China Environmental Statistical Yearbook". On the basis of model (1), the level of government environmental governance (*EN*) and its interaction term with the digital transformation of national auditing (*DID*) is added to construct model (2).

$$GTFP_{it} = \alpha_0 + \alpha_1 DID_{it} + \alpha_2 DID_{it} \times EN_{it} + \alpha_3 EN_{it} + \beta \sum Controls_{it} + \lambda_i + \eta_t + \varepsilon_{it} \quad (2)$$

The empirical analysis results are shown in column (1) of Table 7. It can be seen that the coefficient of the interaction term between national audit digital transformation and government environmental governance level is −0.0037 and is significant at the 10% level. This finding indicates that in regions with relatively low levels of environmental governance, the digital transformation of national audits has a more pronounced effect on promoting regional sustainable development. Thus, it can be concluded that national audit digital transformation will promote local sustainable development by enhancing the environmental governance capacities of local governments.

Table 7. Mechanism test results.

	(1)	(2)
	GTFP	*GTFP*
DID	−0.0023	0.0022 *
	(−1.3496)	(1.9514)
EN	0.0041 **	
	(2.3673)	
DID×EN	−0.0037 *	
	(−1.8863)	
GINN		0.0035
		(0.1229)
DID×GINN		−0.0451 *
		(−1.9013)
FDI	−0.0000	−0.0000
	(−1.3485)	(−1.1993)
INFRA	−0.0014 ***	−0.0015 **
	(−3.0103)	(−2.1313)
INRA	0.0004 **	0.0004 **
	(2.0389)	(2.0505)
INUP	−0.0087 **	−0.0085 **
	(−2.4332)	(−2.4583)
EDU	−0.0387	−0.0366
	(−1.2933)	(−1.2769)
INCOME	0.0000	0.0000 **
	(1.4619)	(2.2820)
EXPEND	−0.0000	−0.0000
	(−0.8570)	(−0.8043)
IN	−0.0000	−0.0000
	(−0.6795)	(−0.4912)
GDP	0.0005	0.0006
	(1.2272)	(1.2225)
Year	YES	YES
Region	YES	YES
Constant	1.0055 ***	1.0022 ***
	(87.8750)	(169.9394)
N	3627	3627
Adj. R2	0.292	0.292

Note: ***, **, and * indicate significance at the 1%, 5%, and 10% levels, respectively.

At the same time, in order to verify the mechanism that the digital transformation of national audit promotes green innovation and subsequently enhances regional sustainable development, this study uses the number of regional green invention patent applications to measure the regional green innovation level. Building upon model (1), model (3) incorporates the regional green innovation level (*GINN*) along with its interaction term with national audit digital transformation (*DID*).

$$GTFP_{it} = \alpha_0 + \alpha_1 DID_{it} + \alpha_2 DID_{it} \times GINN_{it} + \alpha_3 GINN_{it} + \beta \sum Controls_{it} + \lambda_i + \eta_t + \varepsilon_{it} \quad (3)$$

The specific empirical analysis results are presented in column (2) of Table 7. The coefficient of the interaction term ($DID_{it} \times GINN_{it}$), which reflects the relationship between national audit digital transformation and regional green innovation levels, is −0.0451 and is statistically significant at the 10% level. This finding indicates that in regions with low levels of green innovation, the digital transformation of national audit exerts a more pronounced effect on promoting regional sustainable development. Thus, the digital transformation of national audits can enhance regional sustainable development by fostering improvements in regional green innovation levels.

6.2. Heterogeneity Analysis

6.2.1. Dependence on Secondary Industry

A region that is heavily reliant on secondary industry and possesses a singular industrial structure is likely to experience a high degree of resource dependence. This scenario renders the natural resource environment more susceptible to damage and poses significant challenges to the region's sustainable development. Currently, China is undergoing a critical phase of economic transformation, during which the adjustment and upgrading of industrial structures will have a profound impact on regional sustainable development. In regions with a strong reliance on secondary industry, the digital transformation of national auditing has broadened the scope of oversight and enhanced the supervision of both government and market entities. Consequently, these entities are now under increased pressure to maintain environmental legitimacy, which has encouraged them to adopt more positive environmental practices, such as cleaner production, green innovation, and industrial upgrading. These practices, in turn, contribute to the sustainable development of the entire region. Conversely, in areas with low dependence on secondary industry, where large factor inputs are not required, regional development primarily hinges on emerging industries that invest heavily in knowledge and technology. Therefore, while the digital transformation of national auditing holds significant importance for these regions, its effectiveness in promoting sustainable development is constrained.

We posit that in regions with a high degree of dependence on secondary industry, the digital transformation of national auditing can significantly enhance regional sustainable development. Conversely, in regions with a low degree of industrial dependence, this effect may be less pronounced. The degree of industrial dependence is measured using the proportion of secondary industry in GDP. The group regression results are presented in columns (1) and (2) of Table 8. The findings indicate that in areas with high industrial dependence, the coefficient for the digital transformation of national audits is 0.0038, which is significantly positive at the 5% level. In contrast, in regions with weak industrial dependence, the coefficient for *DID* is not statistically significant. The empirical results suggest that the impact of national audit digital transformation on promoting regional sustainable development is more pronounced in areas with high industrial dependence.

Table 8. Heterogeneity analysis: Industry dependence and government competition.

	(1)	(2)	(3)	(4)
	Industry-H	*Industry-L*	*Competition-H*	*Competition-L*
DID	0.0038 **	0.0002	0.0025 *	0.0032
	(1.9922)	(0.1473)	(1.7581)	(0.9269)
FDI	0.0000	0.0000	0.0000	−0.0001 ***
	(0.2645)	(0.1982)	(0.1543)	(−2.8142)
LINFINDEP	−0.0028	0.0013	0.0080	0.0006
	(−0.3131)	(0.3096)	(1.5737)	(0.0674)
INFRA	−0.0022	−0.0020 **	−0.0005	−0.0024 ***
	(−1.3973)	(−2.1214)	(−0.6467)	(−3.5930)
INRA	0.0013 **	0.0005	0.0003	0.0010
	(2.3518)	(1.6134)	(0.9718)	(1.0943)
INUP	−0.0491 **	−0.0028	−0.0072	−0.0138
	(−2.1714)	(−0.8712)	(−1.4856)	(−1.6181)
EDU	−0.0160	−0.0406	−0.0053	−0.0605
	(−0.2857)	(−0.7912)	(−0.1415)	(−0.6531)
INCOME	0.0000	0.0000	0.0000 ***	−0.0000
	(0.4620)	(0.8373)	(2.6738)	(−0.2196)
EXPEND	−0.0000	−0.0000	−0.0000	−0.0000
	(−0.5750)	(−1.5816)	(−0.2467)	(−0.8356)
IN	−0.0000	−0.0000	−0.0000 **	0.0001
	(−0.4916)	(−0.6111)	(−2.1892)	(1.2945)

Table 8. *Cont.*

	(1)	(2)	(3)	(4)
	Industry-H	*Industry-L*	*Competition-H*	*Competition-L*
GDP	0.0002	0.0016 **	−0.0001	0.0021
	(0.2270)	(2.0697)	(−0.2122)	(0.9973)
Year	YES	YES	YES	YES
Region	YES	YES	YES	YES
Constant	1.0331 ***	0.9688 ***	0.8792 ***	0.9684 ***
	(7.6125)	(14.5191)	(11.0199)	(6.0842)
N	1823	1804	1975	1652
Adj. R2	0.3470	0.3799	0.3226	0.4080

Note: ***, **, and * indicate significance at the 1%, 5%, and 10% levels, respectively.

6.2.2. Level of Government Competition

Competition among local governments can hinder the compensation for negative externalities related to resources and the environment, resulting in 'race to the bottom' behavior where local governments strive to lower environmental standards in order to attract funding and stimulate economic development. Furthermore, intense competition among local governments may lead to significant corruption and rent-seeking behaviors. The implicit relationships between governments and enterprises, characterized by such corruption and rent-seeking, can undermine the effectiveness of environmental policies [59–61]. In regions where competition among local governments is intense, the digital transformation of national audits can enhance regulatory intervention, increase the likelihood of detecting rent-seeking in audit activities, and motivate local governments and enterprises to abandon rent-seeking practices. This shift can encourage adherence to relevant environmental protection policies, ultimately contributing to sustainable development. Consequently, we posit that in areas with high levels of local government competition, the digital transformation of national audits can significantly enhance regional sustainable development. However, this effect is less pronounced in regions with lower levels of local government competition.

This study uses the method of Xu et al. [62] to assess the degree of government competition, measured by the number of cities within the province of each city. A higher index indicates greater intensity of government competition faced by the city. The results of the group regression analysis are presented in columns (3) and (4) of Table 8. The findings reveal that in cities characterized by intense local government competition, the coefficient for national audit digital transformation is 0.0025, which is significant at the 10% level. Conversely, in areas with less intense local government competition, the coefficient for *DID* is 0.0032, which does not pass the significance test. These results suggest that the impact of national audit digital transformation on promoting regional sustainable development is more pronounced in areas with higher levels of local government competition.

6.2.3. Level of Public Environmental Concern

If the level of public environmental concern in a certain region is high, it means that issues such as sustainable development have received widespread attention from the public. This also means that there is a higher demand for audits on environmental protection issues from the public. In order to meet the needs of national audits, institutional guidance, resource allocation, and work arrangements for the national audit are tilted towards the field of sustainable development. In regions where public concern for the environment is pronounced, citizens are more likely to engage in environmental governance activities and express their demands for environmental interests. Audit agencies can use digital technologies such as crawlers to monitor pressing environmental issues of public concern and swiftly capture reports and complaints regarding environmental matters. This allows for the identification of legal and regulatory violations within the environmental sector, thereby facilitating the advancement of regional sustainable development. Based on this, we believe that in areas with a high degree of public environmental participation, the digital transformation of national audits can better promote regional sustainable development, while this effect is not obvious in areas with low public environmental concerns.

This study selected the ratio of the total number of environmental petitions and letters from citizens in each region to the total regional population to measure the degree of public environmental participation. The grouping results are shown in Table 9. It can be seen that in cities with higher public environmental concerns, the national audit coefficient of digital transformation is 0.0046 and is significant at the 5% level. However, in areas with low public environmental concern, the coefficient of the national audit digital transformation variable is not significant. This shows that the role of national audit digital transformation in improving regional sustainable development is more obvious in areas with high public environmental concerns.

Table 9. Heterogeneity analysis: public environmental concern.

	(1)	(2)
	Concern-H	*Concern-L*
DID	0.0046 **	0.0016
	(1.9928)	(1.0850)
FDI	−0.0000 ***	0.0000
	(−3.2425)	(0.0251)
LINFINDEP	0.0029	0.0000
	(0.5097)	(0.0027)
INFRA	−0.0009	−0.0016 **
	(−0.7729)	(−2.2968)
INRA	0.0001	0.0007
	(0.2791)	(1.4376)
INUP	−0.0046	−0.0108 *
	(−0.8985)	(−1.7963)
EDU	0.0235	−0.0700 *
	(0.4499)	(−1.6731)
INCOME	0.0000	0.0001 **
	(1.2130)	(2.3813)
EXPEND	−0.0000	−0.0000
	(−0.5098)	(−0.5751)
IN	0.0000	−0.0000 *
	(0.5825)	(−1.6795)
GDP	0.0011 *	−0.0013
	(1.8129)	(−0.9084)
Year	YES	YES
Region	YES	YES
Constant	0.8942 ***	0.9973 ***
	(9.4340)	(11.5689)
N	1824	1803
Adj. R2	0.2547	0.3518

Note: ***, **, and * indicate significance at the 1%, 5%, and 10% levels, respectively.

7. Discussions and Conclusions

Scholars have examined the economic consequences of audit digital transformation, focusing on its impact on the audit process itself. They found that the digital transformation of auditing can enhance the quality of financial analysis and, in better understanding and quantifying audit risks [5], improves the quality of audit data analysis [35], profoundly impacts auditors' behavior [5], and allows auditors to significantly reduce the time spent on data collection, allowing them to concentrate more on analysis and eliminate tedious data-management tasks [36]. Ultimately, audit digital transformation can reduce labor intensity [39], enhance production efficiency [40], and greatly enhance the overall quality of audits [42]. From the perspective of audit practice, the evolution of the audit organization model, along with the formation, storage, and indexing of audit knowledge and models driven by the digital transformation of auditing, has significantly enhanced both audit efficiency and quality. However, due to limitations in the measurement methods of audit digital transformation, relevant empirical research on audit digital transformation remains scarce, and few studies have explored the positive role of national audit digital transformation in promoting regional sustainable development.

This study systematically explored the effects of national audit digital transformation on regional sustainable development and its underlying mechanisms based on the quasi-natural experiment established by audit digitalization departments. The conclusions of this study are as follows. First, national audit digital transformation can significantly promote regional sustainable development levels. Second, the mechanism analysis found that national audit digital transformation facilitates sustainable development through channels such as improving government environmental governance and corporate green innovation. Third, the heterogeneity analysis reveals that the positive impact of national audit digital transformation on regional sustainable development is more pronounced in regions characterized by a high reliance on secondary industries, intense government competition, and elevated public concern for environmental issues.

This article makes a two-fold contribution at the theoretical level. First, our study introduces an innovative measurement approach to audit digital transformation; it provides the impact and mechanism of audit digital transformation on sustainable development. In so doing, we gave a response to calls in the literature for an empirical analysis of national audit digital transformation. Second, by recognizing the established connection between sustainable development and digitalization processes, as highlighted by numerous scholars, we provide a perspective of audit digital transformation to observe and understand the factors influencing sustainable development.

From a practical perspective, this article provides evidence to support the evaluation of the effectiveness of national audit digitalization. China's Supreme Audit Institution has invested large amounts of human resources, financial support, and material guarantees in the process of audit digitization. However, the actual effectiveness of the governance effects of digital transformation in national auditing still needs to be evaluated with objective and conclusive empirical evidence. This article provides empirical evidence on the sustainable development of national audit digital transformation, providing strong support for a comprehensive and objective evaluation of the results of audit digital construction. At the same time, the research conclusions in this article provide the following Chinese experience on how to achieve sustainable development: First, regional national audit digital departments should leverage their critical role in audit digital transformation and accelerate the construction of national audit digitalization. This will enhance audit functioning and provide essential support for achieving sustainable development. Secondly, local governments play a pivotal role in advancing regional sustainable development. National audit departments should closely monitor the performance of local governments in fulfilling their environmental governance responsibilities, actively integrate the latest technologies, such as big data and artificial intelligence, into audit practice, and utilize digital technology to enhance supervision. This will encourage governments to proactively meet their environmental governance obligations. Finally, national audit agencies should use digital technology to create robust incentives and constraints for enterprises, motivating them to prioritize green innovation based on a long-term perspective, thereby fostering improvements in regional sustainable development.

The present exploratory study is limited in that our qualitative sample of national audits comes only from China; we had difficulties in collecting evidence from other countries to see their level of understanding and practice of digital transformation in audits. Therefore, we call for constructing an effective and comprehensive national audit digital transformation evaluation system aimed at measuring the audit digitalization levels across multiple countries. On this basis, comparative research methods can be used to study whether the particularities of each audit institution and the digital transformation method will produce different effects. Moreover, as advancements in measuring the level of audit digitization continue to evolve, future research could delve deeper into the economic ramifications of national audit informatization, extending beyond the realm of sustainable development.

Author Contributions: Conceptualization, Y.Z. (Yuhan Zhang) and Z.W.; Validation, Y.Z. (Yuankang Zhang); Formal analysis, Y.Z. (Yuankang Zhang) and Z.W.; Data curation, Y.Z. (Yuhan Zhang) and Z.W.; Writing—original draft, Y.Z. (Yuhan Zhang); Writing—review and editing, Y.Z. (Yuankang Zhang); Supervision, Y.Z. (Yuankang Zhang) and Z.W.; Project administration, Y.Z. (Yuhan Zhang). All authors have read and agreed to the published version of the manuscript.

Funding: This research received no external funding.

Institutional Review Board Statement: Not applicable.

Informed Consent Statement: Not applicable.

Data Availability Statement: The data presented in this study are available upon request from the corresponding author.

Conflicts of Interest: The authors declare no conflicts of interest.

References

1. United Nations. *Transforming Our World:the 2030 Agenda for Sustainable Development*; United Nations: New York, NY, USA, 2014; pp. 1–41.
2. Manita, R.; Elommal, N.; Baudier, P.; Hikkerova, L. The Digital Transformation of External Audit and Its Impact on Corporate Governance. *Technol. Forecast. Soc. Chang.* **2020**, *150*, 119751. [CrossRef]
3. Warren, D.; Moffitt, K.; Byrnes, P. How accounting records will change with Big Data. *Account. Horiz.* **2015**, *29*, 397–407. [CrossRef]
4. Al-Htaybat, K.; Von Alberti-Alhtaybat, L. Big Data and Corporate Reporting: Impacts and Paradoxes. *Account. Audit. Account. J.* **2017**, *30*, 850–873. [CrossRef]
5. Brown-Liburd, H.; Issa, H.; Lombardi, D. Behavioral Implications of Big Data's Impact on Audit Judgment and Decision Making and Future Research Directions. *Account. Horiz.* **2015**, *29*, 451–468. [CrossRef]
6. Schmitz, J.; Leoni, G. Accounting and Auditing at the Time of Blockchain Technology: A Research Agenda. *Aust. Acc. Rev.* **2019**, *29*, 331–342. [CrossRef]
7. Moffitt, K.C.; Rozario, A.M.; Vasarhelyi, M.A. Robotic Process Automation for Auditing. *J. Emerg. Technol. Account.* **2018**, *15*, 1–10. [CrossRef]
8. Cohen, M.; Rozario, A.; Zhang, C. Exploring the Use of Robotic Process Automation (RPA) in Substantive Audit Procedures. *CPA J.* **2019**, *89*, 49–53.
9. Littley, J. Leveraging Data Analytics and Continuous Auditing Processes for Improved Audit Planning, Effectiveness, and Efficiency. Available online: http://www.kpmg.com/US/en/IssuesAndInsights/ArticlesPublications/Documents/data-analytics-continuous-auditing.pdf (accessed on 4 November 2024).
10. Verhoef, P.C.; Broekhuizen, T.; Bart, Y.; Bhattacharya, A.; Qi Dong, J.; Fabian, N.; Haenlein, M. Digital Transformation: A Multidisciplinary Reflection and Research Agenda. *J. Bus. Res.* **2021**, *122*, 889–901. [CrossRef]
11. Arnold, V.; Collier, P.A.; Leech, S.A.; Sutton, S.G. Impact of Intelligent Decision Aids on Expert and Novice Decision-makers' Judgments. *Account. Financ.* **2004**, *44*, 1–26. [CrossRef]
12. Fang, Q.L.; Zhang, Y.H.; Gao, S.F. Information Empowerment and Regional Environmental Governance—An empirical study based on provincial audit institutions. *Audit. Res.* **2023**, *39*, 27–38. [CrossRef]
13. Zheng, W.; Zhang, L.M.; Cui, W.W.; Xing, C.Y. Information Technology and Government Audit Quality: From the Perspective of the Illegal Amount and GDP of An Area. *J. Audit Econ.* **2020**, *35*, 1–8. [CrossRef]
14. Guo, M.N.; Sun, P.; Wang, X.L.; Du, Z.Y. National Audit Informatization Construction and Audit Fees of State-owned Enterprises: Quasi-natural Experiment Based on China's Golden Auditing Project. *J. Audit Econ.* **2023**, *45*, 54–69. [CrossRef]
15. Porter, M.E.; van der Linde, C. Toward a New Conception of the Environment-Competitiveness Relationship. *J. Econ. Perspect.* **1995**, *9*, 97–118. [CrossRef]
16. Qiu, S.; Wang, Z.; Liu, S. The Policy Outcomes of Low-Carbon City Construction on Urban Green Development: Evidence from a Quasi-Natural Experiment Conducted in China. *Sustain. Cities Soc.* **2021**, *66*, 102699. [CrossRef]
17. Lin, B.Q.; Zhu, J.P. Impact of energy saving and emission reduction policy on urban sustainable development: Empirical evidence from China. *Appl. Energy* **2019**, *239*, 12–22. [CrossRef]
18. Liao, W.L.; Dong, X.K.; Weng, M.; Chen, X.Y. Economic Effect of Market-oriented Environmental Regulation: Carbon Emission Trading, Green Innovation and Green Economic Growth. *China Soft Sci.* **2020**, *35*, 159–173. [CrossRef]
19. Wu, L.; Jia, X.Y.; Wu, C.; Peng, J.C. Impact of Heterogeneous Environmental Regulation on Green Total Factors Productivity. *China Popul. Resour. Environ.* **2020**, *30*, 82–92. [CrossRef]
20. Yu, J.T.; Zhang, R.; Gong, X.Y. How Does Digital Finance Affect Green Total Factor Productivity: Characteristics of Dynamics, Identification of Mechanisms and Spatial Effects. *Mod. Econ. Sci.* **2022**, *44*, 42–56.

21. Zou, J.; Wang, Q.; Yan, H.L.; Deng, X.J. How does the Digital Economy Affect Green Total Factor Productivity? Evidence from Prefecture-level Cities in China. *Soft Sci.* **2024**, *38*, 44–52. [CrossRef]
22. Zhang, Y.H.; Wang, M.F.; Cui, L.M.; Kuang, A.P. Impact of the Digital Economy on Green Total Factor Productivity in Chinese Cities. *Econ. Geogr.* **2022**, *42*, 33–42. [CrossRef]
23. Kwiliński, A.; Lyulyov, O.; Pimonenko, T. Environmental Sustainability within Attaining Sustainable Development Goals: The Role of Digitalization and the Transport Sector. *Sustainability* **2023**, *15*, 11282. [CrossRef]
24. Mensah, C.N.; Long, X.; Dauda, L.; Boamah, K.B.; Salman, M.; Appiah-Twum, F.; Tachie, A.K. Technological Innovation and Green Growth in the Organization for Economic Cooperation and Development Economies. *J. Clean. Prod.* **2019**, *240*, 118204. [CrossRef]
25. Wang, X.; Luo, Y. Has Technological Innovation Capability Addressed Environmental Pollution from the Dual Perspective of FDI Quantity and Quality? Evidence from China. *J. Clean. Prod.* **2020**, *258*, 120941. [CrossRef]
26. Peng, X.H.; Wang, J.Y. High-Speed Rail Construction and Green Total Factor Productivity: Based on Factor Allocation Distortion. *China Popul. Resour. Environ.* **2019**, *29*, 11–19. [CrossRef]
27. Wang, K.L.; Pang, S.Q.; Zhang, F.Q. Can the Opening of High-Speed Rail Improve Urban Green Total Factor Productivity? *Ind. Econ. Res.* **2021**, *20*, 112–127. [CrossRef]
28. Zhang, Y.; Li, L.; Sadiq, M.; Chien, F.S. Impact of a sharing economy on sustainable development and energy efficiency: Evidence from the top ten Asian economies. *J. Innov. Knowl.* **2023**, *8*, 100320. [CrossRef]
29. Zhou, Z.; Lin, S.; Shi, J.; Huang, J.; Han, X. Can Mega Sporting Events Promote Urban Green Transformation? Evidence from China. *Sustainability* **2024**, *16*, 6109. [CrossRef]
30. Li, Z.; Ouyang, X.; Du, K. Does government transparency contribute to improved eco-efficiency performance? An. empirical study of 262 cities in China. *Energy Policy* **2017**, *110*, 79–89. [CrossRef]
31. Dowling, C.; Leech, S.A. A Big 4 Firm's Use of Information Technology to Control the Audit Process: How an Audit Support System Is Changing Auditor Behavior. *Contemp. Account. Res.* **2014**, *31*, 230–252. [CrossRef]
32. Lin, H.; Fan, W. Leveraging Organizational Knowledge through Electronic Knowledge Repositories in Public Accounting Firms: An Empirical Investigation. *Behav. Res. Acc.* **2011**, *23*, 147–167. [CrossRef]
33. Carson, E.; Dowling, C. The Competitive Advantage of Audit Support Systems: The Relationship between Extent of Structure and Audit Pricing. *J. Inf. Syst.* **2012**, *26*, 35–49. [CrossRef]
34. Sajady, H.; Dastgir, M.; Nejad, H.H. Evaluation of the Effectiveness of Accounting Information Systems. *Int. J. Inf. Sci. Manag.* **2008**, *6*, 49–59.
35. Cao, M.; Chychyla, R.; Stewart, T. Big Data Analytics in Financial Statement Audits. *Account. Horiz.* **2015**, *29*, 423–429. [CrossRef]
36. Krahel, J.P.; Titera, W.R. Consequences of Big Data and Formalization on Accounting and Auditing Standards. *Account. Horiz.* **2015**, *29*, 409–422. [CrossRef]
37. Kogan, A.; Alles, M.G.; Vasarhelyi, M.A.; Wu, J. Design and evaluation of a continuous data level auditing system. *Audit. J. Pract. Theory* **2014**, *33*, 221–245. [CrossRef]
38. Zhang, L.-C. Proxy Expenditure Weights for Consumer Price Index: Audit Sampling Inference for Big-Data Statistics. *J. R. Stat. Soc. Ser. A Stat. Soc.* **2021**, *184*, 571–588. [CrossRef]
39. Elliott, R.K. Assurance Services and the Audit Heritage. *CPA J.* **1998**, *68*, 40.
40. Menon, K.; Williams, D.D. Long-Term Trends in Audit Fees. *Audit. J. Pract. Theory* **2001**, *20*, 115–136. [CrossRef]
41. Chan, D.Y.; Vasarhelyi, M.A. Innovation and Practice of Continuous Auditing. In *Continuous Auditing*; Emerald Publishing Limited: Bingley, UK, 2018; pp. 271–283. ISBN 97817874 34141.
42. Lombardi, D.R.; Bloch, R.; Vasarhelyi, M.A. The Current State and Future of the Audit Profession. *Curr. Issues Audit.* **2015**, *9*, P10–P16. [CrossRef]
43. Cunningham, L.M.; Stein, S.E. Using Visualization Software in the Audit of Revenue Transactions to Identify Anomalies. *Issues Account. Educ.* **2018**, *33*, 33–46. [CrossRef]
44. Li, X.; Xu, H. Effect of Local Government Decision-making Competition on Carbon Emissions: Evidence from China's Three Urban Agglomerations. *Bus. Strat. Environ.* **2020**, *29*, 2418–2431. [CrossRef]
45. Popp, D. Induced Innovation and Energy Prices. *Am. Econ. Rev.* **2002**, *92*, 160–180. [CrossRef]
46. Popp, D. International Innovation and Diffusion of Air Pollution Control Technologies: The Effects of NOX and SO2 Regulation in the US, Japan, and Germany. *J. Environ. Econ. Manag.* **2006**, *51*, 46–71. [CrossRef]
47. Lovely, M.; Popp, D. Trade, Technology, and the Environment: Does Access to Technology Promote Environmental Regulation? *J. Environ. Econ. Manag.* **2011**, *61*, 16–35. [CrossRef]
48. Ricci, F. Channels of Transmission of Environmental Policy to Economic Growth: A Survey of the Theory. *Ecol. Econ.* **2007**, *60*, 688–699. [CrossRef]
49. Tiberius, V.; Hirth, S. Impacts of Digitization on Auditing: A Delphi Study for Germany. *J. Int. Account. Audit. Tax.* **2019**, *37*, 100288. [CrossRef]
50. Dangelico, R.M.; Pontrandolfo, P. Being 'Green and Competitive': The Impact of Environmental Actions and Collaborations on Firm Performance: Being "Green and Competitive. *Bus. Strat. Environ.* **2015**, *24*, 413–430. [CrossRef]
51. Rennings, K. Redefining Innovation—Eco-Innovation Research and the Contribution from Ecological Economics. *Ecol. Econ.* **2000**, *32*, 319–332. [CrossRef]

52. Montmartin, B.; Herrera, M. Internal and External Effects of R&D Subsidies and Fiscal Incentives: Empirical Evidence Using Spatial Dynamic Panel Models. *Res. Policy* **2015**, *44*, 1065–1079. [CrossRef]
53. Stiglitz, J.E. Leaders and Followers: Perspectives on the Nordic Model and the Economics of Innovation. *J. Public Econ.* **2015**, *127*, 3–16. [CrossRef]
54. Testa, F.; Iraldo, F.; Frey, M. The Effect of Environmental Regulation on Firms' Competitive Performance: The Case of the Building & Construction Sector in Some EU Regions. *J. Environ. Manage.* **2011**, *92*, 2136–2144. [CrossRef] [PubMed]
55. Vial, G. Understanding Digital Transformation: A Review and a Research Agenda. *J. Strat. Inf. Syst.* **2019**, *28*, 118–144. [CrossRef]
56. Li, H.; Zhang, J.; Osei, E.; Yu, M. Sustainable development of China's industrial economy: An empirical study of the period 2001–2011. *Sustainability* **2018**, *10*, 764. [CrossRef]
57. Quan, T.; Quan, T. A study of the spatial mechanism of financial agglomeration affecting green low-carbon development: Evidence from china. *Sustainability* **2023**, *15*, 965. [CrossRef]
58. Wang, Y.; Li, X. Promoting or inhibiting: The impact of government R&D subsidies on the green innovation performance of firms. *China Ind. Econ.* **2023**, *2*, 131.
59. Welsch, H. Corruption, Growth, and the Environment: A Cross-Country Analysis. *Environ. Dev. Econ.* **2004**, *9*, 663–693. [CrossRef]
60. Cole, M.A. Corruption, Income and the Environment: An Empirical Analysis. *Ecol. Econ.* **2007**, *62*, 637–647. [CrossRef]
61. Zhang, Y.; Jin, P.; Feng, D. Does Civil Environmental Protection Force the Growth of China's Industrial Green Productivity? Evidence from the Perspective of Rent-Seeking. *Ecol. Indic.* **2015**, *51*, 215–227. [CrossRef]
62. Xu, C.; Pang, Y.M.; Liu, D. Local Fiscal Pressure and Government Expenditure Efficiency: The Reform of Income Tax Sharing as a Quasi-natural Experiment. *Econ. Res. J.* **2020**, *55*, 138–154.

energies

MDPI

Article

Rationalization of Energy Expenditure: Household Behavior in Poland

Elżbieta Stańczyk [1,*], Katarzyna Szalonka [1], Małgorzata Niklewicz-Pijaczyńska [1], Wioletta Nowak [1], Piotr Stańczyk [2], Katarzyna Witczyńska [1] and Justyna Ziobrowska-Sztuczka [1,*]

[1] Faculty of Law, Administration and Economics Institute of Economic Sciences, University of Wroclaw, Uniwersytecka 22/26, 50-145 Wroclaw, Poland; katarzyna.szalonka@uwr.edu.pl (K.S.); malgorzata.niklewicz-pijaczynska@uwr.edu.pl (M.N.-P.); wioletta.nowak@uwr.edu.pl (W.N.); katarzyna.witczynska@uwr.edu.pl (K.W.)

[2] Faculty of Economics and Finance, Wroclaw University of Economics and Business, 53-345 Wroclaw, Poland; piotr.stanczyk@ue.wroc.pl

* Correspondence: elzbieta.stanczyk@uwr.edu.pl (E.S.); justyna.ziobrowska@uwr.edu.pl (J.Z.-S.); Tel.: +48-71-3752358 (E.S. & J.Z.-S.)

Abstract: Background: The implementation of the EU climate and energy policy, along with changes in the legal environment, has led to a significant increase in energy prices in Poland. Consequently, energy expenditures are now a larger part of household budgets. These rising energy costs and the evolving legal landscape are compelling households to invest in energy-saving solutions and modify their energy consumption habits. This article aims to identify the activities of households in Poland regarding the rationalization of energy expenditures. It formulates the following research hypothesis: households invest in energy-saving appliances to rationalize energy expenditures and/or change their behaviors to reduce energy consumption. Methods: The paper is based on primary research conducted using an online questionnaire survey on a sample of 331 respondents in Poland in March and April 2023. Results: A classification tree algorithm was used to identify the level of investment activities and behavioral changes made by households to reduce energy expenditures. The authors found that low-income households and people who fear further energy price increases are the first of all to change their behaviors for more energy-efficient ones. Medium- and high-income households take investment measures. They replace household appliances with more energy-efficient ones and install heat pumps and photovoltaic panels. These investments are motivated by responsible consumption, environmental protection, cleanliness, and the ease of use of the appliances.

Keywords: responsible energy consumption; energy innovation; consumer behavior

1. Introduction

In 2014, the European Commission's Communication to the European Parliament, the Council, the European Economic and Social Committee, and the Committee of the Regions outlined the fundamental framework for the future energy policy. According to its content, the EU member states committed to reducing greenhouse gas emissions in the EU by 40% by 2030 compared to the 1990 levels, increasing the share of renewable energy to at least 27% at the EU level, and reforming the existing emissions trading system through a new market stability reserve [1]. The Paris Agreement on climate change, adopted in December 2015 [2], reflected the EU's more radical approach to climate issues and energy transition, which was subsequently confirmed during the European Council summit in December 2019. It was then that the assumption was made that by 2050, the EU should achieve a net-zero balance of greenhouse gas emissions, also known as climate neutrality. Climate neutrality means a situation where the amount of greenhouse gases emitted into the atmosphere is equivalent to the amount that can be absorbed naturally by the environment: forests, water bodies, and soil. Additionally, it aims to decouple economic growth from resource consumption. The planned transformation should be cost-efficient and, from a social perspective, sustainable and fair. In July 2021, the new climate strategy was finally adopted and confirmed through the European Climate Law [3], which came into force in July 2021.

The efforts undertaken at the EU legislative level to promote the cost-effective and economically efficient reduction of greenhouse gas emissions reflected the long-standing belief in the need to improve energy efficiency and transition to sustainable and competitive forms of energy. At the same time, the initiatives aimed at promoting the cost-effective and economically efficient reduction of greenhouse gas emissions resulted in specific consequences. One of these was the increase in energy prices in EU countries, including Poland. The rise in prices in Poland was further exacerbated by the lack of energy transition, heavy dependence on coal, insufficient levels of low-emission investments amid significantly increased energy demand, rising costs of purchasing energy resources and CO_2 emission allowances, growing producer margins, and inadequate infrastructure for storing energy from renewable sources, which is supplied irregularly to the distribution grid [4–6]. The deepening energy crisis caused by the war in Ukraine, which led to the disruption of raw material supplies from Russia and the destabilization of European energy markets, also had a significant impact on the increase in energy prices [7].

However, not only in Poland but also in other EU countries, the increase in prices has been driven by a combination of factors such as rising raw material costs, the geopolitical situation, the EU climate policy, and challenges related to energy transition and infrastructure investments. Meanwhile, energy expenditures represent one of the largest expense categories in the budgets of Polish households. In 2018, the average share of energy expenditures in Polish households was 8.0%. Among the European countries, only Slovakia recorded a higher percentage of energy expenditures at 8.4%, with the average for the EU estimated at 3.9% [8]. It is worth noting that Poland belongs to the group of countries with the highest share of energy costs in total household expenditures among the EU countries [9].

The problems related to the calculation of energy expenditures pose a threat to the global energy system. Short-term periods of high energy and fuel prices create investment uncertainty in this area, while the long-term impacts affect industries and the purchasing power of households. The increase in energy expenses places many households at risk of energy poverty. Energy poverty occurs when energy costs represent too large a share of a household's overall expenses [10], making the access to energy a luxury for many [11]. Energy poverty thus means the inability to meet energy needs at home at a reasonable cost, with these needs being understood as maintaining a certain standard of heating and supplying other types of energy essential for the biological and social functioning of the household members [12]. According to cautious estimates, over 50 million households in the European Union alone experience energy poverty. Energy poverty, varying over time and space, is a multidimensional phenomenon (political, economic, health-related, infrastructural, and social) and is linked to two types of factors: technical, related to the technical conditions of the building, heating system, location of the building, and dwelling; and socioeconomic, related to the household's source of income, household structure, and age [13].

In this situation, households face the urgent need to reassess their previous decisions and take action to optimize their energy expenditures to maximize the utility of the consumed energy [14]. The issue of the rationalization of household energy behaviors has been extensively discussed in the literature. The research in this area comprises selected countries [15–17] or views the problem first of all from a supply-side perspective [18,19]. Many studies focus only on chosen aspects. For example, the impact of the COVID-19 pandemic on energy demands and consumption was addressed in [20], where the authors presented possible directions for further energy development, such as a new lifestyle with lower energy consumption, opportunities for the use of renewable energy sources, and energy storage by households. Meanwhile, the impact of COVID-19-related restrictions on the electricity consumption in Europe was discussed in a comparative article [21]. The rationalization of household energy expenditures in the new geopolitical situation, especially due to the war in Ukraine, has been the subject of numerous analyses [22–25]. Additionally, the issue of household energy demands in terms of minimizing environmental harm was discussed in [26], where an ensemble learning (EL) model for classifying total household energy expenditures was applied.

However, there is a lack of broader analyses concerning the determinants of the behaviors of Polish households. Rationalization, in this context, should be understood as more economical, conscious, and effective action aimed at achieving the highest possible utility while continuously considering the minimal expenditure of financial resources. The rationalization of expenditures can be pursued in various ways, utilizing legal regulations and efficiency measurement methods and introducing new, innovative, and increasingly efficient and cost-effective methods of energy acquisition.

There is a difference between objective and subjective rationalization of energy expenditures, which can be understood through an analysis of how people make decisions regarding energy consumption and how they justify those decisions. In the context of energy consumption and spending, subjective rationalization may involve the different behaviors and attitudes of households. For example, they may justify the use of energy-intensive devices or practices by claiming they are necessary for their comfort, even though more energy-efficient alternatives exist. Another example might be downplaying the impact of personal consumption habits on electricity bills, disregarding energy savings, or avoiding investments in more efficient technologies.

Addressing the issue of rationalizing household energy expenditures requires analysis and research, and it can constitute a further effective financial control system for consumers to prevent inefficiencies in household budgeting.

In light of the above, the aim of this article is to identify the actions taken by households in Poland regarding the rationalization of energy expenditures and the motives of their behavior. The analysis of Polish household energy behaviors from the demand side fills the gap in the literature on the subject.

In order to analyze household energy consumption, different variables are used in the literature. Some studies focus on the influence of income, household size, length of residency, education level, type of occupancy, age [27–31], or building characteristics [32] on gas and electricity consumption. Other papers include occupants' energy-related behaviors within their houses, such as heating, room usage, cooking, household appliance usage, and time spent at home [33–36]. In this paper, financial and non-financial measures are used. Moreover, several motives for energy efficiency measures are considered.

The authors formulated the following research questions: to what extent were households open to new opportunities in the form of energy-saving innovations, and to what extent were they willing to change their lifestyle? What were the main determinants for undertaking investment actions versus lifestyle changes? This article proposes the following research hypothesis: in order to rationalize energy expenditures, households invest in energy-efficient devices and/or change their behaviors to reduce energy consumption.

This paper is based on primary research conducted on a sample of 331 respondents in March and April 2023 in Poland using an online questionnaire survey. An algorithm of classification trees was applied to identify the effects of the levels of investment actions and changes in household behaviors on the reduction of energy expenditures.

2. Materials and Methods

This paper presents the insightful results of a pilot questionnaire survey conducted in Poland in March–April 2023 on a sample of 331 households. Purposive sampling was used to select the participants of the survey. The online survey was sent to persons managing household budgets via email with a link that automatically opened the survey on their device, which lasted 10–12 min. The data were collected subject to Wroclaw University of Economics and Business ethics (application no. 43/2023). Participation in the survey was voluntary. The data were kept strictly confidential and only accessible to the members of the research project team.

The subject scope of the questionnaire survey included, among others, the following crucial issues:

- The types of fuels and energy carriers used in the household;
- A self-assessment of the energy management status of the household, particularly the changes in the monthly energy costs and consumption compared to one year ago;
- The measures taken to rationalize energy expenditures;
- The main motives for taking action to rationalize energy spending;
- The unnecessary energy expenditures incurred due to wasted energy or the failure to use alternative energy-saving options.

The survey was meticulously designed, using both closed and semi-open, single- and multiple-choice questions. Guided by the subject scope of the empirical study, it was assumed that the explanatory (dependent) variables were the selected forms of energy expenditure rationalization measures taken about the individual household investment measures (financial measures) and lifestyle changes (non-financial measures).

The variables $Y_{I(i)}$; i = 1,..., 7 (with the following scale of values: 1—yes, 0—no) were selected to represent the investment measures undertaken in the households to rationalize energy expenditures: replacing old household appliances with more energy-efficient ones ($Y_{I(1)}$); replacing old-type light bulbs with new-generation LED lighting ($Y_{I(2)}$); installing thermostatic valves and heads ($Y_{I(3)}$); saving hot water by installing an aerator in the tap ($Y_{I(4)}$); taking care of the proper insulation of the house (sealing the windows and doors) ($Y_{I(5)}$); changing the energy acquisition system in favor of renewable energy sources (e.g., the installation of heat pumps, photovoltaic panels) ($Y_{I(6)}$); and others, e.g., recuperation and the use of automatic balancing of the central heating system ($Y_{I(7)}$). The coefficient value α-Cronbach = 0.529 (for the binary variables calculated identically to the so-called Kuder–Richardson KR-20 formula for scale reliability) (The Kuder–Richardson Coefficient (KR-20) is a method used to assess the internal consistency of a measure, particularly with dichotomous data. It is based on the variance of the item scores and is the dichotomous equivalent to the coefficient alpha [37].)—Table 1. In addition, the following forms of lifestyle changes were included in the questionnaire (variables $Y_{S(i)}$; i =1,..., 9 with the following value scale: 1—definitely no; 2—rather not; 3—hard to say; 4—rather yes; 5—definitely yes): turning off the TV if it is not being watched ($Y_{S(1)}$); washing only with a full load in the washing machine ($Y_{S(2)}$); lowering the temperature in an unused part of the dwelling and when ventilating ($Y_{S(3)}$); unplugging chargers from the socket if they are not being used for charging a device, e.g., a phone ($Y_{S(4)}$); reducing the temperature in the dwelling to 19–20 degrees Celsius during the day and 17–18 degrees at night ($Y_{S(5)}$); reducing the time spent in the shower or bathing ($Y_{S(6)}$); reducing the use of the oven, e.g., for baking cakes ($Y_{S(7)}$); reorganizing work to work as much as possible during the day using the daylight ($Y_{S(8)}$); and going to bed earlier ($Y_{S(9)}$). The α-Cronbach's coefficient value was 0.856, indicating high measurement reliability and consistency (Table 1).

To avoid excessive detail, when analyzing the seven types of investment activities included in the empirical study ($Y_{I(i)}$; i = 1,..., 7) and the nine forms of lifestyle changes ($Y_{S(i)}$; i = 1,..., 9), synthetic variables were constructed for each respondent:

$$Y_I^{synthetic} = \frac{\sum_{i=1}^{7} Y_{I(i)} - min}{max - min}; \ min = 0 \leq \sum_{i=1}^{7} Y_{I(i)} \leq 7 = max.$$

$$Y_S^{synthetic} = \frac{\sum_{i=1}^{9} Y_{S(i)} - min}{max - min}; \ min = 9 \leq \sum_{i=1}^{9} Y_{S(i)} \leq 45 = max.$$

The variables constructed in this way, indicating the scale of the action taken in terms of investment activities $Y_I^{synthetic}$ and in the scope of the lifestyle changes $Y_S^{synthetic}$, take values in the range [0, 1].

Table 1. Item analysis; Cronbach's alpha coefficient.

Specification	Corrected Item–Total Correlation	Cronbach's Alpha if Item Deleted
Investment activities, 7 items, Cronbach's alpha = 0.529		
$Y_{I(1)}$	0.322	0.531
$Y_{I(2)}$	0.368	0.427
$Y_{I(3)}$	0.384	0.438
$Y_{I(4)}$	0.340	0.490
$Y_{I(5)}$	0.361	0.438
$Y_{I(6)}$	0.294	0.465
$Y_{I(7)}$	0.299	0.557
Lifestyle changes, 9 items, Cronbach's alpha = 0.856		
$Y_{S(1)}$	0.627	0.835
$Y_{S(2)}$	0.649	0.833
$Y_{S(3)}$	0.602	0.838
$Y_{S(4)}$	0.521	0.847
$Y_{S(5)}$	0.547	0.844
$Y_{S(6)}$	0.564	0.842
$Y_{S(7)}$	0.616	0.837
$Y_{S(8)}$	0.575	0.841
$Y_{S(9)}$	0.511	0.847

Source: own elaboration.

The group of independent variables describing and relating to the state of the energy economy and the financial conditions of the respondents were selected from the questionnaire survey:

- The change in the household energy use costs compared to the previous year (variable $X_{S(1)}$ with the following scale of values: 1—costs decreased by more than 50%; 2—decrease between 25% and 50%; 3—decreased by 25% at most; 4—average remained the same; 5—increased by 25% at most; 6—increase between 25% and 49%; 7—increase between 50% and 74%; 8—increase between 75% and 99%; and 9—increased by 100% or more);
- The change in the average monthly level of energy consumption in the household compared to the previous year ($X_{S(2)}$ with the following scale of values: 1—definitely no increase in energy consumption; 2—rather not; 3—hard to say; 4—energy consumption has rather increased; and 5—definitely yes);
- The recognition of the issue of unnecessary energy expenditures incurred as a result of wasted energy or the lack of use of alternative energy-saving options ($X_{S(3)}$ with the following scale of values: 1—definitely not; 2—rather not; 3—hard to say; 4—rather yes; 5—definitely yes);
- $X_{S(4)}$: the average monthly electricity/other heat charges;
- $X_{S(5)}$: the average monthly net income per person in the household;
- $X_{S(6)}$: the primary sources of energy used in the household (grid electricity, district heating, solid fuels, fossil fuels, natural gas, liquefied petroleum gas, fuel oil, and energy from renewable sources).

The main motives for the energy efficiency measures included in the questionnaire survey (the independent variables) were the following $X_{M(i)}$ (i = 1,...6 with values of 1—yes and 0—no) variables: the willingness to save (variable $X_{M(1)}$); conscious, responsible consumption, including being guided by the principles of sustainability ($X_{M(2)}$); limited household budget and the fear of rising energy costs ($X_{M(3)}$); cleanliness and automation in operation ($X_{M(4)}$); information about the positive experiences of family, neighbors, or friends in using innovative energy-saving solutions ($X_{M(5)}$); and the existing legal regulations ($X_{M(6)}$).

The additional control variables characterizing the households were as follows: $X_{H(1)}$—the number of persons in the household; $X_{H(2)}$—the dwelling metric; $X_{H(3)}$—the place of residence (cities with more than 150,000 inhabitants, towns with up to 150,000 inhabitants and more than 50,000 inhabitants, cities with up to 50,000 inhabitants, and rural areas). To construct two models of the relationship between the dependent variables, describing the propensity to undertake investment activities (quantitative variable $Y_I^{synthetic}$) and the propensity to change lifestyle (quantitative variable $Y_S^{synthetic}$), and the independent variables, relating to the state of the respondents' energy economy and the main motives for energy efficiency measures, as well as the control variables characterizing the household, the regression tree method was proposed. Its use in discriminant and regression analysis was presented by Breiman et al. [38]. This method (belonging to the group of non-parametric methods for building discriminant and regression models) is used to predict the value of the explanatory variable measured on a quotient or interval scale.

It consists of a recursive division of the population into disjointed subsets (segments, classes) that are as homogeneous as possible in terms of the explanatory (dependent) variables, analyzing the relationship of each explanatory (independent) variable to the explanatory variables. The forecast error in such a procedure is most often calculated from the sum of the squares of the deviations from the mean at the terminal nodes, most often the RMSE (Root Mean Square Error), R^2 [39]. Nowadays, the method of regression trees (as well as the method of classification trees) is increasingly used in many different research fields, not only within statistics or econometrics but also in microeconomics, e.g., in the segmentation strategies of individuals and companies, to analyze financial health, to assess credit risk, and to achieve success [40,41], or about households, to identify which household characteristics play an important role in income polarization [42], to investigate the additional expenditures caused by the COVID-19 pandemic [43], and to analyze the levels of energy consumption [44] or financial exclusion [45].

Unlike many other statistical methods, regression trees are less sensitive to distributional assumptions than parametric methods like linear regression, and variables measured on weak measurement scales (ordinal or qualitative, including binary) and variables with missing data can also be included in the model. In the case of generating accurate decision trees with manageable complexity and error rates, selecting an appropriate sample size is essential [46].

In this analysis, the regression tree method allowed the determination (and presentation, using the appropriate diagrams) of the affiliation of the object-respondents to the classes of the variable measure of the willingness to invest or the measure of the desire to change lifestyle to rationalize energy spending, i.e., the dependent variables (continuous) based on the above-mentioned independent, explanatory (mainly qualitative) variables and the prediction of the value of the explained variable—the aggregated measure of the willingness to invest or the measure of the desire to change lifestyle.

The final division of the surveyed community is illustrated by the final node leaves, which contain the forecast value of the two analyzed dependent variables (arithmetic mean) and information about the number of respondents assigned to each class. To obtain a relatively simple tree indicating the main significant independent variables, it was necessary to stop the procedure of the recursive division of the community before obtaining complete homogeneity of the segments and classes (at the expense of the quality and reduction of R^2) by performing the so-called pruning of the tree. For this purpose, the direct stopping rule FACT (Fast Algorithm for Classification Trees) was used for a given fraction of objects—5% of the tested community. The calculations were conducted using Statistica software (version 13.3) purchased by the universities under the Site License academic license. The C&RT (Classification & Regression Trees) module was used for the classification tree, a complete implementation of the CART method introduced by Breiman and his colleagues from Berkeley [38].

3. Results

The pilot survey covered 331 households in Poland, with over 82% of the respondents being owner-occupiers. Among the respondents, single-person households accounted for 14% and two-person households 37%, and the remaining 49% were households with three or more persons. Approximately 60% of the dwellings were over 60 m^2, and one-third were over 100 m^2. The respondents lived mainly in cities (75%), with half in cities with more than 150,000 inhabitants. Over half of the respondents had an average monthly net income per person in the household of less than PLN 3500 (EUR 756).

On the other hand, the respondents' average monthly charges for electricity or other heat ranged from 50 PLN to 800 PLN (EUR 11-173), with an average of 507 PLN (EUR 109). Half the respondents were charged, at most, 300 PLN (65 EUR) monthly. Compared to 2022, in the subset of households where the energy use did not increase (or remained the same), the monthly energy expenditures increased for 82% of the respondents. In contrast, for 63% of the households, the monthly energy use and home heating costs increased by 25% or more. The percentage of households declaring an increase in energy use and space heating costs varied according to the primary source and energy carrier. Among the households that used natural gas, liquefied petroleum gas, fuel oil, or fossil fuels, the highest percentage declared an increase in the monthly costs associated with energy use (66% declared an increase in costs by more than one-fourth). Notably, 52% of the households using renewable energy sources also declared increased costs.

The households with renewable energy sources were distinguished by the lowest percentage of respondents who perceived a problem of unnecessary energy expenses incurred from wasting energy or not using alternative energy-saving options (5% of the respondents in the case of electricity charges). The perception of such a problem was much more common among the households using district heating or fossil fuels (26% and 24% of the respondents, respectively—Figure 1).

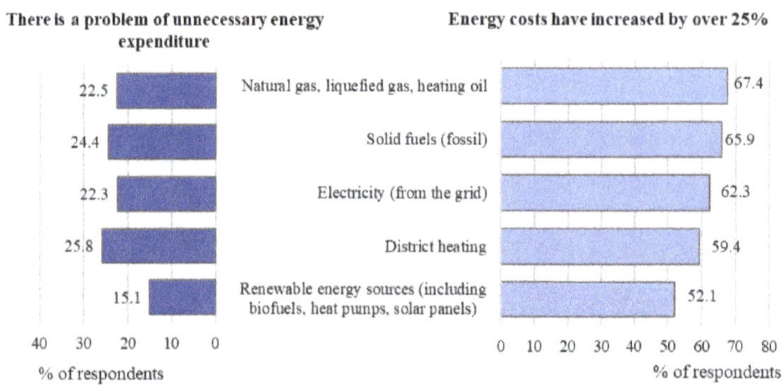

Figure 1. The percentage of households in which the monthly costs related to energy use increased by more than 25% and those in which there was a problem of unnecessary energy expenditure, in groups according to the main energy sources in the households. Source: own elaboration.

Overall, the problem of unnecessary energy expenditures was perceived by 22% of the respondents (answers "yes" or "rather yes"). More than twice as many respondents did not perceive such a problem, with 52% (answers "no" or "rather not")—Figure 2. The perception of the problem of unnecessary expenditures increased in the situation of an observable increase in energy consumption (28% of the respondents in this group, against about 19% among the households in which the energy consumption decreased) (The test statistic $\chi^2 = 15.968$ and the p-value = 0.003. It means that the chance of rejecting a correct H_0 (hypothesis about the independence of the perception of the problem of lost expenses and the increase in the costs related to energy use) was 0.3%. A significant association was found between both variables (p-value < α = 0.05)), and to a lesser extent in the situation of an increase in the costs associated with energy use (24% of the respondents in this group, against 9.5% among the households in which the energy expenditures decreased) (The test statistic χ^2 = 7.820 and the p-value = 0.098. It means that the chance of rejecting a correct H_0 (hypothesis about the independence of the perception of the problem of lost expenses and the increase in the costs related to energy use) was 9.8%).

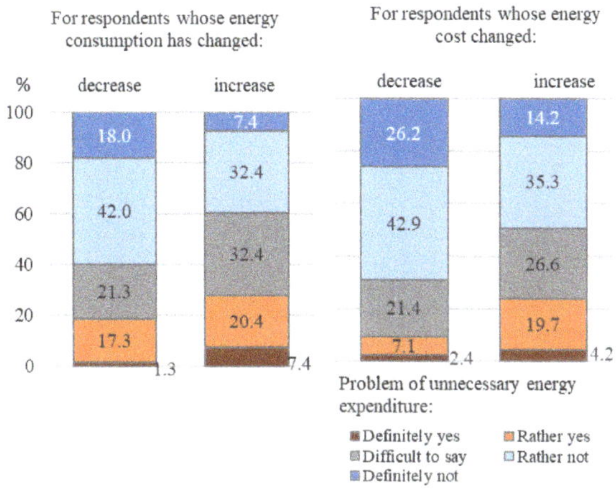

Figure 2. The households with the problem of unnecessary energy expenditure. Source: own elaboration.

According to 46% of the respondents, with more rational energy management, there was a chance to reduce energy charges by more than 5%. For 16% of the respondents, on the other hand, energy expenses could be reduced by more than 10% in this way. The opportunity to reduce the energy expenditures was noted primarily by those who perceived energy wastage in their households or realized that more energy-efficient solutions could be used. In such a subset, 63% of the respondents still had an opportunity to save more than 5% of the current energy charges (compared to 40% of the respondents who did not perceive a wastage problem) (The test statistic χ^2 = 20.984 and the p-value = 0.0003. It means that the chance of rejecting a correct H_0 (hypothesis about the independence of perceiving the problem of lost expenses and the possibility of saving on energy costs) was 0.03%. A significant association was found between both variables (p-value < α = 0.05)).

The most remarkable propensity to reduce wasteful energy consumption was declared by those whose average monthly household income per person was less than PLN 3500.

In 2023, Polish households' most used energy source (primary or supplementary) was electricity (from the grid), particularly for cooking meals. The growing popularity of induction hobs, microwave ovens, and electric ovens may have influenced the importance of the electricity used for cooking meals in Polish households. Natural gas was the households' second most crucial energy source, followed by district heating (the thermal energy supplied by heating companies for space heating and water heating purposes). For the renewable energy households, the energy obtained from photovoltaic installations and the energy from biofuels were most important. Energy obtained from ambient heat was of minor importance. Hard coal was the most important and commonly used solid fuel by households in Poland (Table 2).

In order to rationalize their energy expenditures, Polish households undertook various investment measures. The percentage of such respondents ranged from 10.9% in the category "other" (e.g., recuperation, the use of automatic hydraulic balancing of the central heating system) ($Y_{I(7)}$) to about 55% in the case of replacing old-type light bulbs with new-generation LED lighting ($Y_{I(2)}$). Switching to renewable energy sources, e.g., the installation of heat pumps and photovoltaic panels ($Y_{I(6)}$), was declared by 35% of the respondents (Table 3). A test χ^2 was conducted for each pair of variables; the financial measures for energy efficiency ($Y_{I(i)}$; i = 1,..., 7) and annual changes in energy-related costs, $X_{S(1)}$, showed no significant relationship at a = 0.05 level.

Table 2. Fuels and energy carriers in surveyed households.

Specification	Total	Including in Order to		
		Space Heating	Water Heating	Cooking Meals
	in % of Respondents			
Electricity (from the grid)	74.6	30.8	30.2	60.4
District heating	38.7	36.3	31.4	-
Hard coal	10.9	10.9	6.0	-
Lignite	2.7	2.1	2.4	0.3
Natural gas, liquefied gas	68.0	38.4	36.6	53.8
Heating oil	1.5	0.9	1.2	0.3
Biofuels (e.g., firewood, biogas, bioliquids)	11.2	10.3	5.7	0.9
Water energy	0.3	0.0	0.3	-
Wind energy	0.3	0.3	0.0	-
Solar energy utilization (photovoltaics)	13.6	8.8	10.9	8.8
Use of ambient heat (heat pumps)	3.6	3.6	0.0	0.9

Source: own elaboration.

Table 3. Investment measures taken by Polish households to rationalize their energy expenditures.

Specification $Y_{I(i)}$ i = 1,..., 9	Investment Activities Undertaken *		
	in the Total Number of Holdings Surveyed	on Holdings with Energy-Related Costs $X_{S(1)}$	
		Increased by More Than 25%	Not Increased by More Than 25%
	in %		
$Y_{I(1)}$ replacement of old household appliances, appropriate use of household appliances	39.0	43.3	31.7
$Y_{I(2)}$ replacement of old-style incandescent bulbs with new-generation LED lighting	54.7	54.3	55.3
$Y_{I(3)}$ installation of thermostatic valves and heads	15.4	16.3	13.8
$Y_{I(4)}$ hot water savings by fitting aerator in tap	31.7	33.2	29.3
$Y_{I(5)}$ ensuring proper insulation of the house (sealing of windows and doors)	31.4	28.8	35.8
$Y_{I(6)}$ change of energy generation system (installation of heat pumps or photovoltaic panels)	34.7	34.1	35.8
$Y_{I(7)}$ other (e.g., recuperation, use of automatic hydronic balancing of central heating system)	10.9	9.1	13.8

* $Y_{I(i)}$ = 1 (yes). Source: own elaboration.

Among the different forms of lifestyle changes included in the questionnaire survey that could have had an impact on the energy management of the household, the following were mentioned most frequently by the respondents: switching off the TV if it is not being watched ($Y_{S(1)}$); washing only with a full load in the washing machine ($Y_{S(2)}$); and lowering the temperature in the unused part of the dwelling and when ventilating ($Y_{S(3)}$) (Table 4). The willingness to undertake such household activities depended significantly on the increase in the monthly costs associated with energy use (this conjecture is confirmed by, among others, the chi-square test) (for the pair of variables $Y_{S(1)}$ and $X_{S(1)}$, the test $\chi^2 = 59.1$ and the *p*-value = 0.002; for the pair of variables $Y_{S(2)}$ and $X_{S(1)}$, the test $\chi^2 = 46.6$ and the *p*-value = 0.044; and for the pair of variables $Y_{S(3)}$ and $X_{S(1)}$, the test $\chi^2 = 48.6$ and the *p*-value = 0.030. A significant association was found between both variables (*p*-value < $\alpha = 0.05$).

Table 4. Lifestyle change affecting household energy economy in Poland by selected forms.

Specification i = 1,..., 9	A Decision Has Been Made to Change Lifestyle *		
	in the Total Number of Holdings Surveyed	on Holdings with Energy-Related Costs $X_{S(1)}$	
		Increased by More Than 25%	Not Increased by More Than 25%
	in %		
$Y_{S(1)}$ switching off the TV if it is not being watched	66.8	75.5	52.0
$Y_{S(2)}$ washing only with a full load in the washing machine	66.5	73.6	54.5
$Y_{S(3)}$ lowering the temperature in the unused part of the dwelling and during ventilation	57.1	66.8	40.7
$Y_{S(4)}$ disconnecting chargers from the socket when not in use for charging (e.g., phone)	56.5	60.1	50.4
$Y_{S(5)}$ lowering the temperature in the flat to 19-20 degrees Celsius during the day and 17–18 degrees at night	44.7	54.8	27.6
$Y_{S(6)}$ the reduction in showering and bathing times	41.4	43.8	37.4
$Y_{S(7)}$ reducing the use of the oven (e.g., baking cakes)	31.7	40.4	17.1
$Y_{S(8)}$ the reorganization of work to work as much as possible during the day using the daylight	29.3	32.7	23.6
$Y_{S(9)}$ going to bed earlier	17.5	19.2	14.6

* Responses "yes" or "rather yes," i.e., $Y_{S(i)} = 4$ or 5, i = 1,...9. Source: own elaboration.

The primary statistical parameters of the two aggregate measures testifying to the scale of the energy efficiency measures, with the first synthesizing the information on the investment measures included in the empirical study and the second synthesizing the individual forms of the lifestyle changes, are presented in Table 5. According to the adopted design, both synthetic measures take values in the interval [0, 1], with the synthetic measure of the lifestyle changes equal to 1 if "definitely yes" responses were received for all component forms of lifestyle changes (maximum values). At the same time, it is equal to 0 if "definitely no" responses were received for all component forms of lifestyle changes (minimum values). Based on the empirical survey data, the calculated synthetic measures indicate that, in the context of energy expenditure rationalization, the propensity to change lifestyles was more significant than the propensity to take investment more significant than the arithmetic mean of the variable $\mathbf{Y}_S^{synthetic}$ and greater than the mean of the synthetic variable $\mathbf{Y}_I^{synthetic}$.

Table 5. Basic statistical parameters of synthetic measure of investment activities for energy efficiency and synthetic measure of lifestyle changes.

Specification	Synthetic Measure of Investment Activities $Y_I^{synthetic}$	Synthetic Measure of Lifestyle Changes $Y_S^{synthetic}$
Minimum value (MIN)	0.000	0.000
Maximum value (MAX)	0.889	1.000
Median (ME)	0.333	0.528
First quartile (Q_1)	0.111	0.333
Third quartile (Q_3)	0.444	0.667
Average $\bar{Y}$	0.302	0.496
Standard deviation (S_Y)	0.205	0.230

Source: own elaboration.

The primary motivation for energy efficiency measures most often indicated by the respondents was the desire to save ($X_{M(1)}$), followed by conscious, responsible consumption ($XM_{(2)}$) and limited finances and fears of rising energy costs ($X_{M(3)}$) (Table 6). Those who indicated these motivations were the most likely to make lifestyle changes (the highest values of $Y_S^{synthetic}$). In contrast, the propensity to undertake investment measures to rationalize energy expenditures was highest among those for whom cleanliness and automation in operation ($X_{M(4)}$) and information about the positive experiences of family, neighbors, and friends in the use of innovative energy-saving solutions ($X_{M(5)}$), as well as conscious, responsible consumption ($X_{M(2)}$), were critical motivational criteria.

Table 6. The main motives for the decision to rationalize energy expenditures in the surveyed households.

Main Motivations for Action $X_{M(i)}$ i = 1,..., 6 N = 331	% of Total Respondents	$Y_I^{synthetic}$ Synthetic Measure of Investment Measures	$Y_S^{synthetic}$ Synthetic Measure of Lifestyle Changes
		for $X_{M(i)} = 1$	
$X_{M(1)}$ willingness to save	52.6	0.345	0.531
$X_{M(2)}$ conscious, responsible consumption	28.7	0.381	0.542
$X_{M(3)}$ limited finances and fears of rising energy costs	26.3	0.374	0.613
$X_{M(4)}$ cleanliness and automation in operation	5.1	0.444	-
$X_{M(5)}$ information on positive experiences	4.8	0.438	0.462
$X_{M(6)}$ applicable regulations	4.2	0.341	0.454

Source: own elaboration.

Two regression tree models were constructed to show the impact of the variables included in the questionnaire survey on the levels of investment activities for energy efficiency and lifestyle changes. Only those variables with the highest classification power and those significantly influencing the value of the dependent variables (the measures of investment activities or lifestyle changes) were included. The population size was 305 respondents (the respondents who did not answer the question on average monthly electricity/other heat payments were omitted). A recursive procedure to divide the analyzed set of households by the value of the synthetic measure of investment activities (using appropriate measures of the quality of the division, including the reduction of dispersion around the dependent variable, as well as the tree pruning rule) (RMSE = 0.194; R^2 = 0.305 after so-called tree pruning) identified ten classes of households (Figure 3).

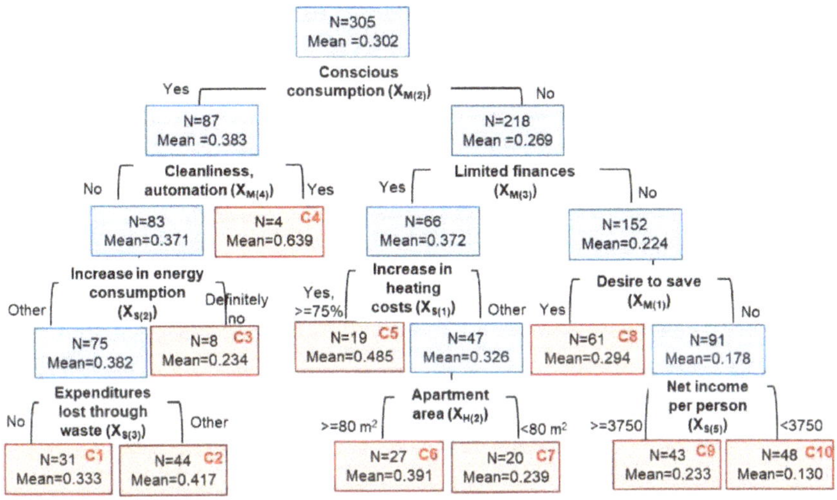

Figure 3. Regression tree for synthetic measure of investment activities to rationalize energy expenditures. Source: own elaboration.

The following variables had the highest discriminatory power in the model for the measure of changes in investment activities: conscious, responsible consumption, including the importance of sustainability ($X_{M(2)}$), and cleanliness and automation in service ($X_{M(4)}$). The following variables, determining the value of the dependent variable and serving to subdivide the household collective further, were limited finances ($X_{M(3)}$); a change in the cost of household energy use compared to the previous year ($X_{S(1)}$); an increase in the average monthly level of household energy consumption compared to the previous year ($X_{S(2)}$); willingness to save ($X_{M(1)}$); perception of the unnecessary energy expenditures incurred as a result of wasted energy or the lack of use of alternative energy-saving options ($X_{S(3)}$); the average monthly net income per person in the household ($X_{S(5)}$); the square meters of the dwelling ($X_{H(2)}$); and information about the positive experiences of family, neighbors, and acquaintances with the use of innovative energy-saving solutions ($X_{M(5)}$).

The value of the average synthetic measure of investment activities for energy expenditure rationalization ranged from 0.13 (C10) to 0.64 (C4) in separate classes. In class C4 (with the highest value of the mean synthetic measure of investment activities, more than twice as high as the average in the whole set of surveyed households), the households that were classified included conscious, responsible consumption and cleanliness and automation in service among the main reasons for the decision to rationalize energy expenditures. The lowest value of the average synthetic measure of investment activities for rationalizing energy expenditures occurred in class C10, covering the households with an average monthly net income per person below 3750 PLN. In addition, these households did not use conscious and responsible consumption.

As a result of the procedure of recursive partitioning of the analyzed set of households concerning the value of the measure of lifestyle changes (using the appropriate measures of the quality of the partitioning, including the reduction of dispersion around the dependent variable, as well as the use of the tree pruning rule), eight classes of households were identified (Figure 4) (RMSE = 0.164; R2 = 0.358 after so-called tree pruning). The value of the synthetic lifestyle measure was most strongly determined by the variable of restricted finances, $X_{M(3)}$. The following variables provided significant classificatory power for further subdivisions of the community: the change in the cost of energy use in the household compared to the previous year ($X_{S(1)}$); the increase in the average monthly level of energy consumption in the household compared to the previous year ($X_{S(2)}$); the average monthly net income per person in the household ($X_{S(5)}$); conscious, responsible consumption, including the importance of sustainability ($X_{M(2)}$); the recognition of the problem of unnecessary energy expenditures incurred as a result of energy waste or the lack of use of alternative energy-saving options ($X_{S(3)}$); and the number of persons per household ($X_{H(1)}$).

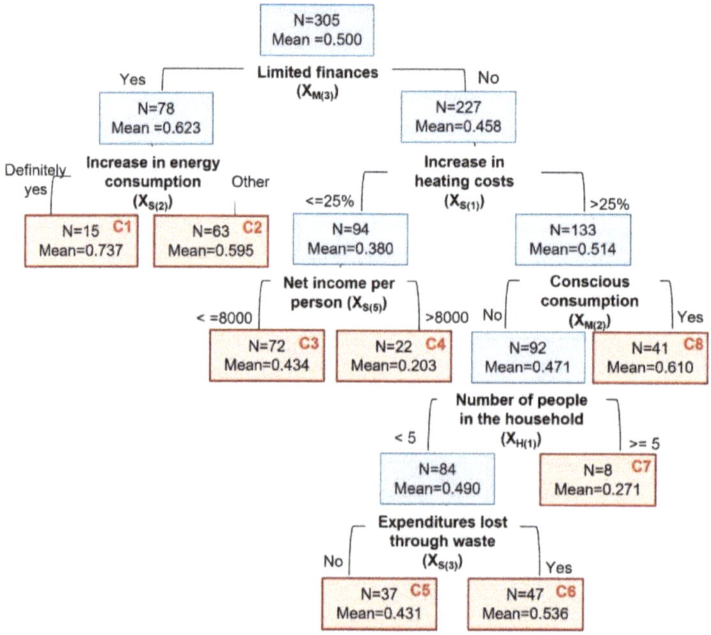

Figure 4. Regression tree for synthetic measure of lifestyle changes to rationalize energy expenditures. Source: own elaboration.

The highest (predicted) value of the mean synthetic measure of lifestyle changes was recorded in the class described by node C1 (equal to 0.74), comprising the households that made decisions to rationalize energy expenditures due to limited finances and noticed an increase in the average monthly level of energy consumption compared to the previous year.

The second highest value for the average lifestyle change measure occurred in class C8 (0.61), comprising the households that followed the principles of conscious and responsible consumption and noted that their average monthly energy use costs had increased by more than 25% compared to the previous year.

The class of respondents described by node C4 (0.20) had the lowest mean (predicted) value of the average measure of synthetic lifestyle changes, i.e., the most miniature scale of the energy expenditure rationalization measures in the form of lifestyle changes. It comprised 7.2% of the respondents with an average monthly net income per person in their household greater than 8000 PLN and declaring that their monthly costs related to energy use decreased, remained unchanged, or increased by 25% at most.

4. Discussion

Restricted finances forced the respondents who were sensitive to electricity price increases to make some changes in their lifestyles and daily household energy-saving behaviors. In order to decrease their average monthly levels of energy consumption, they lowered the temperature in an unused part of the dwelling and when ventilating, and lowered the temperature in the flat to 19–20 degrees Celsius during the day and 17–18 degrees at night. Such behavior is in accordance with the results of some previous studies [35,47].

Turning off the air conditioner when the room is empty and setting the air conditioner at the right temperature, not the lowest or highest, was also observed among urban residents in China who tried to lower electricity consumption [34,36]. Turning off the TV if it was not being watched and pulling the plugs in time are also popular household electricity-saving behaviors in the world [36,48]. Respondents in Poland trying to reduce energy use mainly washed only with a full load in the washing machine, while households in other countries first turned off the light in time and/or in an empty room [34,36,48,49].

If the electricity expenditure is a small proportion of the household consumption expenditures, then respondents are not willing to change their living habits to save electricity. However, they invest in energy-efficient devices. Their electricity-saving consumption is caused by the belief that it contributes to the improvement of the environment. The result is consistent with studies which show that personal moral norms and environmentally friendly habits have a significant effect on people's intentions to reduce electricity consumption [34,36,50]. Respondents in Poland first indicated making investments that required low funding (new-generation lighting, replacing household appliances with more energy-efficient ones, or installing a faucet aerator). Purchasing high-efficiency appliances and equipment as a result of changes in energy prices was shown in many studies [51–54]. One-third of respondents changed their energy generation system (installing heat pumps or photovoltaic panels) and/or improved their home insulation. A recent increase of the share of renewable sources in energy production in Poland was also confirmed by other papers [55,56].

The presented research results have some limitations, since they refer predominantly to a small sample of Polish households. The sample size and measurement scales used limited the possibility of using other multidimensional, parametric statistical models allowing for testing the relationships between variables.

The conducted questionnaire study was exploratory. This pilot study also allowed for testing the possibility of applying the proposed analytical method (regression trees), which could be used in the actual study, with a much larger size.

In the paper, socio-demographic factors were not included in the analysis because the respondents provided answers about their households regardless of the division of the decision-making power in the area of rationalizing energy expenditures. Therefore, in subsequent studies, it is recommended to conduct in-depth research on an extended sample and to modify the survey questionnaire to include the socio-demographic factors of the surveyed households to obtain more reliable measures of internal coherence. We also recommend extending the territorial scope to the international dimension. Moreover, taking into account the importance of the problem under consideration, it is justified to monitor consumers' attitudes towards the rationalization of energy expenditures (in the face of increasing energy prices and institutional requirements) and towards energy innovations.

5. Conclusions

Due to the implementation of the EU's climate and energy policy and ongoing geopolitical changes, a significant increase in energy prices has been observed in Poland. In 2023, the increase in energy expenditures was the highest in households that relied on natural gas or fossil energy sources.

The rising energy costs, along with the changes in legal regulations, have led households to alter their energy management behaviors. To rationalize their energy expenditures, they invested in energy-efficient devices and/or changed their habits to reduce energy consumption.

Pilot studies show that households with low incomes or concerns about further energy price increases were the most inclined to adopt lifestyle changes aimed at reducing energy consumption. The respondents most frequently reported turning off the television when not in use, doing laundry only with a full load, or lowering the temperature in the unused areas of their homes.

New-generation lighting, replacing household appliances with more energy-efficient ones, installing heat pumps or photovoltaic panels, saving hot water by installing a faucet aerator, and improving home insulation were the most commonly cited investment actions by respondents aimed at rationalizing their energy expenditures. Households that undertook more significant investments with higher financial costs were primarily motivated by responsible consumption, environmental protection, and the cleanliness and ease of use of the devices. The positive experiences shared by family, neighbors, or friends regarding the use of innovative energy-saving solutions also had a strong influence on their decisions.

Over 40% of respondents noted that their energy expenditures could be lower if they reduced energy wastage or adopted alternative energy-saving options.

Author Contributions: Conceptualization, E.S., K.S., M.N.-P., W.N., P.S., K.W. and J.Z.-S.; methodology, E.S.; validation, E.S., K.S., M.N.-P., W.N., P.S., K.W. and J.Z.-S.; formal analysis, E.S. and W.N.; investigation, E.S., K.S., M.N.-P., W.N., P.S., K.W. and J.Z.-S.; resources, E.S., K.S., M.N.-P., W.N., P.S., K.W. and J.Z.-S.; data curation, E.S., K.S., M.N.-P., P.S., K.W. and J.Z.-S.; writing—original draft preparation, E.S., K.S., M.N.-P., W.N., P.S., K.W. and J.Z.-S.; writing—review and editing, E.S., K.S., M.N.-P., W.N., P.S., K.W. and J.Z.-S.; visualization, E.S. and J.Z.-S.; supervision, E.S., K.S., M.N.-P., W.N., P.S., K.W. and J.Z.-S.; project administration, E.S. and J.Z.-S.; funding acquisition, K.S. and P.S. All authors have read and agreed to the published version of the manuscript.

Funding: This publication was financed by a subsidy granted to the University of Wroclaw and the Wroclaw University of Economics.

Data Availability Statement: The original contributions presented in the study are included in the article, further inquiries can be directed to the corresponding author.

Conflicts of Interest: The authors declare no conflicts of interest.

References

1. *Communication from the Commission to the European Parliament, the Council, the European Economic and Social Committee and the Committee of the Regions: A Policy Framework for Climate and Energy in the Period from 2020 to 2030*; Document 52014DC0015; European Union: Maastricht, The Netherlands, 2014.
2. *Paris Agreement*; Document 22016A1019(01), OJ L 282; European Union: Maastricht, The Netherlands, 2016.
3. *Regulation (EU) 2021/1119 of the European Parliament and of the Council of 30 June 2021 Establishing the Framework for Achieving Climate Neutrality and Amending Regulations (EC) No 401/2009 and (EU) 2018/1999 ('European Climate Law')*; Document 32021R1119, OJ L 243; European Union: Maastricht, The Netherlands, 2021.
4. Stala-Szlugaj, K.; Grudziński, Z. (Eds.) Zagadnienia surowców energetycznych i energii w gospodarce krajowej. Bezpieczeństwo energetyczne Polski i UE w świetle obecnej sytuacji gospodarczej. In *Zeszyty Naukowe Instytutu Gospodarki Surowcami Mineralnymi i Energią PAN*; Instytut Gospodarki Surowcami Mineralnymi i Energią Polskiej Akademii Nauk: Krakow, Poland, 2024; Volume 1, Available online: https://min-pan.krakow.pl/wydawnictwo/wp-content/uploads/sites/4/2024/03/ZN-2024-nr-112-1.pdf (accessed on 19 October 2024).
5. Brodny, J.; Tutak, M.; Grebski, W. Empirical Assesment of Efficiency of Poland's Energy Transition Process in the Context of Implementing the European Union's Energy Policy. *Energies* **2024**, *17*, 2686. [CrossRef]
6. Bijańska, J.; Wodarski, K. Hard coal production in Poland in the aspect of climate change and energy policy of the European Union and the war in Ukraine. Investment case study. *Resour. Policy* **2024**, *88*, 104390. [CrossRef]
7. Gradzewicz, M.; Jabłonowski, J.; Sasiela, M.; Żółkiewski, Z. The impact of energy prices increases on the Polish economy. *Energy Econ.* **2024**, *140*, 107944. [CrossRef]
8. Piekut, M. Patterns of Energy Consumption in Polish One-Person Households. *Energies* **2020**, *13*, 5699. [CrossRef]
9. Barwińska Małajowicz, A.; Knapková, M.; Szczotka, K.; Martinkovičová, M.; Pyrek, R. Energy Efficiency Policies in Poland and Slovakia in the Context of Individual Well-Being. *Energies* **2023**, *16*, 116. [CrossRef]
10. Bouzarovski, S. Understanding Energy Poverty, Vulnerability and Justice. In *Energy Poverty*; Palgrave Macmillan: Cham, Switzerland, 2018; pp. 9–39. [CrossRef]

11. Conradie, P.; Van Hove, S.; Pelka, S.; Karaliopoulos, M.; Anagnostopoulos, F.; Brugger, H.; Ponnet, K. Why do people turn down the heat? Applying behavioural theories to assess reductions in space heating and energy consumption in Europe. *Energy Res. Soc. Sci.* **2023**, *100*, 103059. [CrossRef]
12. Borowska, A.; Mikuła, A.; Raczkowska, M.; Utzig, M. *Konsumpcja Dóbr i Usług w Gospodarstwach Domowych w Polsce*; Wydawnictwo SGGW: Warszawa, Poland, 2020.
13. Szczygieł, O.; Kalinowski, S.; Łuczak, A. When Energy Becomes a Luxury. Challenges in the Fight Against Energy Poverty. Conclusion from the Research on the Causes of Poverty in the Mazovian Voivodship. *Polityka Społeczna* **2024**, *599*, 21–29. [CrossRef]
14. Piao, X.; Managi, S. Household energy-saving behavior, its consumption, and life satisfaction in 37 countries. *Sci. Rep.* **2023**, *13*, 1382. [CrossRef]
15. Shittu, I.; Latiff, A.R.A.; Baharudin, S.A. Assessing the compensation and reinvestment plans for fuel subsidy rationalization in Nigeria: A dynamic computable general equilibrium approach. *Energy* **2024**, *293*, 130671. [CrossRef]
16. Amores, A.F.; Maier, S.; Ricci, M. Taxing household energy consumption in the EU: The tax burden and its redistributive effect. *Energy Policy* **2023**, *182*, 113721. [CrossRef]
17. Bhattarai, U.; Maraseni, T.; Devkota, L.P.; Apan, A. Facilitating sustainable energy transition of Nepal: A best-fit model to prioritize influential socio-economic and climate perception factors on household energy behaviour. *Energy Sustain. Dev.* **2024**, *81*, 101505. [CrossRef]
18. Blonz, J.A. The costs of misaligned incentives: Energy inefficiency and the principal-agent problem. *Am. Econ. J. Econ. Policy* **2023**, *15*, 286–321. [CrossRef]
19. Jonassen, R.; Andersen, M.S.; Cottrell, J.; Bhattacharya, S. *Carbon Pricing and Fossil Fuel Subsidy Rationalization Tool Kit*; Asian Development Bank: Mandaluyong, Philippines, 2023. [CrossRef]
20. Jiang, P.; Van Fan, Y.; Klemeš, J.J. Impacts of COVID-19 on energy demand and consumption: Challenges, lessons and emerging opportunities. *Appl. Energy* **2022**, *285*, 116441. [CrossRef] [PubMed]
21. Bahmanyar, A.; Estebsari, A.; Ernst, D. The impact of different COVID-19 containment measures on electricity consumption in Europe. *Energy Res. Soc. Sci.* **2020**, *68*, 101683. [CrossRef]
22. Strojny, J.A.; Chwastek, M.S.; Badach, E.; Lisek, S.J.; Kacorzyk, P. Impacts of COVID-19 on Energy Expenditures of Local Self-Government Units in Poland. *Energies* **2022**, *15*, 1583. [CrossRef]
23. Ślusarczyk, B.; Chłąd, M.; Michałek, J.; Dacko-Pikiewicz, Z.; Androniceanu, A. Strategies for supplying enterprises with energy in the context of changing coal prices on the Polish market—The effect of the war in Ukraine. *Resour. Policy* **2023**, *85*, 104028. [CrossRef]
24. Liobikienė, G.; Matiiuk, Y.; Krikštolaitis, R. The concern about main crises such as the Covid-19 pandemic, the war in Ukraine, and climate change's impact on energy-saving behavior. *Energy Policy* **2023**, *180*, 113678. [CrossRef]
25. Colgan, J.D.; Gard-Murray, A.S.; Hinthorn, M. Quantifying the value of energy security: How Russia's invasion of Ukraine exploded Europe's fossil fuel costs. *Energy Res. Soc. Sci.* **2023**, *103*, 103201. [CrossRef]
26. Kesriklioğlu, E.; Oktay, E.; Karaaslan, A. Predicting total household energy expenditures using ensemble learning methods. *Energy* **2023**, *276*, 127581. [CrossRef]
27. Xu, X.; Xiao, B.; Li, C.Z. Critical factors on electricity consumption in residential buildings: An analysis from the point of occupant characteristics view. *J. Clean. Prod.* **2020**, *256*, 120423. [CrossRef]
28. Chen, C.; Xu, X.; Adua, L.; Briggs, M.; Nelson, H. Exploring the factors that influence energy use intensity across low-, middle-, and high-income households in the United States. *Energy Policy* **2022**, *168*, 113071. [CrossRef]
29. Zhang, W.; Calauit, J. Occupancy behaviour and patterns: Impact on energy consumption of high-rise households in southeast China. *Smart Energy* **2022**, *6*, 100072. [CrossRef]
30. Hu, S.; He, J.; Yang, C. Evaluating the impacts of the increasing block tariffs on residential electricity consumption in China. *Sustain. Prod. Consum.* **2022**, *29*, 180–187. [CrossRef]
31. Chen, X.; Gou, Z. Bridging the knowledge gap between energy-saving intentions and behaviours of young people in residential buildings. *J. Build. Eng.* **2022**, *57*, 104932. [CrossRef]
32. Swell, G.; Gauthier, S.; James, P.; Stein, S. Predicting UK Domestic Electricity and Gas Consumption between Differing Demographic Household Compositions. *Energies* **2024**, *17*, 4753. [CrossRef]
33. Ben, H.; Steemers, K. Household archetypes and behavioural patterns in UK domestic energy use. *Energy Effic.* **2018**, *11*, 761–771. [CrossRef]
34. Wang, B.; Wang, X.; Guo, D.; Zhang, B.; Wang, Z. Analysis of factors influencing residents' habitual energy-saving behaviour based on NAM and TPB models: Egoism or altruism? *Energy Policy* **2018**, *116*, 68–77. [CrossRef]
35. Belaïd, F.; Joumni, H. Behavioral attitudes towards energy saving: Empirical evidence from France. *Energy Policy* **2020**, *140*, 111406. [CrossRef]
36. Fu, W.; Zhou, Y.; Li, L.; Yang, R. Understanding household electricity-saving behavior: Exploring the effect of perception and cognition factors. *Sustain. Prod. Consum.* **2021**, *28*, 116–128. [CrossRef]
37. Anselmi, P.; Colledani, D.; Robusto, E. A Comparison of Classical and Modern Measures of Internal Consistency. *Front Psychol.* **2019**, *10*, 2714. [CrossRef]

38. Breiman, L.; Friedman, J.H.; Olshen, R.A.; Stone, C.J. *Classification and Regression Trees*; Chapman and Hall/CRC: New York, NY, USA, 1984.
39. Mitra, A.; Jain, A.; Kishore, A.; Kumar, P. A Comparative Study of Demand Forecasting Models for a Multi-Channel Retail Company: A Novel Hybrid Machine Learning Approach. *Oper. Res. Forum* **2022**, *3*, 58. [CrossRef]
40. Valaskova, K.; Kliestik, T.; Svabova, L.; Adamko, P. Financial Risk Measurement and Prediction Modelling for Sustainable Development of Business Entities Using Regression Analysis. *Sustainability* **2018**, *10*, 2144. [CrossRef]
41. Dudek, A.; Stańczyk, E. Zastosowanie drzew regresyjnych w analizie determinant sukcesu małych i średnich przedsiębiorstw. In *Determinanty Sukcesu Małych i Średnich Przedsiębiorstw*; Woźniak, H., Ed.; Główny Urząd Statystyczny, Urząd Statystyczny we Wrocławiu: Warszawa, Poland, 2022. Available online: https://stat.gov.pl/statystyki-eksperymentalne/badania-i-rozwoj-innowacyjnosc-spoleczenstwo-informacyjne/determinanty-sukcesu-malych-i-srednich-przedsiebiorstw,14,1.html (accessed on 25 February 2024).
42. Mussini, M. On measuring income polarization: An approach based regression trees. *Stat. Transit. New Ser.* **2016**, *17*, 221–236. [CrossRef]
43. Lotfy, N. Regression tree modelling to predict total average extra costs in household spending during COVID-19 pandemic. *Bull. Natl. Res. Cent.* **2021**, *45*, 127. [CrossRef]
44. Nie, P.; Roccotelli, M.; Fanti, M.P.; Ming, Z.; Li, Z. Prediction of home energy consumption based on gradient boosting regression tree. *Energy Rep.* **2021**, *7*, 1246–1255. [CrossRef]
45. Strzelecka, A.; Zawadzka, D. Application of classification and regression tree (CRT) analysis to identify the agricultural households at risk of financial exclusion. *Procedia Comput. Sci.* **2021**, *192*, 4532–4541. [CrossRef]
46. Genç, S.; Mendeş, M. Evaluating performance and determining optimum sample size for regression tree and automatic linear modeling. *Arq. Bras. Med. Vet. Zootec.* **2021**, *73*, 1391–1402. [CrossRef]
47. Kola-Bezka, M.; Leki, K. Household energy behaviour in the times of crisis: Lessons for policy initiatives from peripheral, fossil-dependent regions of the European Union. *Energy Policy* **2024**, *188*, 114113. [CrossRef]
48. Wang, S.; Lin, S.; Li, J. Exploring the effects of non-cognitive and emotional factors on household electricity saving behavior. *Energy Policy* **2018**, *115*, 171–180. [CrossRef]
49. Krupnova, T.G.; Rakova, O.V.; Shefer, E.A.; Semenenko, D.P.; Saifullin, A.F. Domestic energy-saving behavior index as sustainability indicator: Are Russians ready for sacrifices to protect environment? *Environ. Sustain. Indic.* **2022**, *16*, 100209. [CrossRef]
50. Motz, A. Consumer acceptance of energy transition in Switzerland: The role of attitudes explained through a hybrid discrete choice model. *Energy Policy* **2021**, *151*, 112152. [CrossRef]
51. Jacobsen, G.D. Do energy prices influence investment in energy efficiency? Evidence from energy star appliances. *J. Environ. Econ. Manag.* **2015**, *74*, 94–106. [CrossRef]
52. Alberini, A.; Ščasný, M.; Czajkowski, M.; Bigano, A. Volatile energy markets, consumers and energy price expectations. *Energy Econ.* **2023**, *126*, 107039. [CrossRef]
53. Park, J.; Woo, J. Analyzing consumer's willingness to purchase energy-efficient appliances in response to energy price changes: Case study of South Korea. *Energy Econ.* **2023**, *127*, 107088. [CrossRef]
54. Mataloto, B.; Ferreira, J.C.; Pontes Resende, R. Long Term Energy Saving Through User Behavior Modeling in Smart Homes. *IEEE Access* **2023**, *11*, 44544–44558. [CrossRef]
55. Rusin, A.; Wojaczek, A. Trends of changes in the power generation system structure and their impact on the system reliability. *Energy* **2015**, *92*, 128–134. [CrossRef]
56. Rusin, A.; Wojaczek, A. Changes in the structure of the Polish energy mix in the transition period to ensure the safety and reliability of energy supplies. *Energy* **2023**, *282*, 128831. [CrossRef]

sustainability

MDPI

Article

Sustainable Behavior of Generation Z Tourists' Water Consumption

Alvaro Enrique Lima-Vargas [1,*], Oscar Martínez-González [2], Jessica Geronimo-Cruz [1] and Suemi Lima-Vargas [2,3]

1 Playa del Carmen Campus, Autonomous University of Quintana Roo, Playa del Carmen 77710, Mexico; 2026273@uqroo.mx
2 Cozumel Campus, Autonomous University of Quintana Roo, Cozumel 77600, Mexico; oscarmartinez@uqroo.edu.mx (O.M.-G.); 2484434@uqroo.mx or slimav@uaemex.mx (S.L.-V.)
3 Tlalnepantla Professional Academic Unit, Autonomous University of the State of Mexico, Tlalnepantla 54190, Mexico
* Correspondence: aelima@uqroo.edu.mx

Abstract: Tourism is one of the sectors with the highest demand for fresh water. Indicators suggest that water consumption by tourists is more than double that of residents. This phenomenon can be explained by tourists' interest in engaging in unique experiences, which frequently entails putting sustainable practices aside. To mitigate the environmental impact of tourism and promote sustainable practices, the state of Quintana Roo has implemented the Sustainable Tourism Master Plan 2030, linking the government and companies to reduce environmental impacts by reducing tourists' water consumption. Despite these efforts, the use of fresh water by tourists continues to be more than double that of residents. Consequently, tourists' sustainable behavior is not significantly influenced by external factors, such as awareness campaigns, but depends more on the personal aspects of each individual. Generation Z stands out for being more aware of environmental problems and showing a willingness to modify their behavior towards more sustainable practices. The objective of this study was to explain the factors that influence the sustainable behavior related to water consumption among Generation Z tourists. The methodology employed was cross-sectional, with a quantitative approach, focused on attitude variables, subjective norms, and perceived control as determinants of sustainable behavior. This research is also considered nonexperimental, as there was no direct intervention with the subjects of the study, A non-probabilistic convenience sampling method was used to collect data directly from Generation Z tourists in the state of Quintana Roo, specifically from the destinations of Tulum, Playa del Carmen, Bacalar, and Cancún. These locations were selected due to their prominence as major tourist attractions within the data collection areas, without further differentiation or classification. Data were collected through a face-to-face survey conducted over a period of two months, with a total of 408 respondents participating, of whom 57.8% were male. The results revealed that subjective norms and perceived control influence sustainable water consumption behavior, while attitudes do not have a direct impact. These findings will provide governments and businesses with a basis to design more effective strategies that encourage sustainable behavior among Gen Z tourists, thus offering a starting point for understanding the behavior of other generations.

Keywords: sustainable development; environmental; consumer research; Generation Z

1. Introduction

In recent years, the concept of sustainable development has increasingly influenced various sectors, including tourism, leading to the emergence of sustainable tourism [1] This concept emphasizes the development of products and services designed to reduce the environmental footprint and benefit the environment [2], while also promoting pro-environmental behaviors among tourists. Despite these efforts, however, most tourists continue to engage in environmentally irresponsible practices, contributing to ecological damage, environmental problems, and pollution at tourist destinations [3,4]

Academic research on sustainable development has addressed tourism-related issues, focusing on environmental management, energy saving, waste management, and water consumption [5,6]. In tourism, water is used directly in showers, bathrooms, swimming pools, and spas, among other facilities, and indirectly in golf courses, fountains, and laundries [7,8], which places tourism among the sectors with the highest demand for freshwater [9].

Faced with this problem, governments and companies have developed policies and plans to promote sustainable water consumption among tourists [10]. On the one hand, governments have chosen to regulate demand by increasing the price of water in tourist accommodation [11], while hotels have adopted water-saving technologies, implemented conservation programs, and launched awareness campaigns aimed at visitors [12]. As a result, environmental concern among consumers is growing [13], and it is observed that Generation Z shows a greater willingness to modify their consumption habits towards more sustainable practices [14]. In response to this situation, various strategies have been implemented to mitigate the ecological impact of tourism [15].

This is particularly significant given that, in 2023, tourism generated USD 3.3 trillion, representing 3% of the global GDP. This figure indicates an 88% recovery relative to pre-pandemic levels, according to the World Tourism Organization [16]. It highlights the importance of tourism as a major driver of economic prosperity for nations [17] and highlights its crucial role in regional development [18].

In Mexico, the state of Quintana Roo is one of the most popular tourist destinations. It features 80 tourist spots, with the Riviera Maya, Cancun, Cozumel, and Isla Mujeres being the most prominent (Figure 1). The state has welcomed over 21 million tourists, generating an economic impact of USD 20.5 billion [19]. To mitigate the environmental impact of tourism and promote sustainable practices, the Sustainable Tourism Master Plan 2030 has been implemented. This plan unites the government and tourism companies to mitigate environmental impact, with a particular emphasis on reducing water consumption by tourists [20].

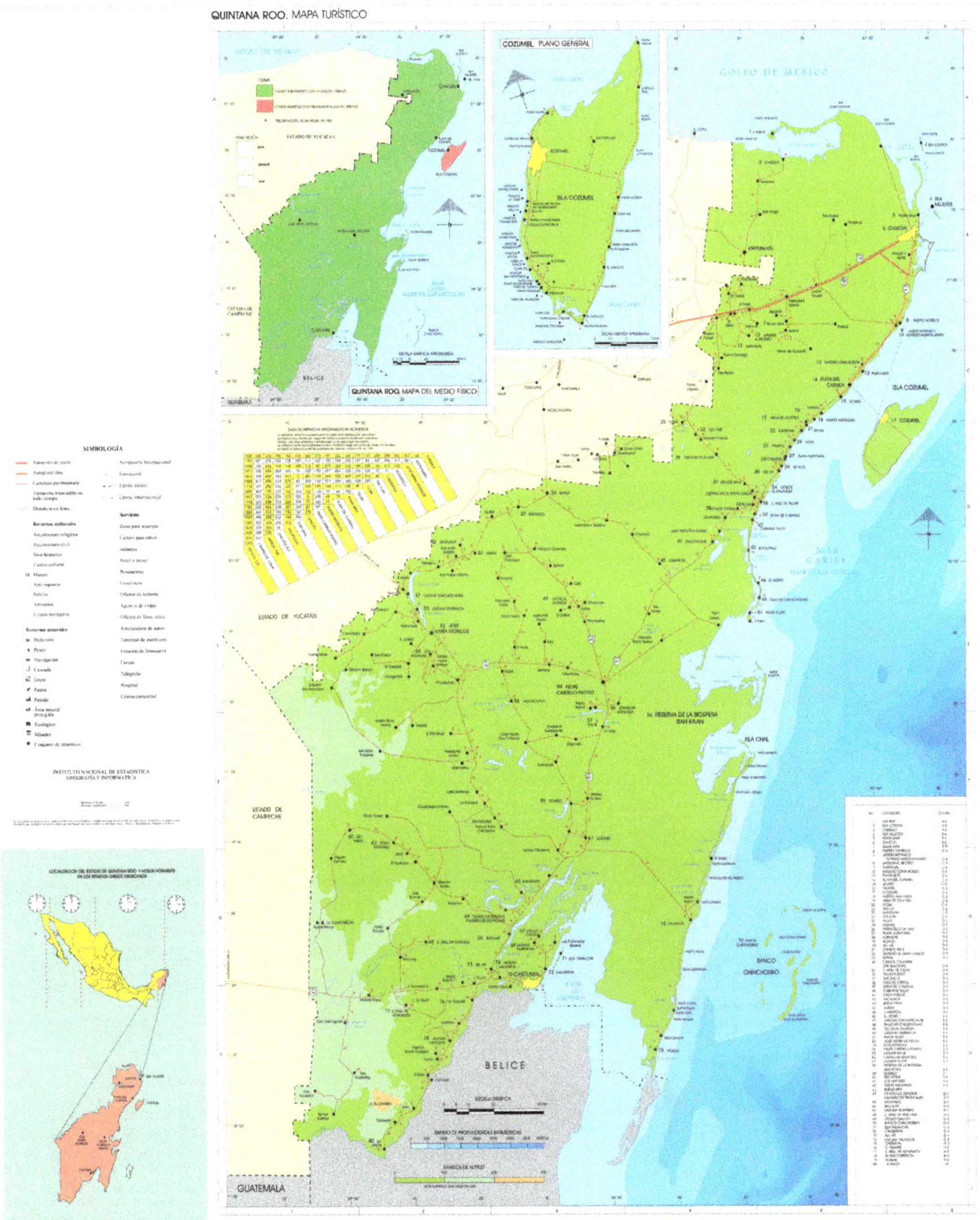

Figure 1. Mapping of tourist areas in Quintana Roo. Source: [21].

Despite these efforts, indicators reveal that tourists' water consumption is more than double that of residents [22]. This phenomenon is attributed to the fact that tourists' environmental behavior is not significantly influenced by external factors such as awareness campaigns, but rather depends on the personal aspects [23,24]. Hence, the following research questions are raised: What determinants motivate Generation Z tourists to engage in sustainable behaviors? and How will these determinants influence sustainable water consumption behavior? This study aims to explain the factors influencing sustainable behavior concerning the water consumption of Generation Z tourists.

This research aimed to examine the water consumption behavior of Generation Z tourists. This focus on Generation Z is due to their heightened awareness of global environmental issues and their potential to challenge existing approaches to managing water depletion. It makes a significant contribution to the literature on sustainable tourist behavior by analyzing the factors (attitudes, subjective norms, and perceived control) that influence sustainable water consumption behavior among Generation Z tourists. This research expects to enhance the understanding of water consumption patterns and apply the Theory of Planned Behavior (TPB) to Generation Z's specific context. This approach will allow us to explore further how these factors affect the water consumption decisions of young tourists, thereby offering a valuable contribution to the academic literature on sustainable tourism.

Thus, the study provides an expanded perspective on the tourist profile and makes a significant contribution through the data collected, its reliability, and its analysis. This research allowed us to reach a deeper understanding of Generation Z tourists' behavior and their preferences for sustainable tourist destinations. The analysis revealed a relationship between sustainable water consumption, the Generation Z cohort, and their pursuit of sustainable alternatives in emerging tourist cities, which are of particular interest to this generation. The main objective of the research was to analyze the sustainable water consumption behavior of Generation Z tourists.

1.1. Theoretical Framework

The Theory of Planned Behavior (TPB) aims to explain human behavior through a model that considers individual factors that must be aligned with planned strategies and mechanisms designed to address specific issues. Within this model, the primary determinants of behavior are the individual's intention and perceived control. Secondary elements include subjective norms and attitudes [25]. In this context, the TPB focuses on behaviors affecting water consumption and the mechanisms for leveraging and managing resources to optimize their use among young populations [26]. However, the research extends beyond organizational processes to address environmental issues and nature conservation, specifically within the context of tourism in cities visited by Generation Z. Hence, this study aims to apply the TPB to tackle the challenges of tourism in cities with economic models impacting rational water consumption from the consumer's perspective [27].

1.2. Theory of Planned Behavior

One of the main theories used to determine individuals' behavior is the TPB, which has been widely referenced in diverse fields, such as management and business, applied psychology, environmental sciences, and environmental studies [28]. The TPB indicates that an individual's behavior will rely on three key individual aspects: attitudes, subjective norms, and perceived control [29]. Concerning tourism, the model has been used to explain environmentally friendly behaviors [30], the preference and choice of eco-friendly hotels [31], and coastal tourism [32], among others.

On the one hand, Generation Z and the Theory of Planned Behavior have been frequently discussed in the literature and used to establish behavior in real estate investments [33] and entrepreneurship [34]. Sustainable behavior, on the other hand, has been used to identify ethical consumption [35] and the consumption of sustainable products [36]. Regarding tourism, it has been used to indicate behavior at heritage sites [37] and the intention to visit eco-friendly hotels [38]. Hence, these studies confirm a link between Generation Z and the Theory of Planned Behavior.

1.3. Gen Z and Sustainable Consumption

Generations have been defined from different stances; however, most postulators agree that each generation will exhibit similar behavior regarding cultural, political, and economic values [39,40], and these will be different from those of other generations [39,41,42]. Generation Z comprises people born between 1997 and 2012 [43] and is characterized by being the generation most aware of environmental problems and looking for ecological products or services that develop pro-environmental strategies [44]. They are also willing to modify their behavior to make it sustainable [45].

1.4. Relationship between TPB and Generation Z

Generation Z includes individuals born between 1997 and 2012 who are characterized by a strong interest in environmental issues and a desire to address them in innovative and sustainable ways [46]). From the perspective of the TPB, it can be argued that this generation is prepared to tackle environmental challenges by planning and implementing actions aimed at attaining sustainability and democratizing decision-making processes [47], having grown up amidst an environmental crisis characterized by climate change, biodiversity loss, and resource scarcity. This generation has also experienced rapid technological advances and the expansion of a global free market, which has led them to access information and connect with others instantly [48]

In this context, Generation Z can make an impact by using the TPB to resolve environmental issues, particularly water depletion. As a crucial resource for human survival, effective water management presents a notable opportunity for this generation [49]. This demographic cohort can lead planned change initiatives that promote new, thoughtful, and humane approaches to interacting with nature and managing natural resources sustainably. Such an approach would focus on water conservation and responsible use rather than focusing exclusively on recreational purposes [50].

Generation Z's ability to address environmental problems innovatively and sustainably can be leveraged through its connectivity and access to information [51]. Their strong interest in environmental issues and their drive to develop novel solutions to tackle them could be crucial for reaching sustainability and democratizing the decision-making process. This is particularly relevant in the context of a polarized global stance on sustainability and ongoing efforts to combat climate change, drought, and extractive economies [52]

1.5. Attitudes towards Sustainable Consumption

Attitudes are psychological tendencies an individual presents towards something specific (people, objects, emotions, situations, etc.). A favorable attitude will generate favorable behavior [29]. For sustainable consumption, attitudes will directly influence consumption behavior [53]. Concerning tourists, studies indicate that their attitudes will become predictors of their behavior during their visit [54].

Diverse studies reaffirm that positive attitudes towards the environment will influence the sustainable consumption behavior of tourists [55,56]; they also indicate that favorable sustainable behavior contributes to protecting the environment and, consequently, achieves a better experience in their visit or some positive return, which stimulates a more sustainable attitude [32,57]. Therefore, it is established that tourists who present a sustainable consumption attitude will generate sustainable behavior, which helps protect the environment. Accordingly, the following hypothesis is established:

Hypothesis 1: *Attitudes towards sustainable consumption have a significant impact on behavioral patterns related to sustainable water consumption.*

1.6. Subjective Standards of Sustainable Consumption

Subjective norms are defined as the social pressure that is exerted on an individual on the behavior that the environment considers appropriate [29]. This social pressure is developed by friends, family, and co-workers [58]. In relation to sustainable behavior, friends, family, and co-workers will influence whether consumers engage in this type of behavior [59].

Concerning the sustainable behavior of tourists, social pressure is also carried out by lodging centers, tourism stakeholders, the government, and other tourists; these will influence behavior. Some studies confirm that the environment directly influences behavior; in relation to tourism, this influence has been identified in visits to ecological hotels [38], in visits to see humpback whales [60] and in visits to wetlands [61], and has been used to measure environmental behavior in general [62]. Therefore, it is established that subjective norms favorable to sustainable consumption generate sustainable behavior towards the environment by tourists. Accordingly, the following hypothesis is established:

Hypothesis 2: *Subjective norms around sustainable consumption exert a significant influence on the behaviors adopted about responsible water consumption.*

1.7. Perceived Control of Sustainable Consumption

Perceived control is defined as an individual's self-perception of their success or failure in performing some activity, and this will determine the individual's behavior [29]. Unlike subjective attitudes and norms, the individual's self-perception of the capabilities and resources available to them will influence the perception of success or failure [63].

In relation to sustainable behavior, perceived control has been established as a critical element in determining sustainable behavior [64]. On the other hand, studies focused on tourism have mentioned that elements such as opportunity cost, time, and money will influence perceived control [65]; this indicates that if a tourist considers adopting a sustainable behavior, they are perceived to have the economic resources and the necessary time, and it will not generate a loss of cost–benefit in their tourist experience [66]. Therefore, it is established that the perceived favorable control towards sustainable consumption generates sustainable behavior towards the environment by the tourists. Consequently, the following hypothesis is established:

Hypothesis 3: *Perceived control over sustainable consumption affects the behaviors adopted in relation to responsible water consumption.*

2. Materials and Methods

2.1. Data Collection Techniques and Study Subjects

This research aims to explain the factors that influence sustainable water consumption behavior among Generation Z tourists. Therefore, this type of research has been established as quantitative, with a correlational–causal objective by seeking to verify the relationship between attitudes, subjective norms, and perceived control with sustainable behavior. In addition, the research is non-experimental since, during the development of the research, the study subjects will not be influenced. It is also called cross-sectional because the sample collection or data collection will be carried out during a specific time (Table 1).

The sources of information for the research were obtained directly from Generation Z tourists in the state of Quintana Roo, through a face-to-face survey. Its application was mainly on Riviera Maya's tourist destinations (Playa del Carmen, Tulum, and Puerto Aventuras), Bacalar, and Chetumal. Consequently, this research is classified as a primary source due to its data collection method, and the information obtained is specific to this research. Limitations include the use of a non-probability convenience sampling method by surveying tourists at the main tourist attractions in the data collection areas without distinguishing or classifying them.

Table 1. Study datasheet.

Title 1	Title 2
Type of research	Quantitative/Non-experimental/Cross-sectional/Correlational–causal
Characteristics of the Universe	Generation Z tourists in the state of Quintana Roo
Data collection	Instrument (surveys)/Primary source
Sampling type	Non-probabilistic for convenience
Trust level and error	95%/5%
Population size	2.5 million
Sample size	n = 384

Source: Own elaboration.

The sample size was calculated (Equation (1)) [67] to determine the number of study subjects.

$$n = \frac{N \times Z_a^2 \times p \times q}{d^2 \times (N-1) + Z_a^2 \times p \times q} \quad (1)$$

where

n = Sample size;
p = Probability of success;
d = Maximum permissible error;
Z = Confidence level;
q = Probability of failure;
N = Population.

The following data were used to calculate the sample size: the sampling error and the confidence level were configured at 5% and 95%, respectively, while the population size was estimated at 2.5 million, corresponding to Generation Z tourists in the state of Quintana Roo (Table 1).

2.2. Data Collection Instrument

The data collection instrument is divided into two sections. The first section contains five questions designed to identify tourist's generation, gender, type of trip, nationality, and type of accommodation. The second section consists of 12 Likert-scale questions, structured on a 5-point scale where 1 denotes "strongly disagree" and 5 denotes "strongly agree", which aim to identify the variables of attitudes, subjective norms, perceived control, and sustainable behavior (Table 2).

Table 2. Data collection instrument.

Variable	Item No.	Question
Attitudes	1	I am aware of the impact of my water consumption, and I care about it.
	2	I am willing to reduce my water consumption to contribute to the protection of the environment.
	3	I am willing to listen to the recommendations of the lodging center and the government authorities to minimize my water consumption.
Subjective Norms	4	The people who accompany me show an active interest in reducing water consumption.
	5	The people who accompany me perceive my level of water consumption in a positive way.
	6	Other tourists consider my water consumption adequate.
Perceived Control	7	I possess the tools, knowledge, and skills necessary to reduce my water consumption if I choose to do so.
	8	I can follow specific recommendations to consume less water, if I choose to do so.
	9	I can effectively reduce my water intake when I set my mind to it.
Sustainable Behavior	10	I have carried out specific actions to reduce water consumption during my stay.
	11	I have listened to the recommendations of the lodging centers on how to save water.
	12	I have adopted behaviors that favor the reduction of water consumption during my stay.

Source: Adapted from Ajzen [29].

2.3. Data Processing

The hypotheses' verification was carried out in two sections. The first, called the validation of the data collection instrument, was conducted in three phases: the first phase, validation of reagents, seeks to ensure that the instruments are 100% complete, eliminating those that still need to be filled. The second phase, the reliability of the data collected, used Cronbach's alpha (Equation (2)) [68], with an estimated acceptance parameter of 0.7. The third phase, estimation errors, includes the evaluation of the multicollinearity of the variables of attitudes, subjective norms, and perceived control by calculating Kendall's Tau (Equation (3)) [68]. The information was processed using SPSS v.22 software.

$$\alpha = \frac{K}{K-1}[1 - \frac{\sum S_i^2}{S_T^2}] \quad (2)$$

where α is the symbol of Cronbach's alpha, K is the number of items, $\sum S_i^2$ represents the sum of the variances of each item, and S_T^2 represents the total variance. This analysis was applied to each of the study constructs.

$$T_b = \frac{P-Q}{\sqrt{(P+Q+X_0)(P+Q+Y_0)}} \quad (3)$$

where Tb is Kendall's Tau symbol, P is the number of concordant pairs, Q is the number of discordant pairs, and X_0 is the number of those related only to the variable X, Y_0.

The second section, called criteria for the validity of the research, was conducted in two phases. The first phase, hypothesis testing, was carried out by calculating the adjusted R^2 with an acceptance parameter of 0.6. The second phase, reaffirmation of the hypothesis, was carried out using Pearson's correlation equation (Equation (4)) [69], with an acceptance parameter of 0.7.

$$r = \frac{S_{xy}}{(S_x)(S_y)} \quad (4)$$

where r is the symbol of Pearson's linear correlation, S_{xy} is the covariance between x and y, S_x is the standard deviation of x, and S_y is the standard deviation of y.

3. Results

Table 3 presents the demographic aspects and specific characteristics of the type of tourism of the 408 subjects studied. The data revealed that of the 408 participants, 57.8% were men and 42.2% were women. Regarding nationality, 73.0% of the respondents were Mexican, 6.9% Spanish, 5.9% Argentinian, 4.9% Chilean and Colombian, 2.0% Italian, 1.5% American, and 1.0% Pakistani. In terms of the type of trip, 55.9% were family holidays, 20.6% were visits with a partner, and 12.7% were trips with friends. Regarding the type of accommodation, 30.4% opted for all-inclusive hotels, 21.6% stayed with family and friends, 20.6% stayed in hotels, and 19.6% stayed in an Airbnb, while 3.9% preferred hostels.

Table 4 shows the reliability of the data collected using Cronbach's alpha, reaching values of between 0.795 and 0.869 for all items. The variables attitudes, subjective norms, perceived control, and sustainable behaviors presented values of 0.862, 0.798, 0.791, and 0.799, respectively, exceeding the acceptance parameter of 0.7, which demonstrates the relevance of the items and variables for their analysis. In addition, the means of the elements ranged from 2.6471 to 4.0221, with standard deviations of between 1.26063 and 1.84710.

Table 5 shows the results of the estimation error in the multicollinearity of the variables attitudes, subjective norms, and perceived control. The relationship between attitudes and subjective norms yielded a value of −0.555, while that between attitudes and perceived control was −0.345, and that between subjective norms and perceived control was 0.525. These values, below the acceptance parameter, confirm the absence of multicollinearity and validate the suitability of these variables for hypothesis testing.

In Table 6, the results of the estimation error analysis, calculating the multicollinearity of the items showed a range between −0.383 and 0.672, placing it below the acceptance parameter of 0.700, indicating that the items are suitable for subsequent analysis. It should be noted that three-item ratios were close to the acceptance limit: the ratio between items 5 and 6 had a value of 0.661, that of items 6 and 7 was 0.672, and that of items 6 and 8 was 0.675. Although these results are below the acceptance threshold, their proximity to the limit merits detailed review in future analyses to determine possible multicollinearities.

Table 3. Descriptive profile of the respondents.

Characteristics	N = 408	Percentage
Gender		
Male	236	57.8%
Female	172	42.2%
Nationality		
Mexican	298	73.0%
Spanish	28	6.9%
Argentinian	24	5.9%
Chilean	20	4.9%
Colombian	20	4.9%
Italian	8	2.0%
American	6	1.5%
Pakistani	4	1.0%
Type of trip		
Family vacation	228	55.9%
Vacation with a couple	84	20.6%
Vacation with friends	52	12.7%
Other	44	10.8%
Type of accommodation		
All-inclusive hotel	124	30.4%
Family and friends	88	21.6%
Hotel	84	20.6%
Airbnb or similar	80	19.6%
Hostel	16	3.9%
Other	16	3.9%

Source: Own elaboration (2024).

Table 4. Reliability analysis.

Item/Variable	Stocking	Standard Deviation	Total Correlation of Corrected Elements	Cronbach's Alpha If the Element Has Been Deleted
Item 1	3.4412	1.84710	−0.308	0.869
Item 2	3.3529	1.62685	−0.203	0.858
Item 3	2.6471	1.63288	−0.440	0.871
Attitudes	**3.5490**	**1.43407**	**−0.371**	**0.862**
Item 4	2.2328	1.53171	0.197	0.834
Item 5	3.5049	1.63349	0.806	0.795
Item 6	3.5098	1.62744	0.891	0.789
Subjective Norms	**3.3431**	**1.38964**	**0.816**	**0.798**
Item 7	3.2500	1.56002	0.819	0.795
Item 8	3.3922	1.56847	0.828	0.795
Item 9	3.4559	1.50609	0.826	0.796
Perceived Control	**3.6078**	**1.49959**	**0.902**	**0.791**
Item 10	3.5613	1.46076	0.803	0.798
Item 11	3.7770	1.26831	0.737	0.805
Item 12	3.7108	1.28100	0.740	0.805
Sustainable behavior	**4.0221**	**1.26063**	**0.854**	**0.799**

Source: Authors' elaboration using SPSS.

Table 5. Variable estimation errors.

	Attitudes (ACT)	Subjective Norms (NS)	Perceived Control (CP)
Attitudes (ACT)	1.000		
Subjective norms (NS)	−0.553	1.000	
Perceived Control (CP)	−0.345	0.525	1.000

Source: Authors' elaboration using SPSS.

Table 6. Item estimation errors.

	Item 1	Item 2	Item 3	Item 4	Item 5	Item 6	Item 7	Item 8	Item 9	Item 10	Item 11	Item 12
Item 1	1.000											
Item 2	0.337	1.000										
Item 3	0.308	0.251	1.000									
Item 4	−0.308	−0.354	−0.315	1.000								
Item 5	−0.321	−0.104	−0.304	0.206	1.000							
Item 6	−0.147	−0.076	−0.289	0.200	0.661	1.000						
Item 7	−0.183	−0.042	−0.264	−0.002	0.643	0.672	1.000					
Item 8	−0.319	−0.118	−0.319	0.190	0.626	0.675	0.639	1.000				
Item 9	−0.221	−0.101	−0.316	0.217	0.575	0.624	0.571	0.600	1.000			
Item 10	−0.359	−0.166	−0.347	0.297	0.555	0.566	0.533	0.557	0.548	1.000		
Item 11	−0.313	−0.172	−0.383	0.216	0.546	0.538	0.524	0.530	0.497	0.548	1.000	
Item 12	−0.307	−0.175	−0.347	0.218	0.521	0.531	0.504	0.537	0.509	0.527	0.542	1.000

Source: Authors' elaboration using SPSS.

Table 7 shows the results of the linear regression analyses, Pearson estimation, estimated error, non-standardized beta coefficient, and error deviation of the items in relation to the variables. It is observed that the relationship of item 1 with attitudes registered an R^2 of 0.625, a Pearson assessment of 0.790, and an estimated error of 0.87776. Similarly, item 2, associated with attitudes, showed an R^2 of 0.601, a rating of 0.775, and an estimated error of 0.90544. On the other hand, item 3, linked to attitudes, reached an R^2 of 0.512, a rating of 0.715, and an estimated error of 1.00132.

Table 7. Checking items and variables.

Item/Variable	Pearson Correlation	Adjusted R Square	Std. Error of the Estimate	Non-Standardized Coefficient Beta
Item 1 → ACT	0.790	0.625	0.87776	0.614
Item 2 → ACT	0.775	0.601	0.90544	0.684
Item 3 → ACT	0.715	0.512	1.00132	0.629
Item 4 → NS	0.633	0.401	1.07577	0.575
Item 5 → NS	0.874	0.765	0.67380	0.744
Item 6 → NS	0.876	0.769	0.66746	0.749
Item 7 → CP	0.901	0.813	0.64765	0.867
Item 8 → CP	0.924	0.854	0.57334	0.884
Item 9 → CP	0.919	0.845	0.59128	0.915
Item 10 → CS	0.905	0.820	0.53456	0.782
Item 11 → CS	0.884	0.783	0.58728	0.880
Item 12 → CS	0.881	0.777	0.59514	0.868

Source: Authors' elaboration using SPSS.

Concerning the relationship between item 4 and subjective norms, the values were an R^2 of 0.401, a compensation of 0.633, and an estimated error of 1.07577. In contrast, item 5, related to subjective norms, obtained an R^2 of 0.765, a rating of 0.874, and an estimated error of 0.67380. Likewise, item 6, also associated with subjective norms, reflected an R^2 of 0.769, a compensation of 0.876, and an estimated error of 0.66746.

For perceived control, item 7 showed an R^2 of 0.813, a compensation of 0.901, and an estimated error of 0.64765. Article 8 linked to it obtained an R^2 of 0.854, a grade of 0.924, and an estimated error of 0.57334. In addition, item 9, related to perceived control, indicated an R^2 of 0.845, a rating of 0.919, and an estimated error of 0.59128.

In relation to sustainable behavior, item 10 achieved an R^2 of 0.820, a rating of 0.905, and an estimated error of 0.53456. Element 11, associated with the same variable, showed an R^2 of 0.783, a compensation of 0.884, and an estimated error of 0.58728. Finally, item 12, linked to sustainable behavior, registered an R^2 of 0.777, a classification of 0.881, and an estimated error of 0.59514.

These results show that items 1 and 2 adequately explain the attitude variables, while items 5 and 6 do the same with subjective norms. On the other hand, items 7, 8, and 9 clarify the perceived control variable, and items 10, 11, and 12 clarify the sustainable behavior variable, all showing values higher than the acceptance parameter of R^2 (0.6) and Pearson's manipulation. On the contrary, the relationship of item 3 with attitudes and item 4 with subjective norms presented values of R^2 below 0.6, indicating that there is no significant relationship between these items and variables.

In Table 8, the results for the hypotheses testing are shown. For Hypothesis 1 (H1), which relates attitudes and sustainable behavior, the data showed an R^2 of 0.324 and a Pearson score of 0.569, with an estimated error of 1.03761. These results do not support H1, indicating that there is no significant relationship between attitudes and sustainable behavior.

Table 8. Hypothesis testing.

Hypothesis	Variables	Pearson Correlation	R Square	Adjusted R Square	Std. Error of the Estimate
1	ACT → CS	0.569	0.324	0.323	1.03761
2	NS → CS	0.844	0.713	0.712	0.67656
3	CP → CS	0.865	0.748	0.748	0.63335

Source: Authors' elaboration using SPSS.

Regarding Hypothesis 2 (H2), which links subjective norms and sustainable behavior, R^2 values of 0.713 and Pearson evaluation values of 0.844 were observed, with an estimated error of 0.67656. These results validate H2, demonstrating that there is a significant relationship between subjective norms and sustainable behavior.

Regarding Hypothesis 3 (H3), which associates perceived control and sustainable behavior, R^2 values of 0.748 and Pearson assessment of 0.865 were recorded, with an estimated error of 0.63335. These data confirm H3, noting that there is a significant relationship between perceived control and sustainable behavior.

4. Discussion and Conclusions

The correlation of the items with the variables' results, specifically the variable of attitude, indicates that Generation Z tourists are aware of the impact of water consumption and are concerned about it (item 1) and are willing to reduce their water consumption to protect the environment (item 2). The results also indicated that they are not willing to listen to the recommendations of the lodging center or the government authorities in terms of the strategies to reduce water consumption. Another critical aspect, in relation to the variable of subjective norms, is that Generation Z tourists consider that the people who accompany them (item 5) and other tourists (item 6) consider their water consumption adequate. However, they affirm that it is not an active interest of the environment to reduce water consumption (item 4).

Regarding perceived control variable, Gen Z tourists indicated that they have the knowledge and skills necessary to reduce their water consumption (item 7), as well as the ability to follow specific recommendations in relation to reducing water consumption (item 8) and that they could effectively mitigate it if they so choose. The behavior-related results indicated that Generation Z tourists have developed activities to reduce water consumption (item 10), that they have heeded the recommendations of lodging centers (item 11), and they exhibit behavior that favors the reduction of water (item 12).

In relation to Hypothesis 1, the results indicated that there is no relationship between attitudes towards sustainable consumption and sustainable water consumption; likewise, Hypothesis 2 was accepted, indicating that subjective norms around sustainable consumption determine the responsible water consumption behavior of Generation Z tourists. Finally, Hypothesis 3 was approved, determining that the perceived control towards water consumption will influence the sustainable water consumption behavior of Generation Z tourists.

These results are partially consistent with those postulated by Ajzen [29], who states that subjective norms and perceived control are determinants in the water consumption behavior of Generation Z tourists, disagreeing with several studies, such as that of Son et al. [54], who indicate that attitudes are predictors of the behavior among tourists. Moreover, Japutra [55] states that positive attitudes among tourists are predictors of sustainable behaviors. This may be justified by considering the fact that the attitudes of Generation Z are favorable towards sustainable behavior, which is why they are not considered predictors, as they are a base element in this generation.

Regarding subjective norms, the studies are consistent with those of Pan et al. [38], Clark et al. [60] and Wang [62], among others, in establishing that social pressure influences sustainable behavior. Finally, in relation to perceived control, the results are correct in agreeing with the extensive literature that relates this variable, which reaffirms the relationship between perceived sustainable control and sustainable tourist behavior [38,63,66,70].

The study identified specific limitations related to the origin of tourists from the region, who come from various countries. While Generation Z is characterized as the first global generation with similar behaviors across individuals, cultural factors continue to exert a significant influence on their actions. Consequently, given the characteristics of this demographic, systematic sampling emerges as a more suitable approach. This consideration, however, introduces new research lines, such as examining the impact of cultural differences on water consumption during trips to tourist destinations. Furthermore, this phenomenon could be classified based on the type of tourism practiced.

In conclusion, this study found that attitudes toward sustainable consumption do not significantly influence the sustainable water consumption behavior of Generation Z tourists. And it was established that subjective norms and perceived behavioral control towards sustainable consumption have a significant impact on the sustainable water consumption behavior of Generation Z tourists. Academically, this study contributes to the literature of the Theory of Planned Behavior by focusing on the sustainable behavior in the water consumption of Generation Z.

Additionally, this research provides empirical evidence on subjective norms and perceived control as determinants of sustainable behavior in water consumption by Generation Z tourists, giving rise to new research on tourist water consumption, which can focus on Millennials, Generation X, and Baby Boomers to understand whether different generations present different predictive elements. On the other hand, the results offer information to lodging centers, tourism stakeholders, and governments on what elements affect the water consumption behavior of Generation Z tourists, allowing them to develop more appropriate strategies to reduce water consumption in this segment.

Author Contributions: Conceptualization, A.E.L.-V. and O.M.-G.; methodology, A.E.L.-V.; software, S.L.-V.; validation, S.L.-V.; formal analysis, A.E.L.-V.; investigation, A.E.L.-V. and J.G.-C.; resources, A.E.L.-V., O.M.-G., S.L.-V. and J.G.-C.; data curation, A.E.L.-V.; writing—original draft preparation, A.E.L.-V. and S.L.-V.; writing—review and editing, O.M.-G.; visualization, J.G.-C.; supervision, A.E.L.-V.; project administration, A.E.L.-V.; funding acquisition, A.E.L.-V., O.M.-G., S.L.-V. and J.G.-C. All authors have read and agreed to the published version of the manuscript.

Funding: This work was supported by "The Sustainable Behavior of Generation Z Tourists in Water Consumption in Quintana Roo" with code CI2464/UNAM-8864. This work was carried out by the Management Research Lab of the Autonomous University of Quintana Roo, Playa del Carmen Campus.

Institutional Review Board Statement: Not applicable.

Informed Consent Statement: Not applicable.

Data Availability Statement: Due to participant privacy and ethics requirements, the data from the current study are not available for sharing.

Conflicts of Interest: The authors have no conflicts of interest to declare.

References

1. Kilipiris, F.; Zardava, S. Developing Sustainable Tourism in a Changing Environment: Issues for the Tourism Enterprises (Travel Agencies and Hospitality Enterprises). *Procedia-Soc. Behav. Sci.* **2012**, *44*, 44–52. [CrossRef]
2. Kollmuss, A.; Agyeman, J. Mind the Gap: Why do people act environmentally and what are the barriers to pro-environmental behavior? *Environ. Educ. Res.* **2002**, *8*, 239–260. [CrossRef]
3. Hines, J.M.; Hungerford, H.R.; Tomera, A.N. Analysis and Synthesis of Research on Responsible Environmental Behavior: A Meta-Analysis. *J. Environ. Educ.* **1987**, *18*, 1–8. [CrossRef]
4. Sisneros-Kidd, A.M.; D'antonio, A.; Monz, C.; Mitrovich, M. Improving understanding and management of the complex relationship between visitor motivations and spatial behaviors in parks and protected areas. *J. Environ. Manag.* **2020**, *280*, 111841. [CrossRef]
5. Li, J.; Coca-Stefaniak, J.A.; Nguyen, T.H.H.; Morrison, A.M. Sustainable tourist behavior: A systematic literature review and research agenda. *Sustain. Dev.* **2023**, *32*, 3356–3374. [CrossRef]
6. Streimikiene, D.; Svagzdiene, B.; Jasinskas, E.; Simanavicius, A. Sustainable tourism development and competitiveness: The systematic literature review. *Sustain. Dev.* **2020**, *29*, 259–271. [CrossRef]
7. Gössling, S.; Hall, C.M.; Scott, D. *Tourism and Water*, 1st ed.; Channel View Publishing: Bristol, UK, 2015.
8. Scott, D.; Hall, C.M.; Gössling, S. Global tourism vulnerability to climate change. *Ann. Tour. Res.* **2019**, *77*, 49–61. [CrossRef]
9. Gabarda-Mallorquí, A.; Palom, A.R.; Daunis-I-Estadella, J. Tourism development and efficient water management. An opportunity for sustainable tourism on the Costa Brava (Girona). *Investig. Turísticas* **2015**, *9*, 50. [CrossRef]
10. Cole, S.K.G.; Mullor, E.C.; Ma, Y.; Sandang, Y. "Tourism, water, and gender"—An international review of an unexplored nexus. *WIREs Water* **2020**, *7*, e1442. [CrossRef]
11. Razumova, M.; Rey-Maquieira, J.; Lozano, J. The role of water tariffs as a determinant of water saving innovations in the hotel sector. *Int. J. Hosp. Manag.* **2016**, *52*, 78–86. [CrossRef]
12. Rico, A.; Olcina, J.; Baños, C.; Garcia, X.; Sauri, D. Declining water consumption in the hotel industry of mass tourism resorts: Contrasting evidence for Benidorm, Spain. *Curr. Issues Tour.* **2019**, *23*, 770–783. [CrossRef]
13. Squires, S. Do generations differ when it comes to green values and products? *Electron. Green J.* **2019**, *1*, 42. [CrossRef]
14. Testa, F.; Pretner, G.; Iovino, R.; Bianchi, G.; Tessitore, S.; Iraldo, F. Drivers to green consumption: A systematic review. *Environ. Dev. Sustain.* **2021**, *23*, 4826–4880. [CrossRef]
15. Madhavan, H.; Rastogi, R. Social and psychological factors influencing destination preferences of domestic tourists in India. *Leis. Stud.* **2013**, *32*, 207–217. [CrossRef]
16. World Tourism Organization. International Tourism Will Reach Pre-Pandemic Levels in 2024. UN Tourism. Available online: https://www.unwto.org/es/news/el-turismo-internacional-alcanzara-en-2024-los-niveles-registrados-antes-de-la-pandemia (accessed on 5 August 2024).
17. Bramwell, B. Tourism development and the environment: Beyond sustainability? *Ann. Tour. Res.* **2010**, *37*, 1194–1196. [CrossRef]
18. Cheng, W.; Zhang, Y. Research on the impact of tourism development on regional ecological environment and countermeasure based on big data. *Technol. Bull.* **2017**, *55*, 464–470.
19. Secretariat of Tourism of the State of Quintana Roo. ¿Cómo Vamos en Turismo? Abril 2022–Abril 2023. 2023. Available online: https://sedeturqroo.gob.mx/ARCHIVOS/comovamos/como_vamos_202304.pdf (accessed on 5 August 2024).
20. Secretariat of Tourism of the State of Quintana Roo. Plan Maestro de Turismo Sustentable 2030. 2022. Available online: https://sedeturqroo.gob.mx/pmts2030/planmaestro.php (accessed on 5 August 2024).
21. Google. Cartografía Zonas Turísticas de Quintana Roo. 2024. Available online: https://www.gifex.com/images/0X0/2009-09-17-5163/Mapa_de_Quintana_Roo.jpg (accessed on 5 August 2024).
22. Gössling, S.; Peeters, P.; Hall, C.M.; Ceron, J.-P.; Dubois, G.; Lehmann, L.V.; Scott, D. Tourism and water use: Supply, demand, and security. An international review. *Tour. Manag.* **2011**, *33*, 1–15. [CrossRef]
23. Imran, S.; Alam, K.; Beaumont, N. Environmental orientations and environmental behavior: Perceptions of protected area tourism stakeholders. *Tour. Manag.* **2014**, *40*, 290–299. [CrossRef]
24. Lee, T.H.; Jan, F.H.; Huang, G.W. The influence of recreation experiences on environmentally responsible behavior: The case of liuqiu island, taiwan. *J. Sustain. Tour.* **2015**, *23*, 947–967. [CrossRef]

25. Ajzen, I. The theory of planned behavior: Frequently asked questions. *Hum. Behav. Emerg. Technol.* **2020**, *2*, 314–324. [CrossRef]
26. Burnes, B. The Origins of Lewin's Three-Step Model of Change. *J. Appl. Behav. Sci.* **2019**, *56*, 32–59. [CrossRef]
27. Alolabi, Y.A.; Ayupp, K.; Al Dwaikat, M. Issues and Implications of Readiness to Change. *Adm. Sci.* **2021**, *11*, 140. [CrossRef]
28. Bosnjak, M.; Ajzen, I.; Schmidt, P. The theory of planned behavior: Selected recent advances and applications. *Eur. J. Psychol.* **2020**, *16*, 352–356. [CrossRef] [PubMed]
29. Ajzen, I. The Theory of Planned Behavior. *Organ. Behav. Hum. Decis. Process.* **1991**, *50*, 179–211. [CrossRef]
30. Gautam, V. Examining environmental friendly behaviors of tourists towards sustainable development. *J. Environ. Manag.* **2020**, *276*, 111292. [CrossRef]
31. Verma, V.K.; Chandra, B. An application of theory of planned behavior to predict young Indian consumers' green hotel visit intention. *J. Clean. Prod.* **2018**, *172*, 1152–1162. [CrossRef]
32. Liu, J.; An, K.; Jang, S. A model of tourists' civilized behaviors: Toward sustainable coastal tourism in China. *J. Destin. Mark. Manag.* **2020**, *16*, 100437. [CrossRef]
33. Gumasing, M.J.J.; Niro, R.H.A. Antecedents of Real Estate Investment Intention among Filipino Millennials and Gen Z: An Extended Theory of Planned Behavior. *Sustainability* **2023**, *15*, 13714. [CrossRef]
34. Setyawati, S.M.; Rosiana, M.; Fauzi, D.R.; Arini, D.V. The Antecedent of Online Entrepreneurial Intention in Generation Z: Planned Behavior Theory Implementation. *Calitatea* **2023**, *24*, 1–9.
35. Djafarova, E.; Foots, S. Exploring ethical consumption of generation Z: Theory of planned behaviour. *Young Consum.* **2022**, *23*, 413–431. [CrossRef]
36. Kurt, B.; Özbaş, N. Generation Z Nursing Students' Perceptions of Sustainable Production and Consumption Based on the Theory of Planned Behavior: A Qualitative Study. *Res. Theory Nurs. Pract.* **2023**, *37*, 468–485. [CrossRef]
37. Siddiqui, S.; Bano, N.; Al Rousan, R. Understanding intention of Gen Z Indians to visit heritage sites by applying extended theory of planned behavior: A sustainable approach. *J. Cult. Herit. Manag. Sustain. Dev.* **2023**. [CrossRef]
38. Pan, J.; Teng, Y.-M.; Wu, K.-S.; Wen, T.-C. Anticipating Z-generation tourists' green hotel visit intention utilizing an extended theory of planned behavior. *Front. Psychol.* **2022**, *13*, 1008705. [CrossRef] [PubMed]
39. Martin, M.A. La teoría de las generaciones de Ortega y Gasset: Una lectura del siglo XXI. *Tiempo Y Espac.* **2008**, *17*, 98–110. [CrossRef]
40. Reisenwitz, T.H.; Iyer, R. Differences in generation X and generation Y: Implications for the organization and marketers. *Mark. Manag. J.* **2009**, *19*, 91–103.
41. Casalegno, C.; Candelo, E.; Santoro, G. Exploring the antecedents of green and sustainable purchase behaviour: A comparison among different generations. *Psychol. Mark.* **2022**, *39*, 1007–1021. [CrossRef]
42. Ham, C.-D.; Chung, U.C.; Kim, W.J.; Lee, S.Y.; Oh, S.-H. Greener than Others? Exploring Generational Differences in Green Purchase Intent. *Int. J. Mark. Res.* **2022**, *64*, 376–396. [CrossRef]
43. Gomes, S.; Lopes, J.M.; Nogueira, S. Willingness to pay more for green products: A critical challenge for Gen Z. *J. Clean. Prod.* **2023**, *390*, 136092. [CrossRef]
44. Dabija, D.-C.; Bejan, B.M.; Pușcaș, C. A Qualitative Approach to the Sustainable Orientation of Generation Z in Retail: The Case of Romania. *J. Risk Financ. Manag.* **2020**, *13*, 152. [CrossRef]
45. Lazányi, K.; Bilan, Y. Generation: Z on the labour market: Do they trust others within their workplace? *Pol. J. Manag. Stud.* **2017**, *16*, 78–93.
46. Dąbrowski, L.S.; Środa-Murawska, S.; Smoliński, P.; Biegańska, J. Rural–Urban Divide: Generation Z and Pro-Environmental Behaviour. *Sustainability* **2022**, *14*, 16111. [CrossRef]
47. Wijaya, D.I.; Kokchang, P. Investigating factors influencing generation Z's pro-environmental behavior to support the energy transition in Jakarta, Indonesia. *BIO Web Conf.* **2023**, *62*, 4005. [CrossRef]
48. Kymäläinen, T.; Seisto, A.; Malila, R. Generation Z Food Waste, Diet and Consumption Habits: A Finnish Social Design Study with Future Consumers. *Sustainability* **2021**, *13*, 2124. [CrossRef]
49. Andruszkiewicz, K.; Grzybowska-Brzezińska, M.; Grzywińska-Rąpca, M.; Wiśniewski, P.D. Attitudes and Pro-Environmental Behavior of Representatives of Generation Z from the Example of Poland and Germany. *Sustainability* **2023**, *15*, 15068. [CrossRef]
50. Cardona-Castaño, J.; Lima-Vargas, A.E.L.; Lamprea-Zona, M.; Morales-Ruano, J. Acompañamiento para la gestión del componente ambiental de las actividades de una organización en Armenia (Quindío), Colombia. *Estud. Gestión Rev. Int. Adm.* **2024**, *16*, 177–202. [CrossRef]
51. Dwidienawati, D.; Abdinagoro, S.B.; Gandasari, D.; Tjahjana, D. Do generation Y and Z really concern about environmental issues? *IOP Conf. Series Earth Environ. Sci.* **2021**, *729*, 12137. [CrossRef]
52. Tsevreni, I.; Proutsos, N.; Tsevreni, M.; Tigkas, D. Generation Z Worries, Suffers and Acts against Climate Crisis—The Potential of Sensing Children's and Young People's Eco-Anxiety: A Critical Analysis Based on an Integrative Review. *Climate* **2023**, *11*, 171. [CrossRef]
53. Vermeir, I.; Verbeke, W. *Sustainable Food Consumption: Exploring the Consumer Attitude-Behavior Gap*; Ghent University, WP: Ghent, Belgium, 2004; Volume 4, p. 268.
54. Song, H.; You, G.-J.; Reisinger, Y.; Lee, C.-K.; Lee, S.-K. Behavioral intention of visitors to an oriental medicine festival: An extended model of goal directed behavior. *Tour. Manag.* **2014**, *42*, 101–113. [CrossRef]

55. Japutra, A.; Loureiro, S.M.C.; Molinillo, S.; Ekinci, Y. Travellers' mindsets and theory of planned behaviour. *Tour. Manag. Perspect.* **2019**, *30*, 193–196. [CrossRef]
56. Tanner, C.; Kast, S.W. Promoting sustainable consumption: Determinants of green purchases by Swiss consumers. *Psychol. Mark.* **2003**, *20*, 883–902. [CrossRef]
57. Giebelhausen, M.; Chun, H.H.; Cronin, J.J.; Hult, G.T.M. Adjusting the Warm-Glow Thermostat: How Incentivizing Participation in Voluntary Green Programs Moderates Their Impact on Service Satisfaction. *J. Mark.* **2016**, *80*, 56–71. [CrossRef]
58. Perugini, M.; Bagozzi, R.P. The role of desires and anticipated emotions in goal-directed behaviours: Broadening and deepening the theory of planned behaviour. *Br. J. Soc. Psychol.* **2001**, *40*, 79–98. [CrossRef] [PubMed]
59. Ting, C.-T.; Hsieh, C.-M.; Chang, H.-P.; Chen, H.-S. Environmental Consciousness and Green Customer Behavior: The Moderating Roles of Incentive Mechanisms. *Sustainability* **2019**, *11*, 819. [CrossRef]
60. Clark, E.; Mulgrew, K.; Kannis-Dymand, L.; Schaffer, V.; Hoberg, R. Theory of planned behaviour: Predicting tourists' pro-environmental intentions after a humpback whale encounter. *J. Sustain. Tour.* **2019**, *27*, 649–667. [CrossRef]
61. Chen, G.; Zou, M.; Ran, N.; Yan, B.; Li, S. The effects of environmental empathy and sustainable intelligence on wetland tourists' revisit intention using an extended model of goal-directed behavior. *J. Clean. Prod.* **2023**, *419*, 138288. [CrossRef]
62. Wang, C.; Zhang, J.; Xiao, X.; Sun, F.; Xiao, M.; Shi, Q. Examining the dimensions and mechanisms of tourists' environmental behavior: A theory of planned behavior approach. *J. Clean. Prod.* **2020**, *273*, 123007. [CrossRef]
63. Yim, B.H.; Byon, K.K. Critical factors in the sport consumption decision making process of millennial fans: A revised model of goal-directed behavior. *Int. J. Sports Mark. Spons.* **2020**, *21*, 427–447. [CrossRef]
64. Patwary, A.K.; Mohamed, M.; Rabiul, K.; Mehmood, W.; Ashraf, M.U.; Adamu, A.A. Green purchasing behaviour of international tourists in Malaysia using green marketing tools: Theory of planned behaviour perspective. *Nankai Bus. Rev. Int.* **2022**, *13*, 246–265. [CrossRef]
65. Brown, T.J.; Ham, S.H.; Hughes, M. Picking up litter: An application of theory-based communication to influence tourist behaviour in protected areas. *J. Sustain. Tour.* **2010**, *18*, 879–900. [CrossRef]
66. Olya, H.G.; Bagheri, P.; Tümer, M. Decoding behavioural responses of green hotel guests: A deeper insight into the application of the theory of planned behavior. *Int. J. Contemp. Hosp. Manag.* **2019**, *31*, 2509–2525. [CrossRef]
67. Fernández, C.; Hernández, R. *Research Methodology*; McGRAW-HILL: Mexico City, Mexico, 2014.
68. Domínguez-Lara, S.; Merino-Soto, C. Why is it important to report Cronbach's alpha coefficient confidence intervals? *Lat. Am. J. Soc. Sci. Child. Youth* **2015**, *13*, 1326–1328. Available online: https://www.redalyc.org/pdf/773/77340728053.pdf (accessed on 5 August 2024).
69. Santabarbara, J. Calculation of the confidence interval for correlation coefficients using syntax in SPSS. *J. Innov. Res. Educ.* **2019**, *12*, 1–14. [CrossRef]
70. Bui, T.B. Applying the extended theory of planned behavior to understand domestic tourists' behaviors in post COVID-19 era. *Cogent Soc. Sci.* **2023**, *9*, 2166450. [CrossRef]

economies

MDPI

Article

Circular Economy: Municipal Solid Waste and Landfilling Analyses in Slovakia

Emese Tokarčíková *, Mária Ďurišová and Terézia Trojáková

Department of Macro and Microeconomics, Faculty of Management Science and Informatics, University of Zilina, 010 26 Zilina, Slovakia; maria.durisova@fri.uniza.sk (M.Ď.); trojakova1@stud.uniza.sk (T.T.)
* Correspondence: emese.tokarcikova@fri.uniza.sk

Abstract: The pursuit of shifting Slovakia towards a circular economy is met with a multitude of obstacles, including the pervasive consumerist mindset among Slovakians. This mindset favors packaged food, leading to its improper disposal in municipal waste instead of being recycled. Furthermore, the inclination towards landfills poses a significant challenge in the management of municipal solid waste (MSW). To address this issue, a quantitative analysis was conducted using developed and validated models, incorporating various factors related to MSW management in Slovakia. Our study confirmed the significance of parameters such as MSW management costs and population size in the amount of MSW generated. Furthermore, our findings include a short-term forecast for MSW generation in Slovakia for the next two years. These results, based on quantitative data, provide valuable insights for policymakers and waste management authorities in Slovakia, emphasizing the urgent need for a transition towards a more sustainable and circular economy.

Keywords: circular economy; municipal solid waste; landfilling; waste management; forecast

1. Introduction

The persistent economic expansion and advancement of human prosperity, particularly in urban areas, inevitably engenders adverse side effects, including disrupted social relations and deteriorating environmental conditions (Grimm et al. 2008). The encroachment upon nature and the prevailing linear production model, which prioritizes human comfort at the expense of destruction and pollution of the natural world, has resulted in the global phenomenon of climate change and the existential threat it poses. This study is situated within the broader discourse on sustainability and urban development, particularly focusing on the transition to a circular economy. According to Geissdoerfer et al. (2017), the circular economy is a restorative framework that aims to minimize resource input and subsequent waste, emissions, and energy dissipation by deliberately decelerating, closing, and restricting material and energy loops. The related literature and research results (Bellemare et al. 2022; Lakatos et al. 2021; Matová et al. 2019) indicate how circular economy practices and R-strategies or frameworks are essential for achieving core sustainability goals. These strategies facilitate the transition towards circularity. This is achieved through technological innovations, developing policies conductive to responsible consumption and efficiency waste management, and engaging and educating individuals on circular economy and waste management principles. As the Slovak Republic is currently transitioning toward a circular economy to promote the reduction in primary resource consumption and encourage material reuse, waste plays a vital role in both production and energy within the circular economy. Despite the increasing volume of research on circular economies, there remains a gap in the identification of specific related parameters and understanding of their interrelations and their impact in the practice. In our contribution we have focused on the research topic particularly regarding municipal solid waste (MSW) management within the context of the Slovak republic. As population and consumption rates continue to rise, the proper management of MSW becomes increasingly important for the protection of human health and the environment. In this context, the country of Slovakia emerges as a particularly interesting case study, with significant efforts and resources being directed towards improving waste management practices. The primary research inquiries driving our investigation pertain to municipal waste generation and state in Slovakia, specifically:

- What is the quantity of municipal waste produced in Slovakia?
- How much of this waste is disposed of in landfills, and what measures have been taken by Slovak authorities in this regard?
- Which particular factors exhibit a noteworthy influence on the volume of municipal solid waste (MSW) in Slovakia?
- What are the short-term projections for MSW generation in the country in the upcoming years?

Our contribution is organized as follows. First, we present a literature review on the current circular economy, municipal waste, and landfill issues in Slovakia. The methodology presents the basic facts about our modeling and variables. Furthermore, we present the main results of our analyses and main empirical findings, and finally, in the discussion section, we offer concluding remarks.

2. Literature Review

The circular economy represents a paradigm shift aimed at reducing waste and promoting the continual use of resources through sustainable production and practices such as recycling and reusing (Korhonen et al. 2018). Effective waste management is crucial for successful implementation of a circular economy, which aims to minimize waste generation and enhance utilization through the cyclical management of materials. This entails the extension and closure of material cycles, thereby facilitating the creation of comprehensive inventories of economic flows (Morseletto 2020; Moraga et al. 2019) Municipal solid waste (MSW), commonly called "trash" or "garbage" refers to the waste generated by households and commercial establishments within the confines of a municipality. This includes waste resulting from the daily activities of individuals as well as waste from recreational facilities. Moreover, MSW also encompasses waste collected by the municipality from sources such as stores, small businesses, offices, and government buildings. Furthermore, the waste produced during the municipality's upkeep of public roads and open spaces (dust, leaf matter, and building debris) and treatment plant residual sludge generated from municipal activities (construction and demolition, street cleaning, landscaping, etc.) are also classified as MSW (Cehlár et al. 2021; Ramachandra 2006; Reddy 2011). Solid waste, referred to as refuse, is also described as any substance that is cast aside or remains from human actions or operations. These substances are deemed to be obsolete or no longer needed and can be perceived as unproductive or undesirable materials (Murray 2002). Nevertheless, they may retain significance or contribute to the well-being of others or other operations. The European Union (EU) has set the goal for all member states to implement closed-loop recycling in their economies by 2023 (CEAP 2020). Following the Slovak waste legislative, Act No. 79/2015 Coll. on Waste and on amendments and additions to certain acts (the "Waste Act"), supplemented and specified by the Act. No 460/2019 Coll. or by Act. No. 372/2021 Coll., waste is defined as a movable substance or object that the holder intends to dispose of, is obligated to dispose of, or is currently disposing of under this Act or specialized regulations. Waste does not include (a) any substance or object that is considered a by-product, (b) specific waste that has been granted the end-of-waste status, (c) waste that has undergone a process of preparation for re-use and meets the requirements for a market-ready product established by specific legislation, or (d) waste that has been designated for household use (LEAP 2024). According to Eurostat's (2022) latest data for 2022, the European Union (EU) has a mean production of 513 kg of municipal waste per capita. Slovakia ranks tenth among the EU member states with the lowest municipal waste generation standing at 478 kg per capita in 2022 (Eurostat 2024b). Despite increasing levels of education and the growing public interest in recycling, Slovakia continues to struggle with its waste management practices. This phenomenon is primarily driven by the escalating consumption rate, where each purchase contributes to the generation of waste. Each European generated 188.7 kg of packaging waste in 2021, a figure that is expected to increase to 209 kg in 2030 without additional measures. Similarly, Slovakia faces challenges in promoting a packaging-free purchasing culture due to the difficulty of altering consumer behavior. The availability of convenient and inexpensive food options in supermarkets further exacerbates this trend, leading to the unnecessary production of waste. The country's municipal waste recovery rate is equally alarming, with recycling rates falling below the standards set by the EU. Achieving the EU's objectives of recycling 65% of municipal waste and limiting landfill to 10% by 2035 appears unrealistic for Slovakia at present. The country's current recycling rate for municipal waste stands at a meager 49.5% in 2022 (calculated based on Eurostat 2024a). The goal of a 50% recycling rate, which Slovakia was supposed to achieve in 2020, was, thus, not achieved. The composition of solid waste is heavily influenced by income levels, as lower-to-middle-income individuals tend to produce a higher proportion of organic waste, while those with higher incomes generate more wastepaper, metals, and glass (Nanda and Berruti 2021; Mor and Ravindra 2023).

The regional distribution of municipal solid waste production per capita (in kilograms) in Slovakia in 2022 displayed a marked disparity (Figure 1). One potential explanation for this finding is the greater availability of employment opportunities in the industrial sector and higher population density in the western region of the country.

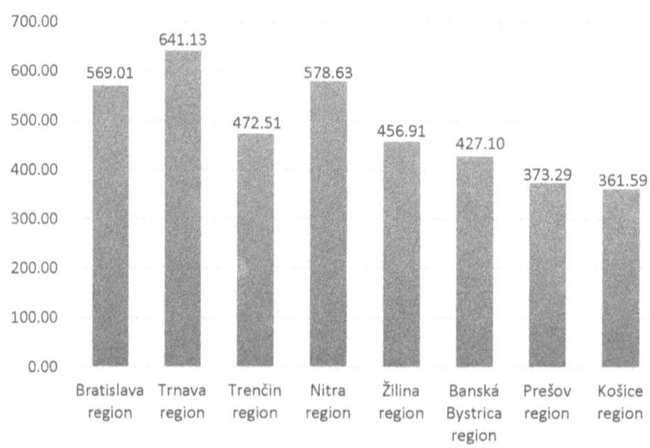

Figure 1. Municipal solid waste production per capita (kg) in Slovakia regions, 2022. Source: https://datacube.statistics.sk/#!/view/sk/VBD_SK_WIN/zp3002rr/v_zp3002rr_00_00_00_sk accessed on 1 May 2024.

However, the country has experienced a rapid increase in municipal waste production in recent years, making it one of the top nations with the swiftest growth in this domain. This is evident in the significant proportion of recyclable materials that are still ending up in landfills, despite the country's efforts to improve its waste management system.

Highly populated areas face challenges when it comes to achieving satisfactory levels of separate collection of municipal solid waste, especially in the context of continuous inflow of new populations. According to Rolewicz-Kalińska et al. (2020), the interplay between bio waste collection and the demands of circular economy in terms of meeting recycling rates and utilizing biogas as a sustainable energy source reveals that none of their proposed scenarios can meet the required municipal solid waste recycling rates by the year 2030 (60%) and 2035 (65%). This clearly points to the pressing need to develop more efficient systems for separate collection, which is also supported by the analytical study conducted by Onungwe et al. (2024) on municipal solid waste management, which focused on conducting comprehensive waste audits.

In recent years, there has been a significant push to make recycling more viable in Slovakia, with many municipalities, cities, and organizations participating in recycling programs. These efforts have been met with some success. Most of the municipalities now have 2–4 separate containers for different types of recyclable materials, distinguished by color (green for glass, yellow for plastic, blue for paper, red for metal, and brown for organic waste). In addition, several legislative measures have been implemented to support and encourage recycling. For instance, since January 2017, it has been mandatory for municipalities to manage biodegradable municipal waste. Furthermore, from 2018 onwards, all residential complexes with green spaces are required to provide residents with brown containers or composters upon request. To further incentivize proper waste management, strict penalties have been introduced for landfill operators who fail to comply with regulations. In addition, lightweight plastic bags are now subject to a charge, metal collection reporting processes have been simplified, and from 2022 the single-use plastic and metal beverage containers are to be collected separately via a deposit-return system).

The management of solid waste in urban areas involves a variety of methods, such as recycling, incineration, waste-to-energy conversion, composting, and landfilling. In many municipalities around the world, landfilling remains the predominant approach to solid waste disposal (Table 1). According to Sondh et al. (2024), only 67% of the globally generated waste is treated by eco-friendly approaches.

Table 1. Amount of generated municipal waste by disposal method (tons)—Slovakia 2017–2022.

	2022	2021	2020	2019	2018	2017
MSW Total	2,597,456.7	2,705,327.1	2,596,725	2,369,725.2	2,325,177.5	2,136,951.8
MSW Landfilled	1,021,583.8	1,099,288.1	1,189,238.5	1,198,249.4	1,250,279.5	1,312,787

Source: https://datacube.statistics.sk/#!/view/sk/VBD_SK_WIN/zp1005rs/v_zp1005rs_00_00_00_sk accessed on 1 May 2024.

The circular economy framework has brought attention to the issue of landfills and calls for the development of novel processes, operational strategies, and intelligent technologies that can mitigate their negative impacts, such as greenhouse gas emissions, methane production, and leachate generation. This push towards a sustainable landfilling approach (depicted in Figure 2) aims to prevent harm to both humans and the environment through systemic changes, the implementation of sustainable practices and technologies, as well as supportive policies. Consequently, this model is driving governments, companies, organizations, communities, and individuals toward a shift to sustainability (Ololade and Orimoloye 2021; Townsend et al. 2015).

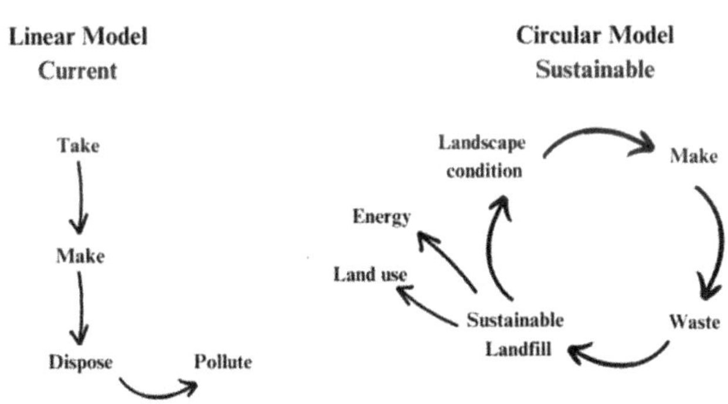

Figure 2. Landfilling in linear and circular economy.

The management of solid waste in urban areas involves a variety of methods, such as recycling, incineration, waste-to-energy conversion, composting, and landfilling. In many municipalities around the world, landfilling remains the predominant approach to solid waste disposal.

Landfilling is the prevalent method of managing municipal solid waste (MSW). However, due to the growing trend in urbanization and increasing consumption patterns, the availability of landfills is becoming scarce. Comparative studies of various waste management methods (landfilling, incineration, composting, etc.) indicate that among the technological options for municipal solid waste (MSW) treatment and disposal, sanitary landfilling or open dumping is prevalent in most countries due to its relatively low cost and minimal technical requirements (Ding et al. 2021; Wang and Nakakubo 2020; Wilson et al. 2012). A recent study by Ichinose (2024) highlights the significant benefits of establishing private landfill sites, not only as a means of managing MSW but also in reducing waste disposal costs. Despite its convenience and cost-effectiveness, the landfilling process poses substantial environmental and human health risks (Vaverková 2019; Das et al. 2019; El-Saadony et al. 2023). The issue is particularly pressing in Slovakia, where the landfill rate greatly surpasses the European average, with over 50% of MSW ending up in landfills, compared to the 24% EU average. This has resulted in severe environmental consequences such as air, soil, and groundwater contamination, with detrimental effects on the health of the population. According to the Institute of Environmental Policy in Slovakia (IEP 2024), there are currently over ten thousand individuals living within a 500 m radius of legally operated landfills in the country. This situation is further compounded by the existence of numerous unregulated "black" landfills, which only exacerbate the issue. Considering these concerning trends, the government of Slovakia has made it a priority to eradicate illegal dumping sites by 2026, as well as implement measures to prevent their recurrence. It is worth noting that marginalized communities, particularly the Roma population, are disproportionately affected by this issue, with thousands of individuals residing within 100 m of larger illegal dumps. The primary aim of the Slovakian municipal waste management strategy for 2025 is to reduce the amount of waste that ends up in landfills. As of 2020, Slovakia has implemented a landfill tax system, where the amount of tax imposed is determined by the extent of municipal waste sorting and separation efforts in each municipality. This approach is geared towards promoting greater recycling and is further supported by a comprehensive program of educational and awareness-boosting initiatives (Ministry of the Environment of the Slovak Republic 2021). This approach is informed by an amalgamation of theoretical scrutiny, pragmatic observations, and scholarly findings regarding the equitable utilization of non-renewable and renewable resources, as well as the evolution towards a circular economic model and complete elimination of emissions (Zhironkin and Cehlár 2022).

3. Data and Methods

Researchers are looking for ways to reuse materials without negative externalities. For this contribution, we utilized data from the Statistical Office of the Slovak Republic. Our focus was on the period between 2008 and 2022, as this is when the tracking of various sectors and production statistics began after Slovakia acceded to the European Union in 2004. Our analysis utilized time series data, specifically annual data, with a total of 15 observations. A significance level of $\alpha = 0.05$ is considered throughout the contribution. The model we developed incorporates a range of variables, namely:

- Year—year of measurement
- Total MunicipalWasteSR—MSW—municipal waste generated each year in Slovakia (tonnes)
- MunicipalWasteManagementCosts—costs associated with MSW (collection, landfilling, energy recovery) have been studied from the years 2008 to 2022, considering the fact that the initial data available in the public database Datacube, is only from the year 2008. Responsibility for the disposal of municipal solid waste (MSW) falls upon the local government in urban areas. To effectively manage the disposal of this waste through techniques such as landfilling or energy recovery, municipalities require adequate funding. This funding is derived from the collection of local taxes and fees

for the provision of waste management services from residents. When constructing models and conducting analysis, it was imperative to identify successful case studies where adjustments in waste costs resulted in a decrease in total KO volume. As a result, municipalities implemented fee structures that served as incentives for residents to accurately pay for their waste output and, thus, encouraged separation and promoted increased recycling efforts. As a logical consequence, cities were able to implement tiered fee structures based on waste production, as the costs associated with collection, transportation, recovery, and disposal were significantly reduced.

- Population—number of inhabitants in the Slovak Republic in the given year. The volume of produced waste is dependent on the number of populations. Furthermore, as waste management costs are covered through fees collected by cities from their residents (both permanent and temporary), the amount, method of payment, and eligibility of payers are determined by the city. This gives the city the power to incentivize citizens to segregate waste more effectively. Fees for waste collection are calculated per person (often a fixed amount based on family size in many municipalities). The more people who participate in waste separation, the less financial resources the city will require for waste management, as the responsible organization for color-coded bins is not the municipalities themselves but rather the producer responsibility organizations (PROs)
- Sales per inhabitant—sales from retail sales per 1 inhabitant
- SalesForOwnServicesAndGood—sales of goods and products in small businesses in thousands of €.
- Similarly, with regard to sales, there is a lack of high-quality relevant data due to the fact that the available data typically includes aggregated information on both performance and goods that do not end up as KO. For instance, the purchase of a car in a given year does not immediately result in waste, as it serves for several years.
- InvestmentsInMunicipalWaste—funds invested in increasing waste management capacity. Continuous data for this parameter are not available in public Slovak databases; therefore, even though it impacts the efficiency of waste management, it has not been included in our models

The determination of the significance of the parameters has been explicated in the findings section, with the size of p-values serving as the metric for evaluation. It is, thus, imperative to select two of these parameters for further examination.

Linear regression was utilized to determine the influence of individual factors on the fluctuation of municipal waste generation. For this contribution, the dependent variable will be referred to as "MunicipalWasteSR", while the remaining variables will act as independent variables, as our focus lies in their impact on the overall quantity of municipal waste produced in the Slovak Republic. It should be noted that the "year" variable is strictly auxiliary and will not be further considered in the linear regression analysis. After verification in several ways, we selected the best-fitting model. Then, we interpreted this resulting model and, thus, tried to find one of the solutions to eliminate the generation of municipal waste. We were inspired by articles and case studies of how it works in other countries or in individual municipalities in the Slovak Republic that have decided to fight against municipal waste individually. We then created a short forecast of waste generation for the next two years, 2023 and 2024, using existing data using Gretl.

4. Results

The Ordinary Least Squares OLS method was employed in the initial model, utilizing the free, open—source Gretl for Windows software (version 1.9.3) to precisely estimate the parameters in model 1.

The selected method for constructing model 1 was based on a sample size of n = 15, as there is limited historical data available for the investment and sales parameters, with data collection beginning only in 2008. As shown in Figure 3, the model is found to be statistically significant, as indicated by the p-value for the F-statistic (9.18×10^{-7}), which falls below the predetermined significance level of $\alpha = 0.05$. Furthermore, the coefficient of determination for this model is calculated to be $R^2 = 0.956335$, indicating that it accounts for 95.63% of the variability in the data. Upon closer inspection, it is evident that only two variables, namely MunicipalWasteManagementCosts and Population, are significant as the *p*-values for the t-statistic (0.0033; 0.0083) are less than 0.05. These results suggest that these two variables have the greatest impact on municipal waste generation, while the influence of the remaining variables can be considered negligible. One crucial aspect of developing a regression model is the validation of the normality assumption regarding the residuals. For optimal parameter values obtained through the ordinary least squares method, it is assumed that the residuals follow a normal distribution. In Model 1, we can observe that the residuals exhibit a normal distribution based on the graphical representation in Figure 4. This is supported by the *p*-value being above our selected significance level, thereby leading to the acceptance of the null hypothesis (H0) that the data indeed stems from a normal distribution.

Model 1: OLS, using observations 2008-2022 (T = 15)
Dependent variable: TotalMunicipalWasteSRton

	Coefficient	*Std. Error*	*t-ratio*	*p-value*	
const	−3,43293e+07	1,06323e+07	−3,229	0,0090	***
SalesForOwnServicesAndGood	5,33089e-05	3,31960e-05	1,606	0,1394	
MunicipalWasteManagementCost	4,90422	1,27909	3,834	0,0033	***
Population	6,50680	1,98341	3,281	0,0083	***
InvestmentsInMunicipalWaste	0,280700	3,46075	0,08111	0,9370	

Mean dependent var	2066139	S.D. dependent var	357022,0
Sum squared resid	7,79e+10	S.E. of regression	88272,44
R-squared	0,956335	Adjusted R-squared	0,938869
F(4, 10)	54,75423	P-value(F)	9,18e-07
Log-likelihood	−189,0658	Akaike criterion	388,1317
Schwarz criterion	391,6719	Hannan-Quinn	388,0940
rho	0,132357	Durbin-Watson	1,538746

Figure 3. Output of linear regression model 1 in Gretl software. Source: own processing, 2024.

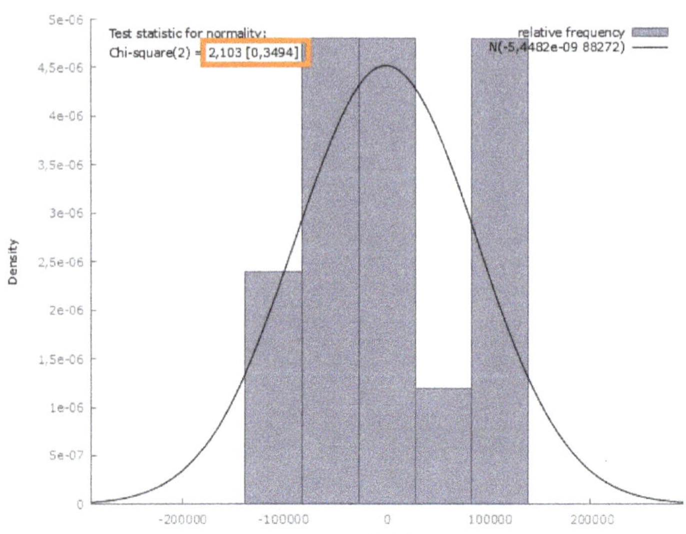

Figure 4. Graphical representation of the normality of model 1 residuals in Gretl software. Source: own processing, 2024.

In the process of amending the initial model 1, the focus was on identifying a set of variables whose values, upon conducting the t-statistics test, would yield a significance level lower than the designated $\alpha = 0.05$. The results depicted in Figure 4 indicate the statistical significance of model 2, with a remarkably low *p*-value in the case of the F-statistic (2.79×10^{-8}), surpassing the chosen significance level. The coefficient of determination, parameter R^2, stands at 0.944937, denoting that the model accounts for 94.49% of the data's variability. Furthermore, all the parameters demonstrate a significant statistical association, evidenced by their *p*-values (0.0007; <0.0001).

Figure 5 also contains text-based information on the residuals' normality. It is observed that the *p*-value, which is 0.46606, exceeds the chosen significance level. As such, it can be concluded that the residuals come from a normal distribution

Model 2: OLS, using observations 2008-2022 (T = 15)
Dependent variable: TotalMunicipalWasteSRton

	Coefficient	*Std. Error*	*t-ratio*	*p-value*	
const	−4,24998e+07	9,57568e+06	−4,438	0,0008	***
Population	8,02948	1,78650	4,495	0,0007	***
MunicipalWasteManagementCosts	6,16925	0,999026	6,175	<0,0001	***

Mean dependent var	2066139	S.D. dependent var	357022,0
Sum squared resid	9,83e+10	S.E. of regression	90489,21
R-squared	0,944937	Adjusted R-squared	0,935760
F(2, 12)	102,9668	P-value(F)	2,79e-08
Log-likelihood	−190,8053	Akaike criterion	387,6106
Schwarz criterion	389,7347	Hannan-Quinn	387,5880
rho	0,117936	Durbin-Watson	1,401697

Test for normality of residual –
Null hypothesis of the normal distribution:
Test statistics: Chi-square (2) = 1,527
with p-value = 0,46606

Figure 5. Output of linear regression model 2 in Gretl software. Source: own processing, 2024.

The developed Model 3 underwent a parameter analysis with modifications to include solely population and total sales of products and goods. The established model exhibits significance through an f-test result of 2.04×10^{-6}, while each parameter demonstrates significance as well. However, in comparison to Model 2, the current model holds a lower R-squared value of 0.887421, indicating a reduced ability to account for variability within the data. Additionally, a thorough examination of the model's residuals reveals a normal distribution (Figure 6).

Based on the methodology employed and the assessment of the adjusted coefficient of determination R_{adj}, it can be determined that model 2 is the most optimal fit. This is due to its significantly higher value of R_{adj} at 0.935760 compared to the other two models. As the data used for constructing the models was oversampled, it is more appropriate to use R_{adj} as a comparative measure, considering the varying sizes of the oversample and the number of independent variables included in each model. Notably, the combination of parameters such as MunicipalWasteManagementCosts and Population can effectively account for 94.49% of the variation in the data, which is slightly lower than the first model. However, all the parameters in model 2 are statistically significant. Moving forward, model 2 will be utilized for further analysis and contributions.

Verification of the model

To accept the selected model, it needs to be tested from several aspects. Regarding the economic verification, a positive unit change in the cost of MSW will increase the collected MSW by 6.16 tonnes. In practice, this means that if the cost of handling MSW increases, i.e., more euros are spent on incineration, regular collection, and collection of MSW, people will put even more of the waste they produce at home in their bins because they will have the opportunity to continuously fill their MSW bins.

```
Model 3: OLS, using observations 2008-2022 (T = 15)
Dependent variable: TotalMunicipalWasteSRton

                         Coefficient    Std. Error     t-ratio    p-value
const                    −4,28350e+07   1,52335e+07    −2,812     0,0157    **
Population                8,14176       2,83747         2,869     0,0141    **
SalesForOwnServices       0,000133472   3,77205e-05     3,538     0,0041    ***
AndGood

Mean dependent var        2066139     S.D. dependent var      357022,0
Sum squared resid         2,01e+11    S.E. of regression      129388,7
R-squared                 0,887421    Adjusted R-squared      0,868658
F(2, 12)                  47,29606    P-value(F)              2,04e-06
Log-likelihood           −196,1691    Akaike criterion        398,3383
Schwarz criterion         400,4624    Hannan-Quinn            398,3157
rho                       0,457146    Durbin-Watson           1,069428
Frequency distribution for residual, obs 1-15
number of bins = 5, mean = -2,08306e-008, sd = 129389
        interval               midpt   frequency   rel.    cum.

           <  − 1,000e+005  − 1,451e+005    3    20,00%   20,00% *******
-1,000e+005 − − 9934 ,      − 5,499e+004    7    46,67%   66,67% ****************
    − 9934 , −  8,017e+004    3,512e+004    1     6,67%   73,33% **
8,017e+004 −    1,703e+005    1,252e+005    2    13,33%   86,67% ****
          >=  1,703e+005      2,153e+005    2    13,33%  100,00% ****

Test for the null hypothesis of the normal distribution:
Chi-square (2) = 2,489
 with p-value = 0,28812
```

Figure 6. Output of linear regression model 3 in Gretl software. Source: own processing, 2024.

The statistical validation of the models was conducted in the preceding subsections, where each model and its associated parameters were scrutinized. Through the utilization of *p*-values and the establishment of a designated level of significance, an evaluation was made on whether the model or parameter is statistically significant.

Since the models are built with temporal data, it is necessary to test for autocorrelation. The selected model contains two variables and for this reason it is also necessary to perform tests for multicollinearity. The autocorrelation coefficient is 0.117936, which represents a low degree of autocorrelation of the residuals. Based on the Durbin–Watson (DW) test of autocorrelation (Figure 7), we cannot know for sure whether the autocorrelation coefficient is significant or not. It is approximated very close to H0, which asserts the statistical insignificance of the autocorrelation coefficient. Since it is not clear whether the autocorrelation is significant or not, in the next step we will proceed to remove it.

Durbin–Watson statistic = 1.4017

H1: positive autocorrelation

p-value = 0.0406227

H1: negative autocorrelation

p-value = 0.959377

Removal of autocorrelation by the Cochrane–Orcutt method.

After removing autocorrelation, all parameters and the model remained statistically significant. The number of observations was reduced by 1 year.

The value of DW is 2.017, indicating the absence of autocorrelation as the autocorrelation coefficient is also close to 0, i.e., autocorrelation is no longer present in the model (Figure 8). Other methods to remove autocorrelation were performed using the Prais–Winsten and Hildreth–Lu methods; however, both the values of rho and DW were very similar, and the significance of the model and the parameters were maintained for all the methods.

Alpha = .05

n\k	1	2	3	4	5	6	7	8	9	10
6	0.610 1.400									
7	0.700 1.356	0.467 1.896								
8	0.763 1.332	0.559 1.777	0.367 2.287							
9	0.824 1.320	0.629 1.699	0.455 2.128	0.296 2.588						
10	0.879 1.320	0.697 1.641	0.525 2.016	0.376 2.414	0.243 2.822					
11	0.927 1.324	0.758 1.604	0.595 1.928	0.444 2.283	0.315 2.645	0.203 3.004				
12	0.971 1.331	0.812 1.579	0.658 1.864	0.512 2.177	0.380 2.506	0.268 2.832	0.171 3.149			
13	1.010 1.340	0.861 1.562	0.715 1.816	0.574 2.094	0.444 2.390	0.328 2.692	0.230 2.985	0.147 3.266		
14	1.045 1.350	0.905 1.551	0.767 1.779	0.632 2.030	0.505 2.296	0.389 2.572	0.286 2.848	0.200 3.111	0.127 3.360	
15	1.077 1.361	0.946 1.543	0.814 1.750	0.685 1.977	0.562 2.220	0.447 2.471	0.343 2.727	0.251 2.979	0.175 3.216	0.111 3.438
16	1.106 1.371	0.982 1.539	0.857 1.728	0.734 1.935	0.615 2.157	0.502 2.388	0.398 2.624	0.304 2.860	0.222 3.090	0.155 3.304
17	1.133 1.381	1.015 1.536	0.897 1.710	0.779 1.900	0.664 2.104	0.554 2.318	0.451 2.537	0.356 2.757	0.272 2.975	0.198 3.184
18	1.158 1.391	1.046 1.535	0.933 1.696	0.820 1.872	0.710 2.060	0.603 2.258	0.502 2.461	0.407 2.668	0.321 2.873	0.244 3.073
19	1.180 1.401	1.074 1.536	0.967 1.685	0.859 1.848	0.752 2.023	0.649 2.206	0.549 2.396	0.456 2.589	0.369 2.783	0.290 2.974
20	1.201 1.411	1.100 1.537	0.998 1.676	0.894 1.828	0.792 1.991	0.691 2.162	0.595 2.339	0.502 2.521	0.416 2.704	0.336 2.885
21	1.221 1.420	1.125 1.538	1.026 1.669	0.927 1.812	0.829 1.964	0.731 2.124	0.637 2.290	0.546 2.461	0.461 2.633	0.380 2.806
22	1.239 1.429	1.147 1.541	1.053 1.664	0.958 1.797	0.863 1.940	0.769 2.090	0.677 2.246	0.588 2.407	0.504 2.571	0.424 2.735
23	1.257 1.437	1.168 1.543	1.078 1.660	0.986 1.785	0.895 1.920	0.804 2.061	0.715 2.208	0.628 2.360	0.545 2.514	0.465 2.670
24	1.273 1.446	1.188 1.546	1.101 1.656	1.013 1.775	0.925 1.902	0.837 2.035	0.750 2.174	0.666 2.318	0.584 2.464	0.506 2.613
25	1.288 1.454	1.206 1.550	1.123 1.654	1.038 1.767	0.953 1.886	0.868 2.013	0.784 2.144	0.702 2.280	0.621 2.419	0.544 2.560
26	1.302 1.461	1.224 1.553	1.143 1.652	1.062 1.759	0.979 1.873	0.897 1.992	0.816 2.117	0.735 2.246	0.657 2.379	0.581 2.513
27	1.316 1.469	1.240 1.556	1.162 1.651	1.084 1.753	1.004 1.861	0.925 1.974	0.845 2.093	0.767 2.216	0.691 2.342	0.616 2.470
28	1.328 1.476	1.255 1.560	1.181 1.650	1.104 1.747	1.028 1.850	0.951 1.959	0.874 2.071	0.798 2.188	0.723 2.309	0.649 2.431
29	1.341 1.483	1.270 1.563	1.198 1.650	1.124 1.743	1.050 1.841	0.975 1.944	0.900 2.052	0.826 2.164	0.753 2.278	0.681 2.396
30	1.352 1.489	1.284 1.567	1.214 1.650	1.143 1.739	1.071 1.833	0.998 1.931	0.926 2.034	0.854 2.141	0.782 2.251	0.712 2.363

Figure 7. DW test for significance of autocorrelation in the model. Source: own elaboration, 2024.

Cochrane-Orcutt, using observations 2009-2022 (T = 14)
Dependent variable: TotalMunicipalWasteSRton
rho = 0,0609573

	Coefficient	Std. Error	t-ratio	p-value	
const	−4,48703e+07	7,66740e+06	−5,852	0,0001	***
Population	8,44921	1,42895	5,913	0,0001	***
MunicipalWasteManagementCost	6,63895	0,805472	8,242	4,91e-06	***

Statistics based on the rho-differenced data:

Sum squared resid	5,41e+10	S.E. of regression	70134,39
R-squared	0,968024	Adjusted R-squared	0,962211
F(2, 11)	148,1298	P-value(F)	1,11e-08
rho	-0,044854	Durbin-Watson	2,017028

Figure 8. Cochrane–Orcutt method (2009–2022) using Gretl software. Source: own processing, 2024.

The model was assessed for multicollinearity using the variance inflation factor (VIF) test. Before the test, it had already been confirmed that the model did not exhibit autocorrelated residuals. Multicollinearity occurs when the VIF values exceed 10 for any variable in the model. As depicted in Figure 9, all variables in the model demonstrated VIF values below 10, leading to the conclusion that multicollinearity was not an issue.

Variance Inflation Factors
Minimum possible value = 1.0
Values > 10.0 may indicate a collinearity problem

Population	1,874
MunicipalWasteManagementCosts	1,874

VIF(j) = 1/(1-R(j)^2), where R(j) is the multiple correlation coefficient between variable j and the other independent variables

Figure 9. VIF method in Gretl software. Source: own processing, 2024.

The developed model (Figure 8), which is not influenced by autocorrelation or multicollinearity, can be elucidated as follows. A growth in population and an escalation in the expenses involved in managing municipal waste would predictably result in a corresponding increase in the production of municipal waste.

It is not possible to determine from the model or the available data the breakdown of costs associated with managing MSW. A different proportional distribution of these costs might be enough to reduce the volume of municipal waste produced. Redirecting some of the existing expenditures on MSW management toward encouraging recycling, instead of solely focusing on collecting municipal waste and distributing bins, could be more efficient.

As part of the modification of the model using the Cochrane method, we also examined a period shorter than one year, i.e., a range 2009–2021 (Figure 10). This method is used to adjust the model in which we have identified autocorrelation. Failure to remove the autocorrelation of the residuals in our model would result in misinterpretation of the conclusions. It would mean that the correlated residuals would have a relationship with each other, and the residual at time t would be somewhat correlated from the residual t-1. Correlation can be positive or negative. After applying the Cochrane–Orcutt, we observed a decrease in the autocorrelation coefficient (ρ/rho/) indicating a decrease in the degree of correlation between the residuals.

In order to detect potential heteroscedasticity in their research, we utilized White's test on a time range spanning from 2008 to 2022 (Figure 11). The established hypotheses were as follows: the null hypothesis (H0) states that the residuals possess a constant variance, indicating there is no heteroscedasticity (homoscedasticity), while the alternative hypothesis

(H1) claims that the variance of the residuals is dynamic, suggesting heteroscedasticity. The interpretation of the test (Figure 11) revealed a p-value of 0.1886, exceeding the commonly accepted significance level of $\alpha = 0.05$. This implies that there is insufficient evidence to reject the null hypothesis, leading to the conclusion that the variances can be considered constant.

We also utilized White's test on a time range spanning from 2008 to 2021 with the same established hypotheses. The interpretation of the test (Figure 12) revealed a p-value of 0.20926, exceeding the commonly accepted significance level of $\alpha = 0.05$. This implies that there is insufficient evidence to reject the null hypothesis, leading to the conclusion that the variances can be considered constant.

Cochrane-Orcutt, using observations **2009-2021** (T = 13)
Dependent variable: TotalMunicipalWasteSRton
rho = 0,173932

	Coefficient	*Std. Error*	*t-ratio*	*p-value*	
const	−3,61280e+07	8,86582e+06	−4,075	0,0022	***
MunicipalWasteManagementCost	8,21751	1,21556	6,760	<0,0001	***
Population	6,79183	1,66014	4,091	0,0022	***

Statistics based on the rho-differenced data:

Sum squared resid	4,38e+10	S.E. of regression	66216,64
R-squared	0,968949	Adjusted R-squared	0,962739
F(2, 10)	113,5489	P-value(F)	1,33e-07
rho	0,002155	Durbin-Watson	1,990325

Statistics based on the original data:

Mean dependent var	2047862	S.D. dependent var	342976,7

Figure 10. Cochrane–Orcutt method (2009–2021) using Gretl software. Source: own processing, 2024.

OLS, using observations **2008-2022** (T = 15)
Dependent variable: TotalMunicipalWasteSRton
HAC standard errors, bandwidth 1 (Bartlett kernel)

	Coefficient	*Std. Error*	*t-ratio*	*p-value*	
const	−4,24998e+07	7,00476e+06	−6,067	<0,0001	***
MunicipalWasteManagementCost	6,16925	0,647125	9,533	<0,0001	***
Population	8,02948	1,29954	6,179	<0,0001	***

Mean dependent var	2066139	S.D. dependent var	357022,0
Sum squared resid	9,83e+10	S.E. of regression	90489,21
R-squared	0,944937	Adjusted R-squared	0,935760
F(2, 12)	117,5471	P-value(F)	1,31e-08
Log-likelihood	−190,8053	Akaike criterion	387,6106
Schwarz criterion	389,7347	Hannan-Quinn	387,5880
rho	0,117936	Durbin-Watson	1,401697

White's test for heteroscedasticity -
Null hypothesis: heteroscedasticity not present
Test statistic: LM = 7,45954
with p-value = P (Chi-square (5) > 7,45954) = 0,188645

Figure 11. White's test for heteroskedasticity (2009–2021) using Gretl software. Source: own processing, 2024.

OLS, using observations **2008-2021** (T = 14)
Dependent variable: TotalMunicipalWasteSRton

	Coefficient	*Std. Error*	*t-ratio*	*p-value*	
const	−3,86727e+07	1,21467e+07	−3,184	0,0087	***
MunicipalWasteManagementCost	6,84169	1,61474	4,237	0,0014	***
Population	7,30494	2,27758	3,207	0,0083	***

Mean dependent var	2028188	S.D. dependent var	337643,7
Sum squared resid	9,57e+10	S.E. of regression	93281,52
R-squared	0,935416	Adjusted R-squared	0,923674
F(2, 11)	79,66075	P-value(F)	2,86e-07
Log-likelihood	−178,3843	Akaike criterion	362,7686
Schwarz criterion	364,6857	Hannan-Quinn	362,5911
rho	0,179080	Durbin-Watson	1,230377

White's test for heteroscedasticity -
Null hypothesis: heteroscedasticity not present
Test statistic: LM = 7,15648
with p-value = P (Chi-square (5) > 7,15648) = 0,20926

Figure 12. White's test for heteroskedasticity (2008–2021) using Gretl software. Source: own processing, 2024.

5. Discussion

The results of our systematic investigation, which combined analytical and practical approaches, offer valuable insights for enhancing decision-making in municipal solid waste management and shaping policies aimed at addressing the mounting waste accumulation issue (Dickson et al. 2023). The significance of this cannot be overstated, particularly in developing economies where, as per Fernando and Zutshi's findings (Fernando and Zutshi 2023), the management of municipal solid waste (MSW) has been deemed inadequate, inefficient, or restricted.

The replacement of traditional municipal waste bins in cities with recycling bins, or the implementation of smart bins equipped with an RFID electronic chip, has been successfully implemented in several European cities, such as Stockholm, Copenhagen, and Zurich. This solution, which requires a strong commitment on the part of city officials, has resulted in a significant reduction in municipal waste. The adoption of smart bins not only enables waste collection workers to optimize their routes by collecting only full bins but also allows for a fairer system of waste collection fees, by ensuring that residents are only charged for the amount of waste they produce (Ahmed et al. 2022; Haque et al. 2020).

For instance, in a lot of Slovak municipalities, a flat fee for municipal waste collection is imposed, which fails to differentiate between residents who produce minimal waste and those who produce a substantial amount. This approach proves to be a hindrance for individuals who are actively seeking to promote a cleaner environment, as they are penalized in the same manner as their neighbors who discard all their waste into communal bins. This causes a disincentive for individuals to actively engage in waste reduction efforts. In many cases, Slovak municipalities operate at a loss as the fees imposed on residents are insufficient to cover the costs associated with municipal waste management (Hrabčák 2011). Therefore, a potential solution to mitigate this issue is to shift some of the costs onto composters (Potočár 2021; Sadovská 2022). In conclusion, the replacement of conventional municipal waste bins with recycling or smart bins is a feasible approach to reducing municipal waste in the Slovak Republic. This in turn not only ensures a fairer system of waste collection fees but also promotes responsible waste management practices, resulting in a cleaner and more sustainable environment for all.

Based on the data presented in Figure 13, it can be observed that the estimated model (represented by the orange line) closely tracks the actual trend in municipal waste generation (represented by the green line). The graph also reveals that until 2021, the curve exhibited an upward trajectory. However, the implementation of backup containers for PET bottles and cans on 1 January 2022 has had a mitigating effect on municipal waste generation, resulting in a decrease in the amount of waste. This finding aligns with the population data for the Slovak Republic, which indicates a similar increasing trend until 2020 but subsequently a decrease due to the impact of the COVID-19 pandemic, leading to a rise in mortality rates within the country.

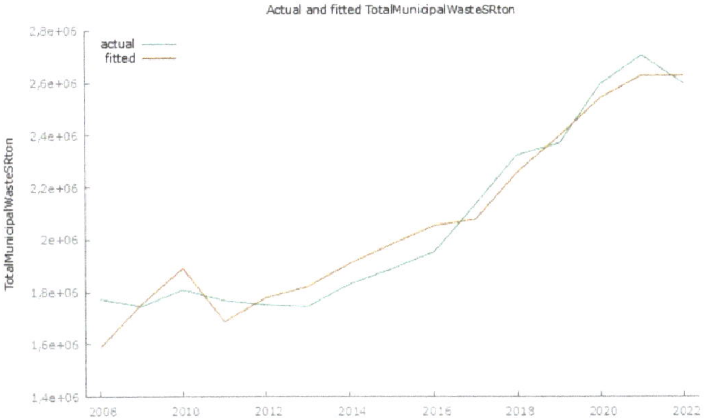

Figure 13. Graph of actual and estimated values in Gretl software. Source: own processing, 2024.

The model can be effectively utilized to examine historical patterns in municipal waste production. As demonstrated in the previous section, the graphical visualization reveals a shift in the municipal waste scenario from 2020 onward due to various factors outlined previously.

As the statistical office does not have data from 2023, we have used our modified model for forecasting purposes. For the forecast, we used the last two years, which record the situation after the pandemic and after the start of the back-up system.

The forecast data (orange) falls within the 95% confidence interval (Figure 14). This means that with many measurements of random municipal waste, 95% of these measurements would include the actual value of municipal waste.

The COVID-19 pandemic period has been found to have a notable impact on the municipal waste volume, a factor not taken into consideration in our contribution. Several research studies have revealed significant modifications in various aspects, such as the consumption of delivered meals, consumer needs, travel habits, corporate culture and the behavior of public and private employees, and education challenges during this period (Hitka et al. 2023; Lorincová et al. 2022; Lizbetinova et al. 2021; Malichová et al. 2023; Staffenova and Kucharčíková 2023). The modified variables mentioned above, in conjunction with various other factors, have the potential to significantly impact the quantity of municipal waste and its implications for environmental sustainability. Additionally, the continuous development of waste management technologies, including disposal, recycling, and landfilling methods and the utilization of advanced information and communication technologies with complex databases for monitoring data and smart technologies (Kvet et al. 2021; Steingartner and Novitzká 2024) can significantly influence forecasting models for future waste volumes.

As we analyze the findings of our study, the generation of municipal waste in Slovakia is impacted by a variety of factors. To accurately model this process, we conducted multiple experiments using different combinations of explanatory variables. Through this process, we have determined that the most effective model includes the variables of costs of creation (KO) and population size, which best describe the fluctuations in municipal waste over time. Our data, gathered from 2008 to 2022, also reveal the impact of external factors such as the pandemic and EU regulations. Of particular note is the 2019 suspension of the sale of disposable plastic plates and cups, which has contributed to a decrease in waste. Additionally, our models illustrate that costs are a key determinant in the creation of municipal waste. As cities and municipalities implement better waste management strategies, including creating a tax burden for residents, it has been observed that individuals are more motivated to separate and sort their waste. This is due to the fact that they are educated on the benefits of waste separation, as they are no longer directly responsible for the costs associated with it. Furthermore, we have also observed a shift from fixed local fees for waste to a more quantitative approach, where fees are determined based on the amount of waste generated. This has led to increased care and concern for the cleanliness of the environment among residents. While it is possible that some waste may end up in landfills outside the city, our data indicate a steady decrease in the amount of waste being landfilled. Ultimately, this suggests a positive trend towards environmental responsibility among residents in Slovakia. Various waste management best practices employed by EU member states, particularly those similar in size and socio-economic conditions to Slovakia, can serve as valuable models for adaptation within Slovakia's waste management framework. In relation to the European Environmental Agency, the implementation of landfill taxes has prompted EU member states to discourage landfilling practices, as exemplified by Poland's adoption of a progressive taxation system for landfilling. Consequently, countries with elevated landfill taxes tend to be more environmentally sustainable and experience fewer issues related to landfilling, primarily due to their increased efforts in waste separation (EEA (European Environment Agency) 2024). However, the recent alterations in the tax burden for residents of Slovakia, including the increase in VAT from 20% to 23% and the introduction of an additional tax on sugary beverages, may unfortunately influence consumer behavior, particularly in border regions. Shoppers may preferentially choose to purchase goods in neighboring countries, where the backup system is less developed than in Slovakia. Consequently, individuals may not return plastic and glass containers to the point of purchase, leading to a rise in waste within Slovakia. After a suitable period, once sufficient data from the backup system has been gathered, it will be possible to analyze whether there has been a notable shift and a decline in individuals' willingness to return recyclable packaging.

For 95% confidence intervals, t(10, 0,025) = 2,228

Obs	TotalMunicipal WasteSRton	prediction	std. error	95% interval
2021	2,70533e+006	2,58635e+006	149572,	(2,25308e+006, 2,91962e+006)
2022	2,59746e+006	2,56972e+006	208249,	(2,10572e+006, 3,03373e+006)

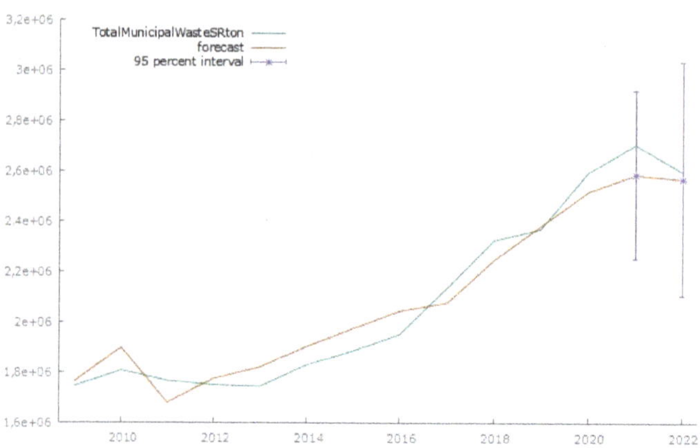

Figure 14. Graph of predicted data in Gretl software. Source: own processing, 2023.

An inspirational case for addressing the issue of municipal waste can be found in the Scandinavian countries, including Norway, Denmark (Sandhi and Rosenlund 2024), and Sweden or the Netherlands (EU Environment Newsletter 2024). In these nations, alongside the implementation of advanced smart technologies (Szpilko et al. 2023) and improved logistics and responsibilities divided into household, producers, and municipalities for waste segregation, there is a significant emphasis on enhancing public awareness regarding the critical nature of effective waste management, like various public awareness campaigns in Finland. Likewise in Poland, the study conducted by Rogowska et al. highlighted the respondents' desire for greater awareness regarding circularity and guidance on waste separation via social media platforms (Rogowska et al. 2024). This approach is exemplified by the initiatives of Bratislava's OLO. Additionally, another method identified was the implementation of social campaigns, as currently being executed by the city of Žilina, which features informative posters on proper waste separation placed at public transportation stops. France (Le Bozec 2008), Greece (Emmanouil et al. 2022), and Italy (Messina et al. 2023) use economic measures such as pay-as-you-throw schemes (Ukkonen and Sahimaa 2021) to encourage waste reduction and promote recycling. Implementing similar financial incentives in Slovakia could motivate households to minimize waste generation. The Zero Waste Cities Certification (Zerowastecities 2024) focuses on reducing waste generation through community engagement in Slovenia and Belgium. A uniform solution is not always effective in addressing the complexities of waste management in diverse regions; however, exemplary practices and innovative strategies—such as economic instruments, comprehensive policies, and advanced treatment methods—can be evaluated to identify the best solutions for Slovakia's unique context.

This article aims to outline the benefits of implementing variable fees for produced waste and to offer insight into the importance of motivating individuals to support waste separation. While many municipalities continue to rely on a tariff system based on the number of family members, this approach lacks the necessary incentives for citizens to engage in waste separation efforts. It is crucial to provide individuals with the opportunity to separate waste, especially in light of the growing concern over environmental issues, such as the climate crisis. By examining case studies from Slovak municipalities, it becomes evident that municipalities have the potential to significantly reduce waste when they prioritize waste management and implement effective separation practices. As a member of the European Union, Slovakia is bound by the EU's goal to recycle at least 65% of municipal solid waste (MSW) and limit landfill usage to 10% by 2035. Therefore, it is imperative for municipal management to prioritize improving their waste management systems by promoting waste separation, implementing fair fees for waste collection, and recognizing the value of waste as a potential raw material. By considering these strategies, municipalities can better achieve their waste reduction targets and contribute to a more sustainable future.

The scope of our study as outlined in this paper is limited by the inadequate data available regarding the economic practices of stakeholders involved in the management of MWS. This lack of information restricts our ability to accurately capture the parameters of our models, particularly those relating to the win-win paradigm (Aguilar-Hernandez et al. 2021) and other circular economy business models. Additionally, the macroeconomic landscape, such as the correlation between the amount of waste and growth of Slovakia's GDP, has been excluded from our models. These limitations hinder a comprehensive understanding of the interplay between economic activities and MWS management.

To achieve progress towards the transition to a circular economy, further research can be directed to a deeper study of the specifics of the implementation of MSW management in various regions of the Slovak Republic, to show advanced methods of waste management and to expand the knowledge of business and the population towards a solution that prevents the development of a circular economy. In this regard, further research is recommended for investigation and giving solutions for decreasing of MSW's landfilling rate. Moreover, the application of Gretl software in our ongoing research efforts will significantly enhance our capability to predict future trends in municipal waste generation by utilizing advanced statistical methodologies such as ARIMA and the Holt–Winters model. Additionally, we plan to compile comprehensive datasets that include daily and monthly waste generation statistics from various targeted regional cities across Slovakia, which is categorized into eight distinct regions. After we have gathered this extensive data, we will perform a detailed comparison of the waste generation trends observed in these regions to derive insightful conclusions.

6. Conclusions

Slovakia is currently facing a pivotal moment in its waste management practices, as it confronts the pressing issues of excessive waste production and an over-reliance on landfill disposal. The country's efforts towards transitioning to a circular economy and promoting recycling have been impeded by a persistent consumerist culture and inadequate avenues for effective waste segregation (Incien 2023).

Our results emphasize critical elements that impact the production of municipal waste, such as the expenses connected with waste generation and the magnitude of the population. These factors can be mitigated through unconventional strategies, such as the implementation of an individualized waste fee system, which has the potential to considerably diminish municipal waste and alleviate the financial strain on local governments.

Furthermore, the financial savings from improved waste management practices can be reinvested into initiatives that restore local biodiversity and promote cultural and tourism activities, fostering a more sustainable community.

In order for Slovakia to successfully transition from a linear production model to a circular economy, prompt and decisive measures are imperative. This necessitates the collaboration of policymakers and waste management authorities to formulate effective strategies that address current challenges while also laying the foundation for a sustainable future. Such efforts to embrace change can serve as a catalyst for a responsible and robust approach to resource management, ultimately yielding benefits for the environment and society at large.

Author Contributions: Conceptualization, E.T. and M.Ď.; methodology, M.Ď.; software, T.T.; validation, M.Ď., E.T. and T.T.; formal analysis, M.Ď.; investigation, E.T.; resources, E.T.; data curation, M.Ď.; writing—original draft preparation, E.T. and T.T.; writing—review and editing, E.T. and M.Ď.; visualization,. E.T. and T.T.; supervision, M.Ď.; project administration, M.Ď.; funding acquisition, E.T. All authors have read and agreed to the published version of the manuscript.

Funding: This research was funded by grant VEGA (Slovakia), grant number VEGA 1/0273/22 Resource efficiency and value creation for economic entities in the sharing economy.

Informed Consent Statement: Not applicable.

Data Availability Statement: Data supporting the reported results can be found on Eurostat and Slovstat.

Conflicts of Interest: The authors declare no conflicts of interest.

References

Aguilar-Hernandez, Glenn A., João F. Dias Rodrigues, and Arnold Tukker. 2021. Macroeconomic, social and environmental impacts of a circular economy up to 2050: A meta-analysis of prospective studies. *Journal of Cleaner Production* 278: 123421. [CrossRef]

Ahmed, Sabbir, Sameera Mubarak, Jia Tina Du, and Santoso Wibowo. 2022. Forecasting the status of municipal waste in smart bins using deep learning. *International Journal of Environmental Research and Public Health* 19: 16798. [CrossRef] [PubMed]

Bellemare, Marie-France, Solen Martin-Déry, Rafael Ziegler, Martine Vezina, Emmanuel Raufflet, and Alex Walsh. 2022. Synergizing Social Economy and Circular Economy. *Canadian Journal of Nonprofit and Social Economy Research* 13: 106–10. [CrossRef]

CEAP. 2020. Circular Economy Action Plan. Available online: https://eur-lex.europa.eu/resource.html?uri=cellar:9903b325-6388-11ea-b735-01aa75ed71a1.0017.02/DOC_1&format=PDF (accessed on 20 May 2024).

Cehlár, Michal, Peter Tauš, Zuzana Šimková, Marcela Taušová, and Igor Šimko. 2021. Analysis of Waste Management Status and Potential of the Circular Economy Model in Slovakia. Available online: https://www.odpady-portal.sk/files/Priloha2/Odborna%20studia_Odpady_FBERG%20final_26082021_.pdf (accessed on 20 May 2024).

Das, Subhasish, S.-H. Lee, Pawan Kumar, Ki-Hyun Kim, Sang Soo Lee, and Satya Sundar Bhattacharya. 2019. Solid waste management: Scope and the challenge of sustainability. *Journal of Cleaner Production* 228: 658–78. [CrossRef]

Dickson, Emmanuel Mela, Astley Hastings, and Jo Smith. 2023. Energy production from municipal solid waste in low to middle-income countries: A case study of how to build a circular economy in Abuja, Nigeria. *Frontiers in Sustainability* 4: 1173474. [CrossRef]

Ding, Yin, Jun Zhao, Jia-Wei Liu, Jizhi Zhou, Liang Cheng, Jia Zhao, Zhe Shao, Iris Çağatay, Bingjun Pan, Xiaonian Li, and et al. 2021. A review of China's municipal solid waste (MSW) and comparison with international regions: Management and technologies in treatment and resource utilization. *Journal of Cleaner Production* 293: 126144. [CrossRef]

EEA (European Environment Agency). 2024. Overview of Landfill Taxes on Municipal Waste Used in EU Member States, 2023. Available online: https://www.eea.europa.eu/en/analysis/maps-and-charts/overview-of-landfill-taxes-on (accessed on 20 October 2024).

El-Saadony, Mohamed T., Ahmed M. Saad, Nahed A. El-Wafai, Hamed E. Abou-Aly, Heba M. Salem, Soliman M. Soliman, Taia A. Abd El-Mageed, Ahmed S. Elrys, Samy Selim, Mohamed E. Abd El-Hack, and et al. 2023. Hazardous wastes and management strategies of landfill leachates: A comprehensive review. *Environmental Technology & Innovation* 31: 103150. [CrossRef]

Emmanouil, Christina, Kalliopi Papadopoulou, Iliana Papamichael, and Antonis A. Zorpas. 2022. Pay-as-You-Throw (PAYT) for Municipal Solid Waste Management in Greece: On Public Opinion and Acceptance. *Sustainability* 14: 15429. [CrossRef]

EU Environment Newsletter. 2024. Circular Wastewater Treatment: Solutions from Industry in the Netherlands and Sweden—European Commission. Available online: https://environment.ec.europa.eu/news/circular-wastewater-treatment-solutions-industry-netherlands-and-sweden-2024-08-29_en (accessed on 18 October 2024).

Eurostat. 2022. Municipal Waste by Waste Operations [env_wasmun]. Available online: https://www.eea.europa.eu/data-and-maps/daviz/municipal-waste-recycled-and-composted-7#tab-chart_7 (accessed on 20 June 2024).

Eurostat. 2024a. Municipal Waste Management Indicators. Available online: https://ec.europa.eu/eurostat/statistics-explained/index.php?title=Waste_management_indicators (accessed on 20 June 2024).

Eurostat. 2024b. Municipal Waste Statistics. Available online: https://ec.europa.eu/eurostat/statistics-explained/index.php?title=Municipal_waste_statistics (accessed on 20 June 2024).

Fernando, Sahan J., and Ambika Zutshi. 2023. Municipal solid waste management in developing economies: A way forward. *Cleaner Waste Systems* 5: 100103. [CrossRef]

Geissdoerfer, Martin, Paulo Savaget, Nancy M. P. Bocken, and Erik Jan Hultink. 2017. The Circular Economy—A new sustainability paradigm? *Journal of Cleaner Production* 143: 757–68. [CrossRef]

Grimm, Nancy B., Stanley H. Faeth, Nancy E. Golubiewski, Charles L. Redman, Jianguo Wu, Xuemei Bai, and John M. Briggs. 2008. Global change and the ecology of cities. *Science* 319: 756–60. [CrossRef]

Haque, Khandaker Foysal, Rifat Zabin, Kumar Yelamarthi, Prasant Yanambaka, and Ahmed Abdelgawad. 2020. An IoT based efficient waste collection system with smart bins. Paper presented at 2020 IEEE 6th World Forum on Internet of Things (WF-IoT), New Orleans, LA, USA, June 2–16; pp. 1–5. [CrossRef]

Hitka, Miloš, Jozef Ďurian, Lenka Ližbetinová, and Eva Nedeliaková. 2023. Crisis COVID-19 and Changes in Motivation in the Public Administraion Sector. *Scientific Papers of the University of Pardubice, Series D: Faculty of Economics and Administration* 31: 1711. [CrossRef]

Hrabčák, Marek. 2011. Municipal Waste Charges and Costs. Available online: https://www.odpady-portal.sk/Dokument/100869/poplatky-a-naklady-miest-na-komunalne-odpady.aspx (accessed on 15 May 2024).

Ichinose, Daisuke. 2024. Landfill Scarcity and the Cost of Waste Disposal. *Environmental and Resource Economics* 87: 629–53. [CrossRef]

IEP. 2024. Environmental Policy Institute. Available online: https://iep.sk/ (accessed on 30 May 2024).

Incien. 2023. Circular Economy. Available online: https://www.incien.sk/cirkularna-ekonomika (accessed on 14 May 2024).

Korhonen, Jouni, Antero Honkasalo, and Jyri Seppälä. 2018. Circular economy: The concept and its limitations. *Ecological Economics* 143: 37–46. [CrossRef]

Kvet, Michal, Emil Kršák, and Karol Matiaško. 2021. Study on Effective Temporal Data Retrieval Leveraging Complex Indexed Architecture. *Applied Sciences* 11: 916. [CrossRef]

Lakatos, Elena Simina, Geng Yong, Andrea Szilagyi, Dan Sorin Clinci, Lucian Georgescu, Catalina Iticescu, and Lucian-Ionel Cioca. 2021. Conceptualizing Core Aspects on Circular Economy in Cities. *Sustainability* 13: 7549. [CrossRef]

Le Bozec, André. 2008. The implementation of PAYT system under the condition of financial balance in France. *Waste Management* 28: 2786–92. [CrossRef]

LEAP. 2024. Law and Enviromental Assistance Platform UN Enviromental Programme. Available online: https://leap.unep.org/en/countries/sk/national-legislation (accessed on 14 May 2024).

Lizbetinova, Lenka, Pavla Lejsková, Eva Nedeliaková, Zdeněk Caha, and Miloš Hitka. 2021. The growing importance of ecological factors to employees in the transport and logicstics sector. *Economic research–Ekonomska Istraživanja* 35: 4379–403. [CrossRef]

Lorincová, Silvia, Milos Hitka, Andrea Novotna, Jozef Durian, and Peter Starchon. 2022. Value preferences supporting company competitiveness in the field of corporate culture. *Polish Journal of Management Studies* 26: 172–88. [CrossRef]

Malichová, Eva, Milan Straka, Ľuboš Buzna, Damiano Scandolari, Mario Scrocca, and Marco Comerio. 2023. Study of traveller's preferences towards travel offer categories and incentives in the journey planning context. *PLoS ONE* 18: e0284844. [CrossRef]

Matová, Hana, Kaputa Vladislav, and Triznová Miroslava. 2019. Responsible Concumer in the context of circular economy. Paper presented at 12th WoodEMA Annual International Scientific Conference on Digitalisation and Circular Economy: Forestry and Forestry based Industry Implications, International House of Scientists Frederic Joliot Currie, Varna, Bulgaria, September 11–13; pp. 69–74.
Messina, Giovanna, Antonella Tomasi, Giorgio Ivaldi, and Francesco Vidoli. 2023. Pay as you own' or 'pay as you throw'? A counterfactual evaluation of alternative financing schemes for waste services. *Journal of Cleaner Production* 412: 137363. [CrossRef]
Ministry of the Environment of the Slovak Republic. 2021. Waste Management Program of the Slovak Republic for the Years 2021–2025. Available online: www.minzp.sk (accessed on 3 May 2024).
Mor, Suman, and Khaiwal Ravindra. 2023. Municipal solid waste landfills in lower- and middle-income countries: Environmental impacts, challenges and sustainable management practices. *Process Safety and Environmental Protection* 174: 510–30. [CrossRef]
Moraga, Gustavo, Sophie Huysveld, Fabrice Mathieux, Gian Andrea Blengini, Luc Alaerts, Karel Van Acker, de Steven Meester, and Jo Dewulf. 2019. Circular economy indicators: What do they measure? *Resources, Conservation and Recycling* 146: 452–61. [CrossRef]
Morseletto, Piero. 2020. Targets for a circular economy. *Resources, Conservation and Recycling* 153: 104553. [CrossRef]
Murray, Robin. 2002. *Zero Waste*. London: Greenpeace Environmental Trust. 211p, Available online: https://www.greenpeace.org.uk (accessed on 25 May 2024).
Nanda, Sonil, and Franco Berruti. 2021. Municipal solid waste management and landfilling technologies: A review. *Environmental Chemistry Letters* 19: 1433–56. [CrossRef]
Ololade, Olusola Oluwayemisi, and Israel R. Orimoloye. 2021. Contribution of a Well-Managed Landfill to Sustainable Development. In *Handbook of Solid Waste Management*. Edited by Chinnappan Baskar, Seeram Ramakrishna, Shikha Baskar, Rashmi Sharma, Amutha Chinnappan and Rasdhmi Sehrawat. Singapore: Springer. [CrossRef]
Onungwe, Ishmael, Dexter V. L. Hunt, and Ian Jefferson. 2024. Transitioning towards circular economy through municipal solid waste analysis and characterisation using SowaCLINK software. *Frontiers in Sustainability* 5: 1321329. [CrossRef]
Potočár, Radovan. 2021. Bulk Waste Collection in the Municipality. Available online: https://www.odpady-portal.sk/Dokument/105895/boli-sme-v-obci-kde-zaviedli-mnozstvovy-zber-a-vazia-vsetky-nadoby-ako-to-funguje.aspx (accessed on 16 April 2024).
Ramachandra, T. V. 2006. *Management of Municipal Solid Waste*. Vancouver: Commonwealt of Learning. Bangalore: Indian Institute of Science. New Delhi: TERI Press.
Reddy, P. Jayarama. 2011. *Municipal Solid Waste Management*. Kolkata: Book Syndicate.
Rogowska, Justyna, Kamila Piątkowska, and Zuzanna Głowacz. 2024. Societal Involvement in Household Waste Sorting Behavior in the Context of the Circular Economy: A Case Study of Poland. *Sustainability* 16: 1841. [CrossRef]
Rolewicz-Kalińska, Anna, Krystyna Lelicińska-Serafin, and Piotr Manczarski. 2020. The Circular Economy and Organic Fraction of Municipal Solid Waste Recycling Strategies. *Energies* 13: 4366. [CrossRef]
Sadovská, Eva. 2022. Analysis of Waste Collection in the Municipality. Available online: https://www.odpady-portal.sk/Dokument/107118/produkcia-komunalneho-odpadu-v-sr.aspx (accessed on 30 April 2024).
Sandhi, Arifin, and Joacim Rosenlund. 2024. Municipal solid waste management in Scandinavia and key factors for improved waste segregation: A review. *Cleaner Waste Systems* 8: 100144. [CrossRef]
Sondh, Sidhartha, Darshit S. Upadhyay, Sanjay Patel, and Rajesh N. Patel. 2024. Strategic approach towards sustainability by promoting circular economy-based municipal solid waste management system—A review. *Sustainable Chemistry and Pharmacy* 37: 101337. [CrossRef]
Staffenova, Nikola, and Alžbeta Kucharčíková. 2023. Digitalization in the Human Capital Management. *Systems* 11: 337. [CrossRef]
Steingartner, William, and Valerie Novitzká. 2024. Operational Semantics in a Domain-Speecific RoboControl Language: A Pedagogical Use Case. *Computer Science and Informational Systems* 21: 1077–95. [CrossRef]
Szpilko, Danuta, Antonio de la Torre Gallegos, Felix Jimenez Naharro, Agnieszka Rzepka, and Angelika Remiszewska. 2023. Waste management in the smart city: Current practices and future directions. *Resources* 12: 115. [CrossRef]
Townsend, Timothy G., Jon Powell, Pradeep Jain, Qiyong Xu, Thabet Tolaymat, and Debra Reinhart. 2015. Planning for Sustainable Landfilling Practices. Sustainable Practices for Landfill Design and Operation. In *Sustainable Practices for Landfill Design and Operation*. New York: Springer, pp. 35–51. [CrossRef]
Ukkonen, Aino, and Olli Sahimaa. 2021. Weight-based pay-as-you-throw pricing model: Encouraging sorting in households through waste fees. *Waste Management* 135: 372–80. [CrossRef]
Vaverková, Magdalena Daria. 2019. Landfill impacts on the environment. *Geosciences* 9: 431. [CrossRef]
Wang, Kehua, and Toyohiko Nakakubo. 2020. Comparative assessment of waste disposal systems and technologies with regard to greenhouse gas emissions: A case study of municipal solid waste treatment options in China. *Journal of Cleaner Production* 260: 120827. [CrossRef]
Wilson, David C., Ljiljana Rodic, Anne Scheinberg, Costas A. Velis, and Graham Alabaster. 2012. Comparative analysis of solid waste management in 20 cities. *Waste Management & Research* 30: 237–54. [CrossRef]
Zerowastecities. 2024. Available online: https://zerowastecities.eu/ (accessed on 18 October 2024).
Zhironkin, Sergey, and Michal Cehlár. 2022. Green Economy and Sustainable Development: The Outlook. *Energies* 15: 1167. [CrossRef]

energies

MDPI

Article

Assessing the Additional Benefits of Thailand's Approaches to Reduce Motor Vehicle Emissions

Pantitcha Thanatrakolsri [1,2,*] and Duanpen Sirithian [1,2]

[1] Faculty of Public Health, Thammasat University, Lampang 52190, Thailand; duanpen.s@fph.tu.ac.th
[2] Thammasat University Research Unit in Environment, Health and Epidemiology, Lampang 52190, Thailand
* Correspondence: pantitcha.o@fph.tu.ac.th

Abstract: Air pollutants and greenhouse gases (GHGs) represent major challenges in our era, contributing to climate change and global health issues. These problems arise from a variety of well-known sources, including motor vehicles. Almost all nations, Thailand included, have formulated and implemented policies to curb greenhouse gas (GHG) emissions in line with the requirements and commitments of the Paris Agreement. The evaluation of specific air pollutants and GHG emissions originating from road vehicles utilises the Thailand database, referencing the year 2019. Data intersections from 2019 to 2022 are grounded in actual data collected from relevant departments in Thailand, while projections for 2023–2030 are forecasted based on the baseline year. The secondary database used in the International Vehicle Emission model is adjusted according to real-world driving data to accurately reflect country-specific emission factors. Dynamic emission factors for specific air pollutants and GHGs are evaluated and integrated with the average Vehicle Kilometres Travelled (VKT) for each vehicle category. The Business-As-Usual (BAU) scenario is then examined, based on existing policies aimed at reducing air pollutants and GHG emissions in Thailand's transport sector. These policies include strategies for the adoption of electric vehicles and the promotion of public transport to reduce VKT. Under the BAU scenario, the overall number of road vehicles in Thailand, including passenger cars, motorcycles, pickups, vans, trucks, and buses, is expected to increase by approximately 6.58% by 2030, leading to a rise in specific air pollutants and GHG emissions compared to the 2019 baseline. However, by adhering to Thailand's strategies and transitioning to new electric passenger cars and buses, greenhouse gas emissions and specific air pollutants from the road transport sector will be significantly reduced.

Keywords: greenhouse gas; particulate matter; mitigation measure; co-benefit; electric vehicle

1. Introduction

Climate change represents the most significant health threat currently facing humanity, and healthcare professionals worldwide are actively addressing the health consequences of this unfolding crisis. The Intergovernmental Panel on Climate Change (IPCC) has conveyed a clear and resolute message: in order to mitigate severe health impacts and avert the potential for millions of climate change-related fatalities, it is imperative for the world to take decisive action to limit the temperature rise to 1.5 °C. Regrettably, historical emissions have already initiated a certain degree of global temperature increase and triggered other climate changes that are now unavoidable. However, it is crucial to recognise that even a global warming of up to 1.5 °C falls short of a safe threshold; every additional tenth of a degree of warming will exact a profound toll on the well-being and lives of people [1–4]. Furthermore, 930 million individuals, representing approximately 12% of the global population, devote a significant portion of their household budgets, at least 10%, to cover healthcare expenses. Among them, the most economically disadvantaged individuals often lack insurance, leaving them susceptible to unforeseen health-related financial burdens. Consequently, an estimated 100 million people are pushed into poverty annually due to health shocks and stresses. The worrying reality is that the impacts of climate change are worsening this troubling trend, adding to the challenges experienced by these vulnerable communities [5–7].

The main driver of climate change is the greenhouse effect caused by greenhouse gas emissions (GHGs). However, it is not just greenhouse gas (GHG) emissions that affect the environment and health; ambient air pollution also poses a significant environmental health challenge, impacting populations across all income levels, from low- to high-income nations [8–11]. The World Health Organization (WHO) has shed light on the specific health impacts of ambient air pollution. In 2019, it estimated that approximately 37% of premature deaths linked to ambient air pollution were due to ischaemic heart disease and stroke, with chronic obstructive pulmonary disease and acute lower respiratory infections contributing 18% and 23%, respectively, to these deaths. Additionally, 11% of the fatalities were associated with cancer of the respiratory tract. This detailed breakdown illustrates the complex and widespread health risks posed by outdoor air pollution. Moreover, the burden of outdoor air pollution falls disproportionately on those living in low- and middle-income countries, where an alarming 89% of the 4.2 million premature deaths occurred [12–20].

The road transport sector, which includes mobile sources, is a major contributor to GHGs, with carbon dioxide (CO_2) being the most significant. CO_2 plays a key role in worsening climate change and global warming. Alongside CO_2, vehicles also emit other GHGs like methane (CH_4) and nitrous oxide (N_2O), released during the production and transportation of fossil fuels. Additionally, mobile sources are significant emitters of air pollutants, including particulate matter (PM), nitrogen oxides (NO_x), carbon monoxide (CO), and sulfur dioxide (SO_2) [21–25].

Globally, the road transport sector accounts for roughly 14% (8.1 $GtCO_2eq$) of total GHG emissions and around 20% of CO_2 emissions as of 2022. Moreover, GHG emissions from transport experienced the largest reduction compared to the pre-COVID-19 period, decreasing by approximately 14.1% in 2020. However, in 2022, this sector saw a significant rebound, with emissions increasing by 4.7% [26,27]. In Thailand, the nation's total GHG emissions, excluding those from land use, land-use change, and forestry, amounted to 372,716.86 $GgCO_2eq$. In 2019, Thailand's primary source of GHG emissions remained the energy sector, contributing 69.96% (260,772.69 $GgCO_2eq$) of the country's overall GHG emissions. Within this sector, a significant share of emissions originated from fuel combustion, particularly in the energy industries, which accounted for roughly 103,356.15 $GgCO_2eq$ (39.63%) of the total. The transport sector was the second largest contributor, representing 76,923.02 $GgCO_2eq$ (29.50%) of the energy sector's total emissions [28].

As previously mentioned, the road transport sector significantly contributes to ambient air pollution in many countries [29–34], including Thailand. The main challenge concerning air pollution revolves around the concentration of particulate matter, including particles smaller than 2.5 μm in diameter ($PM_{2.5}$) and those smaller than 10 μm in diameter (PM_{10}). Emission sources vary across geographic areas, with urban and suburban regions primarily facing air pollution from mobile sources and open burning activities. Notably, open burning is more prevalent during specific periods, particularly before agricultural production cycles, while mobile sources consistently impact air quality throughout the year [35–37]. Therefore, the management of mobile emission sources for reducing GHGs and specific air pollutants should be reviewed nationwide.

Updating the mobile source emission inventory with dynamic emission factors is crucial for improving measures to mitigate GHG and air pollutant emissions. Numerous studies have assessed emission inventories using bottom-up, top-down, or combined approaches to develop inventories of air pollutants. Mobile emission sources offer valuable insights into local pollutant emission characteristics, aiding air quality simulations and the development of pollution control measures, as observed in regions such as South America [38], China [39], India [40], and Ecuador [41].

In addressing climate change, Thailand has outlined its nationally determined contribution (NDC) roadmap for 2021–2030. Thailand submitted its second updated NDC in

2022, demonstrating a strong commitment to reducing GHG emissions by 30% compared to the Business-As-Usual (BAU) level by 2030 [28]. The NDC roadmap prioritises the transportation sector, aiming to improve transportation and logistics efficiency, develop low-carbon transport infrastructure, and incorporate sustainable practices to manage transportation demand. This involves not only reducing GHG emissions but also curbing air pollutants [42–44]. Therefore, to effectively tackle climate change and air pollution resulting from the rapid growth of urban traffic, it is crucial to evaluate the additional benefits of reducing GHG emissions and air pollutants in the urban transport sector.

Currently, co-benefits are often emphasised in discussions about reducing GHG emissions and promoting human health. The co-benefit analysis approach provides a novel method for quantitatively evaluating and developing comprehensive policies related to regional environmental management and climate change. This approach has increasingly become an essential policy tool for promoting sustainable development in the modern era [45–51].

In the coming years, as urbanisation accelerates, achieving sustainable transportation development has become a crucial objective in global urban infrastructure planning. Several studies have evaluated the co-benefits of reducing GHG and air pollutant emissions in the transportation sector, from the national and regional levels down to the city level, in the pursuit of sustainable development goals [46,52,53]. Kim et al. (2020) conducted a comprehensive investigation, merging an air quality model, economic model, and health assessment model to assess the health-related advantages of climate change mitigation across various scenarios in South Korea [47]. Wei et al. (2020) assessed the potential co-benefits of air pollution control and climate mitigation policies within China's electricity sector [50]. In a separate study, Jiao et al. (2020) examined the impact of energy-related measures outlined in the action plan on emissions of major air pollutants and CO_2 within the urban transport sector, focusing on the case of Guangzhou [46].

In Thailand, measures of specific air pollutants and GHGs have been implemented to improve air quality in line with established regulations and to meet the commitments made under the Paris Agreement to combat climate change. It is important to note that the sources of emissions for both GHGs and specific air pollutants are often the same. Therefore, it is essential to implement management strategies concurrently, leveraging co-benefits from mitigation measures to improve air quality and reduce the impacts of climate change simultaneously.

Consequently, this study aims to update and evaluate the dynamic emission factors of both GHGs, such as CO_2, N_2O, and CH_4, as well as specific air pollutants, including CO, VOC, VOC_{evap}, NO_x, SO_2, and PM_{10} from mobile sources using a vehicle emission model. Subsequently, the study examines the emission inventory of each vehicle type by linking the emission factors (g/km) to the annual average vehicle kilometre travel (km/year). Additionally, emissions are modelled for each scenario to evaluate the impact of various measures outlined in the policies. The contributions of different measures within the policies to the co-benefits are further quantified, aiding in the identification of measures with significant potential for reducing emissions of GHGs and specific air pollutants in Thailand.

2. Materials and Methods

To explore efficient tactics for diminishing GHG and specific air pollutant emissions within the urban transportation sector, a vehicle emission model has been formulated as a valuable instrument for conducting assessments of vehicle emission inventories in diverse developing Asian urban areas, such as Hanoi, Vietnam [54]; Bandung, Indonesia [55]; alongside cities in Ecuador [38,56] and India [57].

2.1. Vehicle Emission Model

The research utilised the International Vehicle Emissions (IVE) Model version 2.0 to compile an inventory of emissions estimated from mobile sources in Thailand, using

2019 as the base year. Figure 1 illustrates the overarching methodology employed to assess emissions through the IVE model in the study. To accurately compile inventories of emissions from mobile sources, three key elements are essential: vehicle emission rates, vehicle usage patterns, and the distribution of the vehicle fleet. The IVE model is specifically designed to utilise readily available local data and relevant existing information to quantify these crucial inputs.

The IVE model employs a method to estimate emissions by multiplying the base emission rate for each technology with correction factors adjusted for specific vehicle technologies. These correction factors, defined for each vehicle type, are then multiplied by the amount of vehicle travel associated with each technology, resulting in the total emitted emissions. Equation (1) within the model illustrates the internal calculation process for determining the adjusted emission rate. Equations (2) and (3) further explain the process by multiplying the base emission rate (B) with a series of correction factors (K) to estimate the adjusted emission rate (Q) for each vehicle type. Furthermore, the correction factors can be grouped into distinct categories, encompassing location-related variables (such as ambient temperature, humidity, altitude, inspection or maintenance programmes, and base emission adjustments), fuel quality, and power- and driving-related variables (including vehicle specific power or VSP, road grade, air conditioning usage, and start distribution) [58,59].

$$Q_{(hot/cold)(t)} = B_{(hot/cold)(t)} \times K_{(t)(1)} \times K_{(t)(2)} \times \ldots K_{(t)(n)} \tag{1}$$

$$Q_{(hot,t)} = \sum\nolimits_t \left\{ f_t \times \sum\nolimits_d \left[Q_t \times U_{DC} \times f_{(dt)} \times \ldots K_{(dt)} \right] \right\} / U_c \tag{2}$$

$$Q_{(hot,t)} = \sum\nolimits_t \left\{ f_t \times Q_t \times \sum\nolimits_d \left[f_{dt} \times K_{dt} \right] \right\} \tag{3}$$

where

$Q_{(hot/cold)(t)}$ = the emission factor for each vehicle type and technology, either for hot-running conditions in grams per kilometre (g/km) or cold-start conditions in grams per start;

$B_{(hot/cold)(t)}$ = the base emission factor for each vehicle type and technology t, (hot-running in g/km or cold-start in g/start);

$K_{(1,2,\ldots n),\,t}$ = a set of correction factors specific to each vehicle type and technology t;

f_t = the fraction of travel attributed to a particular technology;

$f_{(dt)}$ = the fraction of time associated with different types of driving or the proportion of soaks attributed to a specific technology;

U_{DC} = the average velocity of the test procedure driving cycle, maintained as a constant value in kilometres per hour (km/h);

U_C = the average velocity derived from the specific driving cycle, as input by the user within the location file, expressed in kilometres per hour (km/h).

Creating input data for the IVE model entails producing the location file, fleet file, and base adjustments file, all pivotal elements for assessing emissions. Guaranteeing precise emissions estimation within a designated area necessitates compiling accurate data pertaining to driving habits and local environmental circumstances. Subsequently, these gathered data are input into the IVE location file, which contains particulars regarding the driving behaviours prevalent in the region. This encompasses variables like the frequency and manner of driving, incorporating details about speed and acceleration profiles.

The driving behaviours within the IVE model are elucidated through the characterisation of VSP and engine stress parameters. These parameters are derived from fundamental vehicle type information and a comprehensive velocity trace recorded on a per-second basis. VSP is determined based on the velocity and acceleration data recorded instantaneously. Engine stress, in tandem with VSP, plays a crucial role in enhancing the accuracy of emission estimates for pollutants such as CO, NO_x, and HC. Research indicates that engine stress demonstrates the strongest correlation with vehicle power load demands over the previous 20 s of operation, implying a close relationship with the inferred engine

RPM [38,60]. Equations (4) and (5) then outline a method for estimating VSP and engine stress, providing a systematic approach to quantify these pivotal parameters [58].

$$\mathrm{VSP} = \mathrm{v}[1.1\mathrm{a} + 9.81(\mathrm{atan}(\sin(\mathrm{grade}))) + 0.132] + 0.000302\mathrm{v}^3 \tag{4}$$

where
grade = $(h_{t=0} - h_{t=-1})/v_{(t=-1 \text{ to } 0\text{ s})}$;
v = velocity (m/s);
a = acceleration (m/s^2);
h = altitude (m).

$$\text{Engine Stress}(\text{unitless}) = \text{RPMIndex} + \left(0.08\frac{\text{ton}}{\text{kW}}\right) \times \text{Preaverage Power} \tag{5}$$

where
Preaverage Power = Average ($VSP_{t=-5\text{ s to }-25\text{ s}}$) (kW/ton);
RPMIndex = $\text{Velocity}_{t=0}$/SpeedDivider (unitless);
Minimum RPMIndex = 0.9.

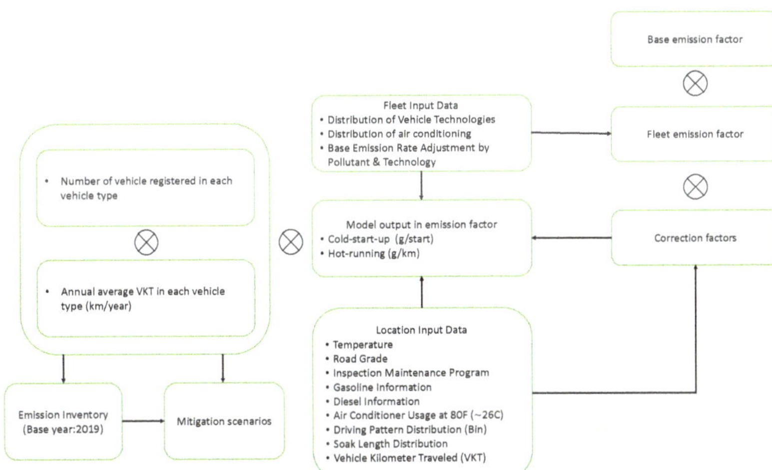

Figure 1. Scope of the study (adapted from [61]).

Stress levels are classified into three categories—low, medium, and high—while VSP is divided into 20 categories, resulting in a total of 60 bins that incorporate both VSP and stress. Low engine stress typically characterises scenarios where the vehicle has experienced low speeds and accelerations in the preceding 20 s of operation, resulting in a low engine RPM. Conversely, high engine stress occurs during periods of high speeds and accelerations in the most recent 20 s, coinciding with a high engine RPM [59,60].

Moreover, the location file logs commencement patterns, requiring information on the frequency of vehicle starts and the duration of engine idle periods prior to restarts. It also meticulously documents essential environmental factors such as altitude, road gradient, and temperature. Finally, it considers fuel attributes like quality and the levels of pollutants and additives. These precise data form the foundation for accurately assessing emissions within the chosen geographical region.

The fleet file serves as a repository for the proportion of vehicle travel associated with each technology. In the IVE model, there are a total of 1372 predefined technologies, supplemented by an additional 45 undefined technologies. These technologies are systematically classified based on diverse parameters, including vehicle size (including trucks), fuel type, vehicle application, fuel supply system, evaporative control system, and exhaust control system/standards. For this study, 71 technologies of specific vehicle types were selected for estimating emission factors.

Furthermore, the IVE model employs base emission rates for both hot-running and cold-start emissions associated with each vehicle technology integrated into the model. These base emissions are determined through dynamometer testing conducted under specific cycles and standard conditions. Most of the current data used to establish these base emission factors originate from research conducted in the United States. Consequently, the base emission rates are modified using vehicle emissions data obtained under the Bangkok driving patterns or driving cycles (345BKKDV) in this study.

Ultimately, the emission factors for conventional air pollutants and GHGs are evaluated, and the emission inventory is calculated based on the vehicle kilometres travelled (VKT) per year and the number of registered vehicles for each specific vehicle type.

2.2. Data Collection

2.2.1. Vehicle Technology and Fuel Type

Improving air quality and protecting public health requires regulating emission sources through the enforcement of emission standards, including those for mobile sources. In Thailand, vehicle emission standards align with European standards and testing procedures. The Euro emission standard includes regulations that govern vehicle emissions for new vehicles in the European Union.

Introduced with the Euro 1 emission standard taking effect in 1992 in Europe [61,62], Thailand adopted the Euro 1, 2, and 3 emission standards in 1998, 2001, and 2005, progressively raising these standards over the years to enforce the Euro 4 standard since 2012 [35,61–63]. These mobile source regulations are complemented by public awareness campaigns promoting the use of public transport and vehicles powered by alternative fuels, such as natural gas and electricity. Additionally, the introduction of the Euro 5 and Euro 6 standards is poised to further enhance emissions control by targeting reductions in nitrogen oxide and hydrocarbon emissions, while also introducing a particulate number standard [64]. Furthermore, Thailand implemented the Euro 5 fuel standard in January 2024 [65].

Additional vehicle data were obtained from the Transportation Statistics Group, Planning Division, Department of Land Transport [66]. Engine technology and emission control for each vehicle type in Thailand are determined based on their specific technological specifications and adherence to emission standards. Passenger cars, vans, and pickups use Euro 1–4 engine standards, equipped with three-way catalysts but lacking exhaust gas recirculation (EGR) technology, while buses and trucks employ EGR technology.

Fuel injection (FI) and multi-point fuel injection (MPFI) technologies are utilised for air and fuel control across all vehicle types except motorcycles, which may feature four-cycle engines. Additionally, gasoline, diesel, natural gas for vehicles (NGV), and liquefied petroleum gas (LPG) fuels are used in all vehicle types except motorcycles, which exclusively use gasoline. In 2019, gasoline was the primary fuel type used in vehicles (61%), followed by diesel (29%), LPG (7%), compressed natural gas (CNG) (2%), and hybrid (1%).

2.2.2. Number of Vehicles and Mobility in Thailand

Thailand has experienced a significant increase in its vehicle population in recent years, reflecting the country's rapid urbanisation and economic growth. Official statistics from the Thai Ministry of Transport reveal that the number of registered vehicles in Thailand has steadily risen over the past decade. This growth includes various vehicle types, with motorcycles making up the largest share, followed by passenger cars, pickups, trucks, vans, and buses [66].

The increase in vehicle ownership has been linked to the significant impact of economic growth on vehicle ownership trends, characterised by two main patterns. Firstly, with rising income levels, there is a general increase in private vehicle ownership, indicative of improved affordability and accessibility. Secondly, once individuals reach a certain level of personal income, there is often a shift from motorcycle ownership to private car ownership [67–69].

The annual breakdown of registered motor vehicle types in Thailand from 2018 to 2021 shows that motorcycles made up the largest proportion (52.25–53.89%), followed by passenger cars (25.36–26.77%), pickups (16.72–17.21%), trucks (2.03–2.89%), vans (1.05–1.07%), and buses (0.31–0.42%). Consequently, the investigation focuses on assessing vehicle emissions for motorcycles, passenger cars, pickups, trucks, vans, and buses. Additionally, there was a consistent increase in the total number of registered vehicles in Thailand from 2008 to 2022.

2.2.3. Vehicle Kilometres Travelled (VKT) Calculation

The data on VKT across various vehicle categories in Thailand have been estimated through comprehensive literature reviews and information from the Pollution Control Department (PCD) database. Estimating emissions from mobile sources in Thailand requires considering the VKT of each vehicle, measured in kilometres per year. This calculation incorporates emission factors derived from model outputs specific to each vehicle type, along with the annual count of registered vehicles.

Furthermore, for the input data required for the vehicle emission model, vehicles are grouped into three categories based on their mileage: up to 79,000 km, 80,000–161,000 km, and over 161,000 km. This classification applies to various vehicle types, including passenger cars, vans, pickups, buses, and trucks. Similarly, motorcycles are segmented into three categories based on their annual VKT: up to 25,000 km, 25,000–50,000 km, and over 50,000 km. Additionally, the size of all vehicles is determined by their gross weight and engine capacity, which are classified into small, medium, and large groups according to the emission model. However, only engine capacity is considered for motorcycles.

The expected mileage for each vehicle type is calculated using vehicle age data provided by the Department of Land Transport (DLT) in Thailand. The mileage categories are then determined by multiplying the vehicle's age (in years) by the average annual VKT in kilometres. For the base year 2019, the breakdown of vehicle ages is broadly as follows: most vehicles are 6 years old, followed by those that are 7 years and 15 years old, respectively. In this research, vehicles older than 20 years are not taken into account, except for buses, as they are still in operation.

2.3. Vehicle Emission Model Input Data

2.3.1. Driving Pattern

Driving patterns, also known as driving behaviours, encompass the various habits and actions of drivers while operating vehicles. These patterns are significantly influenced by factors such as revolutions per minute (RPM), acceleration, deceleration, and speed. Additionally, external factors like traffic congestion and road slope directly impact fuel consumption, air pollutant emissions, and GHG emissions [70,71]. The speed profile measures both speed (km/h) and time (seconds; s).

Figure 2 illustrates the Bangkok driving patterns or driving cycles (345BKKDC) of vehicles used in the study. The average driving speed for buses and trucks is approximately 23.37 km/h, while passenger cars and motorcycles maintain an average speed of 33.47 km/h, and vans and pickups travel at around 35.34 km/h. Additionally, the maximum speeds recorded for these categories are approximately 66.90 km/h, 90.92 km/h, and 86.42 km/h, respectively.

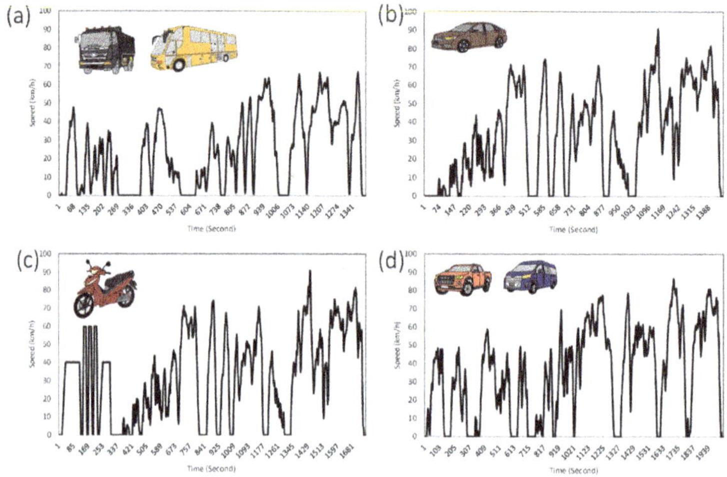

Figure 2. The Bangkok driving cycle (345BKKDC): (**a**) truck and bus driving cycle; (**b**) passenger car and (**c**) motorcycle driving cycle; (**d**) van and pickup driving cycle.

The data align with findings from the Intelligent Traffic Information Center Foundation (iTIC) regarding average speeds in Bangkok and surrounding areas. Between 2011 and 2016, the average speeds, covering both urban and rural regions, declined from 36.29 km/h to 30.92 km/h [72]. Furthermore, the 345BKKDC typically conducts vehicle emissions tests, providing a representation of emissions across a range of vehicle speeds, including low, medium, and high. These tests are conducted at the Automotive Emission Laboratory (AEL) within the Pollution Control Department, Ministry of Natural Resources and Environment in Thailand.

2.3.2. Fleet Characteristics

Table 1 presents the fleet characteristics input for the base year (2019), including fuel type, vehicle technology, and vehicle size. Gasoline is the primary fuel type for passenger cars and motorcycles, while trucks, vans, and pickups predominantly use diesel fuel. Additionally, alternative fuels such as CNG, LPG, and hybrid technologies are also considered.

Table 1. The fleet characteristics input to the IVE model for the base year (2019).

Vehicle Type	Passenger Car	Van	Pick up	Motorcycle	Bus	Truck
1. Fuel type						
• Gasoline	61%	7%	3%	100%	3%	-
• Diesel	29%	86%	94%	-	78%	77%
• NGV/NGV-retrofit	2%	3%	1%	-	17%	3%
• LPG/LPG-retrofit	7%	4%	2%	-	2%	-
• Hybrid	1%	-	-	-	-	-
2. Technology 2.1 Exhaust	Euro 1-/3Way	Euro 1-/3Way	Euro 1-/3Way	Catalyst	Euro 2–3/3Way/EGR	Euro 2–3/3Way/EGR
2.2 Air/fuel control	FI/MPFI/Carburettor	FI/MPFI/Carburettor	FI/MPFI/Carburettor	FI/Carburettor/4 cycle	FI/MPFI/Carburettor	FI/MPFI/Carburettor
3. Vehicle size	Light/Medium	Medium	Medium	Medium	Medium	Medium

Vehicle technology includes both exhaust and air/fuel control systems. The exhaust system encompasses Euro 1–4 engine standards, with the use of exhaust gas recirculation (EGR) varying depending on the specific vehicle technology [73] currently utilised in Thailand. For motorcycles, four-cycle technology is currently employed. Moreover, air/fuel control encompasses fuel injection (FI), multi-point fuel injection (MPFI), and carburettor systems.

Furthermore, vehicles are classified based on size into three distinct groups. The first group comprises light-duty vehicles, such as passenger cars, vans, and pickups, categorised as follows: those weighing between 5000 and 6600 lbs. with engines smaller than 1.5 litres (small size), those weighing between 6601 and 9000 lbs. with engines less than 3 litres (medium size), and those exceeding 9001 lbs. with engines larger than 3 litres (large size).

The second group consists of heavy-duty vehicles, including buses and trucks, categorised as follows: those weighing between 9000 and 14,000 lbs, those weighing between 14,001 and 33,000 lbs, and those exceeding 33,000 lbs, classified as small, medium, and large sizes, respectively.

The last group encompasses motorcycles and is classified as follows: engines with less than 100 cc, engines with less than 300 cc, and engines exceeding 301 cc, also categorised as small, medium, and large sizes, respectively [60].

Additionally, the input data for location vary for each vehicle, but parameters such as temperature, humidity, and fuel properties remain consistent. The average temperature is set at 28 degrees Celsius, with humidity averaging approximately 76%, and the fuel characteristics, for both gasoline and diesel, adhere to the Euro 4 standard [24]. Furthermore, the distribution of technology within each vehicle type is provided as input data, with approximately 25%, 3%, 18%, 18%, 14%, and 10% representing different vehicle technologies allocated to passenger cars, motorcycles, vans, pickups, buses, and trucks, respectively, in the base year.

2.3.3. Base Adjustment

To evaluate vehicle emissions of specific air pollutants and GHGs, a chassis dynamometer test was conducted following the Bangkok driving cycle (345BKKDV), a protocol developed by the Pollution Control Department (PCD) of the Ministry of Natural Resources and Environment in Thailand. These experiments took place at the Automotive Emission Laboratory (AEL) within the PCD.

Vehicle emission factors were established through chassis dynamometer testing, with test conditions closely aligned to regulatory standards, including Thai Industrial Standard 2560-2554 [74]. The emissions database, which records data on CO, NO_x, PM_{10}, CO_2, and CH_4 emissions from vehicles classified by technology and VKT, is used to provide input for the base adjustment rate.

2.4. Emission Inventory Calculation

The specific air pollutants, encompassing CO, VOC, VOC_{evap}, NO_x, SO_x, and PM_{10}, alongside GHGs such as CO_2, CH_4, and N_2O emitted by motor vehicles, are assessed using the IVE model. These emissions are subsequently calculated based on the total annual VKT. The results are presented in Equation (6).

$$\sum EI_{i,j} = EF_{i,j} \times VKT_{i,j} \times N_{i.j} \quad (6)$$

where
i = type of pollutant or GHG;
j = type of vehicle;
EI = emission inventory;
EF = emission factor;
N = total number of vehicles.

2.5. Scenario Analysis

In line with Thailand's policies addressing air pollutants and GHGs from mobile sources, there is a strong emphasis on mitigation strategies aimed at reducing emissions at their source. The National Climate Change Policy Committee (NCCC) has tasked the Energy Policy and Planning Office (EPPO) with leading the development of the national action plan for reducing GHG emissions in the energy sector from 2021 to 2030. This initiative involves collaboration with the Office of Natural Resources and Environmental Policy and Planning (ONEP) and the Thailand Greenhouse Gas Management Organisation (TGO). Moreover, the Office of Transport and Traffic Policy and Planning (OTP) has developed the NDC Sectoral Action Plan for the Transport Sector for the same period, aiming to achieve significant reductions in GHG emissions, equivalent to millions of tonnes of carbon dioxide.

The focus is on creating effective and sustainable transportation systems to reduce energy consumption in the transport sector, alleviate traffic congestion, develop liveable urban areas, and mitigate GHG emissions. To achieve this goal, three key measures are recommended: Firstly, developing policies that discourage unnecessary travel and reduce travel distances by seamlessly integrating urban planning with appropriate transportation planning. Secondly, promoting a shift from private vehicle use to more environmentally friendly and efficient modes of travel, such as public transport, cycling, and walking. Lastly, implementing policies to improve the energy efficiency of motor vehicles and optimise fuel efficiency. This includes adopting automotive energy-saving technologies, exploring alternative energy sources, promoting gasohol use, and popularising electric vehicles (EVs) and bicycles [28].

These action plans include specific measures aimed at reducing air pollution, such as converting diesel bus engines to NGV and Evs [75]. Moreover, promoting public transport [76] and adopting alternative vehicle engines, such as Evs, are seen as strategies for mitigating GHG emissions [77]. Additionally, Evs represent a continuously evolving transportation technology that has emerged over the last decade, characterised by zero emissions [78,79]. Therefore, this study focuses on evaluating scenarios involving encouraging measures such as promoting the use of public transport, transitioning new passenger cars to Evs, and converting buses from diesel engines to Evs.

3. Results and Discussion

3.1. VKT Estimation

The updated PCD database and relevant research [80] provide estimates for the annual average VKT within each primary vehicle category, as shown in Table 2. Buses have the highest average annual VKT per vehicle, followed by trucks, vans, pickups, passenger cars, and motorcycles, respectively. However, the total VKT during the base year was the highest for passenger cars, followed by pickups, motorcycles, trucks, buses, and vans, respectively. This discrepancy can be attributed to the number of vehicles, which affects the assessment of the emission inventory [24,54].

Table 2. The annual average mileage and the total mileage of passenger cars, motorcycles, vans, pickups, buses, and trucks in Thailand.

Vehicle Type	Annual Average VKT (Thousands of Kilometres)	N	Total VKT (2019) (Billions of Kilometres)
Passenger car	23.98	747	210.93
Motorcycle	4.24	687	86.64
Van and pickup	26.85	559	7.20 (Van); 132.92 (Pickup)
Bus	74.53	508	8.81
Truck	62.33	404	38.14

3.2. Emission Factor and Emission Inventory in the Base Year

The IVE model evaluates emission factors for both specific air pollutants and GHGs across all vehicle categories. Table 3 shows the total emission factors for the base year, 2019. Updated emission factors for specific air pollutants and GHGs for each vehicle type and technology in the base year 2019 are also available in the Supplementary Material. Notably, the highest CO emission factors are attributed to passenger cars during cold starts (g/start) and hot running (g/km). The highest VOC emission factors are primarily associated with motorcycles during both cold starts and hot running. The VOC_{evap} emission factor during hot running is predominantly linked to buses, while during cold starts it is highest in passenger cars. Additionally, SO_x and NO_x emission factors are primarily emitted during cold starts by passenger cars and during hot running by buses. The PM10 emission factors are highest in buses during both cold starts and hot running.

Table 3. Emission factors for all vehicle categories in Thailand during the base year of 2019.

Pollutant and GHG	Emission Factor (Unit)	Vehicle Type					
		Passenger Car	Motorcycle	Van	Pickup	Bus	Truck
CO	g/start	28.64	16.83	12.81	6.57	1.70	0.45
	g/km	90.37	33.49	37.73	13.70	31.71	2.78
VOC	g/start	1.75	3.98	0.48	0.32	0.06	0.04
	g/km	0.63	1.73	0.22	0.10	0.64	0.30
VOC_{evap}	g/start	1.19	0.54	0.35	0.13	0.63	0.07
	g/km	0.18	0.15	0.04	0.02	0.20	0.02
NO_x	g/start	1.05	0.78	0.57	0.38	0.76	0.70
	g/km	1.30	0.32	5.59	5.93	6.96	6.12
SO_x	g/start	5.96×10^{-4}	1.04×10^{-4}	4.67×10^{-4}	4.62×10^{-4}	2.81×10^{-4}	2.23×10^{-4}
	g/km	1.62×10^{-3}	4.86×10^{-4}	1.21×10^{-3}	1.19×10^{-3}	2.51×10^{-3}	1.98×10^{-3}
PM_{10}	g/start	0.18	0.14	0.59	0.63	0.64	0.59
	g/km	0.04	0.10	0.09	0.09	0.48	0.44
CO_2	g/start	65.95	21.17	105.05	111.13	87.50	71.61
	g/km	446.86	114.32	1590.40	1714.83	815.22	651.27
N_2O	g/start	0.04	3.34×10^{-3}	0.03	0.03	0.06	0.05
	g/km	0.01	N.A. *	0.01	0.01	0.07	0.08
CH_4	g/start	0.66	0.85	0.63	0.16	0.04	0.01
	g/km	0.50	0.57	0.67	0.15	0.54	0.08

* N.A. = not applicable.

Additionally, significant CO_2 emission factors occur during both starts and running for pickups, while N_2O emission factors during cold starts and running are the primary source of emissions for buses and trucks, respectively. Lastly, regarding CH_4 emission factors, motorcycles exhibit higher emissions than other vehicles during cold starts, and buses have higher emissions during hot running. Subsequently, emission inventories for all vehicle types are compiled based on the average VKT and emission factors.

Moreover, pickups exhibit notable CO_2 emission factors during both starts and running. Cold starts and running are the primary sources of N_2O emission factors for buses and trucks, respectively. Lastly, regarding CH_4 emission factors, motorcycles have higher emissions than other vehicles during cold starts, while vans have the highest emission factor during hot running. Consequently, emission inventories for all vehicle types are established based on the average VKT and emission factors.

The emission inventory for specific air pollutants and GHGs in the base year is outlined in Table 4. The analysis revealed that CO emissions were primarily linked to motorcycles, with passenger cars, trucks, pickups, buses, and vans following in descending order of contribution. Conversely, VOC emissions were predominantly associated with motorcycles, with passenger cars, trucks, buses, pickups, and vans following in sequence. Moreover, passenger cars were identified as the primary sources of VOC_{evap} emissions, with motorcycles, buses, pickups, trucks, and vans subsequently ranking in terms of contribution. Additionally, NO_x emissions were primarily emitted by pickups, with trucks, buses, passenger cars, motorcycles, and vans following, respectively. For SO_x emissions, passenger cars were the main contributors, followed by trucks, pickups, motorcycles, buses, and vans.

Table 4. The emission inventory for conventional air pollutants and greenhouse gases in the base year (2019).

Pollutants and Greenhouse Gas	Emissions (Gg)					
	Passenger Car	Motorcycle	Van	Pickup	Bus	Truck
CO	1065.80	1192.80	4.61	18.94	15.57	20.89
VOC	7.10	52.42	0.05	0.87	0.89	4.01
VOC_{evap}	4.16	3.46	0.01	0.02	0.20	0.02
NO_x	17.00	10.59	7.29	171.08	17.76	80.39
SO_x	0.04	0.02	1.41×10^{-3}	0.03	0.01	0.03
PM_{10}	0.54	3.19	0.12	2.69	1.38	6.47
CO_2	9672.55	3893.96	2029.32	46,707.89	1859.68	8397.76
N_2O	0.25	2.96×10^{-5}	0.01	0.26	0.14	0.77
CH_4	2.16	16.77	0.08	0.09	0.56	0.07

However, the variation observed among vehicle types can be attributed to differences in fuel type, which include gasoline and diesel fuel, all conforming to the Euro 4 standard and containing low sulfur levels (<50 ppm). Regarding PM_{10} emissions, trucks were identified as the primary contributors, followed by motorcycles, pickups, buses, passenger cars, and vans, respectively. This trend can be attributed to older technology and the large VKT of vehicles still in operation within the region.

The analysis revealed that pickups are the primary source of CO_2 emissions, followed by passenger cars, trucks, motorcycles, vans, and buses in that order. N_2O emissions were most significant from trucks, followed by pickups, passenger cars, buses, vans, and motorcycles, respectively. In terms of CH_4 emissions, motorcycles were identified as the largest emitters, with passenger cars, buses, pickups, vans, and trucks following in descending order.

The calculation of GHG emissions in the Fifth Assessment Report (AR5) is expressed in carbon dioxide equivalent (CO_2eq). This standardises the measurement of various gases by their global warming potential (GWP) over a specific timeframe, typically 100 years. GWP serves as an index, with CO_2 assigned a value of 1. Comparatively, other GHGs are evaluated based on how much more warming they induce relative to CO_2. For instance, over a 100-year period, CH_4 has a GWP of 28, while N_2O has a GWP of 265 [45]. Hence, the GHG emissions attributable to pickups amount to 46,779.96 GgCO_2eq, representing the highest emissions category. Subsequently, passenger cars, trucks, motorcycles, vans, and buses display emissions of approximately 9800.10, 8604.74, 4363.43, 2034.94, and 1911.10 GgCO_2eq, respectively, resulting in a cumulative total of 73,494.27 GgCO_2eq. These findings indicating the predominant contribution of air pollutants and GHGs align with previous studies conducted in Myanmar [81], Ecuador [38], and Indonesia [55].

3.3. *Uncertainty Analysis*

Generally, uncertainty in emission inventories primarily arises from data concerning emission factors and activity data [82–84]. The main sources of uncertainty in the emission inventory include variations in the number of registered vehicles, emission factors, the annual average VKT, and the driving pattern. Data on the number of registered vehicles were obtained from official statistics. The dynamic emission factor is determined by multiplying the correction factor with the base emission factor, which has been adjusted using laboratory-measured emission factors for specific air pollutants and GHGs from vehicles. However, vehicle emission testing in Thailand does not encompass all vehicle technologies. Hence, the default correction factors within the vehicle emission model were collectively employed. Regarding the annual average VKT, due to the lack of accurate local statistics, the data used in this study were obtained from official channels, including the PCD and other references.

3.4. BAU Scenario

Vehicle estimates from 2023 to 2030 are based on the growth rates of specific vehicle classifications recorded in the data provided by the DLT from 2010 to 2022. The average annual growth rate for all vehicle types during this period was approximately 3.42%. The specific growth rates for each vehicle category were as follows: 5.27% for passenger cars, 2.93% for vans, −0.36% for pickups, 0.0015% for motorcycles, −0.03% for buses, and 5.36% for trucks. Figure 3 provides a graphical representation of the cumulative vehicle count.

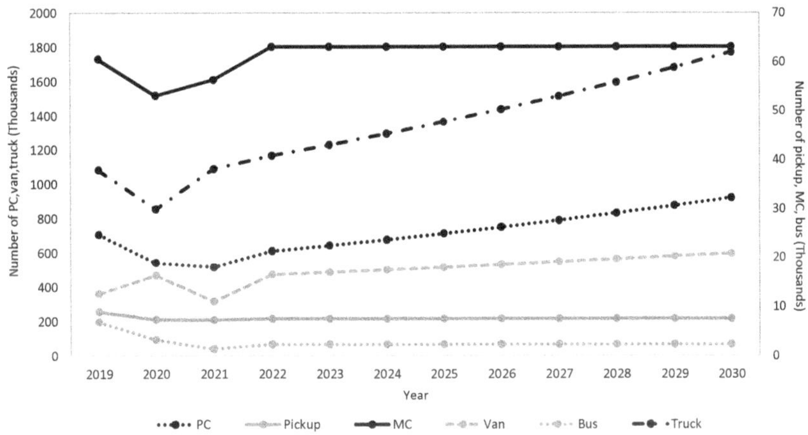

Figure 3. The total number of vehicle types between 2019 and 2030 under the BAU scenario.

The data highlight a notable increase in the cumulative number of trucks, accompanied by rises in the counts of passenger cars and vans, between 2021 and 2030. Conversely, the figures for pickups, motorcycles, and buses remained relatively stable during this period. In 2020 and 2021, all vehicle categories experienced declines due to the impact of the COVID-19 pandemic, except for vans, which saw growth in 2020, and passenger cars and motorcycles, which saw increases in 2021.

The results from the BAU scenario concerning dynamic emission factors and emissions of specific air pollutants from 2019 to 2030 indicate a general rise in CO, VOC, VOC_{evap}, PM_{10}, NO_x, SO_x, CO_2, CH_4, and N_2O across various vehicle categories when compared to the base year. Noteworthy is the decline in emissions from buses across all pollutants. This trend may be linked to the cumulative number of buses deployed across different technologies. Furthermore, motorcycles are recognised as the primary contributors to specific pollutant emissions, constituting 46.58%, followed by passenger cars (40.39%), pickups (7.14%), trucks (4.12%), buses (1.32%), and vans (0.45%), all of which also have a significant impact. Most vehicle emissions are mainly comprised of CO and NO_x, accounting for 85.93% and 10.61% respectively, with VOC (2.33%), PM_{10} (0.65%), VOC_{evap} (0.47%), and SO_x (0.01%) following closely. Motorcycles are the main contributors to CO and VOC emissions, while passenger cars play a significant role in CO, VOC_{evap}, and SO_x emissions. Additionally, PM_{10} emissions primarily originate from a significant portion of diesel engines, with trucks being the largest contributors.

Regarding GHG emissions, CO_2 is the main contributor from vehicles, accounting for 99.92% of the total. Pickups are identified as the largest emitters of CO_2, followed by trucks, passenger cars, motorcycles, vans, and buses, respectively. Moreover, most CH_4 emissions are linked to motorcycles, constituting 86.09% of total CH_4 emissions. This can be attributed to the widespread use of petrol in motorcycles, which have the highest cumulative numbers among vehicle types, making up approximately 52.25% of the total vehicle count. Additionally, N_2O emissions are mainly associated with trucks, albeit relatively low at around 17.57 Gg/year, representing about 0.01% of total GHG emissions.

Furthermore, there has been a general upward trend in GHG emissions from vehicles since the base year. The total GHG emissions ($GgCO_2eq$) are predominantly attributed to pickups (51.81%), with trucks (18.59%), passenger cars (18.13%), vans (2.67%), buses (2.20%), and motorcycles (6.60%) following in descending order of emissions.

3.5. Mitigation Measures Scenario and Co-Benefit

The NDC roadmap outlines strategies to mitigate GHG emissions in the transportation sector from 2021 to 2030, focusing on targeted measures to combat air pollution. Two specific scenarios have been evaluated: bus scenarios and passenger car scenarios. Regarding buses, the scenarios entail transitioning from diesel engines, including those less than 1 year old and phasing out those that are 20 years old, to EVs with adoption rates of both 50% and 100%. Similarly, concerning new passenger cars (less than 1 year old), the focus shifts from reliance on gasoline and diesel fuel to the adoption of EVs at rates around 50% and 100%. This transition is accompanied by a potential 20% reduction in VKT, facilitated by the promotion of public transportation usage.

(1) Passenger car scenarios

The passenger car scenarios include four different plans, each involving the transition of new passenger cars from petrol and diesel engines to EVs. The adoption rates are set at 50%, either with or without a 20% reduction in VKT, and at 100% adoption with the same VKT reduction. In the scenario where 50% of new passenger cars transition to EVs, CO emissions are observed to slightly increase from 2023 to 2030. Specifically, in 2023, the emissions of specific air pollutants and GHGs in this scenario exceed those of the BAU scenario.

Additionally, CO emissions decrease from 2024 to 2030, except for 2025, compared to the BAU scenario. VOC emissions also decrease from 2024 to 2030, except for 2028, while VOC_{evap} emissions exceed the base year from 2023 to 2024, except for 2026. Furthermore, PM10 emissions surpass those of the BAU scenario in 2027 and 2028. Moreover, GHG emissions from 2025 to 2030 decrease compared to the BAU scenario. The variations in emission levels can be attributed to differences in the cumulative number of passenger cars across various technologies.

In the first scenario, the most significant reduction in emissions compared to the BAU scenario is observed in CO at 12.00%, followed by GHG at 10.63%, SO_x at 9.09%, NO_x and VOC at 5.75%, PM_{10} at 3.38%, and VOC_{evap} at 1.74%. In the second scenario, involving the transition of 100% of new passenger cars to EVs, all specific air pollutants and GHG emissions show a slight increase from 2023 to 2030. However, they remain lower than the BAU scenario, except for VOC_{evap} emissions in 2027 and 2028, which exceed those of the BAU case. Furthermore, the most significant reduction in emissions compared to the BAU scenario is observed in GHG at 24.63%, followed by SO_x at 17.66%, CO at 12.44%, NO_x at 10.88%, VOC at 6.59%, PM10 at 5.74%, and VOC_{evap} at 2.85%.

Furthermore, in the final two scenarios, involving the transition of new passenger cars from petrol and diesel engines to EVs at both 50% and 100% adoption rates, with a 20% reduction in VKT, emissions of all types show a slight increase from 2023 to 2030. Nevertheless, these scenarios exhibit a significant decrease compared to the BAU scenario. Therefore, mitigation strategies aimed at reducing GHG emissions require not only the transition from fossil fuel engines to EVs in passenger cars but also initiatives to reduce VKT, such as promoting the use of public transport and minimising the use of private cars.

In the third scenario, which involves transitioning new passenger cars from petrol and diesel engines to EVs at a 50% adoption rate with a 20% reduction in VKT, the most notable reduction in emissions compared to the BAU scenario is observed in CO at 29.60%, followed by GHG at 28.50%, SO_x at 27.27%, NO_x at 24.70%, VOC at 24.60%, PM_{10} at 22.70%, and VOC_{evap} at 21.38%.

Finally, in the last scenario concerning passenger cars, which entails transitioning new vehicles from petrol and diesel engines to EVs at a 100% adoption rate with a 20% reduction in VKT, the most significant reduction in emissions compared to the BAU scenario is observed in GHG at 39.70%, followed by SO_x at 32.12%, CO at 29.95%, NO_x at 28.70%, VOC at 25.27%, PM_{10} at 24.59%, and VOC_{evap} at 22.27%. The emissions of air pollutants and GHGs within the passenger car scenarios from 2019 to 2030 are illustrated in Figure 4.

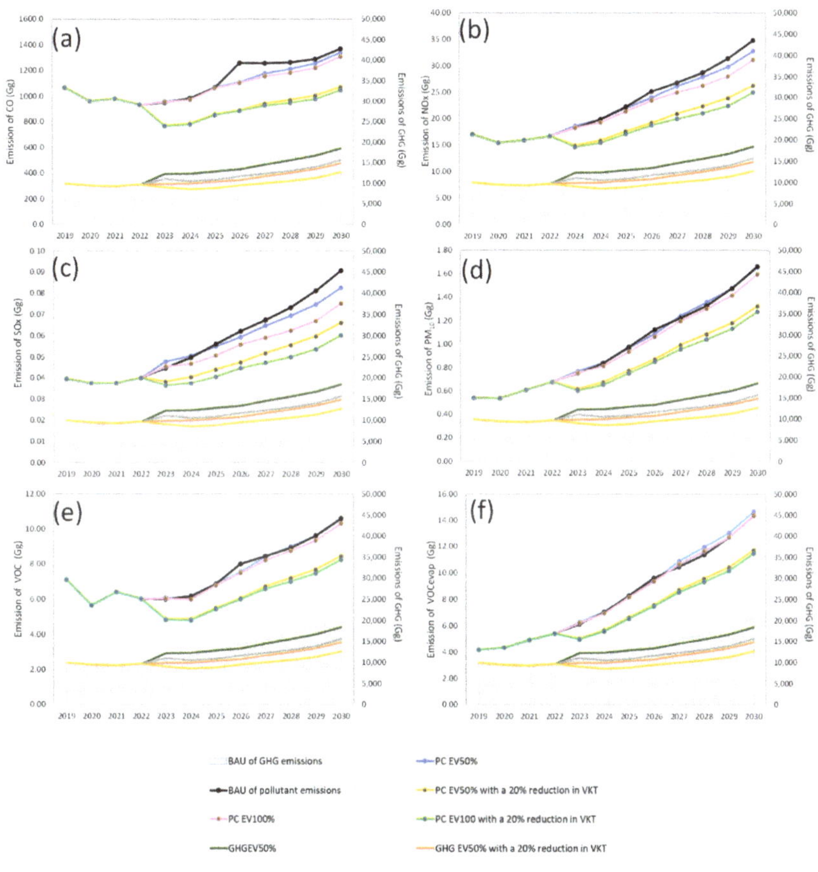

Figure 4. Emissions of air pollutants and greenhouse gases under the passenger car scenarios from 2019 to 2030 include: (**a**) CO emissions; (**b**) NO_x emissions; (**c**) SO_x emissions; (**d**) PM_{10} emissions; (**e**) VOC emissions; (**f**) VOC_{evap} emissions.

(2) Bus scenarios

In the context of bus transportation, two scenarios are considered: one involves replacing existing 20-year-old diesel buses, while the other entails transitioning new diesel buses to electric buses at adoption rates of 50% and 100%, respectively. The outcomes indicate a noticeable decreasing trend from 2023 to 2030. Additionally, emissions of specific air pollutants and GHGs show a significant decrease compared to the BAU scenario. For CO and VOC_{evap}, there is a marginal reduction in emissions, while emissions of other pollutants and GHGs exhibit a significant decrease compared to the BAU scenarios.

In the initial bus scenario, where the shift from new diesel buses to electric buses occurs at a 50% adoption rate, the most significant reduction in emissions compared to the BAU scenario is noticeable in NO_x at 78.72%, followed by PM_{10} at 77.02%, VOC at 73.57%, SO_x at 72.04%, GHG at 70.74%, CO at 26.98%, and VOC_{evap} at 23.18%. In the ultimate bus scenario, involving the complete transition of new diesel buses to electric buses at a 100% adoption rate, the most notable decrease in emissions compared to the BAU scenario is observed in NOx at 80.24%, followed by PM10 at 77.83%, VOC at 75.44%, SO_x at 74.82%, GHG at 74.03%, CO at 39.41%, and VOC_{evap} at 36.17%. This scenario could initially be implemented in specific areas such as the Bangkok Metropolitan Area, Pattaya City, Phuket City, Chiang Mai City, and other urban centres under the authority of the Ministry of Transport of the Thai government [85].

Moreover, the National Electric Vehicle Policy Committee of Thailand has provisionally approved tax incentives to promote the acquisition of electric buses and electric trucks, with the aim of supporting the business sector's efforts to reduce carbon emissions. Under this initiative, companies purchasing domestically manufactured vehicles will qualify for an expense deduction equal to twice the actual price of the vehicles, without a price cap [86]. The emissions of air pollutants and GHGs under the bus scenarios from 2019 to 2030 are illustrated in Figure 5.

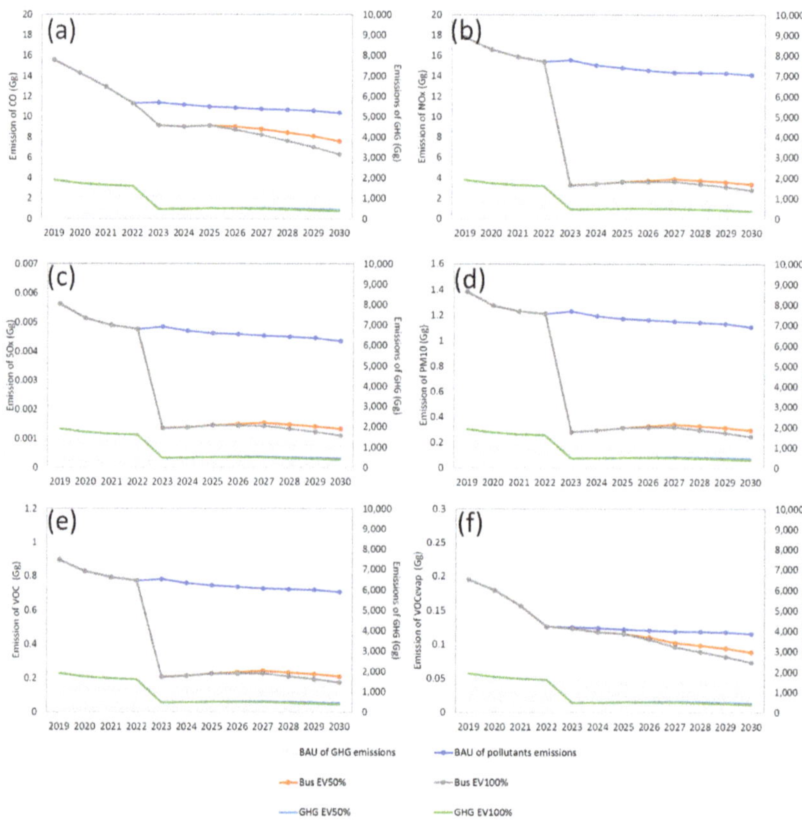

Figure 5. Emissions of air pollutants and greenhouse gases under the bus scenarios from 2019 to 2030 include: (**a**) CO emissions; (**b**) NO_x emissions; (**c**) SO_x emissions; (**d**) PM_{10} emissions; (**e**) VOC emissions; (**f**) VOC_{evap} emission.

The results of the mitigation scenarios, compared to the BAU scenario, indicate that plans incorporating strategies to reduce GHG emissions and specific air pollution, aimed at replacing diesel fuel with EVs in new passenger cars and new buses, led to decreases across all scenarios. In particular, transitioning from new petrol and/or diesel vehicles to EVs at a 100% adoption rate, alongside a reduction in VKT, resulted in the greatest reductions in specific air pollutants and GHG emissions, at 13.65% and 11.11%, respectively.

Table 5 illustrates the percentage reductions of specific air pollutants and GHG emissions for each scenario.

Table 5. GHG emissions reduction (%) of passenger car and bus scenarios from 2023 to 2030.

Pollutant and GHG	Passenger Car Scenarios (A) (% Reduction)				Bus Scenarios (B) (% Reduction)		(A) and (B) (% Reduction)
	(1) EV 50%	(2) EV 50% with a 20% Reduction in VKT	(3) EV 100%	(4) EV 100 with a 20% Reduction in VKT	(5) EV 50%	(6) EV 100%	(4) + (6)
CO	0.31–5.44	8.23–13.41	0.59–5.64	8.37–13.57	0.07–0.10	0.13–0.23	8.48–13.65
VOC	0.02–0.64	1.89–2.73	0.13–0.73	1.95–2.81	0.58–0.95	0.71–1.06	2.90–3.52
VOC_{evap}	0–1.05	9.45–12.92	0.71–1.72	10.72–13.46	0.02–0.11	0.06–0.45	10.87–13.54
NO_x	0.02–0.52	1.12–2.27	0.20–1.02	1.21–2.70	2.74–4.10	3.17–4.38	5.09–6.06
SO_x	0.69–3.51	4.80–10.52	2.05–6.64	6.16–13.03	1.29–2.53	2.26–3.77	8.71–14.41
PM_{10}	0.03–0.18	0.80–1.13	0.11–0.30	0.88–1.49	3.12–5.69	3.54–5.97	4.70–6.58
GHG	0.14–2.65	0.96–7.12	1.45–5.99	2.26–9.79	1.22–1.67	1.32–1.67	3.93–11.11

3.6. Limitations

The limitations of this study require further refinement and exploration. Firstly, to reduce complexity, this paper focused on the co-benefits of GHG emissions and specific air pollutant emissions. However, GHG mitigation also brings additional advantages in terms of health impacts, ecosystems, economic systems, and resource-use efficiency [87,88], which may vary significantly across different areas, thus necessitating further investigations. Future studies should provide a more comprehensive assessment of the human health impacts resulting from emissions reduction.

4. Conclusions

The assessment of specific air pollutants and GHG emissions from road vehicles in Thailand relies on data from the Thailand database, with the reference year of 2019. Actual data from 2019 to 2022 are used, while projections for 2023–2030 are based on the baseline year. The IVE model incorporates secondary database input adjusted to real driving data for emission factors related to Thai vehicle characteristics. Dynamic emission factors for air pollutants and GHGs are evaluated alongside the average VKT for each vehicle type. The BAU scenario, reflecting current policies to mitigate emissions in the transportation sector, considers strategies such as EV adoption and the promotion of public transportation to reduce VKT.

In the BAU scenario, the total number of road vehicles in Thailand, including passenger cars, motorcycles, pickups, vans, trucks, and buses, is projected to rise by around 6.59% by 2030, leading to an increase in specific air pollutants and GHG emissions compared to the reference year. Nevertheless, adopting new EV passenger cars and buses in line with Thailand's outlined strategies will significantly reduce GHG emissions and specific air pollutants from the road transportation sector.

Furthermore, in potential scenarios where the transition to EVs occurs at rates of 50% and 100%, GHG emissions could be reduced by 7940.16 and 7859.03 $GgCO_2eq$, respectively, by the year 2030. These reductions represent decreases of 10.80% and 10.69% compared to the BAU scenario, respectively.

Moreover, the largest emission factors for certain vehicles may not necessarily result in the highest emissions, as this depends on the annual average VKT and the various vehicle technologies each year. Consequently, specific air pollutant and GHG inventories will be crucial factors supporting the sustainable development of Thailand's transportation sector strategies in the future.

Scenario analysis has the potential to reduce GHG emissions and yield additional benefits by lowering local air pollutant emissions. Electric vehicles, particularly passenger cars and buses, demonstrate significant potential for reducing GHG emissions as well as associated air pollutants such as CO, VOCs, VOC_{evap}, NO_x, SO_x, and PM_{10}. However, a more effective strategy for both GHG reduction and air quality improvement involves not only promoting the adoption of EVs but also encouraging the public to use public transport.

Furthermore, the introduction of policies targeting GHG emissions reduction is expected to significantly reduce air pollution emissions, leading to a maximum co-benefit of 11.11% for GHG reduction and 6.58% for PM10 emissions reduction. However, the government must provide support and encouragement for EV adoption. Nevertheless, a high penetration rate of EVs will affect Thailand's energy system, particularly the road transport sector. This impact includes changes in the overall energy demand trend, the load profile of electricity demand, and the indirect GHG emissions generated from energy production for vehicles.

Future scenarios can be explored through long-term strategies, such as setting objectives to achieve carbon neutrality by 2050 and net-zero GHG emissions by 2065 in Thailand. Additionally, other co-benefits, such as health improvements, increased economic outcomes, and the reduction of negative impacts on ecosystems, should be investigated.

Supplementary Materials: The following supporting information can be downloaded at: https://www.mdpi.com/article/10.3390/en17102336/s1. 2019 Baseline Emission Factors for Specific Air Pollutants and GHGs by Vehicle Type and Technology

Author Contributions: Conceptualisation, P.T. and D.S.; methodology, P.T.; software, P.T.; validation, P.T.; formal analysis, P.T.; investigation, P.T.; resources, P.T.; data curation, P.T.; writing—original draft preparation, P.T. and D.S.; writing—review and editing, P.T. and D.S.; visualisation, P.T.; supervision, P.T.; project administration, P.T. and D.S.; funding acquisition, P.T. All authors have read and agreed to the published version of the manuscript.

Funding: This research was funded by the Thailand Science Research and Innovation Fundamental Fund, grant number TUFF15/2564 and the APC was supported by Thammasat University Research Unit in Environment, Health, and Epidemiology.

Data Availability Statement: The original contributions presented in the study are included in the article/Supplementary Material, further inquiries can be directed to the corresponding author.

Acknowledgments: The authors express sincere appreciation to the Automotive Emission Laboratory (AEL) within the Pollution Control Department (PCD), Ministry of Natural Resources and Environment in Thailand. Their invaluable assistance in supplying data and research facilities for evaluating air pollutants and greenhouse gas emissions in this study is deeply valued. This work was supported by the Thammasat University Research Unit in Environment, Health, and Epidemiology, and the Faculty of Public Health at Thammasat University.

Conflicts of Interest: The authors declare no conflicts of interest.

References

1. Apergis, N.; Bhattacharya, M.; Hadhri, W. Health care expenditure and environmental pollution: A cross-country comparison across different income groups. Environ. *Sci. Pollut. Res.* **2020**, *27*, 8142–8156. [CrossRef] [PubMed]
2. Baumgartner, J.; Brauer, M.; Ezzati, M. The role of cities in reducing the cardiovascular impacts of environmental pollution in low-and middle-income countries. *BMC Med.* **2020**, *18*, 1–11. [CrossRef] [PubMed]
3. Badida, P.; Krishnamurthy, A.; Jayaprakash, J. Meta analysis of health effects of ambient air pollution exposure in low-and middle-income countries. *Environ. Res.* **2023**, *216*, 114604. [CrossRef] [PubMed]
4. Ferguson, L.; Taylor, J.; Davies, M.; Shrubsole, C.; Symonds, P.; Dimitroulopoulou, S. Exposure to indoor air pollution across socio-economic groups in high-income countries: A scoping review of the literature and a modelling methodology. *Environ. Int.* **2020**, *143*, 105748. [CrossRef] [PubMed]
5. Bowe, B.; Xie, Y.; Li, T.; Yan, Y.; Xian, H.; Al-Aly, Z. Estimates of the 2016 global burden of kidney disease attributable to ambient fine particulate matter air pollution. *BMJ Open* **2019**, *9*, e022450. [CrossRef] [PubMed]
6. Combes, A.; Franchineau, G. Fine particle environmental pollution and cardiovascular diseases. *Metab.* **2019**, *100*, 153944. [CrossRef]
7. Filippelli, G.; Anenberg, S.; Taylor, M.; van Geen, A.; Khreis, H. New approaches to identifying and reducing the global burden of disease from pollution. *GeoHealth* **2020**, *4*, e2018GH000167. [CrossRef] [PubMed]
8. Kateja, A. Looming Threat of Vehicular Pollution to Human Health. In *Urban Growth and Environmental Issues in India*; Kateja, A., Jain, R., Eds.; Springer: Singapore, 2021; pp. 157–171. [CrossRef]

9. Li, W.W. *Air Pollution, Air Quality, Vehicle Emissions, and Environmental Regulations*; Traffic-Related Air Pollution; Elsevier: Amsterdam, The Netherlands, 2020; Chapter 2; pp. 23–49. [CrossRef]
10. Pompilio, A.; Di Bonaventura, G. Ambient air pollution and respiratory bacterial infections, a troubling association: Epidemiology, underlying mechanisms, and future challenges. *Crit. Rev. Microbiol.* **2020**, *46*, 600–630. [CrossRef] [PubMed]
11. Prasad, R.R.; Paudel, S. *Impacts of Air Pollution on Epidemiology and Cardiovascular Systems*; Air Pollution and Environmental Health; Springer: Singapore, 2020; pp. 179–207. [CrossRef]
12. Tiotiu, A.I.; Novakova, P.; Nedeva, D.; Chong-Neto, H.J.; Novakova, S.; Steiropoulos, P.; Kowal, K. Impact of air pollution on asthma outcomes. *Int. J. Environ. Res. Public Health* **2020**, *17*, 6212. [CrossRef]
13. World Health Organization (WHO) Website. Available online: https://www.who.int/news-room/fact-sheets/detail/ambient-(outdoor)-air-quality-and-health?gclid=CjwKCAjw4P6oBhBsEiwAKYVkq1PnhGQCeb1TGAaP6aourfQo4IVbVojqU10HM4d3-421e55kc2KfiBoC_iIQAvD_BwE (accessed on 10 October 2023).
14. Abbass, K.; Qasim, M.Z.; Song, H.; Murshed, M.; Mahmood, H.; Younis, I. A review of the global climate change impacts, adaptation, and sustainable mitigation measures. *Environ. Sci. Pollut. Res.* **2022**, *29*, 42539–42559. [CrossRef]
15. Ebi, K.L.; Vanos, J.; Baldwin, J.W.; Bell, J.E.; Hondula, D.M.; Errett, N.A.; Hayes, K.; Reid, C.E.; Saha, S.; Spector, J.; et al. Extreme weather and climate change: Population health and health system implications. *Annu. Rev. Public Health* **2021**, *42*, 293–315. [CrossRef]
16. Hoegh-Guldberg, O.; Jacob, D.; Taylor, M.; Guillén Bolaños, T.; Bindi, M.; Brown, S.; Camilloni, I.A.; Diedhiou, A.; Djalante, R.; Ebi, K.; et al. The human imperative of stabilizing global climate change at 1.5 C. *Science* **2019**, *365*, eaaw6974. [CrossRef] [PubMed]
17. Jessel, S.; Sawyer, S.; Hernández, D. Energy, poverty, and health in climate change: A comprehensive review of an emerging literature. *Front. Public Health* **2019**, *7*, 357. [CrossRef]
18. Milner, J.; Hamilton, I.; Woodcock, J.; Williams, M.; Davies, M.; Wilkinson, P.; Haines, A. Health benefits of policies to reduce carbon emissions. *BMJ* **2020**, *368*, l6758. [CrossRef]
19. Romanello, M.; McGushin, A.; Di Napoli, C.; Drummond, P.; Hughes, N.; Jamart, L.; Kennard, H.; Lampard, P.; Rodriguez, B.S.; Arnell, N.; et al. The 2021 report of the Lancet Countdown on health and climate change: Code red for a healthy future. *Lancet* **2021**, *398*, 1619–1662. [CrossRef]
20. World Health Organization (WHO) Website. Available online: https://www.who.int/news-room/fact-sheets/detail/climate-change-and-health?gclid=CjwKCAjwyY6pBhA9EiwAMzmfwYqp1wJnVOIq1Tk3ihKhon54ov0e2rxk5aeIZBDRI4h7LZ3KkrW-1BoC2BQQAvD_BwE (accessed on 15 October 2023).
21. Albuquerque, F.D.; Maraqa, M.A.; Chowdhury, R.; Mauga, T.; Alzard, M. Greenhouse gas emissions associated with road transport projects: Current status, benchmarking, and assessment tools. *Transp. Res. Proc.* **2020**, *48*, 2018–2030. [CrossRef]
22. Cheng, L.; Ji, D.; He, J.; Li, L.; Du, L.; Cui, Y.; Zhang, H.; Zhou, L.; Li, Z.; Zhou, Y. Characteristics of air pollutants and greenhouse gases at a regional background station in southwestern China. *Aerosol Air Qual. Res.* **2019**, *19*, 1007–1023. [CrossRef]
23. World Resources Institute Website. Available online: https://www.wri.org/insights/4-charts-explain-greenhouse-gas-emissions-countries-and-sectors?ref=thesustainableinvestor.org.uk (accessed on 15 December 2023).
24. Sirithian, D.; Thanatrakolsri, P.; Pongpan, S. CO_2 and CH_4 Emission Factors from Light-Duty Vehicles by Fuel Types in Thailand. *Atmosphere* **2022**, *13*, 1588. [CrossRef]
25. United States Environmental Protection Agency (USEPA), Website. Available online: https://www.epa.gov/greenvehicles/greenhouse-gas-emissions-typical-passenger-vehicle (accessed on 15 October 2023).
26. Crippa, M.; Guizzardi, D.; Pagani, F.; Banja, M.; Muntean, M.; Schaaf, E.; Becker, W.; Monforti-Ferrario, F.; Quadrelli, R.; Risquez Martin, A.; et al. *GHG Emissions of All World Countries*; Publications Office of the European Union: Luxembourg, 2023. [CrossRef]
27. United Nations Environment Programme (UNEP) Website. Available online: https://wedocs.unep.org/bitstream/handle/20.500.11822/43922/EGR2023.pdf?sequence=3&isAllowed=y (accessed on 15 December 2023).
28. Ministry of Natural Resources and Environment (MONRE) Website. Available online: https://unfccc.int/sites/default/files/resource/Thailand_BUR4_final_28122022.pdf (accessed on 10 September 2023).
29. Mbandi, A.M.; Malley, C.S.; Schwela, D.; Vallack, H.; Emberson, L.; Ashmore, M.R. Assessment of the impact of road transport policies on air pollution and greenhouse gas emissions in Kenya. *Energy Strategy Rev.* **2023**, *49*, 101120. [CrossRef]
30. Duan, L.; Hu, W.; Deng, D.; Fang, W.; Xiong, M.; Lu, P.; Li, Z.; Zhai, C. Impacts of reducing air pollutants and CO_2 emissions in urban road transport through 2035 in Chongqing. *China Environ. Sci. Ecotechnol.* **2021**, *8*, 100125. [CrossRef]
31. Aslam, A.; Ibrahim, M.; Shahid, I.; Mahmood, A.; Irshad, M.K.; Yamin, M.; Ghazala, T.M.; Shamshiri, R.R. Pollution characteristics of particulate matter (PM2. 5 and PM10) and constituent carbonaceous aerosols in a South Asian future megacity. *Appl. Sci.* **2020**, *10*, 8864. [CrossRef]
32. Bajwa, A.U.; Sheikh, H.A. Contribution of Road Transport to Pakistan's Air Pollution in the Urban Environment. *Air* **2023**, *1*, 237–257. [CrossRef]
33. Anenberg, S.; Miller, J.; Henze, D.A.V.E.N.; Minjares, R. *A Global Snapshot of the Air Pollution-Related Health Impacts of Transportation Sector Emissions in 2010 and 2015*; International Council on Clean Transportation: Washington, DC, USA, 2019; pp. 1–48.
34. Baayoun, A.; Itani, W.; El Helou, J.; Halabi, L.; Medlej, S.; El Malki, M.; Moukhadder, A.; Aboujaoude, L.K.; Kabakian, V.; Mounajed, H.; et al. Emission inventory of key sources of air pollution in Lebanon. *Atmos. Environ.* **2019**, *215*, 116871. [CrossRef]

35. Pollution Control Department (PCD). *Thailand State of Pollutant Report 2022*; Ministry of Natural Resources and Environment: Bangkok, Thailand, 2023.
36. Narita, D.; Oanh, N.T.K.; Sato, K.; Huo, M.; Permadi, D.A.; Chi, N.N.H.; Ratanajaratroj, T.; Pawarmart, I. Pollution characteristics and policy actions on fine particulate matter in a growing Asian economy: The case of Bangkok Metropolitan Region. *Atmosphere* **2019**, *10*, 227. [CrossRef]
37. Chirasophon, S.; Pochanart, P. The Long-term Characteristics of PM10 and PM2.5 in Bangkok, Thailand. *Asian J. Atmos. Environ.* **2020**, *14*, 73–83. [CrossRef]
38. Castesana, P.; Diaz Resquin, M.; Huneeus, N.; Puliafito, E.; Darras, S.; Gómez, D.; Granier, C.; Osses Alvarado, M.; Rojas, N.; Dawidowski, L. PAPILA dataset: A regional emission inventory of reactive gases for South America based on the combination of local and global information. *Earth Syst. Sci. Data* **2022**, *14*, 271–293. [CrossRef]
39. Zhou, M.; Jiang, W.; Gao, W.; Gao, X.; Ma, M.; Ma, X. Anthropogenic emission inventory of multiple air pollutants and their spatiotemporal variations in 2017 for the Shandong Province, China. *Environ Pollut.* **2021**, *288*, 117666. [CrossRef] [PubMed]
40. 40 Prakash, J.; Vats, P.; Sharma, A.K.; Ganguly, D.; Habib, G. New emission inventory of carbonaceous aerosols from the on-road transport sector in India and its implications for direct radiative forcing over the region. *Aerosol Air Qual. Res.* **2020**, *20*, 741–761. [CrossRef]
41. Patiño-Aroca, M.; Parra, A.; Borge, R. On-road vehicle emission inventory and its spatial and temporal distribution in the city of Guayaquil, Ecuador. *Sci. Total Environ.* **2020**, *848*, 157664. [CrossRef]
42. Limmeechokchai, B.; Dul, D. Achievement of Paris Agreement in selected Greater Mekong Sub-region countries: Analyses of renewable electricity and emissions gap. *Energy* **2023**, *1*, 29–66.
43. Meeros, W.; Wichakul, S. Thailand Nationally Determined Contributions (NDCs): Analysis of climate policy for enhancing climate resilience in infrastructure construction sector. In Proceedings of the 2022 International Conference and Utility Exhibition on Energy, Environment and Climate Change (ICUE), Pattaya, Thailand, 26–28 October 2022; pp. 1–8. [CrossRef]
44. Muangjai, P.; Wongsapai, W.; Jaitiang, T.; Ritkrerkkrai, C.; Daroon, S.; Thepsaskul, W. The Pathway to NDC and Carbon Neutrality: Roles of Optimum Degree between Marginal Abatement Cost and Social Cost of Carbon in the Thai Power and Industrial Sector. *Sustainability* **2023**, *16*, 160. [CrossRef]
45. Intergovernmental Panel on Climate Change (IPCC). *Climate Change 2014: Synthesis Report*; Core Writing Team, Pachauri, R.K., Meyer, L.A., Eds.; Contribution of Working Groups I, II and III to the Fifth Assessment Report of the Intergovernmental Panel on Climate Change; IPCC: Geneva, Switzerland, 2014; p. 151.
46. Jiao, J.; Huang, Y.; Liao, C. Co-benefits of reducing CO2 and air pollutant emissions in the urban transport sector: A case of Guangzhou. *Energy Sustain. Dev.* **2020**, *59*, 131–143. [CrossRef]
47. Kim, S.E.; Xie, Y.; Dai, H.; Fujimori, S.; Hijioka, Y.; Honda, Y.; Hashizume, M.; Masui, T.; Hasegawa, T.; Xu, X.; et al. Air quality co-benefits from climate mitigation for human health in Sout Korea. *Environ. Int.* **2020**, *136*, 105507. [CrossRef] [PubMed]
48. Liu, Q.; Gao, J. Co-benefits of reducing greenhouse gas emissions. In *Health of People, Health of Planet and Our Responsibility: Climate Change, Air Pollution and Health*; Al-Delaimy, W.K., Ramanathan, V., Sorondo, M.S., Eds.; Springer: Cham, Switzerland, 2020; pp. 295–307. [CrossRef]
49. Lu, Z.; Huang, L.; Liu, J.; Zhou, Y.; Chen, M.; Hu, J. Carbon dioxide mitigation co-benefit analysis of energy-related measures in the Air Pollution Prevention and Control Action Plan in the Jing-Jin-Ji region of China. *Resour. Conserv. Recycl. X* **2019**, *1*, 100006. [CrossRef]
50. Wei, X.; Tong, Q.; Magill, I.; Vithayasrichareon, P.; Betz, R. Evaluation of potential co-benefits of air pollution control and climate mitigation policies for China's electricity sector. *Energy Econ.* **2020**, *92*, 104917. [CrossRef]
51. Yang, X.; Yang, X.; Zhu, J.; Jiang, P.; Lin, H.; Cai, Z.; Huang, H. Achieving co-benefits by implementing the low-carbon city pilot policy in China: Effectiveness and efficiency. *Environ. Technol. Innov.* **2023**, *30*, 103137. [CrossRef]
52. Bamwesigye, D.; Hlavackova, P. Analysis of sustainable transport for smart cities. *Sustainability* **2019**, *11*, 2140. [CrossRef]
53. Economic and Social Commission for Asia and the Pacific (ESCAP) Website. Available online: https://www.unescap.org/sites/default/files/publications/bulletin87_Fulltext.pdf (accessed on 8 September 2023).
54. Nguyen, Y.L.T.; Nghiem, T.D.; Le, A.T.; Duc, K.N.; Nguyen, D.H. Emission characterization and co-benefits of bus rapid transit: A case study in Hanoi, Vietnam. *Atmos. Pollut. Res.* **2021**, *12*, 101148. [CrossRef]
55. Kim Oanh, N.T.; Huy, L.N.; Didin, D.A.; Zusman, E.; Nakano, R.; Nugroho, S.B.; Lestari, P.; Sofyan, A. Assessment of urban passenger fleet emissions to quantify climate and air quality co-benefits resulting from potential interventions. *Carbon Manag.* **2018**, *9*, 367–381. [CrossRef]
56. Viteri, R.; Borge, R.; Paredes, M.; Pérez, M.A. A high resolution vehicular emissions inventory for Ecuador using the IVE modelling system. *Chemosphere* **2023**, *315*, 137634. [CrossRef]
57. Chaudhry, S.K.; Elumalai, S.P. Assessment of sustainable school transport policies on vehicular emissions using the IVE model. *J. Clean. Prod.* **2024**, *434*, 140437. [CrossRef]
58. International Sustainable Systems Research Center (ISSRC). *IVE Model Users Manual Version 2.0*; International Sustainable Systems Research Center: La Habra, CA, USA, 2008.
59. Wei, Y.; Yu, Y.; Xu, L.; Huang, W.; Guo, J.; Wan, Y.; Cao, J. Vehicle emission computation through microscopic traffic simulation calibrated using genetic algorithm. *J. Artif. Intell. Soft Comput. Res.* **2019**, *9*, 67–80. [CrossRef]

60. International Sustainable Systems Research Center (ISSRC) Website. Available online: http://www.issrc.org/ive/downloads/manuals/AttachC_DA.pdf (accessed on 10 May 2023).
61. Awad, O.I.; Ma, X.; Kamil, M.; Ali, O.M.; Zhang, Z.; Shuai, S. Particulate emissions from gasoline direct injection engines: A review of how current emission regulations are being met by automobile manufacturers. *Sci. Total Environ.* **2020**, *718*, 137302. [CrossRef] [PubMed]
62. Hooftman, N.; Messagie, M.; Van Mierlo, J.; Coosemans, T. A review of the European passenger car regulations–Real driving emissions vs local air quality. *Renew. Sustain. Energy Rev.* **2018**, *86*, 1–21. [CrossRef]
63. Cheewaphongphan, P.; Hanaoka, T.; Chatani, S. Long-term trend of regional passenger road transport demand and emission estimation under exhaust emission regulation scenario in Thailand. *Environ. Res. Commun.* **2020**, *2*, 051009. [CrossRef]
64. Pollution Control Department (PCD) Website. Available online: https://www.pcd.go.th/wp-content/uploads/2020/06/pcdnew-2020-06-05_03-22-03_837433.pdf (accessed on 10 September 2023).
65. Dutta, A.; Chavalparit, O. Assessment of health burden due to the emissions of fine particulate matter from motor vehicles: A case of Nakhon Ratchasima province, Thailand. *Sci. Total Environ.* **2023**, *872*, 162128. [CrossRef] [PubMed]
66. Ministry of Transport Website. Available online: https://web.dlt.go.th/statistics/ (accessed on 15 April 2023).
67. Pongthanaisawan, J.; Sorapipatana, C. Relationship between level of economic development and motorcycle and car ownerships and their impacts on fuel consumption and greenhouse gas emission in Thailand. Renew. *Sustain. Energy Rev.* **2010**, *14*, 2966–2975. [CrossRef]
68. Prabnasak, J.; Taylor, M.A.P.; YUE, W.L. An Investigation of Vehicle Ownership and the Effects of Income and Vehicle Expenses in Mid-Sized City of Thailand. In Proceedings of the 9th International Conference of Eastern Asia Society for Transportation Studies, Jeju, Republic of Korea, 20–23 June 2011; Volume 8. [CrossRef]
69. Thaithatkul, P.; Chalermpong, S.; Laosinwattana, W.; Liang, J.; Kato, H. The relationship between RHA use and car purchases: Trends emerging in Bangkok, Thailand. *Asian Transp. Stud.* **2023**, *9*, 100095. [CrossRef]
70. Giraldo, M.; Huertas, J.I. Real emissions, driving patterns and fuel consumption of in-use diesel buses operating at high altitude. *Transp. Res. D Trans. Environ.* **2019**, *77*, 21–36. [CrossRef]
71. Lois, D.; Wang, Y.; Boggio-Marzet, A.; Monzon, A. Multivariate analysis of fuel consumption related to eco-driving: Interaction of driving patterns and external factors. *Transp. Res. D Trans. Environ.* **2019**, *72*, 232–242. [CrossRef]
72. The Intelligent Traffic Information Center Foundation (iTIC) Website. Available online: https://itic.longdo.com/averagespeed (accessed on 5 September 2023).
73. de Oliveira, A.; Bernardes, A.P.; Ferreira, F. *Reduction of a Diesel Engine NO Emissions Using the Exhaust Gas Recirculation Technique*; SAE Technical Paper, No. 2019-36-0067; SAE: Brasilia, Brazil, 2020. [CrossRef]
74. *TIS.2560-2554*; Exhaust Emissions from Motor Vehicles: Test Method. The Thai Industrial Standards Institute (TISI) Department, Ministry of Industry: Bangkok, Thailand, 2012.
75. Pollution Control Department (PCD) Website. Available online: https://www.pcd.go.th/wp-content/uploads/2021/02/pcdnew-2021-02-18_08-03-46_086635.pdf (accessed on 17 October 2023).
76. Zhao, P.; Tian, B.S.; Yang, Q.; Zhang, S. Influencing Factors and Their Spatial–Temporal Heterogeneity of Urban Transport Carbon Emissions in China. *Energies* **2024**, *17*, 756. [CrossRef]
77. Office of Transport and Traffic Policy and Planning (OTP) Website. Available online: https://www.otp.go.th/uploads/tiny_uploads/PDF/2562-04/25620314-NDCPlan.pdf (accessed on 5 September 2023).
78. Kwan, S.K.; Zakaria, S.B.; Ibrahim, M.F.; Mahiyuddin, W.R.W.; Sofwan, N.M.; Wahab, M.I.A.; Ahmad, R.D.R.; Abbas, A.R.; Woon, W.K.; Sahani, M. Health impacts from TRAPs and carbon emissions in the projected electric vehicle growth and energy generation mix scenarios in Malaysia. *Environ. Res.* **2023**, *216*, 114524. [CrossRef] [PubMed]
79. Aiello, G.; Quaranta, S.; Inguanta, R.; Certa, A.; Venticinque, M. A Multi-Criteria Decision-Making Framework for Zero Emission Vehicle Fleet Renewal Considering Lifecycle and Scenario Uncertainty. *Energies* **2024**, *17*, 1371. [CrossRef]
80. Thanatrakolsri, P.; Siritian, D.; Pongpan, S. Evaluation of Greenhouse Gas Emissions from Motor Vehicles in Bangkok, Thailand. *BUSCIJ* **2023**, *28*, 223–247.
81. Huy, L.N.; Oanh, N.T.K.; Htut, T.T.; Hlaing, O.M.T. Emission inventory for on-road traffic fleets in Greater Yangon, Myanmar. *Atmos. Pollut. Res.* **2020**, *11*, 702–713. [CrossRef]
82. Gong, M.; Yin, S.; Gu, X.; Xu, Y.; Jiang, N.; Zhang, R. Refined 2013-based vehicle emission inventory and its spatial and temporal characteristics in Zhengzhou, China. *Sci. Total Environ.* **2017**, *599*, 1149–1159. [CrossRef] [PubMed]
83. Yang, W.; Yu, C.; Yuan, W.; Wu, X.; Zhang, W.; Wang, X. (2018). High-resolution vehicle emission inventory and emission control policy scenario analysis, a case in the Beijing-Tianjin-Hebei (BTH) region, China. *J. Clean. Prod.* **2018**, *203*, 530–539. [CrossRef]
84. Gao, C.; Gao, C.; Song, K.; Xing, Y.; Chen, W. Vehicle emissions inventory in high spatial–temporal resolution and emission reduction strategy in Harbin-Changchun Megalopolis. *Process Saf. Environ. Prot.* **2020**, *138*, 236–245. [CrossRef]
85. Office of Transport and Traffic Policy and Planning (OTP) Website. Available online: https://www.otp.go.th/uploads/tiny_uploads/ProjectOTP/2561/Project10/FinalReport.pdf (accessed on 20 December 2023).
86. The board of investment of the prime minister (BOI) Website. Available online: https://www.boi.go.th/upload/content/PR20_2567EN.pdf (accessed on 29 February 2024).
87. Wu, P.; Guo, F.; Cai, B.; Wang, C.; Lv, C.; Liu, H.; Huang, J.; Huang, Y.; Cao, L.; Pang, L.; et al. Co-benefits of peaking carbon dioxide emissions on air quality and health, a case of Guangzhou, China. *J Environ Manag.* **2021**, *282*, 111796. [CrossRef]

88. Deng, H.M.; Liang, Q.M.; Liu, L.J.; Anadon, L.D. Co-benefits of greenhouse gas mitigation: A review and classification by type, mitigation sector, and geography. *Environ. Res. Lett.* **2018**, *12*, 123001. [CrossRef]

energies

MDPI

Article

A Composite Index for Tracking the Evolution towards Energy Transition at Urban Scale: The Turin Case Study

Eleonora Desogus [1,2,*], Ettore Bompard [1,2] and Daniele Grosso [1,2]

1 Department of Energy (DENERG), Politecnico di Torino, 10129 Turin, Italy; daniele.grosso@polito.it (D.G.)
2 Energy Security Transition (EST) Lab@energycenter, Politecnico di Torino, 10129 Turin, Italy
* Correspondence: eleonora.desogus@polito.it; Tel.: +39-3342253711

Abstract: Cities play a pivotal role in achieving worldwide carbon neutrality due to their significant contribution to global energy consumption and carbon emissions. Therefore, planning effective strategies and guiding evidence-based policymaking at the city scale becomes even more crucial. Composite indices serve as a valuable tool for monitoring urban energy transition trends. This paper aims to present a novel approach, robust and flexible even under conditions of data scarcity, for tracking the energy transition trend of a city by means of a composite index (UETI). The Turin case study is introduced to test the applicability of the proposed approach. Additionally, to demonstrate the robustness of the composite index framework, the paper includes the findings of correlation and sensitivity analyses. This study reveals a significant improvement in Turin's environmental and energy domains, while the socio-economic domain shows more modest improvement. Furthermore, the study highlights the need to address the shortage of urban data to enhance the accuracy and reliability of metric-based frameworks and to extend the assessment to a larger sample of cities.

Keywords: energy transition; electrification; decarbonization; composite index; city scale; performance indicators

1. Introduction

The global energy transition aims to shift from a fossil-based system to a new smart and sustainable paradigm, intended as a society able to meet the current needs without compromising the ability of future generations to meet their own needs [1]. As nations strive to decarbonize their energy systems, cities have emerged as vital actors in this transition. In fact, currently, they account for 55% of the global population, two-thirds of global energy consumption and more than 70% of global greenhouse gas emissions (GHGs) [2]. Considering the ambitious Climate Law [3] goal of achieving carbon neutrality in Europe-27 (EU-27) by 2050, it is clear that energy transition at an urban scale is crucial in this process. Since the transition involves not only energy aspects but is strictly intertwined with many other dimensions, to achieve a comprehensive and clear understanding of the phenomenon, policymakers need support from science-based tools, able to measure and track the multifaceted reality of energy transition. Among the diverse science-based tools, composite indicators are effective at describing and capturing multi-dimensional phenomena by means of a single metric. In the context of energy transition at the city level, composite metrics can serve as a powerful tool for monitoring the process and for comparative analyses across various cities and timeframes.

A robust index framework can be helpful to policymakers to set medium- and long-term goals and targets, evaluate the effect of ongoing and future policies, highlight criticalities and communicate the progress over time to the stakeholders.

Literature Review

Due to the relevance for policymaking, many studies have focused on building metrics to quantify the impacts and the performance towards energy transition at different spatial scales, although the multitude of proposed approaches and the absence of a commonly recognized framework prove that a single pre-set "one-size-fits-all" method would be inappropriate [4].

At the global scale, the International Renewable Energy Agency (IRENA) defined a composite index (Energy Transition Welfare Index) [5] which includes five sub-domains to measure the multi-dimensional impacts of the energy transition. The World Economic Forum (WEF) [6] and the World Energy Council (WEC) [7] focus instead on the country level; by allocating a score to each country, they rank countries according to their overall energy transition performance. Similarly, the European Commission ranks countries by means of the Transition Performance Index (TPI) [8] and promotes constructive competition by encouraging emulation of countries with the best performance. Due to the emergency of global warming, the European Climate Foundation (ECF) launched the Net Zero 2050 initiative [9] aimed at monitoring the evolution towards global carbon neutrality. The Climate Change Performance Index (CCPI) report ranks and compares the climate performance of 59 countries [10].

By shifting the focus of the study to the urban scale, the International Organization for Standardization (ISO) published the ISO 37120 standard [11], including a set of 100 indicators for measuring the sustainable development of cities. This document is intended to be coupled with ISO 37122 [12], providing further indicators specific to smart cities, and ISO 37123 [13] for the evaluation of the urban resilience. These standards provide a useful directory but give no instruction on how to perform the aggregation and normalization of indicators.

Various European projects share the common objective of developing a comprehensive framework to evaluate city performance in the context of energy transition: the CITYKEYS project [14] (2015–2017) aimed to measure the sustainability and smartness of European cities by means of specific Key Performance Indicators (KPIs), while the REPLICATE project [15] (2016–2021) focused instead on the evaluation of initiatives implemented at the district level of three European "lighthouse cities", namely San Sebastian (Spain), Florence (Italy) and Bristol (United Kingdom). Other projects worth mentioning, funded under the European Union's Horizon 2020 Programme, are POCITYF [16] (2019–2024), IRIS [17] (2017–2023) and SmartEnCity [18] (2016–2022). POCITYF is intended to quantify the impact and effectiveness of adopted strategies in meeting the needs of citizens, whereas IRIS and SmartEnCity are more focused on achieving sustainable, accessible and reliable urban transport and energy supply, by increasing the share of renewable resources in the urban energy mix.

The literature review revealed that many studies do not include the normalization, weighting and aggregation procedures [19–22]: they just focus on the selection of criteria to choose an appropriate set of indicators to describe the phenomenon of interest (e.g., sustainable development, energy transition, etc.).

Other studies instead adopt hybrid methods of aggregation, normalization and weight allocation: the Carbon Neutrality Capacity Indicator System (CNCIS) [23] combines the best–worst method (BWM) to obtain the subjective criteria weights, together with the entropy method (EM), used to compute the objective criteria weights. Furthermore, the Uniform Smart City Evaluation (USCE) Framework [24] adopts a hybrid method of normalization, coupling the distance to a reference method with the categorical-scale method, and uses both the Budget Allocation (BAL) and the equal weight (EW) weighting methods; BAL is adopted to assign the weights to the indicators comprising the three main sub-indexes (Project Performance Index, Sustainability Impact Index and Sustainability Performance Index). Then, EW is implemented in combination with the additive aggregation to quantify the composite USCE index. The Urban Energy Sustainability Index (UESI) [25] proposes a novel framework to build the composite index, assigning different weight according to the type of indicator (i.e., basic, instrumental and complementary indicator), but its results are intricate and difficult to explain to non-experts. While sophisticated methodologies may offer advantages from a scientific perspective, their complexity can impede comprehension among a less specialized audience. Since these methodologies are intended to support policymaking and public engagement, it is not reasonable to develop methodologies beyond the policymakers' comprehension. On the other hand, prioritizing clarity and transparency in methodologies can improve effective communication and interpretation of scientific evidence, enabling broader engagement and fostering greater trust and confidence in the information being conveyed. For this reason, less intricate and sophisticated methods are often preferred: as observed from the literature [6,26–28], the combination of Min–Max normalization with the equal-weight additive aggregation is one of most used methods. When dealing with comparative analyses, rank-based methodologies are widely used, as observed in the Arcadis's Sustainable Cities Index (SCI) [29], the IMD's Smart City Index [30], the Global Cities Report [31], the IESE's Cities In Motion Index (CIMI) [32] and in the Global Power City Index (GPCI). On one hand, these rankings serve as a useful tool to compare cities, but they do not convey the magnitude or extent of differences between them since they only consider the relative position of cities within the ranking. City rankings generally include a subset of global cities (e.g., the Arcadis's ranking [29] includes 100 cities, the IMD's report includes 141 cities, the CIMI's report comprises 174 cities, etc.), often corresponding to the capitals or the main cities of each country. On one hand, rankings provide the opportunity to simultaneously compare many cities at a global level; on the other, they only include a small portion of these cities: for example, as regards Italian cities, several rankings encompass only Milan and Rome [29–31], excluding other important cities like Turin and Naples. In addition, the literature review revealed a limited availability of studies focused on energy transition in Italian cities, except for D'Adamo et al. [33], the Legambiente's Ecosistema Urbano report [34] and the Municipality Transition Index (MTI) [35].

The first study updates the results obtained with the methodology developed by Fondazione Enrico Mattei [36] to assess the performance of 103 Italian cities across all Sustainable Development Goals (SDGs), encompassing not only energy transition but also other aspects such as poverty, quality education and gender equality. Indeed, the final score obtained by aggregating all SDG performance indicators provides the overall percentage of achievement of all the SDGs [37].

The annual report published by Legambiente [34] serves as a valuable reference point for analysing the trends in environmental performance among 104 Italian cities. It encompasses a total of 18 indicators across six components (air, water, waste, mobility, urban environment and energy) included in the composite index (Indice Ecosistema Urbano, IEU). However, the index framework is mainly focused on environmental aspects (e.g., air quality) and resources management (e.g., water, waste and soil). As regards energy indicators, only one indicator (i.e., kW of installed photovoltaic systems per 1000 inhabitants) is included, while various crucial aspects essential for quantifying energy transition performance (e.g., energy intensity by consumption sector, the quality of the power network, the penetration of renewables in urban final consumption and the integration of green vehicles into the urban system) are excluded. Another crucial element missing in this report is the urban CO_2 emissions, a key factor for monitoring the effectiveness of urban decarbonization policies in meeting carbon-neutrality objectives [3]. Furthermore, there is a complete lack of indicators measuring the socio-economic impact of urban energy transition (investments, employment, added value, energy poverty, etc.).

The third study [35] proposes instead the Municipal Transition Index (MTI) to assess energy transition in Italian cities. The MTI is composed of a set of 18 indicators associated with four main dimensions: Digitalization (D); Energy, Climate and Resources (ECR); Sustainable Transport and Mobility (M); and Waste and Materials (W). Although this study provides a valuable overview of the Italian energy transition at an urban scale, covering 7904 municipalities, it does not offer any temporal trends, nor does it specify the reference year for the assessment. In fact, it lacks a consistent reference year for all indicators, using the latest available data, thus resulting in some indicators referring to 2021, others to 2018 and yet others to 2019, thereby creating a discrepancy among indicators and in the overall score of the composite index. In contrast, Shen et al. [28] published in 2023 the assessment of energy transition index of 282 Chinese cities up to 2019, adopting the criterion of extending the analysis up to the year for which all indicators are computable. The literature analysis reveals a prevailing tendency of embracing such compromises to facilitate the application of the proposed method and broaden the study's coverage across a wider spectrum of cities, even in the presence of a scarcity of updated data.

Hence, the literature review has highlighted the absence of a comprehensive framework for monitoring urban-scale energy transition that meets all of the following four requirements: (1) encompassing a broad spectrum of factors crucial for tracking the multidimensional impacts of energy transition (i.e., energy intensity, carbon intensity, green mobility, economic impact, etc.); (2) being easily comprehensible and interpretable by non-experts; (3) demonstrating applicability and adaptability to various urban contexts, even with limited data availability; and (4) including correlation and sensitivity analyses to ensure study transparency.

This paper aims to bridge this gap by proposing a new framework for evaluating urban-scale energy transition in terms of energy, environmental and socio-economic factors (requirement 1) through the Urban Energy Transition Index (UETI) composite index. The choice of preferring more easily understandable methodologies for normalizing, weighting and aggregating indicators (discussed in Section 2) meets requirement 2, while the applicability of the methodology to cities characterized by limited availability of data fulfils requirement 3, as proved by the proposed case study related to the city of Turin (Section 3). Furthermore, in line with requirement 4, both correlation analysis and sensitivity analysis are included (Section 4).

Conclusions and future research are reported in Section 5. Moreover, in line with the aim of enabling the replicability of the proposed approach, the complete list of datasets and data providers are reported in Appendix A.

2. Methodology

Building a composite index is a complex process which requires several intermediate steps which can affect the quality and reliability of the results [38]. Each step is described in the following sub-sections, justifying the choice of adopted methodologies: the definition of the conceptual framework in Section 2.1, the normalization step in Section 2.2 and the weighting and aggregation steps in Section 2.3.

2.1. Conceptual Framework

To evaluate the Urban Energy Transition Index, three main domains have been taken into consideration: Energy, Environmental and Socio-Economic. Each domain is further divided into sub-domains as shown in Figure 1.

The Energy domain encompasses aspects related to the decarbonization of the energy system (energy generation, transmission/distribution and consumption), and it is characterized by four sub-domains: Renewable Energy Resource (RES) penetration, power grid quality, green mobility and energy intensity.

The first sub-domain measures the RES integration in the energy supply; the power grid quality aims to monitor the efficiency of the power grid operations; the green mobility sub-domain is for tracking the electric and hybrid vehicle integration into the traditional automotive fleet as well as the coverage of bike lanes in the urban area; the energy intensity, often used as an approximation of energy efficiency [39], encompasses the energy intensities of the main energy-consuming sectors of the city (residential, industrial, tertiary and transport).

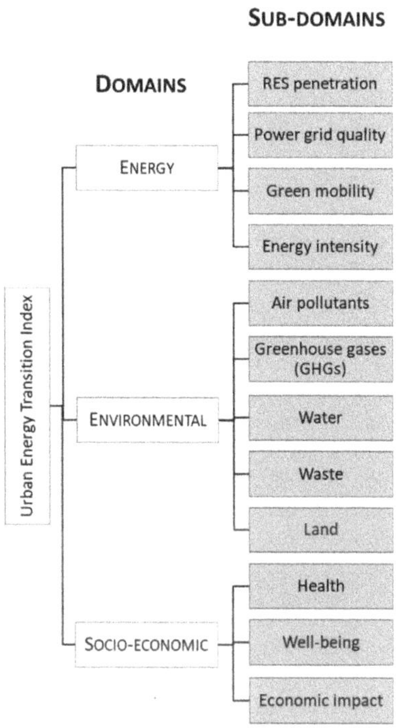

Figure 1. The UETI index conceptual framework: domains and sub-domains.

The Environmental domain includes 5 sub-domains: air pollutants, greenhouse gases, waste, water and land. Air pollutants accounts for gaseous emissions dangerous mainly for human health (e.g., PM_{10} and NOX), greenhouse gases measures the climate-altering emissions in terms of equivalent carbon dioxide (CO_2), waste encompasses waste production and management, whereas the last two sub-domains serve to keep under control the consumption of water and land.

The Socio-Economic domain comprises three main sub-domains: health, well-being and economic impact. The first two sub-domains aim to measure both physical (health) aspects and welfare (well-being) of citizens, whereas the economic impact is specifically focused on tracking the investments and the benefits brought by the energy and environmental sectors in terms of employment and added value.

Once the UETI conceptual framework has been defined, it is necessary to select a set of indicators able to quantitatively assess the impact of energy transition in the city across the 12 sub-domains. However, when dealing with city-scale analyses, data scarcity is a relevant issue which can make calculation of indicators challenging. This aspect emphasized the need for a flexible methodology which provides a robust but also adaptable framework for assessing and monitoring urban energy transition even in the presence of data scarcity.

To maintain flexibility and adaptability of the proposed approach, specific indicators (reported in Appendix A, Table A1) are not included in the conceptual framework. The absence of predetermined specific indicators allows for adjustments and inclusion of relevant metrics based on specific characteristics and data availability on the city under examination. This flexibility ensures that the methodology remains applicable and robust in diverse urban settings, preventing data limitations from impeding the tracking of the energy transition. It acknowledges the dynamic nature of urban data landscapes and strives to provide a comprehensive yet adaptable framework for assessing energy transitions in cities.

2.2. Normalization

Given the dimensional differences among the indicators, it is necessary to perform a normalization before proceeding with the aggregation process. The normalization step involves transforming data into a standard or common scale, typically to remove variations in units or scales, allowing data aggregation into a single metric.

Among the existing normalization methods, z-score, Min–Max, distance to a reference and categorical scale are some of the most common in the literature [40,41]. Z-score normalization is performed by dividing the difference between the raw indicator and the average by the standard deviation. It converts all indicators to a common scale with an average of zero and a standard deviation of one. In case of time-dependent performance indicators, the average and the standard deviation are calculated for a reference year [41]. It is widely used since it is a robust technique, and it is easy to apply and interpret. However, if the data are not normally distributed, the resulting normalized values do not reflect the original data. Moreover, it is not appropriate in the case of small sample sizes, as the values of the mean and standard deviation result in unstable or unreliable estimates [40]. Min–Max normalization, like z-score normalization, is a simple and easy-to-implement method, since it requires just basic mathematical operations (subtraction and division) to rescale the values with respect to the minimum and maximum of the dataset. Compared to the z-score, this method can be applied to a wider range of distributions, including normal and non-normal distributions. Nonetheless, the scale is based on best and worst performances; therefore, just the relative ordering of values is kept after the normalization. Similarly to z-score normalization, outliers could distort the results, and, in addition to this, it requires recalibration of the minimum and maximum when further data are added in the dataset [40]. Distance to reference adopts a different approach, by scaling the raw indicators with respect to a reference benchmark. The disadvantage of this methodology is that the results may be very sensitive to the chosen benchmark, and a significant level of subjectivity is involved when the choice of the benchmark is based on experts' estimations and assumptions rather than universally accepted value from the literature, or normative benchmarks. The same problem occurs in the categorical-scale normalization, which converts the raw indicators into a common scale by assigning categorical (numerical or qualitative) scores, according to a set of reference thresholds. In the case of a lack of universally accepted reference thresholds, they are estimated based on judgements of experts and/or stakeholders [42,43].

The hybrid weighting methods belong to a different category as they combine several subjective and objective methods together: for instance, combining the distance to a reference with the categorical-scale method [24], which allows semi-quantitative (e.g., Likert Scale) and qualitative data (e.g. Boolean logic) to also be included in the analysis. This method could be suitable for cities that do not yet have a robust and organized indicator tracking system, since it works even in the case of a low amount of data [24]; however, it requires the construction of appropriate thresholds, involving a significant degree of subjectivity.

Among these methodologies, the Min–Max normalization method results were the most common and versatile as evidenced by numerous references [26,27,44,45]. Moreover, this normalization allows us to evaluate performance from two perspectives: the performance of a single system or the performance of a set of comparable systems. In the first case, the normalization consists of scaling the values with respect to the minimum and maximum recorded over time by a single system (e.g., country, region, municipality or district), highlighting improvements, worsening and eventual criticalities. As regards the second perspective, Min–Max normalization allows us to compare performances of a group of systems (e.g., European countries, Italian cities or districts of Turin city) by using the minimum and maximum recorded per time unit (e.g., month, quarter or year). This kind of normalization allows us to compare performances of different systems, encouraging healthy competition and emulation of the systems with the highest performance. To assess the energy transition trend in the Turin case study, the first type of Min–Max method has been selected.

2.3. Weighting and Aggregation

As regards weighting and aggregation methods, the OECD's Handbook [40] serves as a useful reference guide to building a composite index. Two main group of weighting approaches can be distinguished: objective (i.e., equal weighting and statistic-based methods) and subjective (i.e., participatory methods) [4]. Equal weighting (objective) involves assigning the same weight to all indicators, with the logic that an indicator cannot be considered more relevant than the others [33]. The statistic-based (objective) weighting derives weights from statistical properties of data, whereas, on the contrary, the participatory (subjective) methods rely on experts' opinions [40]. Furthermore, the OECD's guidelines [40] highlight the limits of application of certain methods: the Min–Max normalization method, for instance, can be used in conjunction with the majority of weighting schemes and aggregation systems, while z-score normalization cannot be used in combination with geometric aggregation or with the Benefit Of Doubt (BOD) weighting method [40].

The Analytic Hierarchy Process (AHP), the conjoint analysis and the Budget Allocation Processes (BAPs) are some of the most common subjective weighting methods [46]. Their main advantage is that they can also be applied in the presence of qualitative input data and in the case of missing or insufficient data. However, they also present several disadvantages such as vulnerability to bias, and personal opinions and experiences of the experts or stakeholders involved can affect the reliability of results; need of many resources (time, effort and expertise) to implement; lack of objectivity; and difficulty in replicating [40]. On the contrary, the objective weighting methods use statistical or data-driven approaches to assign weights to each variable. Examples of objective weighting methods include Principal Component Analysis (PCA), Factor Analysis (FA), data envelopment analysis and regression analysis [40,46]. Similarly to normalization methodologies, some studies adopt hybrid weighting methods by combining subjective with objective weighting methods [23,47]. However, the equal weighting is the most straightforward method among the objective weighting methodologies because of its applicability, ease of communication and simplicity of use. In fact, it assigns the same weight to each variable composing the aggregate metric. The equal-weight method has been selected for this work, both because it is the most common in the literature [5,6,8,26,28,30,32,33,47,48] and because it is often used as reference to perform comparative analyses between different weighting methods [32].

Among the various aggregation approaches observed in the literature, the most straightforward and widely used is the additive aggregation [10,26,27,32,47–49], often combined with the equal weighting. Nevertheless, this methodology is characterized by the so-called "perfect compensability" among performances: underperformance in one component can be perfectly compensated by equivalent overperformance in another component [38,40]. Another methodology is the geometric (multiplicative) aggregation, based on the product of the variables instead of the sum and less sensitive to perfect compensability [50]. The multiplicative aggregation encourages improvements in the weaker components of the composite index since it penalizes mostly the unbalanced performances compared to the additive aggregation.

The aggregation process is performed at three levels:

1. Bottom level: aggregation of normalized indicators belonging to the same sub-domain to calculate the sub-domain performance index;
2. Middle level: aggregation of indexes belonging to the same domain to calculate the domain performance index (Energy, Environmental and Socio-Economic);
3. aggregation of the three indexes to obtain the final UETI.

The preference for additive aggregation over geometric aggregation in the chosen case study (discussed in the Section 2) is justified by its widespread acceptance and frequent use in the existing literature. Furthermore, additive aggregation offers simplicity, versatility and ease of interpretation, and it facilitates effective communication of results to policymakers.

3. Case Study: Energy Transition in Turin

The selection of Turin as the city to test the proposed methodology is rooted in its effort to align with the objectives of the European Green Deal by undertaking measures aimed at advancing its energy transition and carbon neutrality [51,52]. This commitment is prominently demonstrated through Turin's application to the "Climate-Neutral and Smart Cities by 2030" mission (the "100 Cities Mission") [53], launched by the European Commission to speed up the journey of cities towards climate neutrality. Turin has been chosen as one of the 100 European cities participating in the mission, aiming to reach climate neutrality by 2030, through the adoption of ad hoc policy strategies and innovative solutions. Moreover, Turin's case becomes even more intriguing if considering its peculiar situation: due to its geographical location, surrounded by mountains, and due to the high volume of vehicle traffic and fossil-based industrial activities, it is one of the cities in Italy [54] and in Europe [55] with the highest concentration of air pollutants (NO_2, $PM_{2.5}$ and PM_{10}). As it is widely known that polluted air has an adverse effect on human health (e.g., resulting in an increased occurrence of respiratory diseases), the need for decarbonization, especially in Turin, extends beyond the climate change challenge and also encompasses the health of citizens. Thus, by committing to the "Cities Mission" goal, Turin has embraced the challenge of shifting from a traditionally fossil-based industrial city to a sustainable, smart and carbon-neutral city. Due to the magnitude of this challenge, policymaking needs to be provided with reliable and up-to-date information for adapting strategies in each domain involved in the transition process. This study aims to show how the science-based approach can help to address the complexity of the energy transition in the city of Turin, providing a comprehensive view of how the city evolved across various domains (Energy, Environmental and Socio-Economic) and over time (from 2014 to 2019). By using the indicators framework, the progress of the city of Turin towards energy transition can be monitored systematically, providing a real-time understanding of the impact of the implemented measures and facilitating the identification of required adjustments. Moreover, this study strives to promote transparent communication of energy transition progress through indicators to a broader audience, fostering stakeholder and citizen engagement and encouraging collective efforts towards achieving carbon neutrality. Out of over 100 data collected from 34 datasets, provided by 12 data-sources (listed in Appendix A, Table A2), 90 raw data were selected to calculate the UETI. A set of 30 indicators (Appendix A, Table A1), tailored to Turin's context, have been selected to assess the impact of the energy transition across the 12 sub-domains outlined in the Section 2.

Installed capacity (MW) and shares of renewables (RES) in the total final consumption (TFC) are the performance indicators selected to measure the penetration of renewables in the city of Turin. Due to the lack of other renewable resources (wind, hydro and geothermal resources), photovoltaic technology is considered as a benchmark to track the spreading of renewable installations in Turin. On the other hand, due to the lack of data on heat consumption by energy source, the second indicator is obtained by considering the share of renewables on the electricity final consumption, excluding heat consumption. As a result of an increase in RES penetration, power grid quality may suffer from a decrease in grid quality; therefore, three indicators measuring the duration of outages (-), average of disconnections (n.) and power loss (%) are included in the UETI framework. In a city like Turin, characterized by a fossil-fuel-powered transport sector and limited air circulation, air pollution is a critical issue; therefore, PM_{10} exceedances (n. days/year) and NO_2 concentration ($\mu g/m^3$) deserve particular attention. Another aspect strictly connected to air quality is the green mobility trend: it favours transition away from traditional combustion engine vehicles towards electric (EVs) and hybrid vehicles (HVs), allowing for mitigation of air quality issues and a reduction in the overall carbon footprint of the city. Turin's transport sector is expected to increase the share of EVs and HVs in the next decade; to measure this trend, the number of EVs and HVs over 1000 passenger vehicles is included in the framework. However, the growth of EVs and HVs needs to be accompanied by adequate infrastructure (i.e., number of EV charging points over 1000 EVs and HVs), engaging in the exploitation of local renewable energy (i.e., number of RES-EV charging points). In addition to this, tracking the evolution of cycle lanes (km/100 km^2) is incorporated into green mobility: it provides further information about the development of sustainable transportation infrastructure, and it also reflects the commitment of the city to encourage citizens to choose soft mobility (e.g., bikes and e-scooters) over traditional vehicles. Enhancing energy efficiency is considered as a key strategy to achieve carbon-neutrality goals at an urban scale. Tracking the energy consumed per unit of economic activity (i.e., energy intensity) offers valuable insights into the effectiveness of energy efficiency measures across the most energy-demanding sectors: energy intensity residential sector (MWh/inhab), energy intensity tertiary sector (MWh/kEUR), energy intensity transport (MWh/Mpkm), energy intensity industry (MWh/kEUR) and energy intensity municipal service (MWh/m^2). By tracking energy intensity, it is possible to identify trends, evaluate required adjustment and make informed decisions to continually improve efficiency and favour a reduction in the total carbon footprint of the city, specifically measured in terms of tons of CO_2 emitted over the population (carbon intensity: tCO_2/inhab). To obtain a composite indicator providing a comprehensive picture of sustainable urban development, responsible water and land use are essential aspects to be monitored. Including green coverage and pedestrian areas reflects the commitment of Turin to sustainable urban planning, while considering per capita water consumption and the percentage of water recycling provides insights into the city's water management efficiency.

Waste, though historically excluded from the category of local energy resources, can instead serve as a significant example of circular economy if managed effectively while pursuing low waste production coupled with high recycling rates. For this reason, municipal waste production and recycling rate are included in the UETI framework.

Considering city-specific factors such as air quality issues (high concentration of PM_{10} and NO_2), high population density and unique geographical and climatic factors, deaths from respiratory system diseases (%) and mortality rate indicators assume particular importance for assessing the expected positive impact of energy transition process on public health. Moreover, UETI encompasses the economic impact of energy transition by including Energy–Environment Investments (%), Energy–Environment Added Value (EUR) and Energy–Environment Employment (%) as reference indicators. These indicators provide valuable insights into the benefits of energy transition for the urban economy of Turin; indeed, even if historically tied to industrial production like automotive manufacturing, energy transition in the city of Turin presents a unique chance for diversification of the economic landscape, offering the opportunity to be aligned with global sustainability goals, creating new jobs and attracting green investments. As regards the effect on the well-being of citizens, the combination of income (kEUR/inhab) and energy poverty [56] (%) allows for assessing eventual social disparities: income serves as an average indicator of the overall economic status of population; on the other hand, the progress of energy poverty highlights a more nuanced reality, indicating that the vulnerability of certain segments of the population is worsening. Together, these indicators act as a socio-economic barometer, providing insights into an ongoing phenomenon that requires attention and intervention in the context of the energy transition.

Before proceeding to the aggregation of indicators grouped in the same sub-domain (Table 1), Min–Max normalization is adopted to convert into a common scale each of the 30 indicators: 1 corresponds to the best performance and 0 to the worst performance observed over the period considered (2014–2019). Then, three steps of additive aggregation with equal weight are performed. The first aggregation step results in 12 performance indexes (Table 2): 4 for the Energy domain, 5 for the Environment domain and 3 for the Socio-Economic domain. Similarly, the second aggregation calculates the indexes of performance for the Energy domain, the Environment domain and the Socio-Economic domain. These indexes are combined (third aggregation) to obtain the UETI. The overall scores are shown in Figure 2.

Table 1. Selected performance indicators by sub-domain.

Sub-Domains	Performance Indicators
RES penetration	Installed RES capacity (MW) Share RES in electricity consumption (%)
Power grid quality	Duration of outages (-) Average number of disconnections (n.) Power loss (%)
Green mobility	Number of RES-EV charging points (n.) Number of EV charging points over 1000 EV + HV private vehicles (n.) Number of EVs + HVs over 1000 passenger vehicles (-) Cycle lanes (km/100 km^2)
Energy intensity	Energy intensity residential sector (MWh/inhab) Energy intensity tertiary sector (MWh/kEUR) Energy intensity transport (MWh/Mpkm) Energy intensity industry (MWh/kEUR) Energy intensity municipal service (MWh/m^2)
Land use	Green coverage (ha/100,000 inhab) Pedestrian areas (m^2/100 inhab)
Waste	Municipal waste (kg/inhab) Sorted waste (%)
Water	Water consumption (l/day·inhab) Water loss (%)
Air pollutants	PM_{10} exceedances (n. days/year) NO_2 concentration (μg/m^3)
GHGs	Carbon intensity (tCO_2/inhab)
Health	Deaths from respiratory system diseases (%) Mortality rate (n deaths/10,000 inhab)
Economic impact	Energy–Environment Investments (%) Energy–Environment Added Value (EUR) Energy–Environment Employment (%)
Well-being	Income (kEUR/inhab) Energy poverty (%)

Table 2. Trend of Turin's UETI over time (2014–2019).

	2014	2015	2016	2017	2018	2019
RES penetration	0.50	0.37	0.29	0.27	0.44	0.55
Power grid quality	0.62	0.58	0.47	0.50	0.57	0.61
Green mobility	0.00	0.08	0.20	0.34	0.62	1.00
Energy intensity	0.21	0.58	0.65	0.74	0.52	0.83
Energy index	0.33	0.40	0.40	0.46	0.54	0.75
Land use	0.05	0.14	0.27	0.61	0.79	1.00
Waste	0.39	0.43	0.54	0.56	0.41	0.66
Water	0.46	0.12	0.29	0.36	0.33	1.00
Air pollutants	0.37	0.08	0.46	0.13	1.00	0.79
GHGs	0.00	0.46	0.49	0.53	0.85	1.00
Environmental index	0.26	0.25	0.41	0.44	0.68	0.89
Health	1.00	0.56	0.82	0.30	0.03	0.20
Economic impact	0.49	0.30	0.11	0.17	0.68	0.81
Well-being	0.30	0.59	0.59	0.72	0.62	0.50
Socio-economic index	0.59	0.48	0.50	0.40	0.44	0.50
UETI	**0.39**	**0.38**	**0.44**	**0.43**	**0.55**	**0.71**

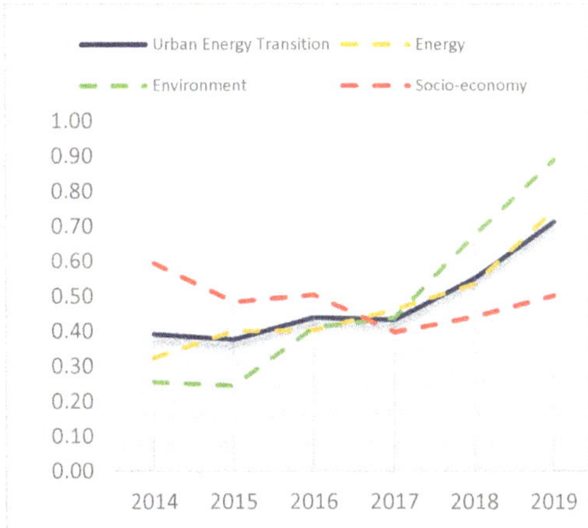

Figure 2. Evolution of UETI and its three domains (2014–2019).

4. Discussion

The results (Table 2) show a clear increase in the overall city performance (UETI) between 2014 and 2019: the Environmental domain shows the best performance, achieving the highest score in 2019, followed by the Energy domain. Despite the other two domains, the Socio-Economic performance decreased over the period of study: in particular, the health performance recorded a significant decrease; indeed, the mortality rate and the deaths from respiratory system diseases increased (Appendix A, Table A1). However, it must be pointed out that this information refers to the province of Turin, because the city-specific data are not available. Similarly, other metrics refer to the regional scale (i.e., energy poverty and Energy–Environment Investment) and provincial scale (i.e., Energy–Environment Added Value and Employment) rather than the city scale due to the lack of data. The Socio-Economic domain lacks updated data for the city of Turin. The utilization of provincial and regional data as an alternative to missing municipal data aligns with the flexibility inherent in the approach proposed in this study, avoiding data constraints to hinder the monitoring of the energy transition. Even though provincial and regional data offer valuable insights, city-specific data provide more accurate representation of the energy transition impact on the specific urban context. For instance, the well-being performance is affected by the low performance of energy poverty at the regional scale (share of people who cannot afford adequate and essential energy services [57]) even if Turin's income performance increased from 16.9 to 18.6 kEUR/inhabitant between 2014 and 2019 (Table A1). Even though these data affect Turin's actual performance, their exclusion in the assessment would lead to an impoverishment of the information content of the composite index. Therefore, to improve the accuracy and reliability of this metric-based assessment, is essential to enhance the current system of data collection at the city level.

As regards the environmental performance, in 2019 GHG emissions, water consumption and land use registered the best performances of the period 2014–2019 (Table 2). In particular, the GHG emission index, resulting from the normalization of CO_2 emissions per inhabitant, shows a continuous and clear trend of improvement. On the contrary, the air pollutants performance shows a fluctuating trend, registering a sharp increase in performance between 2018 and 2019, whereas the waste performance kept a steady, even if moderate, improvement over time. Similarly to the Environmental domain, the Energy performance registered a net increase between 2014 and 2019, especially thanks to the green mobility and the energy intensity trends which demonstrate, on one hand, the effort of the city to integrate the electrical vehicles (EV) and the installations of recharging points, and on the other, the benefit led by the enhanced energy efficiency in the residential buildings (Table 2). Conversely, the RES penetration trend shows in 2019 a limited increase compared to 2014; however, it must be pointed out that the share of RES in TFC omits heat generation due to the lack of data. Moreover, the fact that the score of grid quality performance in 2019 is close to the value in 2014 proves that no significant progress has been made to reduce power losses and disconnection issues.

4.1. Comparative Analysis with Other Studies

Comparing our method with other studies conducted on Italian cities has revealed both differences and commonalities. Since D'Adamo et al. [33] only provide a snapshot of the SDG performance of Italian cities, it was not possible to perform a thorough comparative analysis over the period under study (2014–2019); however, common points with our findings were observed. For instance, their study identifies air pollutants performance as the most critical issue of the city of Turin: in particular, Turin's PM_{10} index is equal to 0.057, significantly below the national average (0.498). On the other hand, coherently with our results, the authors obtain good performance in waste management, especially for municipal waste management with a score equal to 0.695 (above the national average of 0.595) and CO_2 emissions score equal to 0.487 (slightly below the national average of 0.501). Similarly to [33], the MTI's study [35] also provides merely an instant snapshot of the Italian cities' transition situation; therefore, a year-by-year comparison for the city of Turin cannot be performed. Furthermore, the study does not provide the list of performances achieved by each individual city, as the study is aimed at highlighting differences between regions rather than focusing on individual cities' performance; therefore, Turin's performances are deduced from the thematic maps in Figure 5 of the paper [35]. Although it is challenging to deduce precise values for the city of Turin from maps, the comparative analysis appears to confirm that Turin has good performance in waste management and sustainable mobility compared to other Italian cities. On the contrary, the performance of Energy, Climate and Resources (ECR) appears low: the good performance in resource management (e.g., water and soil) and renewable energy penetration is influenced by poor air quality performance. Additionally, the ECR score excludes CO_2 emissions indicator, thereby omitting a crucial component for assessing the city's overall energy, environmental and climate performance.

Finally, the UETI trend has been compared with the evolution of Legambiente's Ecosistema Urbano Index [34] over the period under study (2014–2019). The comparison revealed a similar trend up to 2017, then the trends diverge (Figure 3): the UETI shows continuous growth for two years, reaching its peak in 2019, while the Urban Ecosystem experienced modest growth in 2018 followed by an 11% decrease in 2019. This discrepancy is attributed to intrinsic differences in the composite index frameworks; in particular, the Urban Ecosystem Index mainly addresses environmental and resource management aspects but overlooks critical energy transition indicators like energy intensity, power network quality, renewables penetration, green mobility and CO_2 emissions, all of which experienced significant improvement in 2019 compared to 2018, thus contributing to the overall increase in the UETI.

4.2. Correlation Analysis

While the equal-weight method has been selected for weighting, it does not ensure that every sub-domain contributes equally to the Urban Energy Transition Index (UETI). Therefore, to comprehend the extent to which each sub-domain affects the overall score, it is essential to perform a correlation analysis between the UETI and its individual sub-domains. This analysis provides a more nuanced understanding of the main driving factors shaping the UETI evolution over time. The correlation analysis has been performed by calculating the Pearson correlation coefficient r (Table 3). Indeed, the Pearson correlation is the most common way to evaluate the linear correlation between two variables. The absolute value of the coefficient quantifies the "strength" of the correlation, ranging between 1 (perfect correlation) and 0 (no correlation); the sign ("+" or "−") refers to the orientation (positive or negative) of the correlation: a positive correlation occurs when one variable changes and the other one changes in the same direction (i.e., both variables increase or both variables decrease), while a negative correlation is observed if one variable increases and the other one decreases (they change in opposite directions). When evaluating the correlation coefficient between the composite metric (e.g., UETI) and its components (e.g., Energy, Environmental and Socio-Economic), a strong correlation ($|r| > 0.6$) is desired, because it proves that the composite index represents the behaviour of its components well. On the contrary, a weak correlation is preferred between the components of a composite metric because it means that they are mutually independent; therefore, the redundancy of the information is avoided. Firstly, the correlation analysis has been performed to study the 15 relationships between sub-domains, domains and the composite index UETI. In particular, the Energy and Environmental domains present a very strong correlation ($|r| > 0.80$) with the composite urban index UETI, whereas the Socio-Economic domain has a very weak correlation ($|r| < 0.19$). This means that the overall score of UETI for the city of Turin over time (2014–2019) is less affected by the behaviour of the Socio-Economic domain. Similarly, other relations present weak correlation but still relevant to compose the city index: the power grid quality shows a weak correlation (0.27) with respect to the Energy domain as well as the economic impact with respect to the Socio-Economic domain (0.21). Nevertheless, more than 60% of the 15 analysed relationships have a strong or very strong correlation (Table 3).

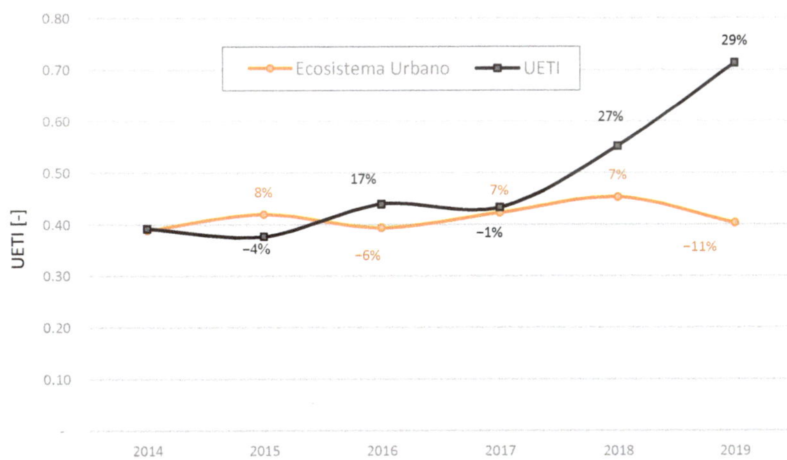

Figure 3. Comparison of UETI's trend with Ecosistema Urbano's trend over time (2014–2019).

Table 3. Pearson correlation coefficient between the UETI, its domains and sub-domains.

Correlation Analysis Ranges			
Very Weak $\mathbf{|r| \leq 0.19}$	**Weak** $\mathbf{|r| = 0.2–0.39}$	**Moderate** $\mathbf{|r| = 0.4–0.59}$	**Strong or Very Strong** $\mathbf{|r| \geq 0.60}$
22%	11%	6%	61%
Correlation of Urban Energy Transition Index (UETI) with its domains:			
DOMAINS			r
Energy			0.97
Environmental			0.98
Socio-Economic			−0.12

Table 3. *Cont.*

Correlation Analysis Ranges			
Very Weak $\lvert r \rvert \leq 0.19$	**Weak** $\lvert r \rvert = 0.2–0.39$	**Moderate** $\lvert r \rvert = 0.4–0.59$	**Strong or Very Strong** $\lvert r \rvert \geq 0.60$
Correlation of Energy Index with its sub-domains:			
SUB-DOMAINS			r
RES penetration			0.49
Power grid quality			0.27
Green mobility			0.98
Energy intensity			0.73
Correlation of Environmental Index with its sub-domains:			
SUB-DOMAINS			r
Land use			0.96
Waste			0.62
Water			0.77
Air pollutants			0.79
GHGs			0.89
Correlation of Socio-Economic Index with its sub-domains:			
SUB-DOMAINS			r
Health			0.75
Economic impact			0.21
Well-being			−0.96

It is important to highlight that the correlation analysis has been performed to provide a complete and transparent view of the proposed methodology, in line with the scope of this paper, which aims to propose a method that is easily replicable; nevertheless, the reliability of the correlation analysis should be further enhanced by improving the data availability with larger time series and constantly updated data.

4.3. Sensitivity Analysis

Conducting sensitivity analysis is crucial for evaluating the robustness and reliability of the composite index, as it systematically explores how variations in input parameters or weights impact the overall results. By subjecting the composite index to sensitivity analysis, one can assess the stability of the index against changes in methodology (e.g., weighting method) and ensure that the final measure is not disproportionately affected by specific variables. Since the proposed methodology is designed to adapt to diverse urban settings, aiming to provide a comprehensive yet adaptable framework, sensitivity analysis is even more relevant to enhance credibility and validity of the composite index. In line with this scope, three cases have been developed to test the UETI robustness (Table 4) [22]. The first case (C1) shows the behaviour of the composite index after varying the weight of those sub-domains characterized by weak correlations (highlighted in pink in Table 3): weights are increased until the moderate level ($\lvert r \rvert = 0.4–0.59$) of correlation is achieved (Table 4). The weight of power grid quality is increased by 80%, while the weight of economic impact is increased by 10% compared to the reference case (equal weighting). The second case (C2) aims to observe the UETI's behaviour when, for each of the three domains (Energy, Environmental and Socio-Economic), the sub-domain with the strongest correlation is excluded from the calculation (i.e., green mobility, land use and well-being). The third case (C3) investigates the effect of excluding a specific domain: the Socio-Economic domain, characterized by the least positive trend (red line in Figure 2), has been excluded from the calculation to better understand how much the perfect compensability of additive aggregation affects the overall UETI trend by compensating the low performance of the Socio-Economic domain with the high performances of the Energy and Environmental domains.

Table 4. Sensitivity analysis scheme.

Case	Objective	Description
C1	▪ |r| ≥ 0.4 between power grid quality and Energy domain ▪ |r| ≥ 0.4 between economic impact and Socio-Economic domain	▪ Power grid quality weight: +80% with respect to EW [1] ▪ Economic impact weight: +10% with respect to EW [1]
C2	Omitting the sub-domains with the strongest correlations to assess how their exclusion impacts the overall composite index	Excluding the sub-domains with the strongest correlation (green mobility, land use, well-being)
C3	Understanding how the exclusion of a specific domain influences the score of UETI	Excluding one specific domain (Socio-Economic) and measuring the perfect compensability effect from additive aggregation

[1] EW: equal-weight allocation.

Figure 4 shows the sensitivity analysis results: the composite index does not vary significantly even in the presence of relevant weight variations (+80% of power grid quality weight with respect to the equal-weight allocation). The trajectory of C1 closely follows the trend of the reference case (EW), with moderate variations ranging between +5.9% (in 2014) and −1.0% (in 2019). On the contrary, C2 and C3 exhibit more pronounced variations compared to C1, as they entail more significant modifications in the structure of the composite index: C2 excludes sub-domains with the highest correlation, and C3 omits an entire domain. C2 variations vary in a range between +26.3% in 2014 and −12.3% in 2017, while C3 varies in a range between −25.3% in 2014 and +14.7% in 2019.

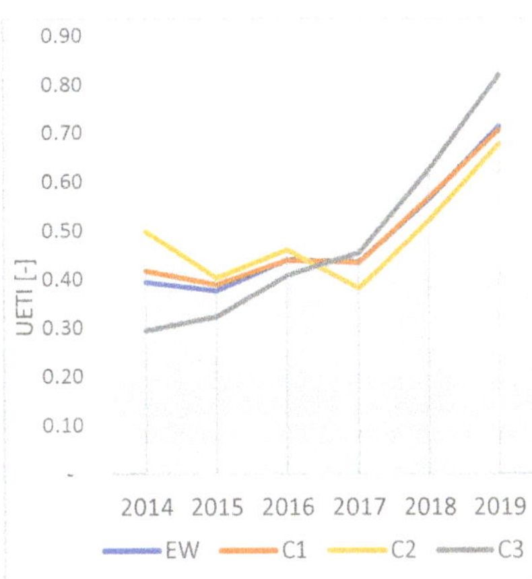

Figure 4. Sensitivity analysis results.

Ultimately, the analysis reveals that even with substantial modifications to weight allocation (C1) and to the structure of the composite index (C2 and C3), the variations in UETI remain limited, demonstrating the robustness of the composite index in the face of significant adjustments.

4.4. Policy Implications

This case study has revealed both strengths and weaknesses in Turin's multi-dimensional evolution towards energy transition. As regards the energy domain, Turin shows a consistent improvement, particularly thanks to the significant penetration of renewables in the urban energy mix, the improvement in green mobility and in energy efficiency leading to a reduced energy intensity in all sectors (e.g., residential, transport, etc.). Power grid quality registered a continuous although modest improvement too. From an environmental perspective, there is a greater disparity between carbon emissions and resources management performances (land use, waste and water) compared to air quality trend. Indeed, air quality remains a persistent issue for the city of Turin, necessitating increased efforts, investments and specific measures to address this challenge effectively. Moreover, this study underscores the need for greater attention to the socio-economic dimension, especially concerning the health and well-being of citizens; often overlooked in favour of technical and economic objectives, social aspects are critical for achieving a truly sustainable transition.

Ultimately, this work underscores the relevance and utility of continuously monitoring the city's evolution towards energy transition through a comprehensive index framework, encompassing not only the energy and environmental aspects but also the economic and social domains. Through consistent monitoring, weaknesses and criticalities can be easily identified and action priorities and investments can be established to advance a balanced energy transition. Therefore, the proposed method can support policymakers in identifying areas of action requiring greater efforts, prioritizing investments and promoting effective and targeted strategies to address the specific challenges faced by the city.

5. Conclusions

Cities will play a crucial role in speeding up the process of energy transition at a wider scale. The quantitative assessment and monitoring of city progress is essential for implementing evidence-informed policymaking, based on decisions grounded in reliable up-to-date information. Indeed, due to the complexity of the process, policymaking should be supported by rigorous science-based tools (methods and technologies) to plan tailored strategies, set targets and track the impacts of policies across the various domains involved in the transition (energy, environmental, society and economy). This study aims to respond to this necessity by proposing a flexible and comprehensive metric-based methodology to assess the energy transition performance at an urban scale. Various metric-based methodologies for monitoring energy transition at an urban scale are available in the literature, but too often the scarcity of urban data constrains the applicability of intricate and sophisticated methods. Furthermore, the intricacy of the methodology could serve as a barrier to effective communication and understanding among a wider and non-expert audience, hindering citizen and stakeholder engagement in the urban transition process.

This case study demonstrates, on one hand, the applicability of the method, and on the other hand, it pointed out the criticality of urban data scarcity.

As highlighted by the preliminary analysis conducted to assess the availability of open-source urban data, the city of Turin, like the majority of Italian municipalities, lacks a systematic data collection and organization system; therefore, the input data necessary to obtain the composite index (UETI) have been collected from 12 different data sources (Appendix A, Table A2) in various formats and time granularity.

The scarcity of urban data required the utilization of provincial- and regional-scale information: to address the lack of socio-economic data, information provided by ISTAT (National Institute of Statistics in Italy) was employed, even if these data primarily pertain to the regional and national scales. This serves as a compromise to overcome data scarcity, but it affects the actual performance of the city (e.g., low performance of energy poverty in Piedmont region impacts the performance of income of Turin).

The overall performance of the city of Turin presents a clear positive trend, mainly thanks to the Environmental and Energy domains, as confirmed by the correlation analysis, which showed that the UETI trend is mainly affected by the Energy and Environmental domains rather than the Socio-Economic trend. The sensitivity analysis proved the robustness of the composite index even in the presence of major modifications in weight allocation (C1) and in the framework structure (C2 and C3). It must be pointed out that to enhance the significance of sensitivity and correlation analyses, it is necessary to extend the sample of data currently limited to five years (2014–2019). Moreover, extending the time series would also improve the overall consistency of time-based Min–Max normalization, which rescales indicators into a 0–1 range according to the best and worst performances recorded in the period under study.

In perspective, developing a standardized data collection system and organizing urban information into a unique database, accessible by means of a simple interface, would be a powerful tool to explore and access municipal data, allowing for comparison among cities and for monitoring over time their UETI trend. Unlike the time-based Min–Max normalization employed in this paper to evaluate the UETI of a single city, when assessing UETI for a set of cities simultaneously, rescaling indicators according to the best and worst performances across cities could serve as incentive for positive competition and continuous collective improvement towards urban energy transition.

Author Contributions: Conceptualization, E.D.; Methodology, E.D.; Writing—original draft, E.D.; Writing—review & editing, E.B. and D.G.; Visualization, E.D.; Supervision, E.B. and D.G.; Project administration, E.B. and D.G. All authors have read and agreed to the published version of the manuscript.

Funding: This research received no external funding.

Data Availability Statement: The complete list of datasets used in the case study is provided in Appendix A, Table A2.

Conflicts of Interest: The authors declare no conflict of interest.

Appendix A

Table A1. Performance indicators of the city of Turin (2014–2019).

Performance Indicators	2014	2015	2016	2017	2018	2019
Installed RES capacity (MW)	18.5	18.9	19.7	21.3	22.4	23.6
Share RES in electricity consumption (%)	4.8	4.2	3.7	3.0	3.3	3.2
Duration of outages (-)	21.5	19.8	31.0	25.4	28.7	25.7
Average number of disconnections (-)	1.2	1.4	1.3	1.7	1.4	1.4
Power loss (%)	6.1	5.9	4.7	3.9	4.2	4.2
Number of RES-EV charging points (-)	0.0	0.0	2.0	2.0	2.0	10.0
Number of EV charging points over 1000 EV + HV private vehicles (-)	4.0	3.8	4.3	15.6	45.9	53.5
Number of EVs + hybrids over 1000 passenger vehicles (-)	2.6	2.9	4.1	6.8	10.3	14.1
Cycle lanes (km/100 km^2)	139.0	146.6	151.6	153.8	159.2	166.1
Energy intensity residential sector (MWh/inhab)	8.9	6.8	6.8	6.7	6.4	6.1
Energy intensity tertiary sector (MWh/kEUR)	0.2	0.2	0.2	0.2	0.2	0.2
Energy intensity transport (MWh/inhab)	3.2	3.2	3.2	3.1	3.2	2.9
Energy intensity industry (MWh/kEUR)	0.1	0.1	0.1	0.1	0.2	0.2
Energy intensity municipal service (MWh/m^2)	3.6	3.3	3.0	2.8	2.8	2.6
Green surface (ha/100,000 inhab)	220.0	219.2	219.8	223.9	225.0	226.2
Pedestrian areas (m^2/100 inhab)	139.0	146.6	151.6	153.8	159.2	166.1
Municipal waste (kg/inhab)	491.4	493.8	482.3	498.0	523.3	510.3
Sorted waste (%)	41.6	42.4	42.1	44.7	46.6	47.7
Consumption (l/day·inhab)	293.0	292.0	288.0	287.0	286.0	282.5
Loss (%)	22.4	24.6	24.7	24.6	25.0	22.2
PM_{10} exceedances (n days/year)	58.5	85.0	63.5	94.5	36.0	45.0
NO_2 concentration (μg/m^3)	39.5	40.5	37.5	38.5	33.0	35.0
Carbon intensity (tCO_2/inhab)	3.9	3.5	3.4	3.4	3.1	3.0
Deaths from respiratory system diseases (%)	14.1	14.2	14.2	14.8	15.3	14.8
Mortality rate (n deaths/10,000 inhab)	103.1	113.6	106.6	113.5	114.7	115.5
Energy–Environment Investments (%)	4.9	3.7	4.9	3.3	9.0	9.7
Energy–Environment Added value (EUR)	1761.4	1795.4	1709.9	1836.2	1961.9	1899.7
Energy–Environment Employment (%)	4.4	4.3	4.2	4.2	4.2	4.3
Income (kEUR/inhab)	16.9	17.3	17.6	17.6	18.4	18.6
Energy poverty (%)	4.8	4.4	4.6	4.3	5.2	5.6

Table A2. List of datasets and data sources.

ID	Dataset Name	Data Source
DT1	Dati di produzione e raccolta differenziata	ISPRA
DT2	Dati sui costi di gestione dei rifiuti urbani (pro capite o per chilogrammo di rifiuto)	ISPRA
DT3	Rifiuti_Produzione rifiuti speciali	ARPA PIEMONTE
DT4	Aria—la qualità dell'aria in Piemonte (Misure)	ARPA PIEMONTE
DT5	Bilancio di sostenibilità SMAT	SMAT
DT6	Ambiente urbano—Verde Urbano	ISTAT
DT7	Consumo del suolo	ISPRA
DT8	Istat_Tavole_Censimento_acque_per_uso_civile	ISTAT
DT9	Catasto Impianti Termici	REGIONE PIEMONTE
DT10	Iren—Bilancio di sostenibilità	IREN
DT11	Ambiente urbano—Energia	ISTAT
DT12	Dichiarazione non finanziaria	GTT
DT13	Autoritratto_2021—Circolante_Copert_2021	ACI
DT14	Open Parco Veicoli	ACI
DT15	Ambiente urbano—Mobilità	ISTAT
DT16	Ambiente urbano—Eco management (dati su illuminazione pubblica)	ISTAT
DT17	Consumi energetici, Impianti e Attestazione di Prestazione Energetica—APE	REGIONE PIEMONTE
DT18	Popolazione residente ricostruita—Anni 2002–2019	ISTAT
DT19	Reddito e principali variabili IRPEF su base subcomunale/comunale	MINISTERO DELL'ECONOMIA E FINANZE
DT20	Mortalità per cause	REGIONE PIEMONTE
DT21	Mortalità per territorio di evento	ISTAT
DT22	AAEP—Anagrafe delle Attività Economiche Produttive—Consultazione	REGIONE PIEMONTE
DT23	Principali aggregati territoriali di Contabilità Nazionale—Valore aggiunto per branca di attività	ISTAT
DT24	Imprese e addetti	ISTAT
DT25	Principali aggregati territoriali di Contabilità Nazionale—Investimenti fissi, lordi, interni e Spesa per consumi finali delle amministrazioni pubbliche	ISTAT
DT26	TAPE	COMUNE DI TORINO
DT27	Torino—Informacasa	COMUNE DI TORINO
DT28	Analisi del potenziale solare per i comuni dell'area metropolitana torinese	PROVINCIA DI TORINO
DT29	Relazione annuale relativa al funzionamento e alla sorveglianza dell'impianto—Termovalorizzatore Gerbido	IREN
DT30	Dichiarazione ambientale—Centrale di cogenerazione Torino Nord	IREN
DT31	Dichiarazione ambientale—Centrale di cogenerazione Moncalieri	IREN
DT32	STATO D'AVANZAMENTO ATTIVITA' DISCARICA E ATTIVITÀ DI GESTIONE DEL BIOGAS	AMIAT
DT33	Annuario Statistico—Settore toponomastica e edilizia	COMUNE DI TORINO
DT34	INDAGINE SULLE SPESE DELLE FAMIGLIE: MICRODATI AD USO PUBBLICO	ISTAT

References

1. United Nations. Sustainability. Available online: https://www.un.org/en/academic-impact/sustainability#:~:text=In%201987,%20the%20United%20Nations,to%20meet%20their%20own%20needs.%E2%80%9D (accessed on 22 May 2023).
2. Science-Based Climate Targets: A Guide For Cities Science-Based Targets For Climate: A Guide For Cities 2 Contents. 2020. Available online: https://sciencebasedtargetsnetwork.org/wp-content/uploads/2021/04/SBTs-for-cities-guide.pdf (accessed on 1 January 2024).
3. EU Parliament and EU Council. European Climate Law. Available online: https://eur-lex.europa.eu/legal-content/EN/TXT/?uri=CELEX:32021R1119 (accessed on 30 June 2021).
4. Gan, X.; Fernandez, I.C.; Guo, J.; Wilson, M.; Zhao, Y.; Zhou, B.; Wu, J. When to use what: Methods for weighting and aggregating sustainability indicators. *Ecol. Indic.* **2017**, *81*, 491–502. [CrossRef]
5. Renewable Energy Agency. World Energy Transitions Outlook 1.5 °C Pathway. 2021. Available online: www.irena.org (accessed on 1 January 2024).
6. World Economic Forum (WEF). Fostering Effective Energy Transition 2021. 2021. Available online: https://www.weforum.org/reports/fostering-effective-energy-transition-2021/in-full (accessed on 1 July 2022).
7. World Energy Council. World Energy Trilemma Index Report. 2019. Available online: www.worldenergy.org (accessed on 1 January 2024).
8. European Commission. *Transitions Performance Inde*; EU Publications: Luxembourg, 2021. [CrossRef]

9. European Climate Foundation. Net-Zero 2050. Available online: https://europeanclimate.org/net-zero-2050/ (accessed on 19 May 2023).
10. Burck, J.; Hagen, U.; Bals, C.; Höhne, N.; Nascimento, L. Climate Change Performance Index 2021: Background and Methodology. Available online: www.newclimate.org (accessed on 1 January 2024).
11. *ISO 37120*; Sustainable Cities and Communities-Indicators for City Services and Quality of Life. ISO: Geneva, Switzerland, 2018.
12. *ISO 37122*; Sustainable Cities and Communities-Indicators for Smart Cities. ISO: Geneva, Switzerland, 2019.
13. *ISO 37123*; Sustainable Cities and Communities-Indicators for Resilient Cities. ISO: Geneva, Switzerland, 2019.
14. Airaksinen, M.; Huovila, A. CITYkeys indicators for smart city projects and smart cities. *CITYkeys Rep.* **2017**, 14–47. [CrossRef]
15. REnaissance of Places with Innovative Citizenship And Technology (REPLICATE EU project). 2017. Available online: https://cordis.europa.eu/project/id/691735 (accessed on 1 January 2024).
16. POCITYF. Available online: https://cordis.europa.eu/project/id/864400 (accessed on 22 May 2023).
17. IRIS Project. Available online: https://cordis.europa.eu/project/id/774199 (accessed on 22 May 2023).
18. SmartEnCity. Available online: https://cordis.europa.eu/project/id/691883 (accessed on 22 May 2023).
19. Angelakoglou, K.; Kourtzanidis, K.; Giourka, P.; Apostolopoulos, V.; Nikolopoulos, N.; Kantorovitch, J. From a comprehensive pool to a project-specific list of key performance indicators for monitoring the positive energy transition of smart cities—An experience-based approach. *Smart Cities* **2020**, *3*, 705–735. [CrossRef]
20. Efkarpidis, N.; Goranović, A.; Yang, C.-W.; Geidl, M.; Herbst, I.; Wilker, S.; Sauter, T. A Generic Framework for the Definition of Key Performance Indicators for Smart Energy Systems at Different Scales. *Energies* **2022**, *15*, 1289. [CrossRef]
21. Anthopoulos, L.; Kazantzi, V. Urban energy efficiency assessment models from an AI and big data perspective: Tools for policy makers. *Sustain. Cities Soc.* **2022**, *76*, 103492. [CrossRef]
22. Zhang, L.; Bai, W.; Xiao, H.; Ren, J. Measuring and improving regional energy security: A methodological framework based on both quantitative and qualitative analysis. *Energy* **2021**, *227*, 120534. [CrossRef]
23. Chun, Y.; Zhang, J.; Sun, B. Evaluation of carbon neutrality capacity based on a novel comprehensive model. *Environ. Sci. Pollut. Res.* **2023**, *30*, 3953–3968. [CrossRef]
24. Kourtzanidis, K.; Angelakoglou, K.; Apostolopoulos, V.; Giourka, P.; Nikolopoulos, N. Assessing impact, performance and sustainability potential of smart city projects: Towards a case agnostic evaluation framework. *Sustainability* **2021**, *13*, 7395. [CrossRef]
25. Marquez-Ballesteros, M.J.; Mora-López, L.; Lloret-Gallego, P.; Sumper, A.; Sidrach-de-Cardona, M. Measuring urban energy sustainability and its application to two Spanish cities: Malaga and Barcelona. *Sustain. Cities Soc.* **2019**, *45*, 335–347. [CrossRef]
26. Molinaro, R.; Najjar, M.K.; Hammad, A.W.A.; Haddad, A.; Vazquez, E. Urban Development Index (UDI): A comparison between the city of Rio de Janeiro and four other global cities. *Sustainability* **2020**, *12*, 823. [CrossRef]
27. Analisi Trimestrale del Sistema Energetico Italiano. Available online: https://www.pubblicazioni.enea.it/le-pubblicazioni-enea/analisi-trimestrale-del-sistema-energetico-italiano.html (accessed on 22 May 2023).
28. Shen, Y.; Shi, X.; Zhao, Z.; Xu, J.; Sun, Y.; Liao, Z.; Li, Y.; Shan, Y. A dataset of low-carbon energy transition index for Chinese cities 2003–2019. *Sci. Data* **2023**, *10*, 1–10. [CrossRef] [PubMed]
29. The Arcadis Sustainable Cities Index. 2022. Available online: https://www.arcadis.com/en/knowledge-hub/perspectives/global/sustainable-cities-index (accessed on 21 February 2024).
30. IMD Smart City Index (SCI). 2021. Available online: https://www.imd.org/smart-city-observatory/home/2021/#:~:text=Smart%20City%20Index%202021,-The%20third%20edition&text=Taking%20the%20top%20three%20places,Geneva%20in%208th%20place.&text=What%20does%20it%20mean%20if%20your%20city%20is%20%E2%80%9Csmart%E2%80%9D? (accessed on 21 February 2024).
31. Global City Index GCI. 2022. Available online: https://www.kearney.com/service/global-business-policy-council/gcr/2022-full-report (accessed on 21 February 2024).
32. IESE Cities in Motion Index. 2022. Available online: https://www.iese.edu/media/research/pdfs/ST-0633-E.pdf (accessed on 21 February 2024).
33. D'Adamo, I.; Gastaldi, M.; Ioppolo, G.; Morone, P. An analysis of Sustainable Development Goals in Italian cities: Performance measurements and policy implications. *Land Use Policy* **2022**, *120*, 106278. [CrossRef]
34. Ecosistema-Urbano-2023. Available online: https://www.legambientefaenza.it/ambiente/2023/10/ecosistema-urbano-2023/ (accessed on 21 February 2024).
35. Muscillo, A.; Re, S.; Gambacorta, S.; Ferrara, G.; Tagliafierro, N.; Borello, E.; Rubino, A.; Facchini, A. An open data index to assess the green transition—A study on all Italian municipalities. *Ecol. Econ.* **2023**, *212*, 107924. [CrossRef]
36. Cavalli, L.; Farnia, L.; Lizzi, G.; Romani, I.; Alibegovic, M.; Vergalli, S. Luglio 2020 Report L'SDSN Italia SDGs City Index per un'Italia Sostenibile: Report di Aggiornamento. Available online: https://econpapers.repec.org/article/femfrepor/2020.07-06.htm (accessed on 21 February 2024).
37. European Commission. Sustainable Development Goals (SDGs). Available online: https://commission.europa.eu/strategy-and-policy/sustainable-development-goals_en (accessed on 21 February 2024).
38. Pareto, A.; Mazziotta, M. Methods for Constructing Composite Indices: One for All or All for One? Methods for Constructing Composite Indices: One for All or All for One? 1. 2013. Available online: https://www.researchgate.net/publication/281106596 (accessed on 1 January 2024).

39. EUROSTAT. Statistics Explained: Glossary. Available online: https://ec.europa.eu/eurostat/statistics-explained/index.php?title=Glossary:Energy_intensity (accessed on 15 September 2023).
40. European Commission; Joint Research Centre; Organisation for Economic Co-operation and Development. *Handbook on Constructing Composite Indicators: Methodology and User Guide*; OECD: Paris, France, 2008.
41. European Commission. Competence Centre on Composite Indicators and Scoreboards—Step 5: Normalisation. Available online: https://knowledge4policy.ec.europa.eu/composite-indicators/10-step-guide/step-5-normalisation_en (accessed on 13 September 2023).
42. Quijano, A.; Hernández, J.L.; Nouaille, P.; Virtanen, M.; Sánchez-Sarachu, B.; Pardo-Bosch, F.; Knieilng, J. Towards Sustainable and Smart Cities: Replicable and KPI-Driven Evaluation Framework. *Buildings* **2022**, *12*, 233. [CrossRef]
43. Becker, W.; Saisana, M.; Paruolo, P.; Vandecasteele, I. Weights and importance in composite indicators: Closing the gap. *Ecol. Indic.* **2017**, *80*, 12–22. [CrossRef] [PubMed]
44. Environmental Perfomance Index. Available online: https://epi.yale.edu/ (accessed on 22 May 2023).
45. Human Development Index. Available online: https://hdr.undp.org/data-center/human-development-index#/indicies/HDI (accessed on 22 May 2023).
46. European Commission. Competence Centre on Composite Indicators and Scoreboards—Step 6: Weighting. Available online: https://knowledge4policy.ec.europa.eu/composite-indicators/10-step-guide/step-6-weighting_en (accessed on 13 September 2023).
47. Antolín, J.; de Torre, C.; García-Fuentes, M.; Pérez, A.; Tomé, I.; Mirantes, M.L.; Hoyos, E. Development of an Evaluation Framework for Smartness and Sustainability in Cities. *Sustainability* **2020**, *12*, 5193. [CrossRef]
48. Bonnet, J.; Coll-Martínez, E.; Renou-Maissant, P. Evaluating Sustainable Development by Composite Index: Evidence from French Departments. *Sustainability* **2021**, *13*, 761. [CrossRef]
49. Prevost, S.; Benavente, D.; Stevenson, A.; Caperna, G.; Panella, F.; European Commission. Directorate-General for Research and Innovation, Transitions Performance index 2021: Towards Fair and Prosperous Sustainability. Available online: https://research-and-innovation.ec.europa.eu/strategy/support-policy-making/support-national-research-and-innovation-policy-making/transitions-performance-index-tpi_en (accessed on 1 January 2024).
50. European Commission. Competence Centre on Composite Indicators and Scoreboards—Step 7: Aggregating indicators. Available online: https://knowledge4policy.ec.europa.eu/composite-indicators/10-step-guide/step-7-aggregating-indicators_en (accessed on 31 August 2023).
51. Piano d'Azione per l'Energia Sostenibile ed il Clima (PAESC) Città di Torino. Available online: https://servizi.comune.torino.it/consiglio/prg/intranet/display_testi.php?doc=A-P202300598:110302 (accessed on 23 January 2024).
52. Bianciardi, G.; Cortassa, C.; Ellena, G.; Iacono, M.; Scarzello, R.; Pagani, R.; Savio, L. Piano D'azione per L'energia Sostenibile. 2010. Available online: https://iris.polito.it/handle/11583/2491508 (accessed on 21 February 2024).
53. European Commission. EU Mission: Climate-Neutral and Smart Cities. Available online: https://research-and-innovation.ec.europa.eu/funding/funding-opportunities/funding-programmes-and-open-calls/horizon-europe/eu-missions-horizon-europe/climate-neutral-and-smart-cities_en (accessed on 14 September 2023).
54. Laurenti, M.; Nuglio, S.; Poggio, A. Mal'Aria di Città. Available online: https://legambiente.campania.it/2024/02/08/malaria-di-citta-2024-i-nuovi-dati-sullinquinamento-atmosferico-nelle-citta-campane/ (accessed on 21 February 2024).
55. European Environment Agency (EEA). European City Air Quality Viewer. Available online: https://www.eea.europa.eu/themes/air/urban-air-quality/european-city-air-quality-viewer (accessed on 14 September 2023).
56. Faiella, I.; Lavecchia, L. La Povertà Energetica in Italia. 2014. Available online: https://www.bancaditalia.it/pubblicazioni/qef/2014-0240/QEF_240.pdf (accessed on 21 February 2024).
57. European Commission. Energy Poverty in the EU. Available online: https://energy.ec.europa.eu/topics/markets-and-consumers/energy-consumer-rights/energy-poverty-eu_en (accessed on 25 September 2023).

economies

Article

Heterogenous Effect of Industrialisation on Environmental Degradation in Southern African Customs Union (SACU) Countries: Quantile Analysis

Teboho Mosikari

Department of Economics, North-West University, Mafikeng 2735, South Africa; teboho.mosikari@nwu.ac.za

Abstract: Southern African Customs Union (SACU) countries are under pressure to commit themselves to sustainable environmental activities. This study employed data from 2007 to 2021 using quantile regression to determine the heterogenous effects of industrialisation on environmental degradation in SACU countries. Prior to the main results, this study investigated and confirmed the existence of a long-run relationship between industrialisation and environmental degradation. This study confirms the heterogeneous effect of industrialisation on environmental degradation. The results through quantile process estimates demonstrated an inverted U-shaped curve. The inverted U-shape suggests that industrialisation at lower and higher quantiles has a minor effect on the environment compared to medium quantiles, where it has a higher effect. This study found that industrialisation increases environmental degradation in the 4th to 6th quantiles, whereas in the 7th to 8th quantiles, industrialisation reduces environmental degradation. Therefore, it is recommended by this study that to mitigate environmental degradation, firms in SACU countries are encouraged to adopt environment-friendly technologies in their production.

Keywords: industrialisation; environmental degradation; SACU; panel quantile regression

JEL Classification: A10; C31; F18; L60

1. Introduction

Africa contributed about 4% of global carbon emissions in 2017 and 3.9% in 2021; however, the pattern in which Africa is developing may mean that Africa's emissions continue to increase undesirably (Ayompe et al. 2021). In the modern world, the global community is desperately seeking ways to mitigate the scale and impact of climate change in many ways. The Southern African Customs Union (SACU) countries, consisting of Botswana, Eswatini, Lesotho, Namibia, and South Africa, has, since 1910, formed a customs union. SACU is one of the oldest unions in Africa, and, recently, the environment they are operating within has significantly changed since its establishment. The fundamental intention of the union is to make trade free or easier for its members. SACU strategic plan for the period 2022–2027 acknowledges that it relates to changes in the environment and the energy transition away from economic activities that harm the environment. The plan further indicates that there are major developments emerging on new methods in industrial production, commodity demand, trade, and investment flows.

SACU regards industrialisation as one of the strategic pillars for its development and growth. However, it can be argued that industrialisation without innovative processes might bring heavy emissions for the region, leading to repercussions of environmental degradation. Industrialisation is the process through which agrarian economies transform into an industrial one through mass manufacturing (Mgbemene et al. 2016). According to Biernacki (2001), industrialisation is described as the process of using mechanical, chemical, and electrical sciences to transform production with inanimate sources of energy. Other scholars such as Dong et al. (2021) mentioned that industrialisation causes the concentration of production and labour in specific areas. Furthermore, they mentioned that it may lead to urbanisation, which may appeal to more production factors, create employment opportunities, and promote new methods of production. Industrialisation can be viewed to play an important part in economic development. Firstly, it stimulates economic activities through value chains, from raw materials to final goods. Also, it tends to create formal employment, which may prevent many social instabilities. In addition, it helps in maintaining a trade surplus by producing goods for export and reducing import dependence. Lastly, it encourages manufacturing and processing capabilities, leading to strong sustainable economic development, in turn creating wealth in the country.

However, it should be mentioned that although industrialisation may bring economic growth and development, it may come with problems of environmental degradation. According to Maurya et al. (2020), environmental degradation is defined as "deterioration of the environment through depletion of resources which includes all the biotic and abiotic element that form our surrounding that is air, water, soil, plant, animals, and all other living and non-living element of the planet of earth". There are two arguments in this process that both natural and human actions are perceived to be participating in increases in environmental degradation. Human activities are one of the reasons for accelerating environmental degradation. These activities can be automobiles and industries increasing poisonous greenhouse gases in the atmosphere. Some critics observe that reducing greenhouse gases is a pointless and prejudicial burden on African low-income countries. However, such critics are clearly climate-blinded and cannot anticipate what would come with high costs in the future. Medinilla and Byiers (2023) suggested that there are four strategies for African countries to consider green industrialisation: (1) decarbonise the existing industries, (2) produce inputs for global green industries, (3) manufacture green goods for African markets, and (4) leverage brown capabilities to jump to green industries. The relationship between industrialisation and the environment involves both threats and opportunities that will impact SACU countries in different ways. It is mentioned in this paper that being capable of foreseeing these adjustments will help these countries to adapt and respond better as part of their regional economic planning.

This study contributes to the literature by being the first to investigate the effect of industrialisation on the environment in SACU countries. Also, this study employs the panel data estimation method of quantile regression, which is powerful to help determine the heterogenous effect of industrialisation on environmental degradation in SACU countries. Secondly, although the work of Opoku and Aluko (2021) investigated the role of industrialisation on environmental degradation in 37 African countries, in their sample, SACU countries were not all included, including Namibia and Lesotho. It is, therefore, from this perspective and the interest of this paper to provide an exclusive study on SACU as one of the oldest customs unions. This attempt will provide an appropriate policy direction to the bureaucracy of SACU. Furthermore, to investigate the effect of industrialisation on environmental degradation in SACU countries, this study estimates its model by controlling for urban population and foreign direct investment (FDI). These control variables were never considered in previous studies such as Opoku and Aluko (2021). These variables are critical, with SACU as an old customs union, and this can be affected by these two variables. According to Darkoh (1997), it is projected that through 1950 to 2025, the urban population in developing economies will have increased from 300 million to 4 billion, which is 14-times more. This statistic provides an interesting claim to control for urbanisation in the current study to understand its heterogenous effect on the environment in SACU to help in policy making. Also, FDI has become a critical cause of private external finance for developing economies. According to the literature, FDI can possibly be channelled from strict environmental economies to divert their production to SACU countries if they have weak environmental policies (Copeland and Taylor (1994) and Nyeadi (2023)). In addition, FDI is considered from the view that SACU countries are still developing countries and to grow their economies, they have weak foreign investment policies that may possibly attract investments that are not environmentally friendly.

Therefore, the overall aim of this study is to examine the heterogenous effect of industrialisation on environmental degradation in SACU countries. This study attempts to answer the following research questions: (i) Is there a long-run relationship between industrialisation and environmental degradation? (ii) Is there a heterogenous effect of industrialisation on the environment? The outline of this study is as follows. Section 2 presents the literature survey. Section 3 describes the method adopted. Section 4 discusses the empirical results. Section 5 provides the conclusion and policy implications of this study.

2. Literature Survey

There has been a reasonably vast literature on the effect of industrialisation and climate change across the world. Environmental economics scholars have hypothesised that as a country develops, environmental degradation will first increase and then decrease in the long run. Therefore, the shape of the relationship between industrialisation and environmental degradation takes the form of an inverted U-shaped curve, and this hypothesis can be explained by the theory of the environmental Kuznets curve (EKC), Capps et al. (2016). Okereke et al. (2019) maintains that transparency and a clear discussion between industries and government are critical in warranting policies to take account of the effects regarding both sustainability and industrialisation agendas for corporations of all sizes. Industries are important to fast track the pace of economic growth, but disorganised industrialisation in the name of development can affect the people and deteriorate its local ecosystem (Mech and Hazarika 2018). Andronie et al. (2021) indicated that industrial big data analytics and sustainable product lifecycle management can assist throughout the decarbonisation process by the use of digital technologies.

In the literature, Mgbemene et al. (2016) descriptively analysed the aftereffects of industrialisation on environmental change. The study found that since industrial development, individuals have extremely enlarged the rate of adjustment in the climate and the environment through moving from agricultural to industrial practices and the pumping of carbon gases into the atmosphere. Majeed and Tauqir (2020) studied a panel of 156 countries for the period 1990 to 2014. This study sorted the countries into different income levels and used the dynamic generalized moments method (GMM) and common correlated effects mean group (CCEMG) to examine the relationships. The study suggested that industrialisation increases carbon emissions in all developmental stages. From industrialised economies, the work of Idowu et al. (2023) investigated whether energy consumption due to industrialisation leads to environmental degradation in OPEC and highly industrialised economies. The findings of their study found that industrialisation on emissions is negative for OPEC countries, and for highly industrialised countries, it is positive and significant. Yusuf et al. (2023) investigated the effect of trade and industrialisation on climate change in Australia. This study found that industrialisation does not have any significant effect on carbon emissions in the long run.

The empirical literature it is not without the work conducted in Asian countries as one of the fast-emerging economies. Firstly, a study by Panayotou et al. (1990) examined the association among industrial growth, structural change, and industrial policy on the environment in Thailand. In their paper, they recommended one aspect of policy transition. This idea emphases that alterations during the conversion phase must be made. They further claimed that the new policy is also possibly adequate for the industries if it is progressively phased in over time. Following this, Wang et al. (2020) examined a panel of Asia-Pacific Economic Cooperation countries for the period 1990 to 2014. This study applied a Westerlund panel cointegration test and dynamic seemingly unrelated co-integrating regression (DSUR) for determining the effect of industrialisation on carbon dioxide emissions. The findings indicated that there was a positive and significant effect of industrialisation on emissions. Similar results were found by Zafar et al. (2020) when they studied a panel of 46 Asian countries for the period of 1991 to 2017. Their paper used various panel techniques, Westerlund cointegration and fully modified OLS, to determine the long-run parameters. This study found that industrialisation has a positive impact on carbon emissions in Asian countries. Furthermore, a study by Ahmed et al. (2022) explored a panel of 55 countries in the Asia-Pacific region. This study used an econometric method, the autoregressive distributed lag model, for the period of 1995 to 2020. The results of their study revealed that industrialisation is positively associated with the environment. Elfaki et al. (2022) examined the impact of economic growth, energy consumption, financial development, and industrialisation on environmental degradation for eight ASEAN countries. Their paper used an autoregressive distributed lag model pooled mean group (PMG/ARDL) method covering the period 1994–2018. The findings from their study concluded that industrialisation has a negative relationship with environmental degradation. Jermsittiparsert (2021) indicated that industrialisation has a positive impact on nitrous oxide emissions in the Association of Southeast Asian Nations (ASEAN). The study used stationary panel models for the period of 1995–2015.

A country-specific study by Yuan et al. (2020) indicated that industrialisation has brought China both prospects and challenges since 1978. The results indicate that rapid industrialisation has put Eastern China under serious pollution stress. The most tremendous effects of industrialisation on environmental degradation are reflected in aquatic and soil ecosystems. Their paper recommends that China needs to enhance the incorporation of environmental observation and conservation monitoring systems. Wang et al. (2011) found that heavy industries have a positive impact on promoting China's carbon emissions. The study concluded that a 1% increase will lead to a 0.27% increase in China's carbon emissions in the long run. A study by Liu and Bae (2018) explored the effects of energy intensity, real GDP, industrialisation, urbanisation, and renewable energy on the environment for China for the period from 1970 to 2015. The results show that the coefficients of industrialisation have a positive and significant effect on carbon emissions. This study showed that a 1% increase in industrialisation accelerates carbon emissions by 0.3%. It is clear from the above review in Asian economies that there is clear evidence of a detrimental effect of industrialisation on environmental quality.

Bekabil (2020) pointed out that in Africa, small and growing industries' development requires subsidies from their governments to stimulate GDP and create job opportunities. However, governments in these developing countries should encourage green production techniques. Aladejare and Nyipute (2022) and Lin et al. (2015) are among the authors who studied the relationship between industrialisation and climate change. The work of Aladejare and Nyipute (2022) examined a panel of 32 African countries for the period 1991 to 2019. This study used the generalised least squares mixed effect model and dynamic common correlated effect to investigate the phenomena at hand. The findings of the study suggest that industrialisation has an adverse effect on the environment, whereas a single country study by Lin et al. (2015) indicated that there is no evidence that industrialisation does not increase carbon emissions in Nigeria. This study applied the technique of Johansen cointegration, and the study period was between 1980 and 2011. An independent study in BRICS countries by Voumik and Sultana (2022) investigated how various types of industrialisation relate to environmental degradation in the BRICS region. Their study applied the CS-ARDL for the period spanning from 1972 to 2021. This study discovered that industrialisation contributes negatively to the environment. Throughout the literature, especially related to African countries, there has been no study conducted in SACU countries. Furthermore, the existing literature still lacks evidence on the relationship between industrialisation and climate change based on quantile analysis. The advantage of this econometric technique is that it provides the researcher with evidence of heterogeneous effects on the phenomena under study.

3. Empirical Model and Data

This study aims to examine the heterogenous effect of industrialisation on environmental degradation in SACU countries. In relation to the other control variables, this study makes use of the following variables: foreign direct investment, manufacturing value added, and urban population. Therefore, the econometric model of this paper is derived as follows:

$$LED_{it} = \gamma_i + \delta_1 LFDI_{it} + \delta_2 LMVA_{it} + \delta_3 LUP_{it} + V_t \quad (1)$$

where the following is the description of the notations in the above equation.

LED_{it} is carbon emissions from SACU countries to capture environmental degradation.

$LFDI_{it}$ measures the inflow of foreign direct investment in SACU countries.

$LMVA_{it}$ measures industrialisation in SACU countries.

LUP_{it} captures the urban population in SACU countries.

V_t captures the residuals.

To capture the behaviour of the model in Equation (1), this study employs five countries of the SACU (i.e., Botswana, Lesotho, Namibia, Eswatini, and South Africa) for the panel time series of 2007 to 2021. The selection of this time span is purely based on the availability of data. Therefore, the description of the variable under study is presented in Table 1.

Table 1. Description of variables.

Abbreviation	Unit of Measure	Source
$LFDI_{it}$	Foreign direct investment, net inflows (% of GDP)	International Monetary Fund (IMF)
LED_{it}	CO_2 emissions (kt)	World Development Indicators (WDI)
$LMVA_{it}$	Manufacturing, value added (% of GDP)	World Development Indicators (WDI)
LUP_{it}	Urban population (% of total population)	World Development Indicators (WDI)

Modelling procedure: This study first assesses the description of the data, by applying the correlation analysis, normality test of the variables, panel unit root testing, and finally conducting the panel cointegration and estimating the quantile regression.

Panel unit root: This study applies two types of panel unit root tests, Levin et al. (2002) (LCC) and Im et al. (2003) (IPS). The LCC test works by pooling cross-sectional time series as a way of creating more powerful unit root tests. The test processes are created to assess the null hypothesis that the unit cross-section in the panel has in the integrated time series versus the alternative hypothesis that all cross-sections time series are stationary. In 2003, IPS proposed a panel stationarity test for dynamic heterogeneous panels based on the mean of cross-sectional stationarity statistics. The test was also applied in the literature by many authors such as Mosikari et al. (2016) and Fowowe (2012).

To assess the long-run relationship between industrialisation and environmental degradation, this study applies the Pedroni test. Pedroni (2004) pioneered the test of seven statistics that assess the null hypothesis of "no cointegration in the panels". The seven test statistics are categorised into two dimensions: group-mean statistics, which provides the average for the results of cross-country test statistics, and panel statistics that pool the statistics along the within dimension.

After determining the long-run equilibrium between industrialisation and environmental degradation, the following, according to the interest of the current study, is to heterogeneously determine the effect of industrialisation on the environment using the quantile panel analysis. Due to the unequal heterogeneity in SACU countries, an interaction between industrialisation and environmental degradation is likely to perform differently across different levels of emissions. The advantage of quantile regression is that it permits the parameters to vary with different quantiles. It also has a unique advantage of identifying the variation in the effect of industrialisation on the distribution of carbon emissions. Following the work of Koenker (2004), the quantile regression for the study can expressed as follows:

$$ED_{yit}(\tau x_{it}) = x_{it}'\beta(\tau) + \alpha_i + \varepsilon_{it} \quad (2)$$

where $ED_{yit}(\tau x_{it})$ indicates the τth quantile of the dependent variable, and x_{it} captures the vector of the explanatory variables. α_i stands for the sectional effect, τ represents the quantile, and $\beta(\tau)$ denotes the regression coefficients of the τth quantile and can be computed through the following function:

$$\beta(\tau) = \frac{argmin}{\beta(\tau)} \sum_{k=1}^{q} \sum_{t=1}^{T} \sum_{i=1}^{N} (|y_{it} - \alpha_i - x_{it}'\beta(\tau)|w_{it}) \quad (3)$$

where q presents the number of quantiles, T denotes the number of years and N for cross-countries in the panel, and w_{it} is the weight of the ith countries in the tth year.

4. Empirical Analysis

The table below presents the correlation results of the study. The purpose of this is to examine if there is a possibility of high correlation among the independent variables.

Table 2 presents the correlation analysis for the variables under study. It is evident from the table that there is a negative relationship between environmental degradation (LED)and foreign direct investment (LFDI) also, there is a negative correlation between environmental degradation and industrialisation (LMVA). The descriptive results also show that there is a positive correlation between environmental degradation and urban population. Lastly, it can be observed among the independent variables, i.e., foreign direct investment, urban population (LUP), and industrialisation, that there is no high correlation (about 0.9), which might be considered a sign of high multicollinearity.

Table 2. Correlation analysis results.

	LED	LFDI	LUP	LMVA
LED	1	−0.187	0.865	−0.512
LFDI	−0.187	1	−0.186	−0.048
LUP	0.865	−0.186	1	−0.796
LMVA	−0.512	−0.048	−0.796	1

Table 3 presents the normality test results. This study applied the Kolmogorov–Smirnov and Anderson–Darling tests developed by Smirnov (1939) and Anderson and Darling (1954), respectively. According to these two tests, they assume the null hypothesis that the data are normally distributed. According to the results, it can be observed considering the p-values that this study rejects the null hypothesis since all the p-values are significant. This implies that environmental degradation, foreign direct investment, industrialisation, and urban population are not normally distributed. Therefore, Cheng et al. (2021) suggested that panel quantile regression is likely suitable in this regard.

Table 3. Normality test results.

Variables	Kolmogorov–Smirnov Test	Anderson–Darling Test
LED	0.245 (0.000)	6.567 (0.000)
LFDI	0.170 (0.000)	3.626 (0.000)
LMVA	0.136 (0.001)	2.035 (0.000)
LUP	0.182 (0.000)	4.761 (0.000)

Notes: p-values in brackets.

Before this study performed the empirical investigation, it was crucial to investigate the unit root in panel data.

Table 4, above, presents the panel unit root results for variables considered in this paper. This study employed the LCC and IPS pioneered by Levin et al. (2002) and Im et al. (2003), respectively. It can be observed from the findings that the variable foreign direct investment is not stationary at levels, whereas environmental degradation, industrialisation, and urban population are found to be integrated in the order of 1. In using the IPS test, only industrialisation was found to be integrated in the order of zero, whereas variables environmental degradation, foreign direct investment, and urban population were found integrated in the order of one. The results from these two tests demonstrate, overall, that the variables in hand are in a mixture in the order of 1 and 0. Therefore, this study proceeds to determine the long-run equilibrium.

Table 4. Panel unit root test results.

Variables	LCC Test (Intercept and Trend)	ΔLCC Test (Intercept and Trend)	IPS Test (Intercept and Trend)	ΔIPS Test (Intercept and Trend)
LCO2	−1.402 (0.080) *	−8.375 (0.000) ***	−0.664 (0.253)	−6.268 (0.000) ***
LFDI	0.232 (0.591)	−12.353 (0.000) ***	0.896 (0.815)	−5.964 (0.000) ***
LMVA	−3.285 (0.000) ***	−4.942 (0.000) ***	−1.521 (0.064) *	−2.999 (0.000) ***
LUP	−1.503 (0.066) *	−1.637 (0.050) *	−0.719 (0.236)	0.178 (0.570)

Notes: 1% ***/5% **/10% *.

Table 5 presents the Pedroni cointegration. The method provides the three test statistics, which are Modified Phillips–Perron, Phillips–Perron, and Augmented Dickey–Fuller, to determine the possible existence of cointegration. The results show that with a significant *p*-value at 5%, this study rejects the null hypothesis of "no cointegration". This implies that there is a long-run equilibrium existence between foreign direct investment, environmental degradation, and urban population. After the confirmation of cointegration in the system, this study estimates the parameters using quantile regression.

Table 5. Pedroni test for cointegration results.

	Test Statistics	*p*-Value
Modified Phillips–Perron	1.6613	0.048 *
Phillips–Perron	−2.8340	0.002 **
Augmented Dickey–Fuller	−2.5783	0.005 **

Notes: 1% ***/5% **/10% *.

Since the data series in this paper does not follow the assumption of normal distribution, the study applies panel quantile regression. In comparison to the existing methods, such as pooled effect, fixed effect, and random effect regression methods, the current paper uses the panel quantile regression method, which provides the regression results for each quantile. Interestingly, it can offer more evidence regarding the heterogenous effect of industrialisation on environmental degradation in SACU countries.

To offer a comprehensive presentation for different quantiles, this study adopted 10 quantiles (i.e., 1st quantile to 9th quantile) throughout the regression process. The quantile regression results are shown in Table 6 and Figure 1 below. As can be observed from the results, environmental degradation was explained with variables LFDI, LMVA, and LUP. As can be observed, the effect of industrialisation on the environment is positive, demonstrating that industrialisation is detrimental to the environment in SACU countries. In terms of observation, the effects are heterogeneous across different quantiles, with an inverted U-shape (see Figure 1 for LMVA curve). The results suggest that industrialisation at lower and higher quantiles has a small effect on the environment compared to medium quantiles, where it has a higher effect. This confirms the heterogeneous effects of industrialisation across various distributions of emissions in SACU economies.

Table 6. Quantile regression results.

Variables	Quantile	Coefficients	*t*-Statistics
LFDI	0.1	−0.015	−0.178
	0.2	−0.025	−0.339
	0.3	0.009	0.130
	0.4	0.082	0.757
	0.5	0.079	0.499
	0.6	−0.170	−0.495
	0.7	−0.029	−0.105
	0.8	−0.132	−0.501
	0.9	−0.049	−0.216
LMVA	0.1	1.365	1.929 ***
	0.2	1.290	1.610 **
	0.3	1.234	1.758 **
	0.4	1.633	1.770 **
	0.5	2.165	1.722 **
	0.6	4.088	2.409 ***
	0.7	2.804	2.302 ***
	0.8	1.295	1.619 **
	0.9	1.159	1.723 ***
LUP	0.1	3.745	3.865 ***
	0.2	3.890	3.478 ***
	0.3	3.898	3.921 ***
	0.4	4.425	3.374 ***
	0.5	5.226	2.924 ***
	0.6	8.832	7.212 ***
	0.7	8.203	10.176 ***
	0.8	7.254	15.241 ***
	0.9	7.210	18.015 ***
Intercept	0.1	−9.919	−1.835 ***
	0.2	−10.094	−1.634 **
	0.3	−9.910	−1.822 ***
	0.4	−12.833	−1.797 ***
	0.5	−17.014	−1.754 ***
	0.6	−34.435	−3.670 ***
	0.7	−28.454	−4.449 ***
	0.8	−20.426	−5.451 ***
	0.9	−19.864	−6.313 ***

Notes: 1% ***/5% **/10% *.

Specifically, the highest positive effect was at the 6th quantile at 4.088, decreasing to 1.159 at the 9th quantile. The result shows that there is a negative relationship between foreign direct investment and environmental degradation. The negative effects are observed at the 1st, 2nd, 6th, 7th, 8th, and 9th quantiles, whereas the positive effect is observed at the 3rd, 4th, and 5th quantiles. The effects in all quantiles are not significant. For the urban population variable, the effect is positive in all quantiles, and they are statistically significant. It can be observed (see Figure 1 for LUP curve) that the effects are heterogeneous across the different quantiles with an inverted U-shape, specifically peaking at the 6th quantile and becoming relatively stable throughout.

Table 7 presents the Chi-square statistic value of 106.298, which is statistically significant at the 1% level. Therefore, the null hypothesis of "slope equality" across quantiles is rejected. This result confirms the conclusion imposed by Figure 1 and serves as proof that the relationship between industrialisation, FDI, urban population, and the dependent variable (environmental degradation) is heterogenous across quantiles. These findings are critical because they demonstrate that in some cases, the study might model industrialisation and environmental relationships with only linear models; this can lead to inappropriate conclusions. Also, the Chi-square statistic value of 34.746 results for the test of "symmetry". The test assumes the null hypothesis of symmetric quantiles. This study rejects the null hypothesis at the 1% significance level. These findings confirm the heterogeneous effect of industrialisation on environmental degradation in SACU countries.

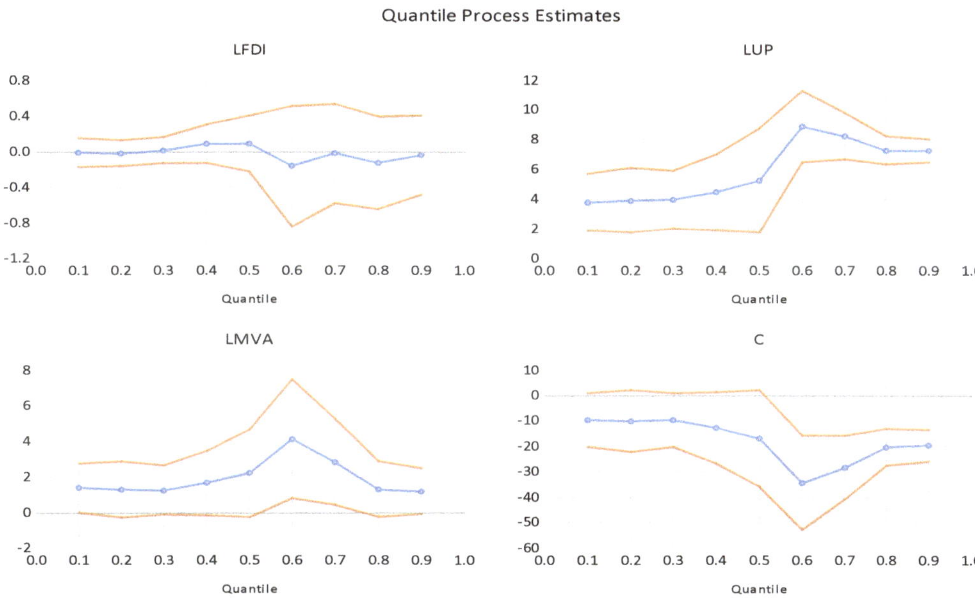

Figure 1. Graphical quantile process.

Table 7. Slope equality and symmetric quantile test.

Tests	Chi-Square Statistics	*p*-Value
Slope equality test	106.298	0.000 ***
Symmetric quantile test	34.746	0.004 ***

Notes: 1% ***/5% **/10% *.

5. Conclusions and Policy Implications

The empirical research in this study explored the heterogenous effect of industrialisation on environmental degradation in SACU countries. This study used the method of quantile regression to examine the proportions of industrialisation on environmental degradation for the period 2007 to 2021. This study also used the urban population and foreign direct investment as control variables. However, prior to quantile results, this study employed the Pedroni cointegration test, and the results confirmed the long-run existence of a relationship between industrialisation and environmental degradation in SACU countries. Furthermore, this study explored the parameter estimates through quantile effects. The results show that for FDI and environmental degradation, there is a negative relationship existing between them, but it is not statistically significant. These results are in line with Voumik and Ridwan (2023). The result for urban population shows that there was a positive relationship. This finding implies that as the urban population is increasing, there is a tendency to impact negatively on the environment. This can be explained as the behaviour of urban people to the increasing demand for food, construction (building houses), and rapid transportation. These activities might put pressure on the environment. These findings are consistent with a study by Mosikari and Eita (2020). The results between industrialisation and environmental degradation show a significant positive relationship. These results are consistent with the work of Ahmed et al. (2022).

The policy implications based on the findings of this study are as follows: On the positive relationship between urban population and environmental degradation, policy makers may consider encouraging the urban population to use local transportation that is environmentally friendly, such as electric cars, and for household lighting may encourage renewable lights. Also, on the positive relationship between industrialisation and environment, it is suggested by this study that SACU firms should start to think of greener innovative measures in their production processes to mitigate environmental degradation. There are several critical ways to ensure environmental friendliness: Firstly, industries should have treatment plans for the waste released by them. Waste management should be an integral part of the production process in each industry. Secondly, financial institutions need to broaden the area of green financing and incentives. Lastly, industries should reduce the use of fossil fuels and invest in renewable energy, including energy generated by wind, the sun, and rain. The suggestion for further research is that future studies should consider technological innovation in the manufacturing sector and consider the spatial effect of industrialisation in SACU countries.

Funding: This research received no external funding.

Institutional Review Board Statement: Not applicable.

Informed Consent Statement: Not applicable.

Data Availability Statement: The data can be available on a reasonable request.

Conflicts of Interest: The author declares no conflict of interest.

References

Ahmed, Farhan, Imtiaz Ali, Shazia Kousar, and Saira Ahmed. 2022. The environmental impact of industrialization and foreign direct investment: Empirical evidence from Asia-Pacific region. *Environmental Science and Pollution Research* 29: 29778–92. [CrossRef]

Aladejare, Samson Adeniyi, and Ishaku Rimantanung Nyipute. 2022. Ecological Response to Industrialisation Drivers in Africa. Research Square Working Paper. Available online: https://www.researchsquare.com/article/rs-2076419/v1 (accessed on 5 January 2024).

Anderson, Theodore Wilbur, and Donald Allan Darling. 1954. A Test of Goodness-of-Fit. *Journal of the American Statistical Association* 49: 765–69. [CrossRef]

Andronie, Mihai, George Lăzăroiu, Mariana Iatagan, Iulian Hurloiu, and Irina Dijmărescu. 2021. Sustainable Cyber-Physical Production Systems in Big Data-Driven Smart Urban Economy: A Systematic Literature Review. *Sustainability* 13: 751. [CrossRef]

Ayompe, Lacour M., Steven J. Davis, and Benis N. Ego. 2021. Trends and drivers of African fossil fuel CO_2 emissions 1990–2017. *Environmental Research Letters* 15: 124039. [CrossRef]

Bekabil, Urgessa Tilahun. 2020. Industrialization and Environmental Pollution in Africa: An Empirical Review. *Journal of Resources Development and Management* 69: 18–21.

Biernacki, Roland. 2001. Industrialization. *International Encyclopedia of the Social & Behavioral Sciences*, 7356–60.

Capps, Krista A., Catherine N. Bentsen, and Alonso Ramírez. 2016. Poverty, urbanization, and environmental degradation: Urban streams in the developing world. *Freshwater Science* 35: 429–35. [CrossRef]

Cheng, Cheng, Xiaohang Ren, Kangyin Dong, Xiucheng Dong, and Zhen Wang. 2021. How does technological innovation mitigate CO_2 emissions in OECD countries? Heterogeneous analysis using panel quantile regression. *Journal of Environmental Management* 280: 111818. [CrossRef]

Copeland, Brian, and Scott Taylor. 1994. North-South trade and the environment. *The Quarterly Journal of Economics* 109: 755–87. [CrossRef]

Darkoh, Michael Bernard Kwesi. 1997. Environmental Impacts of Urban-Industrial Development in The Developing World. *Journal of African Research and Development* 27/28: 127–37.

Dong, Hanmin, Minggao Xue, Yujia Xiao, and Yishuang Liu. 2021. Do carbon emissions impact the health of residents? Considering China's industrialization and urbanization. *Science of The Total Environment* 758: 143688. [CrossRef] [PubMed]

Elfaki, Khalid Eltayeb, Zeeshan Khan, Dervis Kirikkaleli, and Naveed Khan. 2022. On the nexus between industrialization and carbon emissions: Evidence from ASEAN + 3 economies. *Environmental Science and Pollution Research* 29: 31476–85. [CrossRef] [PubMed]
Fowowe, Babajide. 2012. Energy consumption and real GDP: Panel co-integration and causality tests for sub-Saharan African countries. *Journal of energy South Africa* 23: 8–14. [CrossRef]
Idowu, Ayodele, Obaika Micheal Ohikhuare, and Munem Ahmad Chowdhury. 2023. Does industrialization trigger carbon emissions through energy consumption? Evidence from OPEC countries and high industrialised countries. *Quantitative Finance and Economics* 7: 165–86. [CrossRef]
Im, Kyung So, Hashem Pesaran, and Yongcheol Shin. 2003. Testing for Unit Roots in Heterogeneous Panels. *Journal of Econometrics* 115: 53–74. [CrossRef]
Jermsittiparsert, Kittisak. 2021. Does Urbanization, Industrialization, and Income Unequal Distribution Lead to Environmental Degradation? Fresh Evidence from ASEAN. *International Journal of Economics and Finance Studies* 13: 253–72. [CrossRef]
Koenker, Roger. 2004. Quantile regression for longitudinal data. *Journal of Multivariate Analysis* 91: 74–89. [CrossRef]
Levin, Andrew, Chien-Fu Lin, and Chia-Shang James Chu. 2002. Unit Root Tests in Panel Data: Asymptotic and Finite Sample Properties. *Journal of Econometrics* 108: 1–24. [CrossRef]
Lin, Boqiang, Oluwasola E. Omoju, and Jennifer U. Okonkwo. 2015. Impact ofindustrialisationon CO_2 emissions in Nigeria. *Renewable and Sustainable Energy Reviews* 52: 1228–39. [CrossRef]
Liu, Xuyi, and Junghan Bae. 2018. Urbanisation and industrialisation impact of CO_2 emissions in China. *Journal of Cleaner Production* 172: 178–86. [CrossRef]
Majeed, Muhammad Tariq, and Aisha Tauqir. 2020. Effects of urbanization, industrialization, economic growth, energy consumption, financial development on carbon emissions: An extended STIRPAT model for heterogeneous income groups. *Pakistan Journal of Commerce and Social Sciences (PJCSS)* 14: 652–81.
Maurya, Pradip Kumar, Sk Ajim Ali, Anwar Ahmad, Qiaoqiao Zhou, Jonatas Da Silva Castro, Ezzat Khane, and Hazrat Ali. 2020. An introduction to environmental degradation: Causes, consequence, and mitigation. In *Environmental Degradation: Causes and Remediation Strategies*. Edited by Vinod Kumar, Jogendra Singh and Pankaj Kumar. Haridwar: Agro Environ Media, Agriculture and Environmental Science Academy, vol. 1, pp. 1–20.
Mech, Annesha, and Parinita Hazarika. 2018. A Study on the Impact of Industrial Effluents on Local Ecosystem and Willingness to pay for its Restoration. *Amity Journal of Economics* 3: 61–74.
Medinilla, Alfonso, and Bruce Byiers. 2023. The political economy of green industrialisation in Africa, ECDPM, DISCUSSION PAPER No. 363. Available online: https://ecdpm.org/application/files/1917/0263/7204/The-political-economy-green-industrialisation-Africa-ECDPM-Discussion-Paper-363-2023.pdf (accessed on 10 January 2024).
Mgbemene, Chigbo A., Chidozie C. Nnaji, and Chekwubechukwu Nwozor. 2016. Industrialization and its Backlash: Focus on Climate Change and its Consequences. *Journal of Environmental Science and Technology* 9: 301–16. [CrossRef]
Mosikari, Teboho Jeremiah, and Joel Hinaunye Eita. 2020. CO_2 emissions, urban population, energy consumption and economic growth in selected African countries: A Panel Smooth Transition Regression (PSTR). *OPEC Energy Review* 44: 319–33. [CrossRef]
Mosikari, Teboho Jeremiah, Mmamontsho Charlotte Senosi, and Joel Hinaunye Eita. 2016. Manufactured Exports and Economic Growth in the Southern African Development Community (SADC): A Panel Cointegration Approach. *Acta Universitatis Danubius* 12: 266–78.
Nyeadi, Joseph Dery. 2023. The impact of financial development and foreign direct investment on environmental sustainability in Sub-Saharan Africa: Using PMG-ARDL approach. *Economic Research-Ekonomska Istraživanja* 36: 2106270. [CrossRef]
Okereke, Chukwumerije, Alexia Coke, Mulu Geebreyesus, Tsegaye Ginbo, Jeremy J. Wakeford, and Yacob Mulugetta. 2019. Governing green industrialisation in Africa: Assessing key parameters for a sustainable socio-technical transition in the context of Ethiopia. *World Development* 115: 279–90. [CrossRef]
Opoku, Eric Evans Osei, and Olufemi Adewale Aluko. 2021. Heterogeneous effects of industrialization on the environment: Evidence from panel quantile regression. *Structural Change and Economic Dynamics* 59: 174–84. [CrossRef]
Panayotou, Theodore, Phanu Kritiporn, and Krerkpong Charnpratheep. 1990. Industrialization and Environment in Thailand: A NIC at What Price? *TDRI Quarterly Review* 5: 20–26.
Pedroni, Peter. 2004. Panel Cointegration: Asymptotic and Finite Sample Properties of Pooled Time Series Tests with an Application to the PPP Hypothesis. *Econometric Theory* 20: 597–625. [CrossRef]
Smirnov, Nikolai Vasil'evich. 1939. Estimate of deviation between empirical distribution functions in two independent samples. *Bulletin Moscow University* 2: 3–16.
Voumik, Liton Chandra, and Mohammad Ridwan. 2023. Impact of FDI, industrialization, and education on the environment in Argentina: ARDL approach. *Heliyon* 9: e12872. [CrossRef]
Voumik, Liton Chandra, and Tasnim Sultana. 2022. Impact of urbanization, industrialization, electrification and renewable energy on the environment in BRICS: Fresh evidence from novel CS-ARDL model. *Heliyon* 8: e11457. [CrossRef]
Wang, Zhaohua, Yasir Rasool, Bin Zhang, Zahoor Ahmed, and Bo Wang. 2020. Dynamic linkage among industrialisation, urbanisation, and CO_2 emissions in APEC realms: Evidence based on DSUR estimation. *Structural Change and Economic Dynamics* 52: 382–89. [CrossRef]
Wang, Zhongping, Qiang Li, Changliang Shi, and Gang Wang. 2011. Impact of Heavy Industrialization on the Carbon Emissions: An Empirical Study of China. *Energy Procedia* 5: 2610–16. [CrossRef]

Yuan, Jingjing, Yonglong Lu, Chenchen Wang, Xianghui Cao, Chunci Chen, Haotian Cui, Meng Zhang, Cong Wang, Xiaoqian Li, Andrew C. Johnson, and et al. 2020. Ecology of industrial pollution in China. *Ecosystem Health and Sustainability* 6: 1779010. [CrossRef]
Yusuf, Mohamud Said, Hammed Oluwaseyi Musibau, Khadar Ahmed Dirie, and Waliu Olawale Shittu. 2023. Role of trade liberalization, industrialisation and energy use on carbon dioxide emissions in Australia: 1990 to 2018. *Environmental Science and Pollution Research* 30: 79481–96. [CrossRef]
Zafar, Ayaz, Sana Ullah, Muhammad Tariq Majeed, and Rizwana Yasmeen. 2020. Environmental pollution in Asian economies: Does the industrialisation matter? *OPEC Energy Review* 44: 227–48. [CrossRef]

economies MDPI

Article

Impacts of the COVID-19 Pandemic on the Production Costs and Competitiveness of the Brazilian Chicken Meat Chain

Luiz Clovis Belarmino [1,*], Margarita Navarro Pabsdorf [2] and Antônio Domingos Padula [3]

[1] Embrapa Temperate Climate, Brazilian Agricultural Research Corporation, POB 403, Pelotas 96010-971, Brazil
[2] Faculty of Economic and Business Sciences, University of Granada, P. de Cartuja, 18011 Granada, Spain; pabsdorf@ugr.es
[3] School of Administration, Federal University of Rio Grande do Sul, Porto Alegre 90010-460, Brazil; antonio.padula@ufrgs.br
* Correspondence: luiz.belarmino@embrapa.br

Abstract: Sanitary requirements, geopolitical crises, and other factors that increase price volatility have an impact on the organization of markets and changes in investment policies and business strategies. The COVID-19 pandemic interrupted the trade of chicken meat, due to the drastic reduction in the circulation of goods, interrupted the supply of production chains, changed consumption habits, and made it difficult to reorganize business due to the slow resumption of operations by suppliers of inputs and in distribution logistics. The magnitude of these impacts has not been studied despite the high relevance of this economic dimension and the managerial implications for sector governance and trade management. The purpose of this study was to evaluate the economic impact of the COVID-19 pandemic on the production costs and competitiveness of the Brazilian chicken meat production chain. The methodology consisted of the detailed collection of information and data on private and social prices carried out using the Policy Analysis Matrix (PAM) method. The competitiveness coefficients and policy effects in the Brazilian broiler production chain before (2015) and during (2022) the COVID-19 pandemic were quantified and compared. Generally, the significant increases in the production costs of chicken meat (30.49%) caused a decrease in total factor productivity (−19.54%), a reduction in gross revenue, and lower tax collection. The pandemic has reduced the profitability of the chicken production chain in Brazil by 32.31%, reduced the competitiveness of exports, and worsened other economic indicators of the production chain. To the best of our knowledge, no other study has investigated the impacts of the COVID-19 pandemic on the competitiveness of the Brazilian chicken meat production chain. The PAM method allows for prices paid and received to be updated in real terms in projects representative of Brazil, the world leader in exports. This information is important for both national and international stakeholders. Additionally, this model is applicable to other meats traded in the international market, as it provides greater precision in business management and can estimate the impacts of risks on the availability or quality of food and health crises with robust results.

Keywords: poultry costs; production chain efficiency; politic; competitiveness; Brazil

1. Introduction

A close and complex link exists between domestic food production, international trade, and price impacts (Nonnenberg et al. 2021). This link is created via the interconnections between aggressive competition and industrial concentration associated with the commercial diplomacy of governments, in which increases in import tariffs and cultural issues associated with demand are predominant (Nkgadima and Muchopa 2022; Yeong et al. 2021). In contrast, animal products are the main sources of protein and energy available for human consumption, and territories with serious increases in hunger persist worldwide (Wijerathna-Yapa and Pathirana 2022). Therefore, it is expected to increase the supply of quality food, either via local production or imports (Soendergaard et al. 2023), where the global consumption of poultry meat will increase by 16% and represent 41% of all meat protein by 2031 (OECD-FAO. Organisation for Economic Co-operation and Development-Food Agricultural Organization 2022).

In this sense, in times of deep globalization with capital movements (Clapp 2019), specifically foreign direct investment (Bargoni et al. 2022; Chen 2017; Manning and Baines 2004), the availability of useful information for the formulation, monitoring, and evaluation of investment and governance policies, both public and private, is scarce, outdated, and poorly parameterized. In the last four decades, the chicken meat supply chain has been consolidated. Concurrently, it was necessary to assess factors that affect development, such as exchange rate fluctuations, technology transfer and innovation, the evolution of transnational companies, new clusters (Chen 2017), technological growth, digitalization (Qi and Chu 2022), lower prices, greater diversity and convenience (Chen et al. 2023; Martindale and Schiebel 2017; Bargoni et al. 2022), new investment patterns, and anti-competitive policies (Clapp 2017).

Competitiveness is a broad and complex concept (Constantin et al. 2023) but generally expresses the economic condition of an industry, company, product, or cluster to dispute markets or attract investments from competitors (Clapp 2017). The competitiveness of agricultural commodities can be conceptualized as the ability of organizations to gain, maintain, or expand their market share compared to competitors (Belarmino et al. 2022). The main factors determining this market condition are the use of universal production standards, permanent investments in innovation, and steady increases in productivity and efficiency (Almeida et al. 2020).

Competitiveness coefficients and policy effects on the Brazilian broiler production chain before (2015) and during (2022) the COVID-19 pandemic were identified, quantified, and compared. The economy of agri-food systems, management of the Food Supply Chain, and Global Value Chain (GVC) were prioritized.

2. Theoretical Background

2.1. The Production and Export of Meat Worldwide and in Brazil

Fruits, vegetables, and meat are typically the foods most recommended and valued by nutritionists. Their supply is conditioned by seasonality, the availability of productive resources, consumer income levels, and items of access and entry into markets. Generally, the production and trade of these foods are facing greater sanitary and quality controls, less regular supply, local or short chains, and greater price volatility (Aday and Seckin-Aday 2020). In this dynamic of linking prices and costs in the chains, economic impacts and supply crises arise, as in the case of agricultural fertilizer prices, which recently rose, thus indirectly increasing the costs of producing grains and, consequently, animal feed, all of which impacts meat competitiveness (CEPEA. Centro de Estudos Avançados em Economia Aplicada 2023; MAPA. Ministério da Agricultura 2023; USDA-FAS. United State Department of Agriculture-Foreign Agriculture Service-GAIN 2022).

Figure 1 presents the most significant countries in the international market in terms of quantity produced, yield/carcass weight, and quantity and value of exported chicken meat. Brazil is the third-largest producer of chicken meat (14.38% of the total) and the largest exporter (34.61%) and domestically consumes 10.03% of the global production.

Figure 2 presents additional information on the main market variables. This trend is relevant for the growth of the Brazilian economy and in other countries that produce and sell chicken meat, especially, for example, those that value this economical source of protein to supply low-income populations. Domestic production amounted to 14.705 million tons in 2022, and exports reached 4.6 million tons (FAOSTAT. Food and Agriculture Organization of the United Nation-Statistics Division 2023; MAPA. Ministério da Agricultura 2023). In 2020, the top exporters of poultry meat were Brazil (USD 5.59 billion), the United States (USD 3.93 billion), Poland (USD 2.61 billion), the Netherlands (USD 2.36 billion), and Thailand (USD 921 million). That year, the top importers of poultry meat were China (USD

2.99 billion), Germany (USD 1.79 billion), the United Kingdom (USD 1.34 billion), France (USD 1.27 billion), and Hong Kong (USD 1.08 billion).

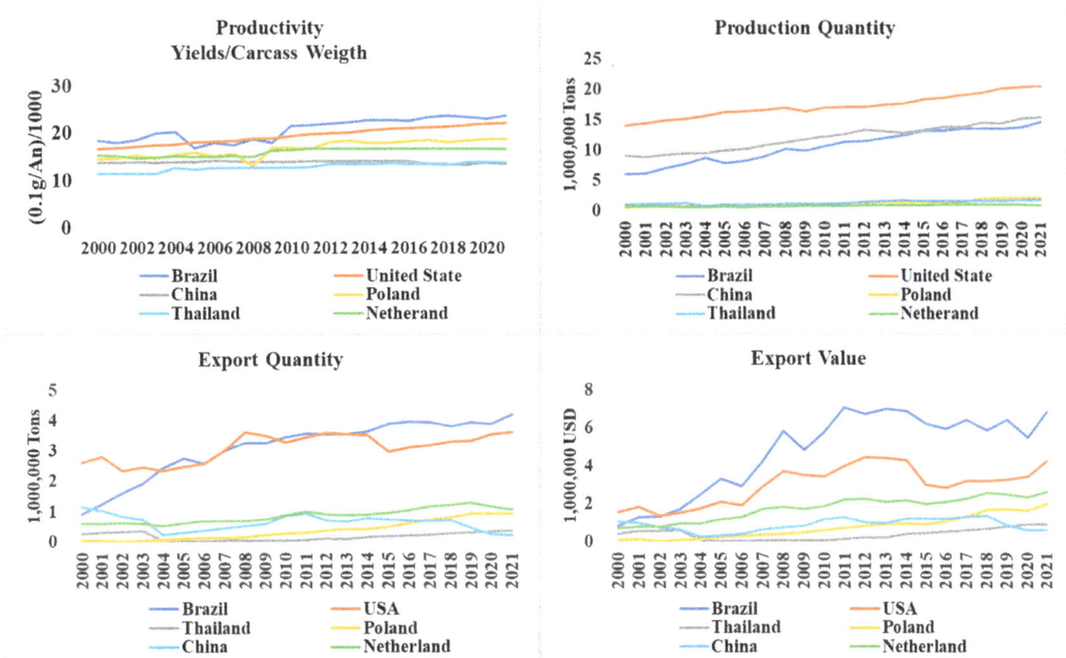

Figure 1. Main countries by the quantity produced, yield/carcass weight, and quantity and values exported of chicken meat. Source: FAOSTAT. Food and Agriculture Organization of the United Nation-Statistics Division (2023).

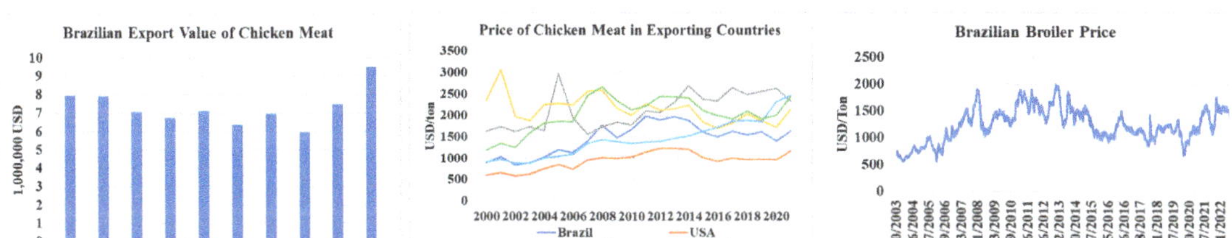

Figure 2. Brazilian export values and observed prices in Brazil and in competitors for chicken meat. Sources: FAOSTAT. Food and Agriculture Organization of the United Nation-Statistics Division (2023), MAPA. Ministério da Agricultura (2023), and CEPEA. Centro de Estudos Avançados em Economia Aplicada (2023).

The foreign trade of chicken in the USA represents 17% of national production (USDA-FAS. United State Department of Agriculture-Foreign Agriculture Service-GAIN 2022). The main destinations for Brazilian exports are China (USD 1.27 billion), Saudi Arabia (USD 690 million), Japan (USD 660 million), the United Arab Emirates (USD 426 million), and Hong Kong (USD 236 million) (CEPEA. Centro de Estudos Avançados em Economia Aplicada 2023; Associação Brasileira de Proteína Animal (ABPA) 2023).

2.2. Global Competitiveness of Poultry Meat

The recent dynamics of these economic variables and competitiveness factors express important indicators and coefficients for understanding the state of the art in the production and international trade of chicken meat in Brazil and other leading countries in this global value chain. Thus, for example, they indicate the competitive advantage revealed by the respective market share, the current levels of productivity (increase in weight/animal carcass), different prices or costs per ton of the national offers, and also the revenue obtained. These cost and revenue indicators in each country indicate the profitability or private profit margin, which can be transformed into a competitive advantage and also form the basis for calculating social profit (when taxes and other market failures are removed), which corresponds to the comparative advantage of each production chain studied. Therefore, the difference between private and social profitability monetarily scales the effects of national policies on the ability of exported commodities to compete in international trade.

Brazilian leadership in chicken exports is mainly due to the lower cost of production and marketing compared to competitors such as the United States (USDA-FAS. United State Department of Agriculture-Foreign Agriculture Service-GAIN 2022) and the European Union (EC 2021). Yields observed as weight gain (Figure 1), expressed in grams divided by the weight of gutted chicken, revealed that Brazil has the best conversion rate among the main exporters. In 2021, it surpassed the United States (−6.4%), Poland (−21.0%), the Netherlands (−29.5%), Thailand (−41.7%), and China (−43.4%). Generally, international meat prices should remain high in the short term, and the export price of chicken meat should closely follow grain prices, given the high share of feed costs in production (MAPA. Ministério da Agricultura 2023).

Poultry meat consumption has increased in virtually all countries and regions and is expected to reach 154 million tons, as consumers are attracted to lower prices, consistency, the adaptability of the product, and higher protein/low fat content (USDA-FAS. United State Department of Agriculture-Foreign Agriculture Service-GAIN 2022). The price of chicken should stabilize above the pre-COVID level after peaking in the first half of 2022 and stabilize at approximately EUR/USD 2000/ton by 2032, mainly owing to sustained demand in the European Union (EC 2021). Consequently, the price difference with Brazil (which produces EUR 1500/ton) will continue (USDA-FAS. United State Department of Agriculture-Foreign Agriculture Service-GAIN 2022), making it almost impossible for Europe to compete on the same ground (FAOSTAT. Food and Agriculture Organization of the United Nation-Statistics Division 2023; Associação Brasileira de Proteína Animal (ABPA) 2023). As shown in Figure 2, the average price in 2021 in Poland was 30.6% higher than in Brazil, and in the Netherlands, prices were 49.5% higher than in Brazil, which recorded USD 1629.22/ton this year, slightly above the average for this millennium (USD 1454.56/ton). In the United States, this average was lower (USD 937.75/ton) since most exports were whole chicken, which receives a lower price than chicken meat cut into specific pieces.

Contrarily, in Brazil, approximately 2/3 of sales are of cuts (Associação Brasileira de Proteína Animal (ABPA) 2023), and prices are higher than those paid for the whole chicken. These fluctuations in value were also observed by a 6.3% increase in food prices in general, which occurred in the first half of 2022 (World Bank 2023) but showed consecutive declines in the second half, as confirmed by the FAO Food Price Index (FAO FFPI. Food and Agriculture Organization of the United Nations 2023). In Brazil, the Index of Producer Prices of Agricultural Products Groups of CEPEA. Centro de Estudos Avançados em Economia Aplicada (2023) accumulated a nominal increase of 10.1% in 2022 (Figure 2), whereas the Index of Prices of Industrial Products (FGV IBRE. Fundação Getúlio Vargas-Instituto Brasileiro de Economia 2023) increased by 10.7% during the same period. Between 2021 and 2022, international food prices increased by 14.3% (FAO FFPI. Food and Agriculture Organization of the United Nations 2023). For the remainder of 2022 and early 2023, USDA-FAS. United State Department of Agriculture-Foreign Agriculture Service-GAIN (2022) predicted that domestic food prices in Brazil would continue to rise owing to rising inflation.

The costs, revenues, and national and international trade in Brazil are monitored and disclosed periodically by Embrapa Suínos e Aves (2023), MDIC COMEX STAT. Ministério do Desenvolvimento (2023), CEPEA. Centro de Estudos Avançados em Economia Aplicada (2023), IBGE PPM. Instituto Brasileiro de Geografia e Estatística (2023), CONAB. Companhia Nacional de Abastecimento (2023), and Associação Brasileira de Proteína Animal (ABPA) (2023). State agencies such as CEPA EPAGRI-SC. Centro de Socioeconomia e Planejamento Agrícola-Empresa de Pesquisa de Santa Catarina (2023), DERAL-PR. Departamento de Economia Rural-Secretaria de Agricultura e Abastecimento do Paraná (2023), and IEA APTA. Instituto de Economia Agrícola-Agência Paulista de Tecnologia Agrícola de São Paulo (2023) also monitor the price fluctuations in the Brazilian chicken meat chain.

By 2030, the global demand for chicken meat is expected to increase by 47% (CEPEA. Centro de Estudos Avançados em Economia Aplicada 2023). However, collectives of producers and other stakeholders have warned about the challenges posed by higher costs, which concern the industry, especially after corn and soybean meal prices rose by 3.4 and 2.5 times, respectively, in the last seven years (CEPEA. Centro de Estudos Avançados em Economia Aplicada 2023). Supply and demand may drive prices more significantly in 2023, stimulating production with continued shipments at high levels, in addition to domestic demand in Brazil's remaining firm (GEF-WFP. Global Economic Forum-United Nations World Food Programme 2023).

The impact of these foreign trade variables and domestic policies on the production, distribution, and consumption of chicken meat allows for the evaluation of the competitive positioning of each national production chain, the results of which increase knowledge on the subject and, thus, support decision making on investments and adjustments in the sector's governance, as well as supporting management measures in companies (Constantin et al. 2023).

2.3. *Impacts of COVID-19 on the Production Processes and GVC of Chicken Meat*

Generally, isolation measures and restrictions on the movement of people affected all sectors of the economy, creating serious difficulties in industrial operations. The primary sector was not the target of these activity restrictions, as it does not involve crowds of people. Even so, production typically takes place in open, isolated, and ventilated environments with other contamination difficulties. Aviaries are not work intensive and do not require significant manpower. Agri-industries concentrate many people on slaughter and cutting lines, but the interruption was partially adopted only at the beginning of the first contamination peaks, as these environments are regulated via hygiene standards and personal safety, strict surveillance of workers' health, and constant public inspection. Table 1 summarizes information from the bibliographic review, which selected the events and consequences of the pandemic in chicken meat GVC.

The export quantities in Figure 1 provide evidence that the COVID-19 pandemic did not affect the flow of Brazilian chicken meat export production chains. The prices of the main grains (corn and soybeans) experienced a period of high inflation at the global level. This was confirmed via the statistics of the volumes generated (IBGE PPM. Instituto Brasileiro de Geografia e Estatística 2023; MAPA. Ministério da Agricultura 2023; Associação Brasileira de Proteína Animal (ABPA) 2023) and the constancy of exports (MDIC COMEX STAT. Ministério do Desenvolvimento 2023; FAOSTAT. Food and Agriculture Organization of the United Nation-Statistics Division 2023). Figure 1 shows the positive growth in production (7.56%) and exports by volume (22.25%) and value (35.25%) during the COVID-19 pandemic and between 2015 and 2022. The average price of exports of chicken meat between 2017 and 2022 was USD 1604.19/ton, increasing by 10.63% in the same period. However, some countries reported problems in operation due to the pandemic, especially at the beginning of the outbreak. Interruptions occurred in food supplies, vaccines, medicines, and equipment in Bangladesh, with losses estimated at USD 825 million in the poultry sector (Sattar et al. 2021). In Indonesia, while the demand and price of chicken remained unaffected, economic growth decreased from 4.97% to 2.97%

(Surni et al. 2021). The COVID-19 pandemic also impacted the consumption, transport, and trade of chickens in Saudi Arabia (Hafez and Attia 2020). In India, losses exceeded USD 3.035 million, among other dramatic impacts on specific territories (Biswal et al. 2020). Similar effects of the pandemic have been reported in Nigeria (Fafiolu and Alabi 2020), Egypt (Abu Hatab et al. 2021), Ghana (Obese et al. 2021), and Myanmar (Fang et al. 2021). The impact of the pandemic has also been reported on energy supplies in China (Wu and Ma 2021) and other G7 countries (Awan et al. 2021), on macroeconomic indicators such as inflation in China (Feng et al. 2021), and on the supply chain for chicken feed (Attia et al. 2022).

Table 1. The COVID-19 pandemic and associated consequences on chicken meat production and markets worldwide and in Brazil.

Impacts of the COVID-19 Pandemic on the Chicken Meat Chain and Causes or Solution Strategies		
	Impacts	**Consequences**
WORLDWIDE	Initial reduction in consumption and trade Poor and reduced logistics Macroeconomic changes Biggest price increase in history	Uncertainties about harmlessness Blockages in production Lower economic growth of countries Shortage of production inputs Decreased business Prices dropped initially and then rose sharply Discrepancy between supply and demand Billions in losses in India Decrease in imports, followed by recovery
BRAZIL	Uninterrupted production and slightly lower exports Consumption and retail changed at the beginning of forced isolation Shortages in the supply of imported agricultural inputs in 2020 and 2021 Assistance and social security policies and relative inflation control	Ephemeral reductions in exports in 2020 (volume and value) and in consumption (see Figures 1 and 2) Increase in online purchases, adaptation to circulation control systems, and full recovery of exports in 2021 Adaptation of aviaries and agri-industries Substantial replenishment of exports from Ukraine

Source: Constructed by the authors, with updates from the literature cited.

In the United States and Europe-27, online shopping has increased sharply in the last decade; at the start, it accounted for only 11% of total retail trade (OECD-FAO. Organisation for Economic Co-operation and Development-Food Agricultural Organization 2022). According to Zeballos et al. (2023), the COVID-19 pandemic altered the entire food sector and induced a large increase in online food purchases (Ellison et al. 2021; Muresan et al. 2022; Todua and Jashi 2015). Another important aspect to consider is that in Brazil, the consumption of chicken meat should remain unchanged owing to low economic growth and high inflation, which can keep purchasing power weak. In this context, the USDA-FAS. United State Department of Agriculture-Foreign Agriculture Service-GAIN (2022) estimated that per capita consumption in Brazil would increase by 2.8% from 2022 to 2023 and pointed to a 3.8% increase in Brazilian chicken meat exports in 2023. The Associação Brasileira de Proteína Animal (ABPA) (2023) predicted an increase of 8.5% in international sales, reaching 5.2 million tons. CEPEA. Centro de Estudos Avançados em Economia Aplicada (2023) predicted that chicken meat production could reach 15.1 million tons in 2023, 2.3% above the forecast for 2022, with 6.2 billion animals slaughtered and a growth of 2.4% in the period.

The economic analysis of these variables and factors underlies several studies on the ability of companies to generate profits, monitor business results, predict the development or otherwise of the organization, practice business intelligence, monitor the trend of the segment in which the company operates as well as the economic situation and the country's fiscal policy, participate in the definition of credit sources and conditions, and also select the best logistics with suppliers and distributors (Lopes et al. 2012; FAO RLC. Organización de las Naciones Unidas para la Agricultura y Alimentación-Oficina Regional para América Latina y Caribe 2007).

3. Methods

Data collection for analysis was carried out at a representative establishment (RE) (Lopes et al. 2012; Monke and Pearson 1989), which was prioritized in the southern region of Brazil, the oldest and most traditional production center of chicken farms and agri-industries. This region represents a stabilized industry with consolidated productive and commercial experience and is recognized as a competitive agri-export cluster. Production costs, physical yields, revenues, and taxes were collected directly from RE accounting in the form of expenses and receipts actually incurred and not average estimates from the literature or databases. This ensured greater fidelity to market values and provided more credibility to the results of the competitiveness metrics (FAO RLC. Organización de las Naciones Unidas para la Agricultura y Alimentación-Oficina Regional para América Latina y Caribe 2007; Lopes et al. 2012).

To compare the data collected in 2015 with those obtained in 2022, the effect of the pandemic was isolated by selecting the same RE, repeating the export corridor and the same mode of transport (road) between the links in the chain, in addition to using the same metrics as in the Policy Analysis Matrix (PAM) method. Furthermore, the production chain in Brazil continued to be free of avian flu and to participate in the global chicken meat value chain; the macroeconomic policies to control exchange rates and interest rates did not change, nor were there interruptions in the use of credit and insurance instruments; the feeding and sanitary control processes were not modified; commercialization strategies followed the same transaction practices, and no innovations in operations occurred in the periods; and the levels of taxes and other charges on intermediate inputs and sold products also remained the same in the two periods.

Moreover, the structures and functions of the technology transfer organizations followed the same recommendations for handling the fattening batches in the aviaries and conducting the production lines in the slaughterhouse processing; the genetics of the birds and the content of the different types of diets did not differ between the two collection periods; the coordination of chain development and governance standards continued to be exercised by the third link (meatpacking plant); and the integration structure and quality standards remained unchanged.

Thus, it was understood that there was no change in the use of productive resources (land, capital, and labor), as there were no changes in the use of physical and human capital. The effects of the war in Ukraine, which started in 2022 (Sohag et al. 2023), did not affect the isolation implemented for measuring the effects of the COVID-19 pandemic, as the collection for 2022 was made with data and information prior to the outbreak of the conflict.

Based on these experimental conditions, the effect of the pandemic was isolated to assess costs, prices, and revenues, as the PAM method allows the maintenance of the physical yield indices in the chain, such as feed conversion rate per chicken fed, use of meat from the weight of the finished chicken at the slaughterhouse, performance of physical inputs, and maintenance of labor productivity. Thus, by collecting only private price variations between the two periods, the other chain competitiveness variables were isolated. Therefore, the products and services traded on the four links changed due to restrictions on the movement of goods and people. This segregation of price variation and guarantee of no change in other conditions allowed for a comparative analysis before and after COVID-19.

3.1. Economic Analyses with the Policy Analysis Matrix (PAM) Method

The PAM method generates the product of two economic identities. The first horizontally defines profit as the difference between revenues and costs, and the second vertically determines the effects of divergence and the impacts of both distorted policies and market failures (Table 2). To build the PAM spreadsheets, a production system based on territory or RE and the respective logistics corridor where transactions occur was previously defined (Lopes et al. 2012).

Table 2. Structure of the accounting matrix of the Policy Analysis Matrix method.

Price	Revenue	Costs		Profits
		Inputs Tradable	Domestic Factors	
Private	A	B	C	D (1)
Social	E	F	G	H (2)
Differences	I (3)	J (4)	K (5)	L (6)

Source: Monke and Pearson's (1989) striking indicators; FAO RLC. Organización de las Naciones Unidas para la Agricultura y Alimentación-Oficina Regional para América Latina y Caribe (2007) and Lopes et al. (2012). PAM's accounting results: (1) private profits (D = A–B–C); (2) social profits (H = E–F–G); (3) revenue transfers (I = A–E); (4) transfers of inputs (J = B–F); (5) factor transfers (K = C–G); and (6) net transfers (L = D–H or L = I–J–K).

Revenues and costs with domestic factors (land, labor, and fixed capital) in the PAM and marketable inputs (variables) were calculated at private and social prices. The values at private prices were determined from the actual prices charged in the market. These private values include the effects of policies, market failures, and shortcomings. Values at social prices corresponded to the economic valuation. However, this prevails in markets without policies or failures.

3.2. *Indicators of Comparative Advantage and Competitiveness in the Chain*

The first line of the PAM (Table 2) represents private profit (D). This is achieved by excluding sales revenues (A), the costs of marketable inputs to the market (B), and costs of domestic factors (C). A positive result indicates that the system is profitable and competitive because it exceeds input costs and positively remunerates capital, labor, and land. The second line estimates the profit at social prices (H), which is equal to social income (E) minus the social costs of marketable production factors (F) and internal factors (G). When H is positive, an economically efficient system with a comparative advantage exists. However, a negative result for social profits indicates that the system cannot survive without (in)direct payments, which wastes scarce resources and will make the value obtained for revenue exceed the import price (Lopes et al. 2012).

In this model, the concept of efficiency considers the use of resources in activities that provide higher levels of production and revenue, reflecting the cost of social opportunities (FAO RLC. Organización de las Naciones Unidas para la Agricultura y Alimentación-Oficina Regional para América Latina y Caribe 2007). The indicators and coefficients were obtained for each of the four links in the chain, starting with the prices paid and received in the production of chickens in the aviaries, the transport of live chickens from the poultry farmer to the slaughterhouse, expenses, and revenues in agri-industrialization, and the transport of processing and packaging to the port of embarkation. All amounts are expenses incurred effectively, taken directly from the accounting of the RE previously chosen in the sector, with the consolidation of data made with market agents and in the records of official and reliable databases. Therefore, data and information represent the area that best employs productive resources and has the best management of organizational innovation (Lopes et al. 2012).

3.3. *Coefficients of Productivity, Profitability, and Protection or Subsidy in the Chicken Meat Chain*

Table 3 presents the main performance coefficients of the competitiveness analyses used to assess the economic impact of the COVID-19 pandemic on the chicken meat chain in Brazil. They are separated into coefficient types: performance, formula, interpretation, and importance.

Table 3. Productivity, profitability, and taxation coefficients were used in the evaluation of economic performance of the chicken meat chain after the COVID-19 pandemic in Brazil.

Coefficient of Performance	Formula	Interpretation	Importance
		1. Profit Sharing in Revenue (PPR)	
–Private –Social	(D/A)*100 (H/E)*100	Share of profit in revenue How much of the revenue is profit	Rate of return Continuity of the chain
		2. Share of Added Value in Revenue (PAVR)	
–Private –Social	((A–B)/A)*100 ((E–F)/E)*100	Percentage of value addition Value added	Value created in the chain Capacity for innovation
		3. Share of Domestic Factors in Added Value (PDFAV)	
–Private –Social	(C/(A–B))*100 (G/(E–F))*100	Domestic factors' remuneration Efficiency gain/loss	Tendency is to reduce Aggregation performance
		4. Total Factor Productivity (FTP)	
–Private –Social	A/(B + C) E/(F + G)	Overall revenue result minus costs Growth of productive efficiency	Chain performance measure Ability of the chain to grow
5. Nominal Product Protection Coefficient (NPCP)	A/E	Calculates the taxation of chicken meat	Assesses the economic distortions to be corrected
6. Nominal Entry Protection Coefficient (NPCI)	B/F	Evaluates the taxation incident on the inputs used in the chain	Higher taxation reduces the competitiveness of the chain
7. Effective Protection Coefficient (EPC)	(A–B)/(E–F)	General measure of taxation that burdens gains in the chain	The weight of public policies in reducing profits
8. Vulnerability of Policy Chains (VCP)	((H–D)/H)*100	Measures the increase in profitability by removing taxation	Greater technification generates less vulnerability
9. Profitability Coefficient (PC)	D/H	Estimates the value of all policies in the profitability of the chain	Interventional terms shuttle income from the chain
10. Chain Taxation Level (CTL)	(L/E)*(−1)*100	Total amount of taxation levied on chain transactions	Excessive taxation reduces the supply chain competitiveness

Source: Monke and Pearson's (1989) striking indicators; FAO RLC. Organización de las Naciones Unidas para la Agricultura y Alimentación-Oficina Regional para América Latina y Caribe (2007) and Lopes et al. (2012).

3.4. Collection of Primary Data and Analysis

The lots consisted of an average of 39,000 broilers distributed over 2.970 m^2 (13 chickens/m^2), resulting in 250,965 chickens/aviary/year or 752,895 tons/aviary/year, with an average production cycle of 45 days and 6.5 cycles or lots/year. A mortality rate of 3.0% results in the effective delivery of 37,830 finished chickens/batch with a median weight of 3039 g. The agri-industry processes 75,000 chickens per day, with standards of international technology, using refrigerated road transport to take the whole chicken to the Rio Grande-RS port (648 km away). Although the agri-industry selected as the RE has 149 officially enabled chicken cuts, the present study established whole gutted and frozen chicken as the base product for comparison with international reference prices. The average yield of the chickens varied according to the cuts and, in the case of whole chickens, was 85.63%. However, most exports from Brazil's agri-industry (70%) are composed of different chicken cuts, such as thighs, breasts, wings, and feet, and other preparations (MAPA. Ministério da Agricultura 2023). The international prices used in this study were obtained from the values observed for social prices, which were generated by converting the private prices. The option of using conversion factors has been consolidated in national publications (Lopes et al. 2012; Torres et al. 2013) and accepted internationally (FAO RLC. Organización de las Naciones Unidas para la Agricultura y Alimentación-Oficina Regional para América Latina y Caribe 2007).

4. Results

The description of the results consists of a socioeconomic overview and analyses in micro-, meso-, and macroeconomic dimensions, according to the theoretical bases of the competitiveness of GVC used in the PAM method.

4.1. Results of the Analyses of Brazilian Chicken Meat Competitiveness after the COVID-19 Pandemic

The results are divided into the accounting and economic coefficients of competitive performance. The most affected factors were the magnitude of revenue sources and cost items (variable and fixed), in addition to the business environment surrounding the companies, especially tax encumbrances, which reached 40% of the total gross amount of revenue. This influence occurred via macroeconomic policies such as interest rates, exchange rates, and various taxes. Also relevant were credit and insurance policies, the promotion of research, innovation, and fiscal deregulations to expand investments in basic infrastructure (transport, communications, and energy), the promotion of exports, and other interventions in the market via reforms for greater economic freedom, simplification, and tax justice.

4.2. Accounting Indicators of the GVC of Chicken Meat Competitiveness and Comparative Advantages

The first row in Table 4 presents the accounting matrix for the GVC of the chicken meat corridor in southern Brazil. Letter D, in the years 2015 and 2022, shows that the existing economic performance proves competitiveness in international trade, as private profits were USD 950.39 and USD 564.24 per ton of frozen whole chicken, respectively. Social profit (letter H) amounted to USD 1312.75/ton in 2015 and USD 792.65/ton in 2022, values that prove the productive efficiency and comparative advantage of this chain in Brazil compared to international market prices. On the other hand, these figures also show that competitiveness reduced during the COVID-19 pandemic, owing to the significant differences in profitability between the figures for 2015 and 2022.

Table 4. Competitiveness indicator results in the Brazilian chicken meat chain, in dollars (USD) per ton before (2015) and after (2022) the COVID-19 pandemic.

Prices	Revenues	Costs		Profits
		Tradable Inputs	Domestic Factors	
Private	A	B	C	D
2015	2.234, 62	1.055, 98	228, 25	950, 39
2022	1.960, 13	1.193, 28	202, 62	564, 24
Social	E	F	G	H
2015	2.252, 13	792, 25	147, 13	1.312, 75
2022	1.979, 89	1.073, 66	113, 58	792, 65
Divergence	I	J	K	L
2015	(−17, 51)	263, 85	81, 12	(−362, 36)
2022	(−19, 76)	119, 62	89, 04	(−228, 41)

Source: based on search results in Lopes et al. (2012). Note: In September 2015, one US dollar was quoted, on average, at BRL3.92, and in March 2022, it had a change in exchange equal to BRL4.74.

4.3. Economic Coefficients of the Competitiveness of Brazilian Broiler Chickens during the COVID-19 Pandemic

The PAM method generated several economic performance coefficients for the chicken meat chain, such as the returns for the production factors and the inputs used, expressed in terms of efficiency (productivity and profitability) and taxation encumbrances on inputs and chicken meat (Table 5) as a result of existing policies.

Table 5. Competitiveness coefficients and effects of policy on the Brazilian production chain of broilers before (2015) and during (2022) the COVID-19 pandemic.

Coefficients, Formulas, and Results for Brazilian Chicken Meat Competitiveness	2015	2022
1. Profit Sharing in Revenue (PPR)		
–Private, PPR = (D/A)*100	42.53%	28.79%
–Social, PPR = (H/E)*100	58.29%	40.03%
2. Share of Added Value in Revenue (PAVR)		
–Private, PAVR = ((A–B)/A)*100	52.74%	39.12%
–Social, PAVR = ((E–F)/E)*100	64.82%	45.77%
3. Share of Domestic Factors in Added Value (PDFAV)		
–Private, PDFAV = (C/(A–B))*100	19.37%	26.42%
–Social, PDFAV = (G/(E–F))*100	10.08%	12.53%
4. Total Factor Productivity (FTP)		
–Private, FTP = A/(B + C)	1.74	1.40
–Social, FTP = E/(F + G)	2.40	1.67
5. Nominal Product Protection Coefficient (NPCP), NPCP = A/E	0.99	0.99
6. Nominal Entry Protection Coefficient (NPCI) B/F, NPCI = B/F	1.33	1.11
7. Effective Protection Coefficient (EPC), EPC = (A–B)/(E–F)	0.81	0.85
8. Vulnerability of Policy Chains (VCP), VCP = ((H–D)/H)*100	27.60%	28.82%
9. Profitability Coefficient (PC), PC = D/H	0.72	0.71
10. Chain Taxation Level (CTL), CTL = (L/E)*(−1)*100	16.09%	11.54%

Source: results obtained using the PAM method, FAO RLC. Organización de las Naciones Unidas para la Agricultura y Alimentación-Oficina Regional para América Latina y Caribe (2007), Lopes et al. (2012), and Torres et al. (2013).

The results for the broiler chain are shown in Table 5, and the coefficients for September 2015 and March 2022 were obtained using the same technologies and tax policies.

5. Discussion and Managerial Implications

5.1. Transformations in Brazilian Chicken Meat Prices and Competitive Performance after the COVID-19 Pandemic

This study found relevant effects of the pandemic on chicken meat prices in Brazil, with a strong reduction in competitiveness, and revealed numerous opportunities to increase organizational and technical innovation capabilities in industrial poultry farming. This knowledge about the transformations that have occurred can support the formulation of production and trade policies that seek the competitiveness of companies and the sustainability of the productive sector, as well as suggest the promotion of more management and governance changes such as, for example, prioritizing new assessments of reducing production costs and prioritizing value addition, as increasing the technical performance indices and improving the financial indicators can determine new economic advances and enable the reorganization of companies and the sector to increase performance.

The results that support these implications were divided into micro-, meso-, and macroeconomic dimensions, as the PAM method generated new information that allows us to understand the relevant points of the competitiveness of chicken meat and quantify the impacts for management purposes.

5.1.1. The Microeconomic Dimension: Changes in Costs, Revenues, and Profitability in the Brazilian Chicken GVC

The results come from a chain comprising an aviary and high-performance slaughterhouse with a high standard of management and governance. In the evaluations using the data of each cost and revenue item of each of the four links, which constituted Table 4, the items with the greatest impact on the cost of chicken meat were recognized as the cost of live chicken (finished chicken) coming from the farm, whereas, in poultry, the highest weights were attributed to intermediate inputs, such as the purchasing of chicks, feed, and medication, as also observed by other authors in Brazil (CONAB. Companhia Nacional de Abastecimento 2023; Santos Filho et al. 2018). It was observed that in 2015, the cost of producing chicken meat on the farm was USD 739.13 per ton of live chicken and was based on the price of BRL 2.92 per dollar, while in 2022, the cost was USD 963.46/ton and the per-dollar price was BRL 5.13.

Therefore, there was an increase of 30.49% in the cost of chicken meat production in Brazil. The largest increase in expenses occurred in the purchase of intermediate inputs, which represented 94.19% and 89.18% of the cost of production in the farm in 2022 and 2015, respectively. Day-old chick prices increased, as did grain expenditure. These went from USD 165.21/ton (25.06%) in 2015 to USD 241.44/ton (25.06%) in 2022. Generally, the percentage shares of these items in expenses are similar to those in other studies found in the literature (CEPEA. Centro de Estudos Avançados em Economia Aplicada 2023). In the agri-industry, the processing cost to obtain a ton of chicken meat was USD 94.06/ton in 2015, which increased to USD 137.16/ton in 2022.

During the pandemic, intermediate inputs in the agri-industry contributed 29.70% of total costs/ton of chicken meat, while labor accounted for 47.85%. In the logistical analysis, the fixed cost represented 3.87% of the total cost from the farm (USD 0.43/ton) and 6.37% of the total cost from the slaughterhouse to the port (USD 0.88/ton) in 2015. In 2022, the respective percentages were 9.69% (USD 0.26/ton) and 1.24% (USD 1.05/ton). Diesel was the main component of the transport cost in 2022, representing 27.22% of the total cost for the transfer of the whole frozen chicken to the port, whereas this participation was 39.75% for the live chicken transferred from the farm to the port. This is likely due to the high prices of this fuel in the international market following the pandemic.

The relationship between gross revenue and total expenses in the Brazilian poultry chain also changed owing to the COVID-19 pandemic. In 2015, in the agri-industry, this ratio, calculated using the formula "revenue/expenditure*100," showed a result of 218.02%, while in 2022, this fell to 129.74%. The causes of this reduction were variations in exchange and interest rates, as reported by several Brazilian authors (Associação Brasileira de Proteína Animal (ABPA) 2023; CEPEA. Centro de Estudos Avançados em Economia Aplicada 2023). This generated information on new costs, revenues, and profits proves that the pandemic has increased expenses and reduced revenues and profits. These microeconomic changes represent a new theoretical framework and describe how firms can optimize production and cost efficiency given existing technologies and input prices.

5.1.2. The Mesoeconomic Dimension: Sectorial Variations That Impacted Prices in the Chicken GVC during the COVID-19 Pandemic

The international demand for meat is vigorous, and the market in Brazil has experienced sustained growth in recent years (USDA-FAS. United State Department of Agriculture-Foreign Agriculture Service-GAIN 2022) despite discussions about the possible threats of climate change (Lara and Rostagno 2013) and the emergence of alternative proteins (Andreoli et al. 2021). The national per capita consumption of chicken meat (45.27 kg) and pork (17.58 kg) increased by approximately 3 kg per capita in 2021, whereas that of beef (32.69 kg) reduced by 2 kg (Talamini and Martins 2022). Chicken and pork meat have much lower consumer prices than beef, in values that can reach up to 20% of the average prices of cuts, despite the large increase in animal feed costs (UNCTAD. United Nations for Conference on Trade and Development 2022; USDA-FAS. United State Department of Agriculture-Foreign Agriculture Service-GAIN 2022; World Bank 2023).

The production cost of a ton of live chicken in the aviary at the end of the process increased during the pandemic period, as it was 71.49% of a ton of chicken eviscerated and frozen in the agri-industry in 2022, while the share in 2015 was 38.32%. The percentage share of the cost of chicken at the end of the process in a ton of meat transported to the port was 59.34% in 2022, higher than in 2015 (31.13%).

Similarly, increases in the costs of intermediate inputs also occurred during this period of the pandemic because the expenses added up throughout the production chain; the values per ton of chicken conducted to the port of embarkation were USD1055.98/ton in 2015 and USD1193.28/ton in 2022 (Table 4), which indicates a 13% increase in the variable spending of the poultry chain during this period. Similarly, the costs of domestic production factors, such as capital and labor, especially in the Brazilian currency BRL/ton, increased in value and weight in the actual RE expenses, which impacted 10.43% of the total costs of the value chain in 2015, and 14.52% in 2022. This result is linked to an increase in the exchange rate, as reported by numerous authors (Talamini and Martins 2022; Zylbersztajn et al. 2015). Brazilian public policies in force since 2005 have considered several sectorial measures, such as aggressive trade policy and the valorization of family farms (USITC. United State International Trade Commission 2012).

Nevertheless, Brazilian logistics are considered uncompetitive in international trade (World Bank 2017). In Brazil, although logistics is the last frontier for reducing expenditure in the supply chain, further studies are lacking (Lepchak and Voese 2020). The reduction in the competitiveness of Brazil's chicken meat chain was due to a shortage of inputs, which increased prices (Bairagi et al. 2022; Yu et al. 2020). This information is also supported by the fluctuations that occurred in Brazil's gross domestic product (IBGE POF. Instituto Brasileiro de Geografia e Estatística 2023), which fell by 3.9% in 2020 but grew by 4.6% in 2021 and 3.0% in 2022 (FGV IBRE. Fundação Getúlio Vargas-Instituto Brasileiro de Economia 2023).

5.1.3. The Macroeconomic Dimension: Post-Pandemic Modifications to the Chicken GVC's Comparative and Competitive Advantages

The estimates in Table 5 were divided into the following competitive capabilities.

(a) Competitive advantage (PPR: profit sharing in revenue, PAVR: value-added share in revenues, and VCP: profitability coefficient): The pandemic decreased the total profitability of the chicken chain by 32.31%, measured using the share of profit on revenue (PSRP). This means that the chain's private revenue continued to outperform the costs of domestic inputs and factors, but with a sharp drop, likely due to an economic recession of 5% in 2020 (OECD-FAO. Organisation for Economic Co-operation and Development-Food Agricultural Organization 2022). The drop in levels of value aggregation over total revenue (PAVR) was 25.83% in 2022 compared to 2015.

(b) Comparative advantage (PDFAV: participation of domestic factors in the added value and total TFP: productivity): The total factor productivity (TFP) decreased from 1.74 to 1.40 from 2015 to 2022, a significant reduction of 19.54%. Therefore, the production chain obtained lower revenue from the expenses of intermediate inputs and the use of domestic factors. For a company, productivity indicates the possibility of increasing employees' salaries, making new investments, or even continuing to operate (Lopes et al. 2012). On a national scale, productivity can be defined as the difference between quality-of-life standards. In comparative advantage analyses, stagnant or contracted productivity reveals future problems for individuals, organizations, and nations. This implies that management and governance measures should be promoted to value this coefficient.

(c) Impacts of protection and subsidy policies (NPNC: nominal product protection coefficient, nominal protection coefficient of the inputs, and EPC: effective protection coefficient): The policy vulnerability of Brazilian aviculture increased during the COVID-19 pandemic because the effective protection coefficient (the sum of the effects on the product and intermediate inputs) increased from 0.81 in 2015 to 0.85 in 2022. Similarly, the level of overall profitability of the chain fell slightly from 0.72 in 2015 to 0.71 in 2022, as the effects of policies increased slightly from 27.60% to 28.82%. This means that there was interference from policies that distorted the prices paid in the chain, as there was no change in the specific coefficient of the price received by the product. This result's implication is related to the tax incident, which motivates producers to seek solutions to increase the added value to production because it is lower than the value added in economic terms (Lopes et al. 2012).

5.2. Conclusions and Future Research

This study evaluated the socioeconomic impact of the COVID-19 pandemic on the competitiveness of the Brazilian chicken meat production chain. The economic analysis of private and social prices in the four links used the PAM method. Competitiveness coefficients and the effects of policy on the chain before 2015 and during the 2022 COVID-19 pandemic were characterized and compared. The results revealed that there were significant increases in the production costs of chicken meat (30.49%). In the agro-industry, the cost of processing more than doubled, from USD 94.06/ton in 2015 to USD 137.16/ton in 2022. This caused a drop in total factor productivity (−19.54%) and gross revenue and lowered the tax collection volume. The pandemic reduced the chain's profitability by 32.31% and reduced the competitiveness of exports.

The political vulnerability of Brazilian poultry farming increased during the pandemic due to high percentages of taxes, fluctuations in input prices, and significant distortions in production costs. This had an important impact on the Brazilian chicken meat chain's competitiveness, with useful managerial and financial implications for stakeholders (Caetano 2022; Valdes 2022). This new condition revealed in the results of this study indicates that investments in the chain are viable and can be continued (Miller et al. 2022; Yuzaria et al. 2021). These results match market trends that indicate that chicken meat will be the most consumed livestock product in the world in the coming decades (Miller et al. 2022), especially in emerging and developing countries (UNCTAD. United Nations for Conference on Trade and Development 2022).

Another implication of the results was the perception that it is essential to continue and accelerate innovation and the exploration of the successful combination of physical capital and human capital with new and transparent business practices. This can be undertaken by incorporating more automation and digitalization via affordable financing and better basic infrastructure, such as with new agile and loss-reducing transport and storage logistics. Therefore, even with a historically competitive Brazilian chicken meat production chain, albeit given the reduction in its profitability, as demonstrated in this article, it is believed that future diagnoses of priority challenges will be essential and, therefore, listing the fundamental increases in productivity and continued investments in technological and organizational innovation capabilities, such as the remodeling of impacted businesses and strategic planning for the recovery of competitiveness indices prior to the COVID-19 pandemic event, will also be crucial.

To the best of our knowledge, no other study has investigated the impacts of the COVID-19 pandemic on the competitiveness of the Brazilian chicken meat production chain. The MAP method has proven to be a verifiable metric and can be reported as suitable for measuring the effects of crises and related policies that affect food markets. Additionally, methodological originality allows for periodic updates in real terms and in support of the formulation, monitoring, and evaluation of projects in agri-food systems. Specifically, this model is applicable to other meats traded on the international market, as it provides greater precision in business management and can estimate the impacts of risks on the availability or quality of food in health or geopolitical crises with robust results.

The limitations of this study involve the data collection time, as the economic condition in the Brazilian chicken meat chain was measured and compared in 2015 and 2022. The pandemic changed prices and competitiveness, as shown in this study; however, the initial disorganization of transactions in the chain is gradually recovering to the pre-pandemic levels. Therefore, despite this likely recovery of normality in the markets occurring, it was not studied. Another limitation of this study involves the use of the option of the internationalization of prices, which can be expanded in new studies, especially regarding the faithful capture of values transacted in international centers of price formation. This option could qualify the obtaining of social prices, enrich the formulation of public policies of interest to other countries, and subsidize new economic strategies in companies, in addition to generating information and insights for new studies.

Studies with the PAM method use the general equilibrium model of international trade and in this study they demonstrated how and how much divergence (distorting police and market failures) affects the values of outputs and inputs associated with global chicken meat systems. Future studies could focus on assessing the impacts of restrictive and distortionary non-tariff measures.

Author Contributions: Conceptualization, L.C.B.; methodology, L.C.B.; software, L.C.B.; validation, A.D.P. and M.N.P.; formal analysis, A.D.P. and M.N.P.; investigation, L.C.B.; resources, L.C.B. and A.D.P.; data curation, L.C.B.; writing—original draft preparation, L.C.B.; writing—review and editing, A.D.P. and M.N.P.; visualization, A.D.P. and M.N.P.; supervision, M.N.P.; project administration, L.C.B. and A.D.P.; funding acquisition, L.C.B. and A.D.P. All authors have read and agreed to the published version of the manuscript.

Funding: This research received no external funding.

Informed Consent Statement: Not applicable.

Data Availability Statement: Data are available upon reasonable request.

Acknowledgments: This research had institutional support from the Faculty of Economic and Managment Palmeira das Missões Campus-Federal University of Santa Maria, Embrapa-Brazilian Agricultural Research Corporation, School of Administration at the Federal University of Rio Grande do Sul (UFRGS) and the Department of International and Spanish Economics at the University of Granada, and the CNPq-National Council for Scientific and Technological Development (Brazil).

Conflicts of Interest: The authors declare no conflict of interest.

References

Abu Hatab, Assem, Zhen Liu, Asmaa Nasser, and Abourehab Esmat. 2021. Determinants of SARS-CoV-2 impacts on small-scale commercial broiler production systems in Egypt: Implications for mitigation strategies. *Animals* 11: 1354. [CrossRef] [PubMed]

Aday, Serpil, and Mehmet Seckin-Aday. 2020. Impact of COVID-19 on the food supply chain. *Food Quality and Safety* 4: 167–80. [CrossRef]

Almeida, Mario Augusto Gouvêa, Hoyêdo Nunes Lins, and Eva Yamila da Silva Catela. 2020. Cadeias globais de valor, inovação e uppgrading: Estudo sobre empresas industriais Argentinas com base em microdados. *Revista de Economia Contemporânea* 24: 1–33. [CrossRef]

Andreoli, Vania, Marco Bagliani, Alessandro Corsi, and Vito Frontuto. 2021. Drivers of protein consumption: A cross-country analysis. *Sustainability* 13: 7399. [CrossRef]

Associação Brasileira de Proteína Animal (ABPA). 2023. Mercados, Estatísticas Setoriais e Relatório Anual. Available online: https://abpa-br.org/mercados/ (accessed on 19 January 2023).

Attia, Youssef A., Md Tanvir Rahman, Md Jannat Hossain, Shereen Basiouni, Asmaa F. Khafaga, Awad A. Shehata, and Hafez M. Hafez. 2022. Poultry production and sustainability in developing countries under the COVID-19 crisis: Lessons learned. *Animals* 12: 644. [CrossRef]

Awan, Tahir Mumtaz, Muhammad Shoaib Khan, Inzamam Ul Haq, and Sarwat Kazmi. 2021. Oil and stock markets volatility during pandemic times: A review of G7 countries. *Green Finance* 3: 15–27. [CrossRef]

Bairagi, Subir, Ashok K. Mishra, and Khondoker A. Mottaleb. 2022. Impacts of the COVID-19 pandemic on food prices: Evidence from storable and perishable commodities in India. *PLoS ONE* 17: e0264355. [CrossRef]

Bargoni, Augusto, Bernardo Bertoldi, Chiara Giachino, and Gabriele Santoro. 2022. Competitive strategies in the agri-food industry in Italy during the COVID-19 pandemic: An application of K-means cluster analysis. *British Food Journal* 124: 4782–99. [CrossRef]

Belarmino, Luiz Clovis, Antonio Domingos Padula, and Margarita Navarro Pabsdorf. 2022. Economic sustainability in emerging agro-industrial systems: The case of Brazilian olive cultivation. *Agriculture* 12: 2085. [CrossRef]

Biswal, Jyotsnarani, Kennady Vijayalakshmy, and Habibar Rahman. 2020. Impact of COVID-19 and associated lockdown on livestock and poultry sectors in India. *Veterinary World* 13: 1928–33. [CrossRef] [PubMed]

Caetano, Gerardo. 2022. Analysis and foresight of the European Union-Mercosur Association Agreement. Madrid, Fundación Carolina. Fundación EU-LAC, Documentos de Trabajo. Occasional Paper, No. 4, p. 56. Available online: https://eulacfoundation.org/sites/default/files/2022-04/Especial_FC_EULAC_4_EN.pdf (accessed on 12 September 2023).

CEPA EPAGRI-SC. Centro de Socioeconomia e Planejamento Agrícola-Empresa de Pesquisa de Santa Catarina. 2023. Boletim Agrícola. Available online: https://cepa.epagri.sc.gov.br/index.php/publicacoes/boletim-agropecuario/ (accessed on 19 January 2023).

CEPEA. Centro de Estudos Avançados em Economia Aplicada. 2023. Frango. Série de Preços. Available online: https://cepea.esalq.usp.br/br/indicador/frango.aspx (accessed on 19 January 2023).

Chen, Lurong. 2017. Globalization and trade liberalization in supporting GVCs upgrade: The case of the Republic of Korea. *Journal of Korea Trade* 21: 161–70. [CrossRef]

Chen, Xiangfeng, Chenyu Wang, and Shuting Li. 2023. The impact of supply chain finance on corporate social responsibility and creating shared value: A case from the emerging economy. *Supply Chain Management* 28: 324–46. [CrossRef]

Clapp, Jennifer. 2017. Concentration and power in the food system: Who controls what we eat? *Global Environmental Politics* 17: 151–52. [CrossRef]
Clapp, Jennifer. 2019. The rise of financial investment and common ownership in global agrifood firms. *Review of International Political Economy* 26: 604–29. [CrossRef]
CONAB. Companhia Nacional de Abastecimento. 2023. Preços Agrícolas. Available online: https://www.conab.gov.br/info-agro/precos (accessed on 19 January 2023).
Constantin, Marius, Juan Sapena, Andreea Apetrei, and Simona Roxana Pătărlăgeanu. 2023. Deliver smart, not more! Building economically sustainable competitiveness on the ground of high agri-food trade specialization in the EU. *Foods* 12: 232. [CrossRef]
DERAL-PR. Departamento de Economia Rural-Secretaria de Agricultura e Abastecimento do Paraná. 2023. Estatísticas Básicas. Available online: https://www.agricultura.pr.gov.br/Boletins-Informativos-Atuais (accessed on 19 January 2023).
EC. 2021. *EU Agricultural Outlook for Markets, Income and Environment, 2021–31*. Brussels: European Commission, DG Agriculture and Rural Development, 83p. Available online: https://agriculture.ec.europa.eu/system/files/2023-01/agricultural-outlook-2021-report_en_0.pdf (accessed on 30 June 2023).
Ellison, Brenna, Brandon McFadden, Bradley J. Rickard, and Norbert L. W. Wilson. 2021. Examining food purchase behavior and food values during the COVID-19 pandemic. *Applied Econonomis Perspectives and Policy* 43: 58–72. [CrossRef]
Embrapa Suínos e Aves. 2023. Centro de Inteligência em Aves e Suínos. Available online: https://www.embrapa.br/suinos-e-aves/cias (accessed on 19 January 2023).
Fafiolu, A.O., and Joal Alabi. 2020. Beyond COVID-19 pandemic period: Strategies for sustainable livestock feed and food production. *Nigerian Journal of Animal Science* 22: 107–21.
Fang, Peixun, Ben Belton, Xiaobo Zhang, and Hnin Ei Win. 2021. Impacts of COVID-19 on Myanmar's chicken and egg sector, with implications for the sustainable development goals. *Agricultural Systems* 190: 103094. [CrossRef]
FAO FFPI. Food and Agriculture Organization of the United Nations. 2023. FAO Food Price Index. Available online: https://www.fao.org/worldfoodsituation/foodpricesindex/en/ (accessed on 19 January 2023).
FAO RLC. Organización de las Naciones Unidas para la Agricultura y Alimentación-Oficina Regional para América Latina y Caribe. 2007. *Competitividad de la agricultura em América Latina y el Caribe, matriz de analisis de política: Ejercicios de cómputo*. Santiago: FAO-RLC, 113p. Available online: https://www.cepal.org/sites/default/files/courses/files/03_3_map_manual_fao.pdf (accessed on 23 January 2023).
FAOSTAT. Food and Agriculture Organization of the United Nation-Statistics Division. 2023. Food and Agriculture Data. Available online: https://www.fao.org/faostat/en/#home (accessed on 25 February 2023).
Feng, Yanhong, Shuanglian Chen, Xuan Wang, and Aaron Tan. 2021. Time-varying impact of U.S. financial conditions on China's inflation: A perspective of different types of events. *Quantitative Finance and Economics* 5: 604–22. [CrossRef]
FGV IBRE. Fundação Getúlio Vargas-Instituto Brasileiro de Economia. 2023. Índice de Preços. Available online: https://portalibre.fgv.br/indices-de-precos (accessed on 19 February 2023).
GEF-WFP. Global Economic Forum-United Nations World Food Programme. 2023. Food Security. Available online: https://intelligence.weforum.org/topics/a1Gb0000000pTDPEA2 (accessed on 23 January 2023).
Hafez, Hafez M., and Youssef A. Attia. 2020. Challenges to the poultry industry: Current perspectives and strategic future after the COVID-19 outbreak. *Frontiers in Veterinarian Science* 7: 516. [CrossRef]
IBGE POF. Instituto Brasileiro de Geografia e Estatística. 2023. Pesquisa de Orçamentos Familiares (POF). Available online: https://www.ibge.gov.br/estatisticas/sociais/saude/24786-pesquisa-de-orcamentos-familiares-2.html (accessed on 19 January 2023).
IBGE PPM. Instituto Brasileiro de Geografia e Estatística. 2023. Produção da Pecuária Municipal. Available online: https://www.ibge.gov.br/estatisticas/economicas/agricultura-e-pecuaria/9107-producao-da-pecuaria-municipal.html?=&t=resultados (accessed on 19 January 2023).
IEA APTA. Instituto de Economia Agrícola-Agência Paulista de Tecnologia Agrícola de São Paulo. 2023. Valor de Produção. Available online: http://ciagri.iea.sp.gov.br/bancodedados/valorproducao (accessed on 19 January 2023).
Lara, Lucas J., and Marcos H. Rostagno. 2013. Impact of heat stress on poultry production. *Animals* 3: 356–69. [CrossRef] [PubMed]
Lepchak, Alessandro, and Simone Bernardes Voese. 2020. Evaluation of the efficiency of logistics activities using data envelopment analysis (DEA). *Gestão & Produção* 27: e3371. [CrossRef]
Lopes, Mauro de Resende, Luiz Clovis Belarmino, Antônio Jorge Oliveira, Joaquim Raimundo de Lima, Daniela Parente Torres, Dirceu João Duarte Talamini, and Franco Müller Martins. 2012. Matriz de análise de política. Metodologia e análise. Brasília-DF, Embrapa, 227 pp. Available online: https://ainfo.cnptia.embrapa.br/digital/bitstream/item/197254/1/Matriz-analise-de-politica.pdf (accessed on 23 January 2023).
Manning, Louise, and R.N. Baines. 2004. Globalization: A study of the poultry-meat supply chain. *British Food Journal* 106: 819–36. [CrossRef]
MAPA. Ministério da Agricultura, Pecuária e Abastecimento-Estatísticas de Comércio Exterior do Agronegócio Brasileiro. 2023. Exportação/Importação e Tabelas de Agrupamentos. Available online: https://indicadores.agricultura.gov.br/agrostat/index.htm (accessed on 19 January 2023).
Martindale, Wayne, and Walter Schiebel. 2017. The impact of food preservation on food waste. *British Food Journal* 119: 2510–18. [CrossRef] [PubMed]

MDIC COMEX STAT. Ministério do Desenvolvimento, Indústria e Comércio-Estatísticas de Comércio Exterior. 2023. Exportação e Importação Geral. Available online: http://comexstat.mdic.gov.br/pt/geral (accessed on 19 January 2023).

Miller, Matthew, Adam Gerval, James Hansen, and Grace Grossen. 2022. *Poultry Expected to Continue Leading Global Meat Imports as Demand Rises*. Washingthon, DC: USDA-ERS/Amber Waves-Features: Poultry & Eggs. Available online: https://www.ers.usda.gov/amber-waves/2022/august/poultry-expected-to-continue-leading-global-meat-imports-as-demand-rises/ (accessed on 19 January 2023).

Monke, Eric A., and Scott R. Pearson. 1989. *Policy Analysis for Agricultural Development*. Ithaca: Cornell University, p. 294.

Muresan, Iulia C., Rezhen Harun, Anca Monica Brata, Vlad Dumitru Brata, Daniel I. Chiciudean, Olivia Paula Tirpe, Andra Porutiu, and Diana E. Dumitras. 2022. Factors affecting food consumers' behavior during COVID-19 in Romania. *Foods* 11: 2275. [CrossRef]

Nkgadima, Kgothatso, and Chiedza L. Muchopa. 2022. Do import tariff adjustments bolster domestic production? Analysis of the South African-Brazilian poultry market case. *Economies* 10: 318. [CrossRef]

Nonnenberg, Marcelo José Braga, Uallace Moreira Lima, Mateus Azevedo Araujo, Fernanda Pedrosa, and Scarlett Queen Almeida Bispo. 2021. Agribusiness trade between Brazil and China: Pillars and opportunities. *IPEA-CAITEC, Brasília-DF. Discussion Paper 259*. [CrossRef]

Obese, Frederick Y., Richard Osei-Amponsah, Eric Timpong-Jones, and Edwin Bekoe. 2021. Impact of COVID-19 on animal production in Ghana. *Animal Frontiers* 11: 43–46. [CrossRef] [PubMed]

OECD-FAO. Organisation for Economic Co-operation and Development-Food Agricultural Organization. 2022. *OECD-FAO Agricultural Outlook 2022–31*. Paris: OECD Publishing. [CrossRef]

Qi, Yudong, and Xi Chu. 2022. Development of the digital economy, transformation of the economic structure and leaping of the middle-income trap. *China Political Economy* 5: 14–39. [CrossRef]

Santos Filho, Jonas Irineu dos, Dirceu João Duarte Talamini, Gerson Neudi Scheuermann, and Teresinha Marisa Bertol. 2018. Impacto da logística brasileira nas cadeias produtivas de aves e suínos. *Revista de Política Agrícola* 27: 48–64. Available online: http://www.alice.cnptia.embrapa.br/alice/handle/doc/1097745 (accessed on 23 January 2023).

Sattar, Abdullah Al, Rashed Mahmud, Md Abu Shoieb Mohsin, Nurun Nahar Chisty, Md Helal Uddin, Nusrat Irin, Tony Barnett, Guillaume Fournie, Eve Houghton, and Md Ahasanul Hoque. 2021. COVID-19 impact on poultry production and distribution networks in Bangladesh. *Frontiers in Sustainable Food Systems* 5: 714649. [CrossRef]

Soendergaard, Niels, Camila Dias de Sa, and Ana Flavia Barros Platiau, eds. 2023. *Sustainability Challenges of Brazilian Agriculture. Governance, Inclusion, and Innovation*. Cham: Springer Nature. [CrossRef]

Sohag, Kazi, Md Monirul Islam, Ivana Tomas Žiković, and Hoda Mansour. 2023. Food inflation and geopolitical risks: Analyzing European regions amid the Russia-Ukraine war. *British Food Journal* 125: 2368–91. [CrossRef]

Surni, Nendissa, Doppy Roy, Muhaimin Abdul Wahib, Maria Haryulin Astuti, Putu Arimbawa, Maximilian MJ Kapa, and Evi Feronika Elbaar. 2021. Socio-economic impact of the COVID-19 pandemic: Empirical study on the supply of chicken meat in Indonesia. *AIMS Agricultural and Food* 6: 65–81. [CrossRef]

Talamini, Dirceu João Duarte, and Franco Müller Martins. 2022. A avicultura brasileira e o mercado mundial de carnes. *Anuário 2022 da Avicultura Industrial* 114: 14–21. Available online: https://www.infoteca.cnptia.embrapa.br/handle/doc/1150191 (accessed on 23 January 2023).

Todua, Nugzar, and Charita Jashi. 2015. Some aspects of social media marketing (Georgian case). *Journal of Business, Human and Social Sciences* 9: 1160–63. [CrossRef]

Torres, Danielle Alencar Parente, Joaquim Raimundo Lima Filho, and Luiz Clovis Belarmino. 2013. Competitividade das cadeias agroindustriais Brasileiras. Brasília, DF, Embrapa, 191p. Available online: https://ainfo.cnptia.embrapa.br/digital/bitstream/item/197253/1/Competitividade-de-cadeias-agroindustriais.pdf (accessed on 23 January 2023).

UNCTAD. United Nations for Conference on Trade and Development. 2022. World Investment Report 2022. Available online: https://unctad.org/publication/world-investment-report-2022 (accessed on 11 November 2022).

USDA-FAS. United State Department of Agriculture-Foreign Agriculture Service-GAIN. 2022. Poultry and Products Annual September. Available online: https://www.fas.usda.gov/data/brazil-poultry-and-products-annual-8 (accessed on 19 January 2023).

USITC. United State International Trade Commission. 2012. Brazil: Competitive factors in Brazil affecting U.S. and Brazilian Agricultural Sales in Selected Third Country Markets. Washington, DC, Investigation No. 332-524, USITC Publication 4310, 422p. Available online: https://www.usitc.gov/publications/332/pub4310.pdf (accessed on 11 November 2022).

Valdes, Constanza. 2022. *Brazil's Momentum as a Global Agricultural Supplier Faces Headwinds*. Washinthon, DC: USDA-ERS. Available online: https://www.ers.usda.gov/amber-waves/2022/september/brazil-s-momentum-as-a-global-agricultural-supplier-faces-headwinds/ (accessed on 11 November 2022).

Wijerathna-Yapa, Akila, and Ranjith Pathirana. 2022. Sustainable agro-food systems for addressing climate change and food security. *Agriculture* 12: 1554. [CrossRef]

World Bank. 2017. *Brazil-Multimodal Freight Transport: Selected Regulatory Issues*. Washington, DC: World Development Sources, WDS 1998-1, 61p. Available online: http://documents.worldbank.org/curated/en/931541468224701179/Brasil-Transporte-multimodal-de-carga-questoes-regulatorias-selecionada (accessed on 19 February 2023).

World Bank. 2023. World Bank Commodities Price Data (the Pink Sheet). Available online: https://www.worldbank.org/en/research/commodity-markets (accessed on 19 January 2023).

Wu, Yilin, and Shiyu Ma. 2021. Impact of COVID-19 on energy prices and main macroeconomic indicators-evidence from China's energy market. *Green Finance* 3: 383–402. [CrossRef]
Yeong, Siew-Wei, Mukvinder Kaur Sandhu, and Hiram Ting. 2021. The future of food: Responsible production, acquisition, consumption and disposition. *British Food Journal* 123: 2953–58. [CrossRef]
Yu, Xiaohua, Chang Liu, Hanjie Wang, and Jan-Henning Feil. 2020. The impact of COVID-19 on food prices in China: Evidence of four major food products from Beijing, Shandong and Hubei Provinces. *China Agricultural Economic Review* 12: 445–58. [CrossRef]
Yuzaria, D., A. R. Wahyuni, and M. Fajrin. 2021. Competitiveness and effect of government policies on laying hens farming business in West Pasaman Regency, West Sumatra Province. Indonesia. *IOP Conference Series: Earth and Environmental Science* 757: 012010. [CrossRef]
Zeballos, Eliana, Xiao Dong, and Ergys Islamaj. 2023. A Disaggregated View of Market Concentration in the Food Retail Industry. ERR-314, U.S. Department of Agriculture, Economic Research Service, 21p. Available online: https://www.ers.usda.gov/publications/pub-details/?pubid=105557 (accessed on 11 November 2022).
Zylbersztajn, Decio, Marcos Fava Neves, and Silvia Morales de Queiroz Caleman. 2015. *Gestão de sistemas de agronegócios*. São Paulo: PENSA-FIA-FEA/USP, 328p. Available online: https://edisciplinas.usp.br/mod/resource/view.php?id=3064690 (accessed on 11 November 2022).

energies

MDPI

Article

A Study on the Spatial Distribution and Gradient Transfer of Atmospheric Pollution Intensive Industries Such as the Thermal Power Industry in Hebei Province

Jingkun Zhou [1], Juan Tian [2,*], Xiaoyan Wang [3] and Xu Bai [4]

1 School of Business, Ludong University, 186 Hongqizhong Road, Yantai 264025, China; zhoujingkun@163.com
2 School of Business Administration, Hebei University of Economics and Business, Shijiazhuang 050061, China
3 School of Sports and Arts, Hebei Sport University, Shijiazhuang 050041, China; xiyanli713@163.com
4 School of Public Administration, Hebei University of Economics and Business, Shijiazhuang 050061, China
* Correspondence: tiansiqi0505@126.com; Tel.: +86-152-2716-9812

Abstract: Owing to a long-term, extensive development model and inadequate industrial development planning, cases of atmospheric environment pollution frequently occur in Hebei province. By using such approaches as the Spatial Gini Coefficient, the Herfindahl–Hirschman Index and location entropy, this paper analyzes the spatial-distribution characteristics of atmospheric pollution-intensive industries such as the thermal power industry in Hebei province. As shown, atmospheric pollution-intensive industries, such as the thermal power industry in Hebei province, excessively cluster. As industrial agglomeration continuously intensifies, the spatial imbalance becomes increasingly prominent. Taking the number of days with excellent air quality as a benchmark, this paper divides prefecture-level cities of Hebei province into four types of industrial management and designs targeted strategies for the optimization of atmospheric pollution-intensive industries, such as the thermal power industry. In terms of policies, Type I and Type II cities are advised to strengthen the transfer of atmospheric pollution-intensive industries such as the thermal power industry, and Type III and Type IV cities are advised to improve capacities in atmospheric self-purification and green-technology innovation in a bid to help government departments to scientifically manage atmospheric pollution-intensive industries such as the thermal power industry.

Keywords: spatial distribution; gradient transfer; atmospheric pollution-intensive industries such as the thermal power industry(APIISTPI)

1. Introduction

As the 2021 China Ecological Environment Status Bulletin released by the Ministry of Ecology and Environment of the People's Republic of China suggested, among 337 prefecture-level cities nationwide, the average proportion of cities with excellent air quality reached 87.5%, and 121 cities failed to meet the standards of excellent air quality. In cities in the Beijing–Tianjin–Hebei Region and the neighboring regions, the average number of days with excellent air quality was 67.2%, which is 19.5% lower than that in the Yangtze River Delta. In the treatment of the atmospheric environment, Hebei province should give priority to the prevention of fog-haze pollution in the autumn and winter. Because of the excessive agglomeration of atmospheric pollution-intensive industries such as iron and steel, petrochemical and thermal power in Hebei province, its atmospheric pollution features soot-type particulate matter. Atmospheric pollutants are immensely emitted. Hebei ranks first in the emission of nitrogen oxides and particulate matter (PM) and second in the emission of sulfur dioxide in China. Simultaneously, with the coordinated development of the Beijing–Tianjin–Hebei Region and the continuous adjustment of regional industrial structure, Hebei undertakes the transfer of atmospheric pollution-intensive industries such as steel and iron and thermal power in Beijing and Tianjin, which aggravates atmospheric pollution. Tian et al. [1] analyze the status quo of industrial transfer in Beijing–Tianjin–Hebei urban agglomeration by using the model of the 'transfer-receive-dynamic process', arguing that industrial transfer has become an inevitable trend and that the overall capacity of Hebei to undertake industrial transfer has obvious regional heterogeneity, which has different effects on regional economic development.

The existing scholarship on atmospheric pollution-intensive industries, such as the thermal power industry (hereinafter called 'APIISTPI'), mainly centers on three aspects. The first aspect is the relationship between industrial agglomeration and environmental pollution. He et al. [2] examined the spatial location and development trend of the agglomeration of pollution-intensive industries and environmental pollution, revealing that they form an inverted U-shaped nonlinear relationship. Chen et al. [3] discovered the spatial effect in the agglomeration of the manufacturing industry. As suggested, the agglomeration of both local and non-local manufacturing industries has a non-linear effect on local environmental pollution, reducing the emission of pollutants in the long term. Chen and Huang et al. [4] emphasized that industrial agglomeration promotes industrial cooperation and enhances the innovation capacity of technological systems, thus improving the treatment of environmental pollution. Chen et al. [5] argued that industrial agglomeration significantly optimizes end-of-pipe treatment technology among industries. Gai and Zhou et al. [6] noticed that industrial agglomeration leads to the homogenization of industrial structures and the waste of resources, which intensifies environmental pollution. Qiu et al. [7] stressed that presently, the industrialization agglomeration index has peaked in China, and therefore, environmental pollution deteriorates with the expansion of industrial agglomeration. These studies show a nonlinear relationship between industrial agglomeration and environmental pollution.

The second aspect is the study of the spatial distribution of industries. Tang and Dou et al. [8] regarded the spatial-transfer pattern of pollution-intensive industries as a key issue for local and national sustainable development. Wen and Su et al. [9] confirmed that local pollutant-emission standards remarkably affect the spatial distribution of pollution-intensive industries. He and Wu et al. [10] believed that the positive externality of technological spillover and the negative externality of environmental pollution that arise from the agglomeration of economic activities have an important impact on the behavior of spatial choice of economic entities and then on the structural balance of industrial spatial distribution. Wang et al. [11] concluded that industrial pollution constitutes the main source of urban atmospheric pollutants and that coal-fired energy-intensive industries have the greatest impact on sulfur dioxide concentration. The above research shows that atmospheric pollution-intensive industries have significant spatial effects.

The third aspect is the study of industrial transfer. Kaname A. [12], a Japanese economist, first proposed the Flying Geese Paradigm for industrial transfer, which originally intended to summarize the development trajectory of Japanese industrial transfer. Ma et al. [13] initiated a 'multi-center radiation' industrial transfer model, which achieves large-scale gradient progress via small-scale multi-center radiation. Chen et al. [14] proposed the 'marginal penetration' transfer model, which requires the coordination and integration of economic development in developed and developing regions that are geographically connected, and to boost the full flow of economic resources in various regions. In this way, some industries in economically developed regions, which gradually lose comparative advantages with the rise of marginal cost, eventually transfer to economically developing regions. Jiang and Sun et al. [15] suggested the holistic transfer model, which involves the overall transfer of technology, equipment, management and talents from industries in developed regions to developing regions so as to establish new industrial production bases. Fu and Zeng et al. [16] summed up two models of industrial transfer, i.e., selective transfer and replicable transfer. Xiang and Hu et al. [17] noted that developed regions could transfer non-core-value production chains without competitive advantages to developing regions by means of outsourcing and OEM while promoting industrial integration and divestiture.

In the research on industrial transfers between eastern and western China, Huang et al. [18] proposed six models, i.e., factor flow and direct investment, corporate internal integration, virtual corporate integration, the new Flying Geese Paradigm, industrial cluster transfer and industrial transfer parks. Ma and Hu et al. [19] stated that inter-regional industrial transfer models in China basically include cost-oriented transfer, market development transfer, diversified operation transfer, competition follow-up transfer, supply-chain connection transfer, economies-of-scale-oriented transfer and policy-oriented transfer. Li and Li et al. [20] measured the industrial transfer via the transfer of embodied carbon. The above shows that industrial transfers are primarily oriented toward the government and the economy. Developing regions often undertake the industrial transfer of developed regions. Additionally, the development conditions remain inadequate in some transfer-in regions, which exacerbates the negative externalities of industrial agglomeration.

In summary, a few researchers have studied the industrial gradient transfer based on the time–space distribution of industrial agglomeration, and the time–space distribution of industrial agglomeration can objectively reflect the quality of the industrial transfer and provide a basis for the scientific design of industrial gradient transfer. This paper takes APIISTPI in Hebei province as a research subject. In the 'Twelfth Five-Year Plan', six major industries, including thermal power, steel and iron, non-ferrous metal, petrochemical, cement and chemical industries, are the key industries in the prevention and treatment of atmospheric pollution [21]. This paper chooses the electric and thermal power production and supply industry, the ferrous metal smelting and calendering processing industry, the petroleum processing and coking industry, the chemical raw material and chemical product manufacturing industry, the non-metallic mineral product industry and the non-ferrous metal smelting and calendering processing industry as index parameters for APIISTPI. By using such approaches as the Spatial Gini Coefficient, Herfindahl–Hirschman Index and location entropy, this paper analyzes the spatial-distribution characteristics of APIISTPI in Hebei province, designs a mechanism for the gradient transfer of APIISTPI, and provides recommendations for achieving sustainable economic development in Hebei province.

2. Research Methods and Data Sources

2.1. Research Methods

Traditional measures of industrial agglomeration include the Spatial Gini Coefficient, location entropy index, Herfindahl–Hirschman Index, regional entropy index and Krugman Specialization Index [22]. The location entropy index is a measurement of relative agglomeration, while the Spatial Gini Coefficient Herfindahl–Hirschman Index is the measurement of absolute agglomeration. The measurement of absolute agglomeration requires more detailed industry data. The Spatial Gini Coefficient only considers the degree of agglomeration between industries and ignores the differences in corporate size and concentration among different industries. The Herfindahl–Hirschman Index(HHI) focuses on reflecting the differences in industrial concentration between regions [23].

Spatial balance refers to the spatial allocation model of various production factors in industrial development, as well as the layout of regional production capacity that proves compatible and coordinated with natural resources and the environmental endowment of one country or region and meets the needs of sustainable development. As an important technological approach, the Gini Coefficient is widely used to describe the distribution differences between various spatial factors and compare regional distribution and differences between research subjects and cities in order to reveal regional distribution and change trends. The comprehensive calculation formula of the Gini Coefficient is expressed as follows:

$$G = \sum_{i}^{N} (S_i - x_i)^2 \quad (1)$$

In particular, G stands for the Spatial Gini Coefficient, S_i stands for the proportion of relevant indexes of APIISTPI in region i in Hebei province, x_i stands for the proportion of relevant indexes of region i in Hebei Province, and N stands for the number of enterprises in Hebei Province. Among them, the value of G is between 0 and 1. If the average value of

G is closer to 0, the distribution of the industry is more balanced; if the average value of G is closer to 1, the imbalance enlarges, or industrial agglomeration intensifies.

Restricted by the availability of data, this paper analyzes the degree and trend of agglomeration in APIISTPI in Hebei province by using location entropy, calculates the regional agglomeration of APIISTPI in Hebei province by using the Spatial Gini Coefficient, and measures the industrial agglomeration of APIISTPI in Hebei province by using the Herfindahl–Hirschman Index (HHI), with an attempt to further test the analysis results.

2.2. Data Sources

The panel data in this paper are mainly derived from China Statistical Yearbook, China Industrial Statistical Yearbook and China Environmental Statistical Yearbook. The area data are derived from China Statistical Yearbook. The PM10 data are derived from local Environmental Statistics Bulletins. The PM2.5 data are derived from the Chinese PM2.5 Concentration Data of Columbia University. In terms of emission scale and intensity, the data on the emissions of industrial sulfur dioxide, industrial nitrogen oxide and industrial particulate matter, as well as the total industrial output value and the number of enterprises, are derived from China Industrial Statistical Yearbook.

3. Research Analysis

Using the emission scale and intensity of industrial atmospheric pollutants and the industrial data of various provinces in China, this paper first calculates the atmospheric pollution intensity index of different industries in China, and proceeds to choose six industries with higher atmospheric pollution intensity index as atmospheric pollution-intensive industries, i.e., the electric and thermal power production and supply industry, the ferrous metal smelting and calendering processing industry, the petroleum processing and coking industry, the chemical raw material and chemical product manufacturing industry, the non-metallic mineral product industry and the non-ferrous metal smelting and calendering processing industry [24]. The spatial quantile-quantile plot can classify corresponding indicator observations of various spatial units according to a numerical value and represent the spatial distribution of the research indicators. In order to detail the overall spatial distribution and change of APIISTPI, this paper classifies the gross industrial output value of APIISTPI in Hebei province and draws quantile-quantile (Q-Q) plots of APIISTPI in various regions in 2006, 2009, 2012, 2015 and 2019. By comparing the Q-Q plots, this paper investigates the regional agglomeration of APIISTPI. A larger gross industrial output value of APIISTPI in one region means a larger number or scale of enterprises in APIISTPI. Or the region has a dense distribution of APIISTPI. Otherwise, the region has a sporadic distribution of APIISTPI.

3.1. The Overall Space-Time Distribution of APIISTPI

In Figure 1, 11 prefecture-level cities in Hebei Province are categorized into six levels in line with the gross industrial output value of APIISTPI. The deepest color represents the densest distribution of APIISTPI. As the color becomes lighter, the intensity of API-ISTPI weakens. In 13 years, from 2006 to 2019, Tangshan produced the highest gross industrial output value of APIISTPI, and Shijiazhuang and Handan stayed in the second gradient. Tangshan adjoins the suburbs of Beijing. By dint of favorable geographical location and abundant natural resources, Tangshan undertakes the transfer of a large number of APIISTPI from Beijing, with a relatively high gross industrial output value. Since 2010, Tangshan has accelerated the adjustment of its industrial structure. Concerning the industrial added value of enterprises above designated size, six industries, i.e., steel and iron, building material, energy, chemical engineering, coking and thermal power, have achieved an added value of 137.832 billion yuan, with a year-on-year increase of 8.7%. As the capital of Hebei Province, Shijiazhuang attracts a good deal of funds and talent. In addition, advanced railway transportation ensures an adequate supply of raw materials and fuels for APIISTPI. Therefore, Shijiazhuang has stood in the second gradient. Located in the south of Hebei Province, Handan possesses rich natural resources for relevant industries. It has been an old resource-based industrial base in Hebei Province and an important production base for metallurgy, coal, power, textiles, ceramic and building material industries in China.

From 2006 to 2019, Baoding, Qinhuangdao, Zhangjiakou and other cities stayed in the end-of-pipe gradient. The main reason lies in that Baoding, Langfang, Zhangjiakou and other places encircle Beijing and Tianjin. To guarantee air quality in Beijing, these cities curb the development of APIISTPI.

3.2. The Spatial Distribution of APIISTPI

3.2.1. The Spatial Distribution of the Chemical Raw Material and Chemical Product Manufacturing Industry

As Figure 2 shows, the overall distribution of the industry roughly forms a radial pattern centered on Shijiazhuang. From 2006 to 2019, the overall development level of the industry in Hebei Province continued to rise, and Shijiazhuang stayed in the first quantile. In 2006, Cangzhou and Tangshan stayed in the second quantile; in 2009, Cangzhou and Langfang stayed in the second quantile, and in 2015, Cangzhou and Tangshan stayed in the second quantile.

There are multiple reasons. In 2006, Shijiazhuang maintained rapid industrial development and realized the simultaneous growth of economic benefit and production. In addition, Shijiazhuang has attained remarkable achievements in building a 'medical city' and has become one of three national bases for the biological industry, with the gross industrial output value of the pharmaceutical industry continuing to grow. The high degree of industrial correlation substantiates that the chemical industry (mainly the pharmaceutical industry) produces scale economies effect and considerable economic benefit. As an economic mainstay in Shijiazhuang, the chemical industry enlarges relative industrial agglomeration of the chemical raw material and chemical product manufacturing industry.

In 2006, Baoding deepened the reform of industrial enterprises and achieved an overall upturn in economic performance. Based on the IPOs of Swan Chemical Fiber, Lucky Film and Baoshuo Group, Baoding completed the capital increase and equity allocation of Swan Chemical Fiber and Baoshuo Group, as well as the shareholding system reform of large enterprises or groups, such as Tianwei Group, Fengfan Group and Lingyun Electronic. In 2006, the competitive industries in Baoding were mainly textiles, the production and supply of electricity, steam and hot water, chemical raw material and chemical product manufacturing, electrical machinery and equipment manufacturing, and non-metallic mineral product. In 2009, transportation equipment manufacturing with strong capital and technological strengths and high economic benefits developed rapidly in the '16th Five-Year Plan', whose overall strength ranked first in the entire industry. Meanwhile, the non-ferrous metal smelting and calendering processing industry developed fast, yet, the chemical raw material and chemical product manufacturing industry dropped to the seventh place, whose industrial agglomeration decreased.

Tangshan slipped from the second gradient in 2006 to the third gradient in 2009. The reason was that in 2009, hammered by the global financial crisis, Tangshan vigorously implemented the 'Five Decisive Campaigns' and 'Eight Major Projects', furthered the adjustment of industrial structure, and significantly improved the level of industrial intensification. Nevertheless, the salt chemical industry, the coal chemical industry, the chemical fertilizer industry and the petrochemical industry were buffeted. The elimination of outdated production capacity was another reason.

In 2006, in Qinhuangdao, the primary processing and low-value-added raw material industry dominated the secondary industry. Enterprises that engaged in basic raw material products (e.g., glass, cement, paper-making, fertilizer and medium-thick steel sheet), primary processing products and labor-intensive products accounted for a high proportion of all manufacturing enterprises. Owing to technological progress, these long-term products, as well as high energy consumption and high pollution products, were sifted out, resulting in a decrease in industrial agglomeration.

3.2.2. The Spatial Distribution of the Non-Ferrous Metal Smelting and Calendering Processing Industry

As Figure 3 shows, in 2006, Shijiazhuang stayed in the first quantile; from 2009 to 2019, Baoding stayed in the first quantile, with a significant change in the spatial distribution of the non-ferrous metal smelting and calendering processing industry. In terms of industrial output value, Qinhuangdao also stayed in the second quantile. The reason is that Baoding possesses rich natural resources. The western mountainous areas contain abundant mineral resources, including over 50 varieties of metallic and non-metallic mineral resources, such as iron, copper, lead, zinc, granite, limestone and ceramic raw material, with vast potential markets and superior geographical location. With the implementation of the 'Beijing Economic Circle' strategy, Baoding expands outwardly. Within a radius of 150 km, a multi-level high-capacity consumption market takes shape, which covers a population of over 40 million and comprises Beijing, Tianjin, Shijiazhuang, Baoding and other cities. In the non-ferrous metal smelting and calendering processing industry, Baoding has a market share of up to 47% of Hebei, with great technological innovation strength and a prominent development advantage.

In 2005, Qinhuangdao was beset by an imbalanced industrial structure. As the demand outstripped the supply, the prices of major production materials skyrocketed. The grade of raw ore in mines decreased, the supply of resources remained severely inadequate, and structural contradictions worsened. In most varieties of non-ferrous metals, the processing capacity exceeds the smelting capacity, and the smelting capacity exceeds the concentrate guarantee capacity, with a prominent problem of blind investment in aluminum oxide, copper and lead-zinc smelting industries. In the aluminum processing industry, industrial agglomeration proves low, and product structure proves unreasonable. The high price of electrolytic copper causes business distress to copper processing enterprises. In lead-zinc smelting enterprises, environmental pollution accidents occur frequently. These factors reduce the industrial agglomeration of the non-ferrous metal smelting industry in Qinhuangdao.

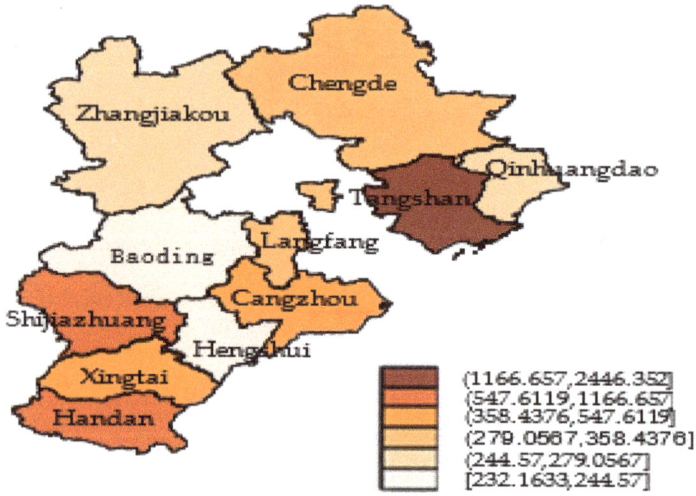

(**a**) 2006

Figure 1. *Cont.*

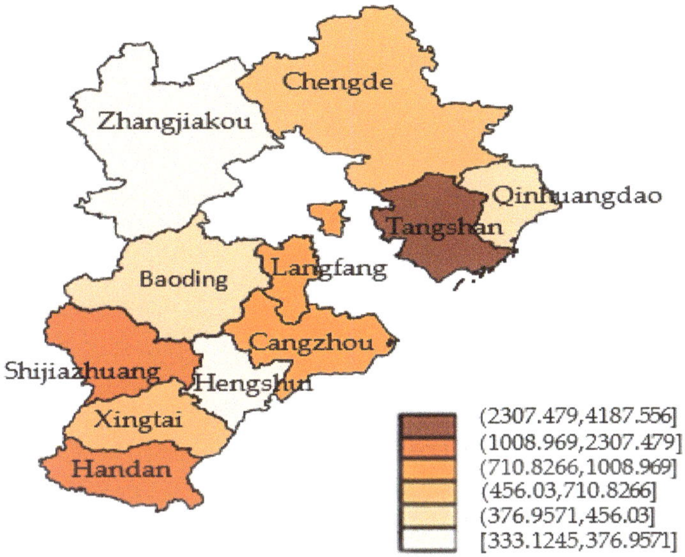

(**b**) 2009

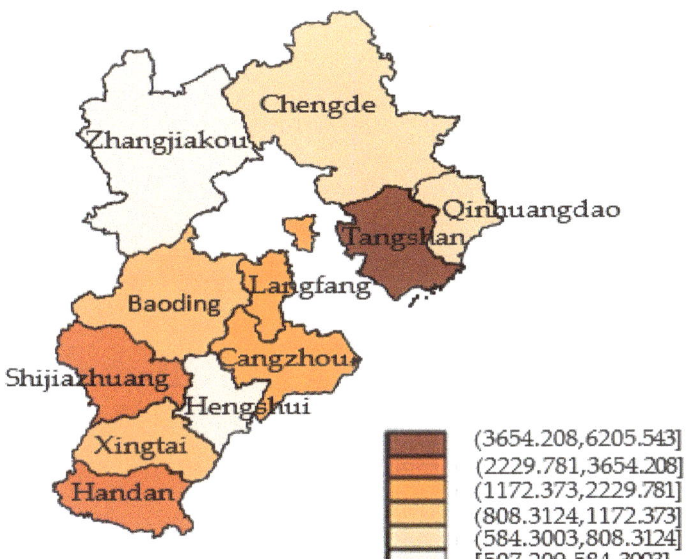

(**c**) 2012

Figure 1. *Cont.*

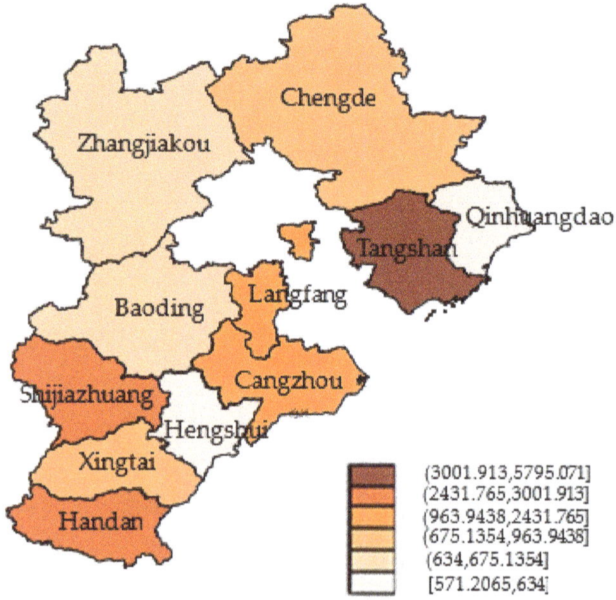

(**d**) 2015

(**e**) 2019

Figure 1. The Spatial Quantile–Quantile (Q-Q) Plot of the Gross Industrial Output Value of APIISTPI in Hebei Province.

3.2.3. The Spatial Distribution of the Electric and Thermal Power Production and Supply Industry

As Figure 4 shows, from 2006 to 2019, Tangshan stayed in the first quantile, and Shijiazhuang and Handan stayed in the second quantile. Douhe Power Plant, a national first-class thermal power plant, is in Tangshan. Since 2000, the industrial agglomeration of the electric and thermal power industry has remained high in Tangshan. In order to provide powerful support for other industries and secure industrial profits, Shijiazhuang and Handan enthusiastically developed the thermal power industry.

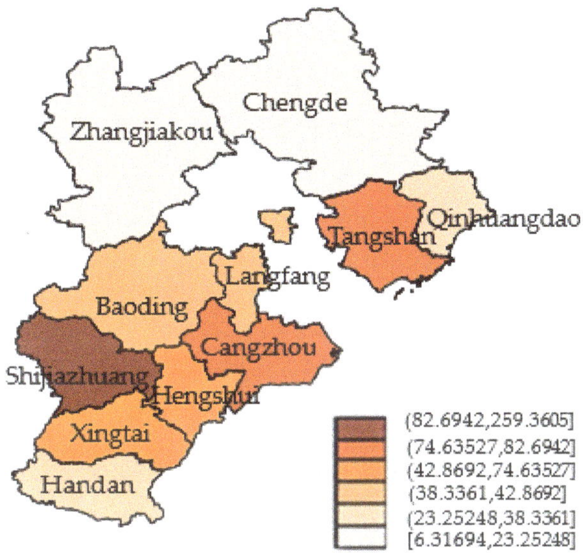

(**a**) 2006

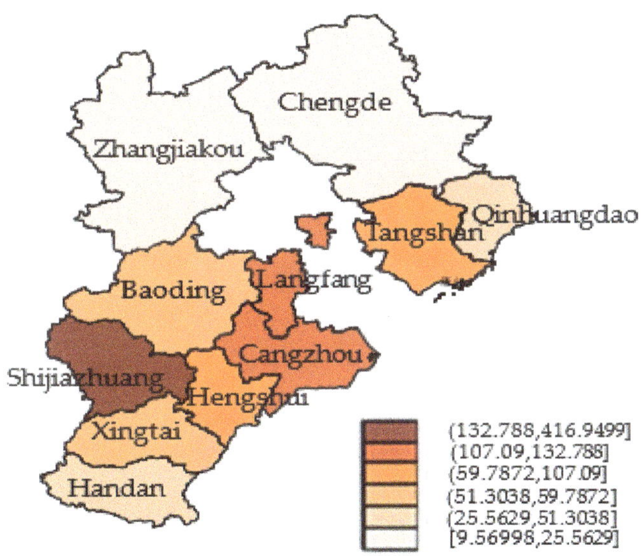

(**b**) 2009

Figure 2. *Cont.*

(**c**) 2012

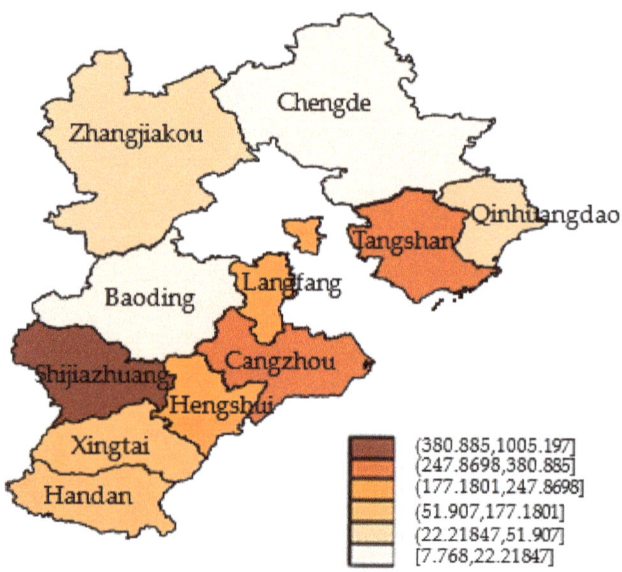

(**d**) 2015

Figure 2. *Cont.*

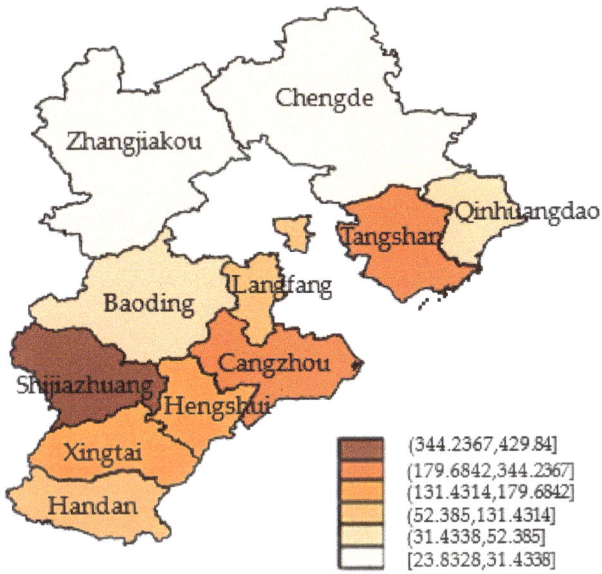

(e) 2019

Figure 2. The Spatial Quantile-Quantile (Q-Q) Plot of the Gross Industrial Output Value of the Chemical Raw Material and Chemical Product Manufacturing Industry in Hebei Province.

3.2.4. The Spatial Distribution of the Petroleum Processing and Coking Industry

As Figure 5 shows, from 2006 to 2019, the spatial distribution color of Cangzhou embodied the deepest color. The main reason was that in 2000, the petroleum processing industry developed slowly in Hebei. Noticeably, the advantage of the port facilitated the development of the petroleum industry in Cangzhou. According to the data, in 2015, leading industries in Cangzhou, e.g., petrochemical, pipeline equipment, metallurgy, mechanical manufacturing, textiles, clothing and food processing, achieved an industrial added value of 107.21 billion yuan, accounting for 85.3% of enterprises above designated size in Cangzhou, with an increase of 7.2%. Particularly, the petrochemical industry realized an industrial-added value of 34.37 billion yuan, accounting for the highest proportion of 27.3%. In 2017, the industrial added value increased by 3% vis-à-vis 2015. That is mainly because, in the mid-and long-term development plan for the oil refining industry, China advocates and guides the transfer of resource-based industries from resource-based cities to coastal areas with huge market demands and favorable conditions for the import of raw materials.

Additionally, large resource-based enterprises, especially large transnational corporations, draw their strategic blueprints in China, which form an important factor in promoting the distribution of large petrochemical and steel and iron projects in coastal areas. Their strategic choices are grounded in comprehensive factors in coastal areas, such as market size, transportation conditions, policy orientation, and proximity to international markets and raw material sources. The petrochemical industry in Cangzhou enjoys a long history. After several years of development, a complete production system matures, which covers the petrochemical, coal chemical and fine chemical industries. Generally speaking, the petrochemical industry shows a trend of chain and group-based development, becoming the foremost pillar industry in Cangzhou.

3.2.5. The Spatial Distribution of the Non-Metallic Mineral Product Industry

As Figure 6 shows, from 2006 to 2019, the spatial distribution of Shijiazhuang and Tangshan embodied deeper colors. Since 2006, the spatial distribution of Hengshui has displayed a trend of industrial agglomeration. After 2005, under the influence of national macro-control, the entire cement industry fluctuated wildly. In Tangshan, the industrial agglomeration of the cement industry diminishes. FRP products in Hengshui have a high market share nationwide. At the end of 2004, the production capacity was discontinued for the sake of environmental protection in Hengshui, resulting in a decrease in marginal supply in 2005. As environmental protection was upgraded and environmental supervision was tightened, the industrial agglomeration of the glass industry continued to decrease. In the '10th Five-Year Plan', industrial resources were optimized with policy support.

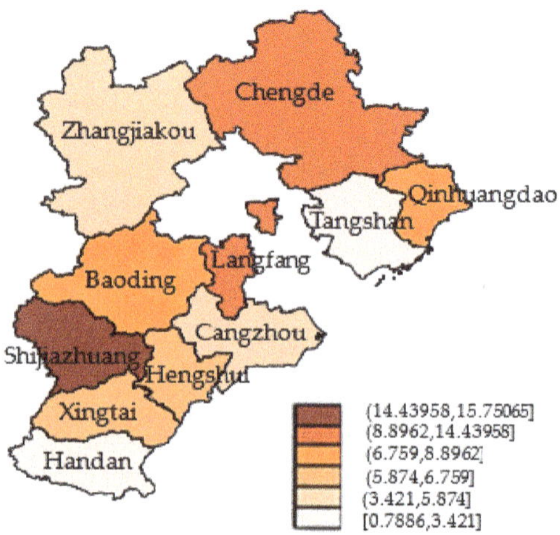

(**a**) 2006

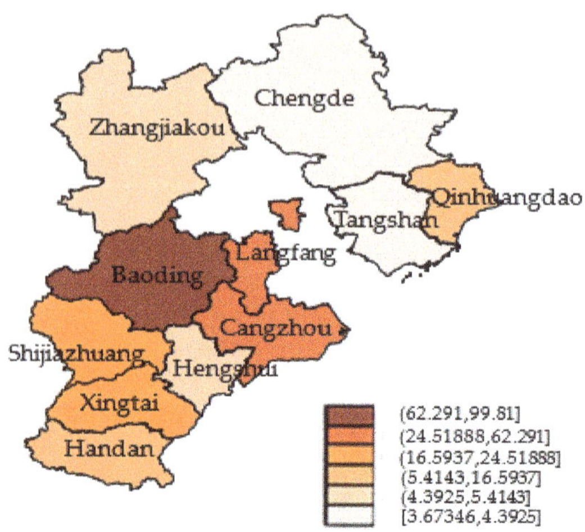

(**b**) 2009

Figure 3. *Cont.*

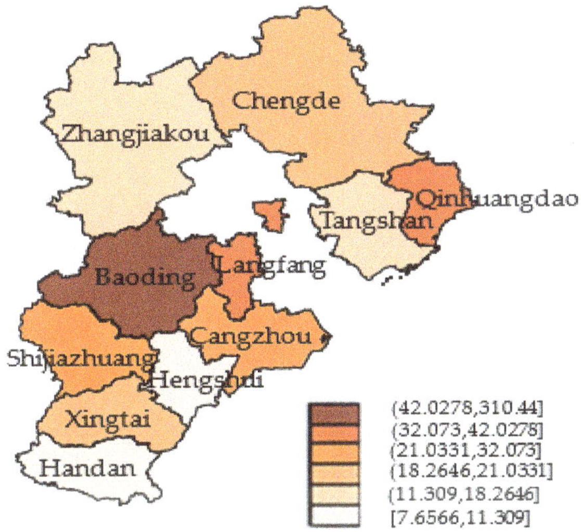

(c) 2012

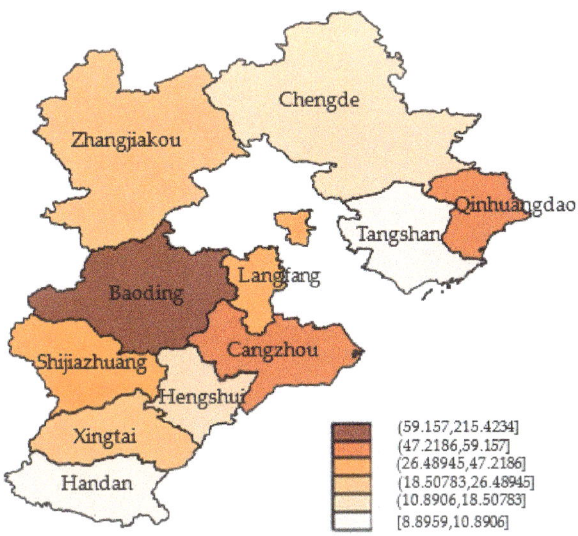

(d) 2015

Figure 3. *Cont.*

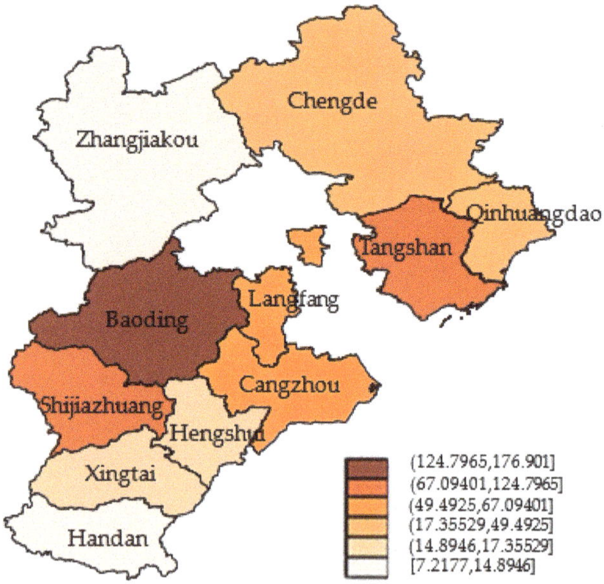

(**e**) 2019

Figure 3. The Spatial Quantile-Quantile (Q-Q) Plot of the Gross Industrial Output Value of the Non-Ferrous Metal Smelting and Calendering Processing Industry in Hebei Province.

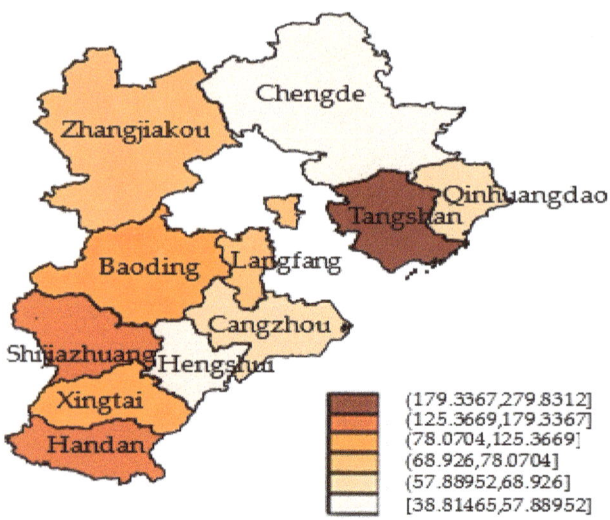

(**a**) 2006

Figure 4. *Cont.*

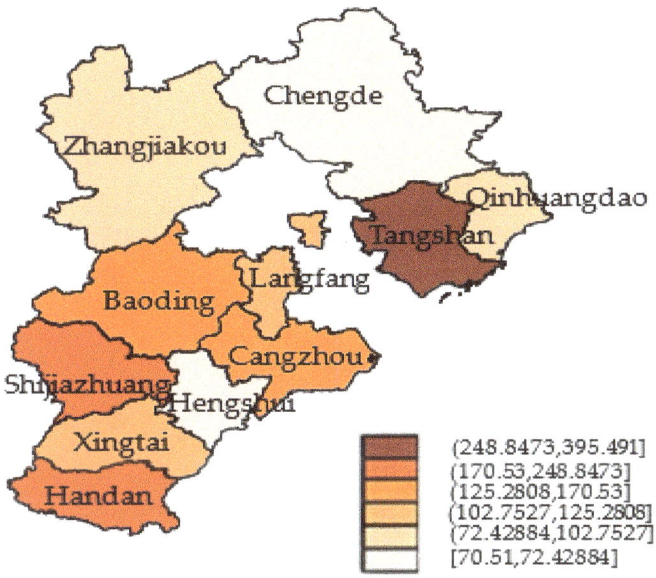

(**b**) 2009

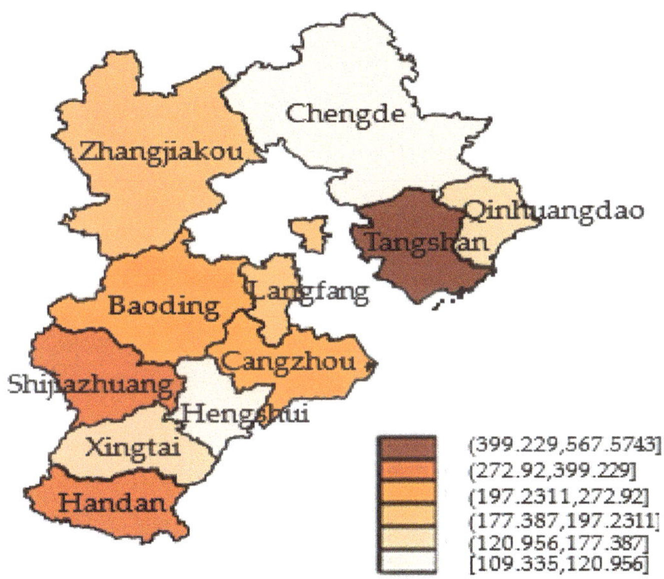

(**c**) 2012

Figure 4. *Cont.*

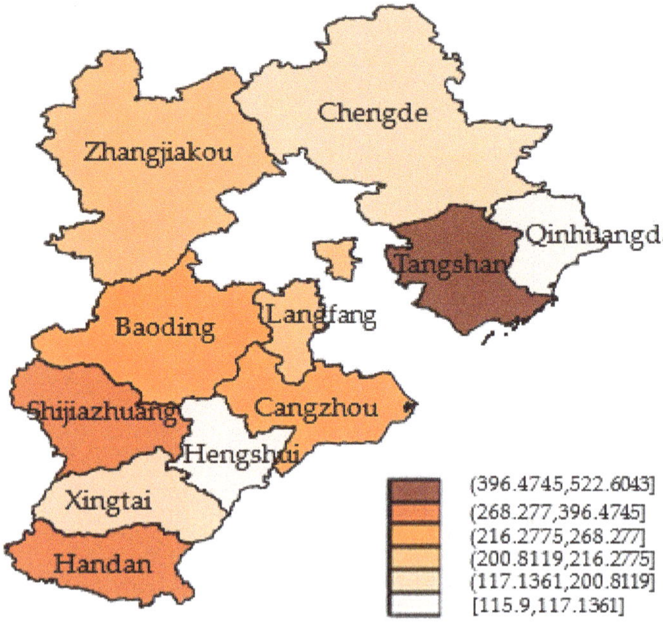

(**d**) 2015

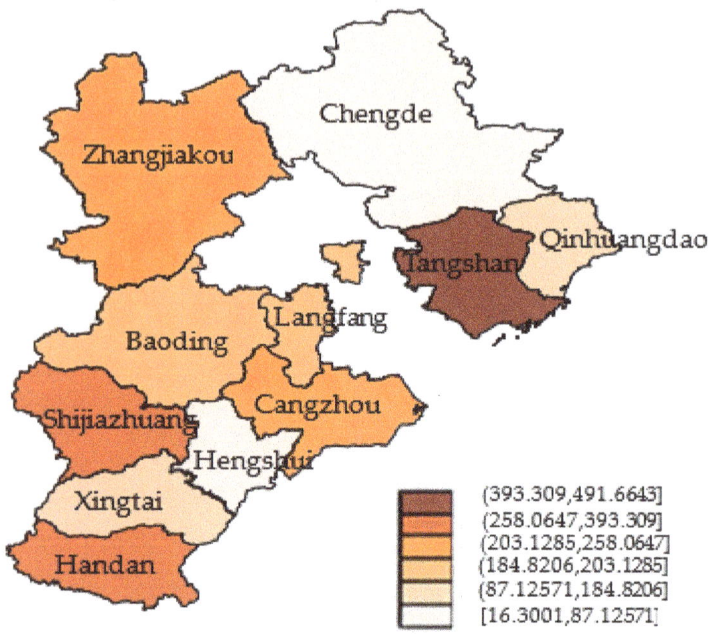

(**e**) 2019

Figure 4. The Spatial Quantile-Quantile (Q-Q) Plot of the Gross Industrial Output Value of the Electric and Thermal Power Production and Supply Industry in Hebei Province.

In 2011, the BBMG Group restructured Hebei Taihang Cement, which tremendously increased its production capacity. The production capacities of Dingxin Group and Quzhai Group in Shijiazhuang rivaled that of Tangshan Jidong Cement in 2001. This solved the problems of high energy consumption per unit output value, systematic drawback, insufficient investment and weak functionality of the cement industry in Shijiazhuang in the '9th Five-Year Plan'. Taking the Liulihe Cement Plant as the benchmark, the BBMG Group quickened the transformation of its cement plant. By the end of the '13th Five-Year Plan', the BBMG Group had realized the transformation and upgrading of 42 enterprises, invested in 58 environmental-protection industrial projects, and further improved the industrial agglomeration of the non-metallic mineral product industry in Shijiazhuang.

3.2.6. The Spatial Distribution of the Ferrous Metal Smelting and Calendering Processing Industry

As Figure 7 shows, from 2006 to 2019, the spatial distribution of Zhangjiakou changed from deep to light. The spatial distribution of Handan embodied deeper color with a high level of industrial agglomeration. The spatial distribution of Tangshan embodied the deepest color, showing a high level of industrial agglomeration. In 2005, the spatial distribution shifted from Zhangjiakou and Handan to Tangshan and formed industrial agglomeration. This trend continued until 2017. There are multiple reasons. In the '10th Five-Year Plan', the domestic steel and iron market continued to thrive, and China introduced a series of policies so that the craze for the export of steel material would wear off. Urged by The Policies on the Development of the Steel and Iron Industry, Zhangjiakou swiftly eliminated outdated production capacity, formulated the standards for steel and iron energy consumption, and supervised energy conservation and emission reduction. Many small and medium-sized enterprises had to be closed, converted or restructured.

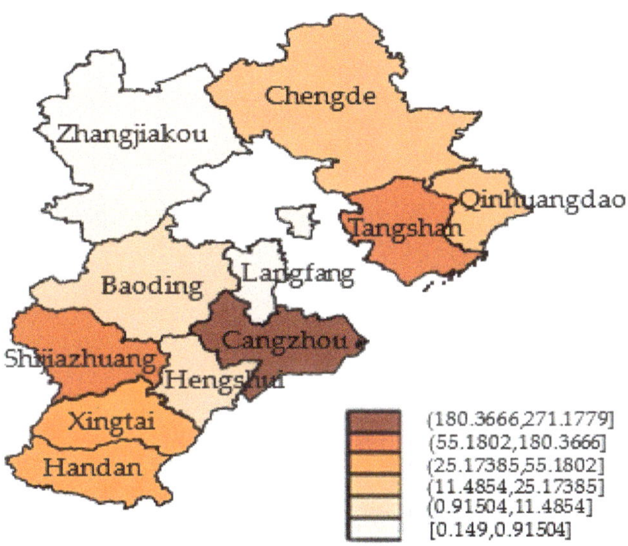

(**a**) 2006

Figure 5. *Cont.*

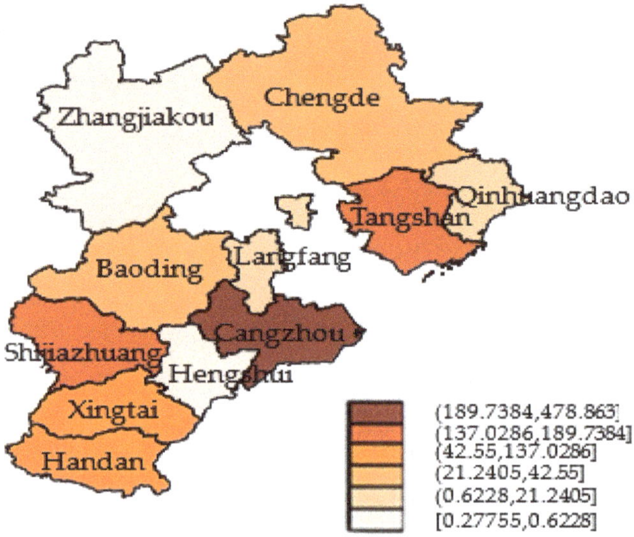

(**b**) 2009

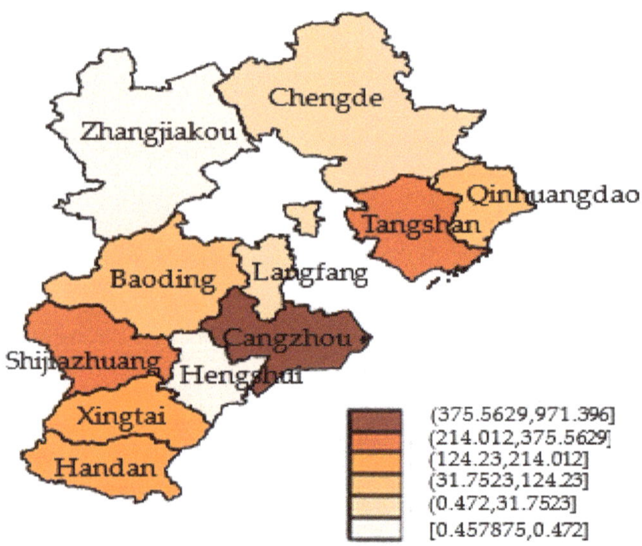

(**c**) 2012

Figure 5. *Cont.*

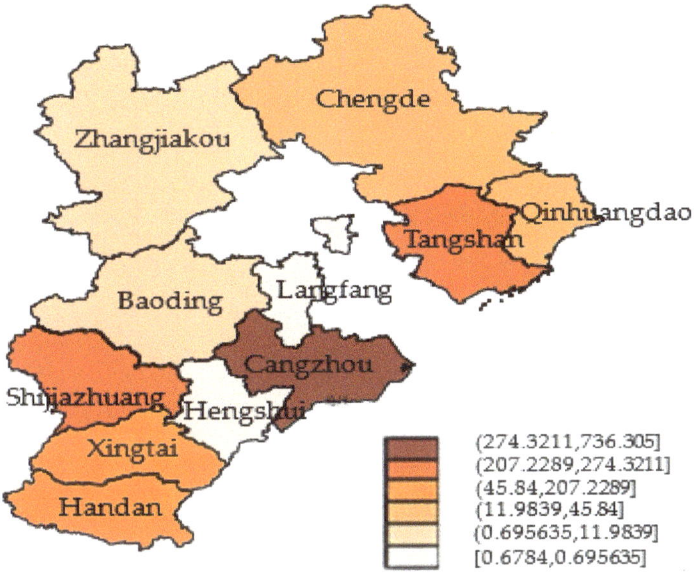

(d) 2015

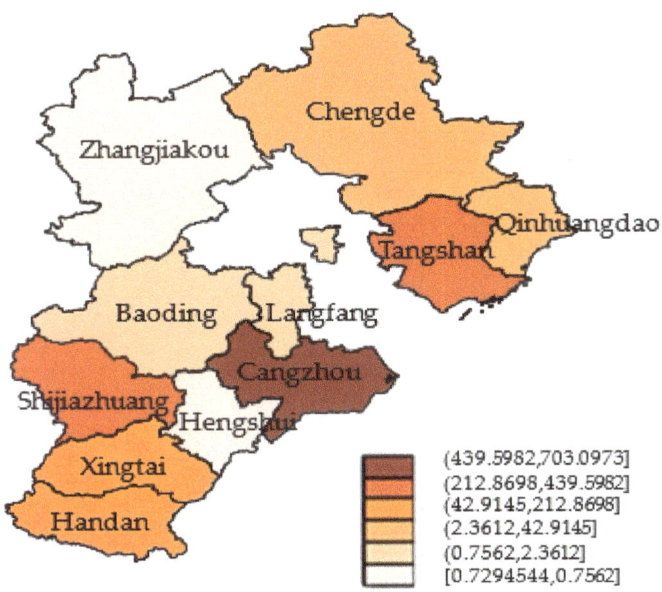

(e) 2019

Figure 5. The Spatial Quantile-Quantile (Q-Q) Plot of the Gross Industrial Output Value of the Petroleum Processing and Coking Industry in Hebei Province.

In 2005, in Handan, the supply-demand relationship in the steel and iron industry presented a landscape of 'excessive supply and structural imbalance'. Small and medium-sized steel and iron enterprises in Handan developed quickly. With no substantial progress in industrial mergers and acquisitions, the industrial agglomeration of the steel and iron industry tended to decline. Restricted by the existence of regional barriers, the deterioration of market performances and the absence of investors, the industrial agglomeration of the steel and iron industry in Handan can hardly be improved in the short term. After 2004, Hebei prioritized the optimization of industrial layout and the adjustment of product structure in the development of the steel and iron industry and re-distributed the newly added production capacity along the coastal cities. Based on the Caofeidian Port for large ores, Hebei organized a steel and iron complex and built Tangshan into the largest steel and iron base in China [21].

3.3. The Analysis of Spatial Balance of APIISTPI

The spatial quantile–quantile plot basically shows the spatial distribution pattern of APIISTPI in Hebei Province. However, it fails to effectively explain the intrinsic spatial correlation of APIISTPI among cities in Hebei Province. Therefore, this paper uses such approaches as Spatial Gini Coefficient to further analyze the issue. Tables 1 and 2 show the calculation results.

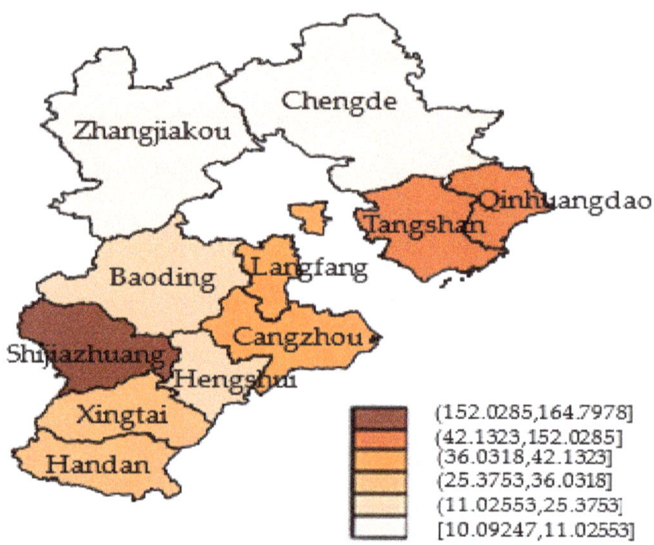

(a) 2006

Figure 6. *Cont.*

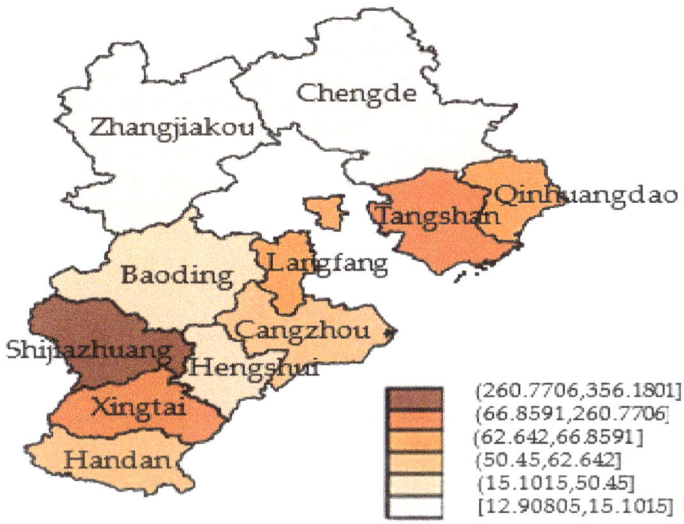

(**b**) 2009

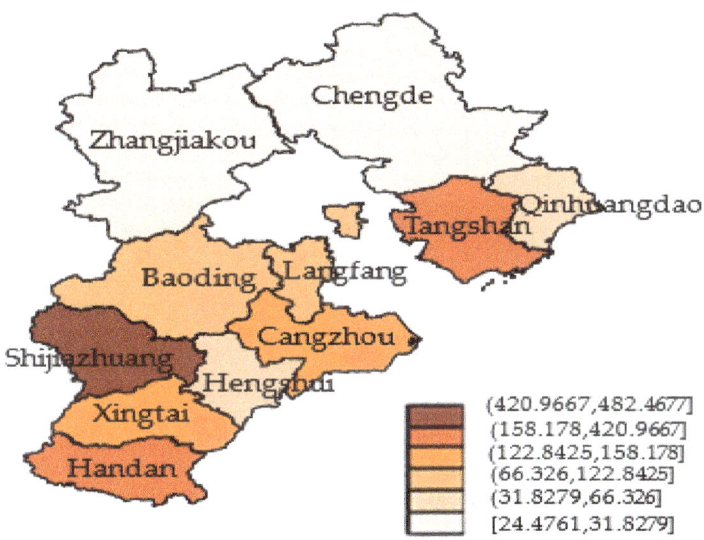

(**c**) 2012

Figure 6. *Cont.*

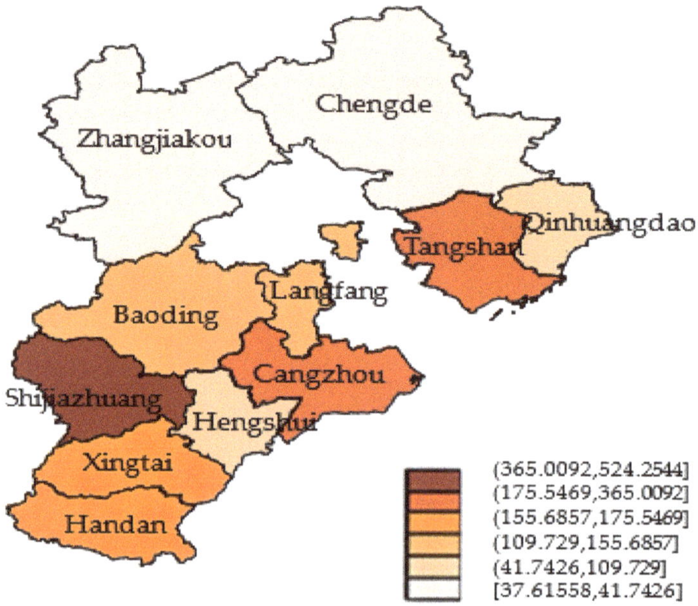

(**d**) 2015

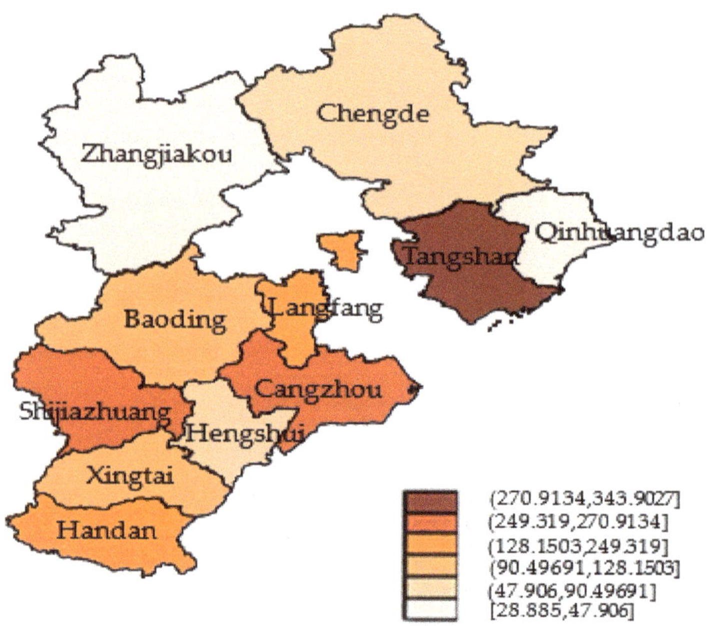

(**e**) 2019

Figure 6. The Spatial Quantile–Quantile (Q-Q) Plot of the Gross Industrial Output Value of the Non-Metallic Mineral Product Industry in Hebei Province.

In Table 1, the emissions of industrial dust and the balance degree of industrial sulfur dioxide in APIISTPI in Hebei province are relatively high. From 2006 to 2009, the G-values were higher than 0.3. The mean value of the emission and the balance degree tended to be close to 1, with an overall trend of imbalanced development. The G-values of PM2.5 and PM10 tend to be close to 0, indicating that the distribution of PM2.5 and PM10 is more balanced. The atmospheric pollution comprehensive index mainly refers to the measurement of the level of industrial agglomeration by using the Spatial Gini Coefficient, Herfindahl–Hirschman Index and location entropy [24]. The results show that atmospheric pollution in Hebei province presents a trend of balance. Hebei relies heavily on the petroleum, coal and other fuel processing industry, the ferrous metal smelting and calendering processing industry and the chemical raw material and chemical product manufacturing industry, which emit lots of industrial smoke and dust and industrial sulfur dioxide. Particularly, since the implementation of the Beijing–Tianjin–Hebei Coordinated Development, Hebei has undertaken the transfer of the heavy industries in Beijing and Tianjin and increased the degree of industrial agglomeration. Therefore, the spatial imbalance of pollutant emission in APIISTPI in Hebei became more prominent after 2010. As Table 2 verifies, the spatial imbalance of APIISTPI in Hebei province enlarges. Notably, in the petroleum, coal and other fuel processing industry and the ferrous metal smelting and calendering processing industry, the Spatial Gini Coefficients reach higher than 0.5 and approach 1. This means that as industrial agglomeration intensifies and spatial imbalance arises, the industrial agglomeration has a significant impact of negative externality on the environment. Under such circumstances, Hebei must scientifically control or regulate industrial agglomeration so as to balance the spatial distribution of industrial agglomeration and achieve the positive externality of industrial agglomeration on the environment.

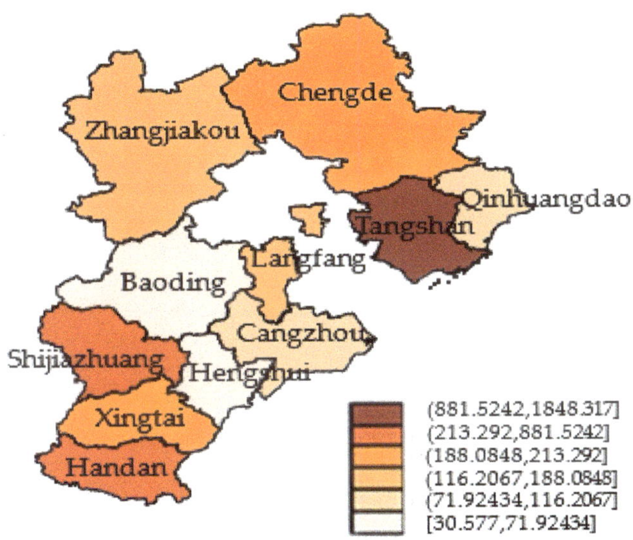

(**a**) 2006

Figure 7. *Cont.*

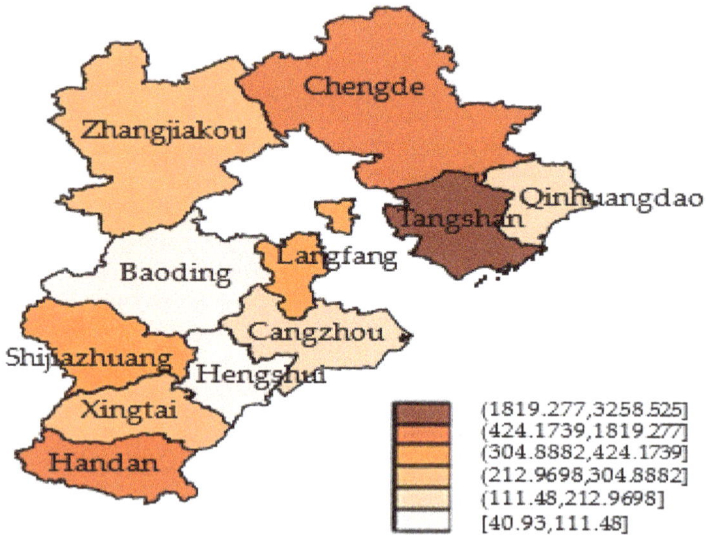

(**b**) 2009

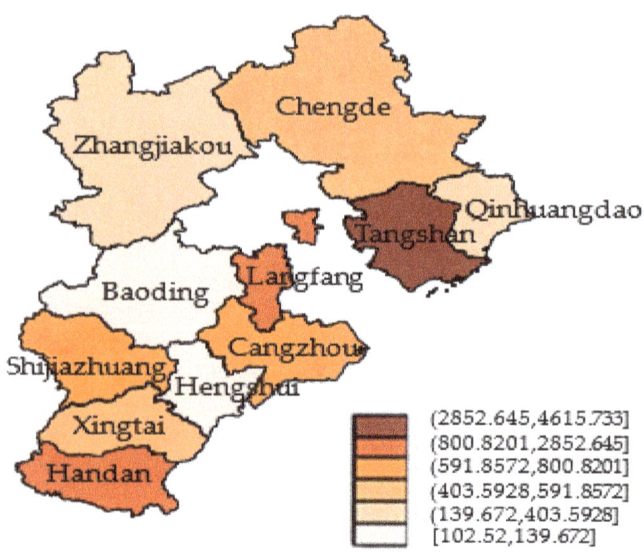

(**c**) 2012

Figure 7. *Cont.*

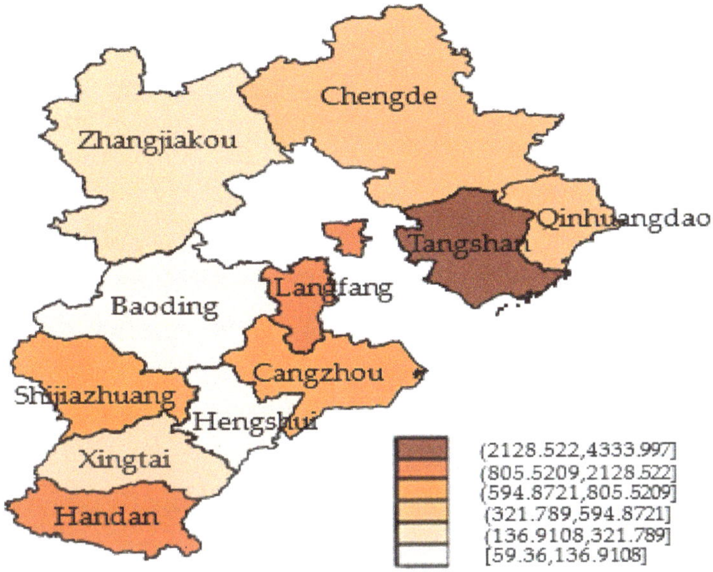

(**d**) 2015

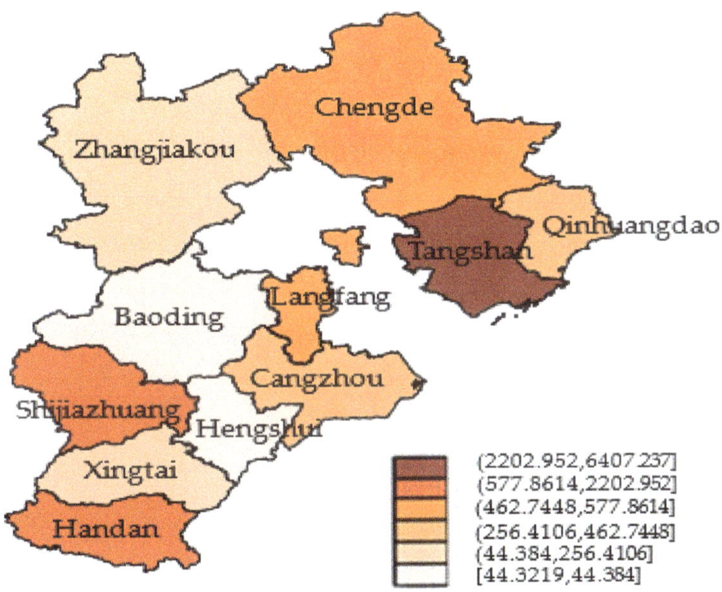

(**e**) 2019

Figure 7. The Spatial Quantile–Quantile (Q-Q) Plot of the Gross Industrial Output Value of the Ferrous Metal Smelting and Calendering Processing Industry in Hebei Province.

Table 1. The Balance Degree of Pollutants in APIISTPI in Hebei Province.

	The Total Emission of Industrial Sulfur Dioxide (Ton)	The Total Emission of Industrial Particulate Matter (PM) (Ton)	PM 2.5 ($\mu g/m^3$)	PM 10 ($\mu g/m^3$)	Atmospheric Pollution Comprehensive Index
	Gini_ya1	Gini_ya2	Gini_ya3	Gini_ya4	Gini_y
2006	0.366	0.389	0.230	0.110	0.198
2007	0.363	0.443	0.221	0.105	0.205
2008	0.368	0.435	0.217	0.098	0.235
2009	0.364	0.419	0.210	0.096	0.237
2010	0.369	0.384	0.216	0.085	0.228
2011	0.375	0.521	0.227	0.085	0.174
2012	0.364	0.521	0.234	0.195	0.189
2013	0.354	0.530	0.202	0.207	0.199
2014	0.333	0.502	0.183	0.168	0.199
2015	0.336	0.526	0.170	0.137	0.190
2016	0.382	0.595	0.166	0.134	0.201
2017	0.490	0.550	0.165	0.142	0.313
2018	0.466	0.559	0.137	0.117	0.209
2019	0.506	0.597	0.111	0.104	0.235

Table 2. The Balance Degree of APIISTPI in Hebei Province.

	The Gross Industrial Output Value of APIISTPI	The Petroleum, Coal and Other Fuel Processing Industry	The Chemical Raw Material & Chemical Product Manufacturing Industry	The Non-Metallic Mineral Product Industry	The Ferrous Metal Smelting and Calendering Processing Industry	The Non-Ferrous Metal Smelting and Calendering Processing Industry	The Electric and Thermal Power Production & Supply Industry
	Gini_x	Gini_xa1	Gini_xa2	Gini_xa3	Gini_xa4	Gini_xa5	Gini_xa6
2006	0.443	0.639	0.439	0.456	0.592	0.317	0.315
2007	0.420	0.625	0.445	0.477	0.565	0.564	0.297
2008	0.455	0.629	0.475	0.476	0.589	0.495	0.298
2009	0.454	0.607	0.475	0.486	0.601	0.530	0.299
2010	0.419	0.625	0.451	0.429	0.580	0.511	0.293
2011	0.434	0.629	0.478	0.476	0.566	0.582	0.282
2012	0.432	0.625	0.479	0.452	0.558	0.588	0.280
2013	0.409	0.672	0.499	0.415	0.536	0.544	0.259
2014	0.430	0.672	0.507	0.417	0.572	0.458	0.262
2015	0.424	0.657	0.538	0.408	0.564	0.496	0.256
2016	0.439	0.646	0.544	0.412	0.595	0.495	0.250
2017	0.483	0.658	0.579	0.421	0.656	0.486	0.267
2018	0.490	0.642	0.435	0.331	0.650	0.495	0.310
2019	0.502	0.641	0.436	0.352	0.659	0.473	0.316

4. Conclusions

As 2021's Ecological Environment Status Bulletin of Hebei Province discloses, the average number of days with excellent air quality in Hebei Province reaches 269 days. This paper categorizes cities with less than 250 days into Type I (including Handan, Shijiazhuang, Xingtai and Baoding), cities with more than 250 days yet less than 269 days into Type II (including Cangzhou, Langfang, Hengshui and Tangshan), cities with more than 269 days yet less than 300 days into Type III (including Qinhuangdao), and cities with more than 300 into Types IV (including Chengde and Zhangjiakou). From 2006 to 2019, the distribution characteristics of APIISTPI in Hebei Province were observed every three years. In combination with the characteristics of spatial distribution and balance analysis of APIISTPI in Hebei province, this paper draws the following conclusion.

4.1. APIISTPI in Type-I Cities

Type-I cities mainly include Handan, Shijiazhuang, Xingtai and Baoding, with the highest level of the agglomeration of atmospheric pollution-intensive industries such as the thermal power industry and the most serious atmospheric pollution. The economies of these four cities hinge on steel and iron, thermal power, chemical raw material and chemical product manufacturing and petroleum processing. In Handan, the steel and iron industry develops fast. With an unreasonable industrial layout, atmospheric pollution-intensive enterprises stand upwind. In Shijiazhuang, thermal power plants and steel mills flourish. Take the Jingye Group in Pingshan County, Shijiazhuang, as an example. Pingshan County lies in the northwest of Shijiazhuang and faces the wind. This directly pollutes the atmospheric environment in Shijiazhuang. In Xingtai, the coking industry develops quickly, with the sheet material industry in the east, the coking industry in the north, the steel and iron enterprises in the west, and power plants, cement plants and glass plants in the south. Coupled with the lack of airflow, these industries exacerbate atmospheric pollution in Xingtai. In Baoding, the power production industry thrives. In addition, the printing, dyeing and paper-making industries in Baoding account for 70% of coal consumption. As a consequence, in the non-heating period of summer and autumn, large-scale fog-haze weather often appears. To sum up, Type-I cities mainly depend on the petroleum, coal and other fuel processing industry, and the mean value reaches higher than 0.6 and approaches 1, which embody the most imbalanced development compared with other industries. In Type-I cities, in addition to the above-mentioned industries, weak atmospheric self-purification capacity that arises from less rainfall and wind occasions serious, atmospheric pollution and attenuates the bearing capacity of these cities.

4.2. APIISTPI in Type-II Cities

Type-II cities mainly include Cangzhou, Langfang, Hengshui and Tangshan, with a high level of agglomeration of atmospheric pollution-intensive industries such as the thermal power industry and serious atmospheric pollution. These cities develop the steel and iron, and oil industries, which emit massive atmospheric pollutants and cause serious, atmospheric pollution. For instance, in Cangzhou, APIISTPI covers the petroleum processing industry, which produces motor vehicle exhaust and dust and coal combustion, forming the sources of atmospheric pollution. In Hengshui, pillar industries are chemical pharmaceuticals, metal products and food. For instance, Hengshui Laobaigan Factory emits lots of atmospheric pollutants in production. Tangshan developed the steel and iron, and thermal power industries represented by Tangshan Steel Group. The steel and iron industry constitutes a source of sulfur dioxide, which poses great harm to the environment. Noteworthily, the thermal power industry has a significant impact on atmospheric pollution. To sum up, the Type-II cities mainly rely on the ferrous metal smelting and calendering industry. The mean value reaches higher than 0.5 and approaches 1, with imbalanced development. To mitigate environmental pollution, the industry can be transferred.

4.3. APIISTPI in Type-III Cities

Type-III cities only include Qinhuangdao, with the average level of the agglomeration of atmospheric pollution-intensive industries such as the thermal power industry and less serious atmospheric pollution. In Qinhuangdao, low-value-added industries have been eliminated in recent years. Qinhuangdao mainly develops the non-ferrous metal smelting and calendering industry and chemical raw materials and chemical products manufacturing industries. The mean value reaches 0.4, with relatively balanced development. In addition, Qinhuangdao is a coastal area with a certain environmental self-purification capability. Therefore, Qinhuangdao can realize industrial gradient transfer by adjusting the industrial structure.

4.4. APIISTPI in Type-IV Cities

Type-IV cities mainly include Chengde and Zhangjiakou. In these cities, the level of economic development plays a minor role because Zhangjiakou and Chengde are important water sources and ecological protection areas for Beijing and Tianjin, with the lowest level of atmospheric pollution. The industrial agglomeration of APIISTPI in Type-IV cities proves the lowest level. Hebei province attaches importance to the eco-environmental construction in Chengde and Zhangjiakou and mainly develops the energy industry and the electric and thermal power production and supply industry. The mean value reaches 0.3, with an overall trend of balanced development. The construction of an ecological environment also strengthens environmental self-purification capacity in Type-IV cities, where industrial transfer focuses on heavy industry.

Through analyzing the spatial distribution characteristics of Hebei APIISTPI, the paper finds that the main reason for atmospheric pollution in Hebei is that the concentration of APIISTPI exceeds the capacity of environmental self-purification. The higher the degree of excess, atmospheric pollution is more serious. In addition, there are reasons, such as the low self-cleaning capacity of the Hebei atmosphere and the unreasonable distribution of APIISTPI.

5. Recommendations

After the goals of carbon peaking and carbon neutrality are proposed, industrial gradient transfer plays an important role in promoting regionally balanced development and optimizing the spatial distribution of productivity [25]. Huang and Xiao et al. [26] hold that industrial gradient transfer narrows the regional economic gap, with different effects of different dimensions on the regional economic gap. As evinced, the regions with a high degree of industrial transfer play a positive role in reducing the regional economic gap. Accordingly, the following recommendations are made.

5.1. Recommendations for Optimizing Type-I Cities

Practical recommendations are made to optimize APIISTPI in Type-I cities. Firstly, the transfer-in regions of APIISTPI should have strong capacities for atmospheric self-purification and green technology innovations. For example, APIISTPI can be transferred to coastal areas, which fully utilizes strong atmospheric self-purification capacity in coastal areas and augments the self-purification of pollutants emitted by polluting industries. Secondly, Type-I cities can transfer APIISTPI to open and downwind terrains and consider the wind direction to avoid the effect of pollution spillover and the pollution of the neighboring environment. Thirdly, Type-I cities can strengthen industrial transformation and upgrading, enhance green-technology innovation, and quicken the desulfurization and denitrification in industries to reduce the emission of atmospheric pollution and eliminate the production equipment of enterprises that have difficulties in completing the technological transformation.

5.2. Recommendations for Optimizing Type-II Cities

Firstly, Type-II cities are advised to reduce atmospheric pollution by industrial transfer. Tangshan and Cangzhou, two cities with densely distributed APIISTPI, should conduct further market research and choose better cities for the transfer of APIISTPI. For example, Tangshan can transfer the steel and iron industry to the coastal areas of Guangxi, which have strong atmospheric self-purification capacity with abundant rainfall and strong wind. This effectively hedges against the impact of pollutants in APIISTPI. In Cangzhou, the petroleum processing industry produces pollutants in the atmospheric environment. However, Cangzhou possesses a unique geographical advantage. Cangzhou can reduce the pollution generated by the petroleum processing industry by accelerating the high-quality economic development of the coastal city, fostering the port transformation and upgrading (from a coal export port to a modern comprehensive service port) and building a new business form of modern port transportation. Langfang adjoins Beijing. With less APIISTPI, Langfang can easily achieve industrial transformation. Hengshui mainly relies on the manufacturing of alcoholic beverages. By controlling production processes and crafts, Hengshui reduces environmental pollution and pollutant emission. Moreover, in industrial transfer, various cities are advised to clarify specific divisions of labor and task, reasonably optimize industrial layout, and realize coordinated development, according to resource endowments, economic conditions and atmospheric self-purification capacity. Various cities can scientifically choose the transfer-in regions with strong capacities in environmental self-purification and green-technology innovation, e.g., the coastal areas of Guangxi or other southern areas, to ensure smooth industrial transfer and lessen the impact of APIISTPI on the environment. Notably, green production technologies can tremendously reduce pollutant emissions, save energy consumption and protect the environment. Therefore, transforming the production model and improving green-technological innovation also reduces pollutant emissions.

5.3. Recommendations for Optimizing Type-III Cities

Control measures can be taken for APIISTPI in Qinhuangdao, and industrial transfer can be carried out if necessary. The main reasons lie in that Qinhuangdao shifted from a 'light industry-preferred' development model to a 'heavy industry-driven development model and that atmospheric self-purification capacity reduces atmospheric pollution caused by coal transportation. In addition, large wetlands, abundant rainfall and strong wind effectively alleviate the accumulation and deposition of pollutants. From 2016 to 2020, in urban planning, Qinhuangdao implemented a municipal gasification project, Qinhuangdao Port coal dust treatment project, Qinhuangdao Port ship fuel-pollution treatment project, Beidaihe atmospheric environment monitoring, warning and big data decision and support system project, urban heavy-polluting-enterprise relocation and renovation project, as well as the desulfurization, denitrification and dust-removal project in key industries such as power, steel and iron, cement and glass. In this way, Qinhuangdao improves atmospheric pollution.

5.4. Recommendations for Optimizing Type-IV Cities

The roles of Zhangjiakou and Chengde in environmental protection are strictly controlled. Geographically, Zhangjiakou and Chengde surround Beijing and Tianjin, and the development level of high-pollution industries remains low. Since the '10th Five-Year Plan', spurred by national policies on environmental protection, the Zhangjiakou government has actively eliminated outdated production capacity and formulated energy-consumption standards for high-pollution industries to regulate their emissions. Chengde vigorously develops agriculture and tourism. Surrounded by mountains, Chengdu has a high green coverage rate and good air quality. In addition, in order to protect the ecological environment of Beijing and Tianjin, China takes administrative measures on environmental regulation in Zhangjiakou and Chengde, such as returning farmland to forests, restricting the pollution emissions of local enterprises, and shutting down some enterprises on chemical manufacturing and pesticide production. To sum up, catalyzed by national and municipal administrative policies on environmental regulation, Zhangjiakou and Chengde effectively control the deterioration of the ecological environment and achieve satisfactory ecological effects (for example, good air quality).

Author Contributions: J.Z. and J.T. conceived and completed the writing of the manuscript. X.W. made key modifications to important knowledge parts. X.B. assisted with the code design and experimental analysis. All authors have read and agreed to the published version of the manuscript.

Funding: This research was funded by the Hebei Social Science Foundation for Spatial distribution and gradient transfer of air pollution-intensive industries in Hebei Province, China (HB19GL002).

Data Availability Statement: The data presented in this study are available upon request from the corresponding author. The data are not publicly available because the supporting project has a confidentiality agreement.

Acknowledgments: The authors would like to thank the editor and reviewers for their valuable comments and suggestions, which helped them improve the manuscript.

Conflicts of Interest: The authors declare no conflict of interest.

References

1. Tian, Y.Y. Regional industrial transfer in the Jingjinji urban agglomeration, China; An analysis based on a new transferring area-undertaking area-dynamic process model. *J. Clean. Prod.* **2019**, *235*, 751–766. [CrossRef]
2. He, Z.X.; Cao, C.S.; Wang, J.M. Spatial Impact of Industrial Agglomeration and Environmental Regulation on Environmental Pollution-Evidence from Pollution-Intensive Industries in China. *Appl. Spat. Anal. Policy* **2022**, *15*, 1525–1555. [CrossRef]
3. Chen, Z.Y.; Ding, X.S.; Han, C.; Yu, H. The Correlation Mechanism of Manufacturing Agglomeration and Pollution, and the Practical Path of Green Development: A Study Based on Spatial Spillover Model. *Stat. Res.* **2022**, *39*, 46–61.
4. Chen, L.Q.; Huang, P. The Analysis of Industrial Cluster and Urbanization: From the Perspective of Externality. *J. Xihua Univ. Sci.* **2007**, *2*, 16–20.
5. Chen, Y.Y. The Impact of Industrial Agglomeration on Industrial Clean Production and End-of-Pipe Treatment. *South China J. Econ.* **2011**, *5*, 17–27.
6. Gai, W.Q.; Zhu, H.S. Flexible Industrial Agglomeration and Regional Competitiveness. *Econ. Theory Bus. Manag.* **2001**, *10*, 25–30.
7. Qiu, Z.Y. The Impact of International Vertical Industrial Agglomeration on China's Environment. *Acad. Forum* **2012**, *3*, 121–124+178.
8. Tang, C.H.; Dou, J.M. The Impact of Heterogeneous Environmental Regulations on Location Choices of Pollution-Intensive Firms in China. *Front. Environ. Sci.* **2021**, *9*, 626. [CrossRef]
9. Wen, X.Q.; Su, R.X. The Impact of Environmental Regulation on the Spatial Distribution of Polluting Industries in the Beijing-Tianjin-Hebei Region. *J. Tianjin Norm. Univ. Sci.* **2015**, *6*, 58–62.
10. He, X.L.; Wu, H. Technological Spillover, Environmental Pollution and Industrial Spatial Distribution: A New Extension of Krugman's Core and Periphery Theory. *West Forum* **2021**, *31*, 34–51.
11. Wang, Y.Z.; Duan, X.J.; Liang, T.; Wang, L.; Wang, L.Q. Analysis of spatio-temporal distribution characteristics and socioeconomic drivers of urban air quality in China. *Chemosphere* **2022**, *291*, 132799. [CrossRef]
12. Kaname, A. The Synthetic Principles of the Economic Development of our country. *J. Econ.* **1932**, *6*, 179–220.
13. Ma, H.X. Comparative Analysis of Two Spatial Models of Regional Transmission: On the Selection Trend of Spatial Model of Regional Transmission in China. *Gansu Soc. Sci.* **2001**, *2*, 30–32.
14. Chen, J.J.; Ye, W.Y. The Research on Industrial Transfer to Developing Regions in Zhejiang Province. *J. Bus. Econ.* **2002**, *4*, 28–31.
15. Jiang, W.J.; Sun, H.G. Coordinating Industrial Transfer: An Important Measure for the Rapid Development of Small and Medium Enterprises in Developing Regions. *Manag. J.* **2001**, *7*, 15–16.
16. Fu, Z.P.; Zeng, S.Y. Transfer Models and Action Characteristics in the Transfer of Clustered Industries: An Analysis from the Perspective of Corporate Social Network. *Manag. World* **2008**, *12*, 83–92.
17. Xiang, G.E.; Hu, P. The Mechanisms and Characteristics of Regional Industrial Transfer from the Perspective of Global Value Chain: A Case Study of the Pan Yangtze River Delta. *East China Econ. Manag.* **2011**, *2*, 45–49.
18. Huang, Z.Y. Industrial Transfer: Trends in the East and Choices in the West: Taking Chongqing as an Example. *Econ. Issues* **2009**, *7*, 117–120.
19. Ma, Z.H.; Hu, H.B. An Exploration of the Major Models of Inter-regional Industrial Transfer in China. *Product. Res.* **2009**, *13*, 141–143.
20. Li, Y.Y.; Li, H. China's inter-regional embodied carbon emissions: An industrial transfer perspective. *Environ. Sci. Pollut. Res.* **2022**, *29*, 4062–4075. [CrossRef]
21. Zhou, J.K.; Li, Y.T. Research on Spatial Distribution Characteristics of High Haze Pollution Industries Such as Thermal Power Industry in the Beijing-Tianjin-Hebei Region. *Energies* **2022**, *15*, 6610. [CrossRef]
22. Wei, H.K. *China's Industrial Cluster and Cluster Development Strategy*; Economic Management Publishing House: Beijing, China, 2008.
23. Zhang, J.Y. Spatial Difference and Convergence of Innovative Development in Yangtze River Delta: Based on Dagum Gini Coefficient and Decomposition. *Resour. Environ. Yangtze Basin* **2023**, *32*, 235–249.

24. Zhou, J.K.; Tian, J.; Zhang, D.D. Pollution Effect of the Agglomeration of Thermal Power and Other Air Pollution-Intensive Industries in China. *Int. J. Environ. Res. Public Health* **2023**, *20*, 1111. [CrossRef] [PubMed]
25. Zhang, Z.Q. Problems and Countermeasures for China's Industrial Gradient Transfer Under Carbon Peak and Carbon Neutrality Target. *Environ. Prot.* **2023**, *51*, 47–50.
26. Huang, H.; Xiao, Y.L. Three Dimensional Economic Openness Industrial Gradient Transfer and Regional Economic Disparity. *J. Tech. Econ. Manag.* **2023**, *4*, 111–117.

water

MDPI

Article

Unveiling the Value of Nature: A Comprehensive Analysis of the Ecosystem Services and Ecological Compensation in Wuhan City's Urban Lake Wetlands

Juan Du [1,2,3], Rajendra Prasad Shrestha [1,*], Vilas Nitivattananon [1], Thi Phuoc Lai Nguyen [1] and Amar Razzaq [4,5]

1 Department of Development and Sustainability, School of Environment, Resources and Development, Asian Institute of Technology, Pathumthani 12120, Thailand; st121584@ait.ac.th (J.D.); vilasn@ait.ac.th (V.N.); phuoclai@ait.ac.th (T.P.L.N.)
2 Sixth Geological Brigade of Hubei Geological Bureau, Xiaogan 432000, China
3 Hubei Key Laboratory of Resources and Eco-Environmental Geology, Xiaogan 432000, China
4 Business School, Huanggang Normal University, Huanggang 438000, China
5 College of Economic and Management, Huazhong Agricultural University, Wuhan 430070, China
* Correspondence: rajendra@ait.ac.th

Abstract: Urban lake wetlands play an essential role in providing ecological services, promoting urban sustainability, and enhancing the quality of urban life. This study quantitatively assesses the ecosystem services value (ESV) of the Zhangdu, East, and Ziyang urban lake wetlands in Wuhan, China, based on primary survey data and methodologies, including the market price, shadow engineering, and travel cost methods. The ESV is categorized into direct use value (DUV), indirect use value (IUV), and non-use value (NUV). Our findings reveal that the non-use value proportion is significant, amounting to 1.569×10^8 CNY $\times$ yr^{-1} for Zhangdu Lake, 1.527×10^8 CNY $\times$ yr^{-1} for East Lake, and 1.060×10^8 CNY $\times$ yr^{-1} for Ziyang Lake. This indicates a high willingness to pay among respondents, reflecting a recognition of the value of wetland services. In addition to the non-use value, this study underscores the considerable material production, water conservation, and leisure tourism value that these urban lake wetlands provide. The assessment of ESV delivers a scientific basis for the management and protection of urban lake wetlands. It also highlights the challenges faced, such as pollution and fragmented management approaches due to unclear property rights and insufficient funding. This study concludes by emphasizing the need for future research to explore mechanisms that promote social participation in wetland management, with the aim of enhancing the overall ecological health of urban lake wetlands.

Keywords: urban lake wetland; ecosystem service value; ecological compensation; sustainability

1. Introduction

Wetlands refer to natural or artificial, permanent or temporary swamps, wetlands, peatlands, or water areas with static or flowing, or fresh, brackish, or salty water bodies, including shallow waters with a depth of not more than 6 m at low tide [1]. Wetlands are distributed between terrestrial and aquatic ecosystems, but wetlands have many features that distinguish them from terrestrial and marine ecosystems. The most notable features include the appearance of stagnant water during parts of the growing season, unique soil conditions, vegetation tolerance, and adaptability to saturated soil [2].

Wetlands are one of the most critical environments for human survival, with biologically important habitats, ecosystem types, and natural landscapes with the richest biodiversity [3]. They have a variety of service functions, such as providing food and fiber, regulating floods, conserving water sources, improving climate, consolidating soil and fertilizing, purifying the environment, and maintaining biodiversity [4]. They are also the "gene pool of species" and "paradise for birds" [5].

The evaluation of the wetland ecosystem service function value is the foundation of wetland protection and rational utilization [6]. A scientific and rational assessment of the functions of various wetlands is conducive to improving the level of wetland research, monitoring, protection, and utilization and is the basis for wetland protection planning [7]. The rational use of wetland ecological resources provides a scientific basis, which is of great significance for improving the quality of the ecological environment and ensuring regional ecological security [8].

The global wetland area is about 7–9 million km^2, which is 4–6% of the earth's land surface [9]. However, wetlands have suffered the most serious damage from human activities in modern history. Wetlands have become the last frontier in terms of attention and conservation efforts, lagging behind other ecosystems such as agriculture, forestry, and deserts. In many developed and developing countries, wetland areas are disappearing at an alarming rate. About 50% of wetlands in the continental United States have disappeared, and wetlands in Europe, parts of Australia, Canada, and Asia are disappearing much faster [10]. In the 30 years from 1978 to 2008, 33% of wetlands in China disappeared [11] due to reasons such as reclaimed land for settlements. This indicates that their function and value have not been properly recognized by humans for a long time. However, the gradual recognition of the importance of wetlands has attracted worldwide attention, and the research on wetland ecosystems has gradually increased [12].

China has a large wetland area, about 384,900 km^2, including 362,000 km^2 of natural wetlands, ranking first in Asia and third in the world after Canada and Russia [13]. Hubei Province is known as the "province of a thousand lakes" and is rich in wetland resources. The existing wetland covers an area of 1.6169 million hectares in five categories: river wetlands, lake wetlands, swamps, swampy meadow wetlands, and artificial wetlands (reservoir ponds) [14]. Wuhan, the capital of Hubei Province, is known as the city of rivers and the city of hundreds of lakes. A total of 165 rivers are crisscrossed; 16 lakes are dotted; the wetland area is 12,000 km^2, accounting for 1.9% of Wuhan's land area; and the wetland resources rank among the top 3 inland cities in the world [15]. Wuhan achieved a major milestone in June 2022, as it was successfully selected as part of the second cohort of international wetland cities. Additionally, the city is home to a renowned national wetland park in China. With the most surface water per capita, Wuhan ranks first in the world [16].

Wetland ecosystems, renowned for their unparalleled biodiversity, are undeniably fragile and necessitate diligent protection to preserve their unique characteristics [17]. Despite their richness and vital role, these wetlands confront numerous challenges, including pollution, fragmented management approaches, unclear property rights, and insufficient funding, all of which threaten their sustainability. In light of these issues, it becomes imperative to understand the ecosystem services provided by various wetlands [18]. This understanding is instrumental in establishing ecological compensation standards that aid in the protection and restoration of wetlands, thereby improving wetland ecology through rigorous scientific assessment. However, current studies have seldom applied a comprehensive approach to assessing the ecosystem service function values of urban wetlands.

There is an immediate need for research that sheds light on the value of these wetlands and elucidates strategies for their protection and enhancement. A noticeable research gap exists in the comprehensive evaluation of these wetlands' ecosystem service function values and the calculation of their ecological compensation. This study aims to fill this void by contributing to the literature by applying the theories of eco-economics and resource economics to provide a comprehensive assessment of the ecosystem service function values of the Zhangdu, East, and Ziyang urban lake wetlands, including their resource function, environment function, and social function. Furthermore, we conducted a survey to analyze respondents' environmental consciousness, the significance of the non-use value component, the frequency distribution of respondents' willingness to pay (WTP), the per capita WTP value, and the total payment value. We also analyzed the characteristics of respondents' WTP and the correlation between WTP and societal factors. These findings will provide valuable insights into the development of strategies to manage and protect these crucial wetlands.

2. Materials and Methods

2.1. Study Area

This study focuses on Wuhan, the capital city of Hubei province, located at the intersection of the Yangtze River and the Han River. Wuhan's climate is subtropical monsoon with abundant rainfall and distinct seasons. Geomorphologically, the city is characterized by a flat center surrounded by hills and ridges in the south and low bristly mountains in the north (Figure 1) [19].

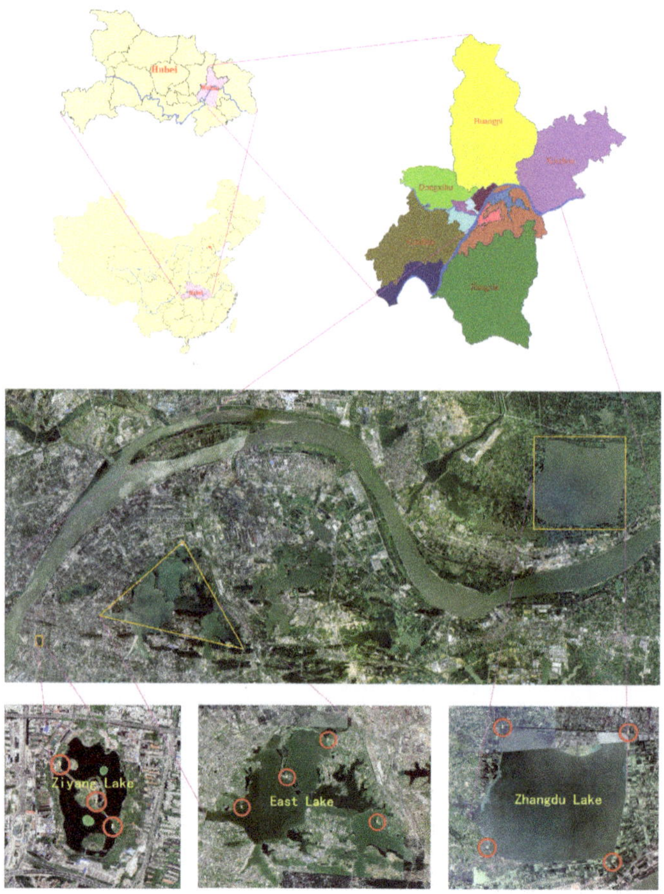

Figure 1. Surface water sampling site of selected lakes.

Wuhan city boasts a rich variety of wetland resources, including rivers, lakes, swamps, marsh meadows, pools, and paddy fields [20]. According to Pan [21], it mainly includes six types of wetlands: rivers, lakes, swamps, marsh meadows, pools, and paddy fields, among which shallow lake wetland systems (including lake wetlands, swamps, and marsh meadow wetlands). However, these wetlands have faced significant challenges due to erosion, reclamation, pollution, and residential encroachment, leading to a reduction in their area, biodiversity, and overall ecosystem function [22].

For this study, we chose three typical urban lake wetlands: Zhangdu Lake, East Lake, and Ziyang Lake, based on their size, location, and type. Surface water samples were collected from 11 sites across these lakes, with four in Zhangdu Lake, four in East Lake, and three in Ziyang Lake (Figure 1 and Table 1).

Table 1. Characteristics of selected urban lake wetlands in Wuhan [23,24].

Lake Name	Lake Type	Geographic Location	Primary Use	Administrative District
Zhangdu Lake	Natural	113°34′–114°52′ E, 30°32′–30°45′ N	Large ecological land	Huangpi
East Lake	Semi-natural	114°20′–114°55′ E, 30°33′–30°34′ N	Lake park/natural scenic area	Wuchang and Hongshan
Ziyang Lake	Artificial	114°21′ E, 30°32′ N	City lake park	Wuchang and Hongshan

2.2. Data Collection

Our study combined primary and secondary data collection with several analytical approaches.

2.2.1. Primary Data Collection

Primary data were collected through key informant interviews and observations through field research. The details are described below:

Key Informant Interviews: Ten key informants (Table 2), including local government officials, local leaders, and general citizens, were interviewed to gain insights into the strengths and weaknesses of the study areas, the major threats to urban lake wetlands, and the benefits derived from them.

Table 2. List and number of key informants interviewed.

Key Informant Category	Number of Informants
Study Area Leaders	3
Government Officers	2
Resident Representatives	3
NGO Officers	2
Total	10

Observation/Field Research: Fieldwork was conducted to observe and collect information on people's behavior; the actual condition of the field; the value of aquatic resources such as fish, crabs, and shrimp derived from the lakes; the benefits imparted by the wetlands; and the cost of living for the local populace.

Socio-Economic and Willingness to Pay (WTP) Surveys

We conducted a two-part survey from August 2021 to April 2022 to gather socio-economic data and evaluate the non-use value of lake wetlands. It should be noted that the response rate for this survey was high, at 96%, emphasizing the reliability of the collected data.

A structured questionnaire was administered through face-to-face interviews at the household level in 13 administrative districts under the jurisdiction of Wuhan. Using the equation suggested by Yamane [25], we required a sample size of 372 households (Table 3).

Table 3. Study lakes and corresponding sample sizes.

Study Lake	Number of Households	Sample Size
Zhangdu Lake	1974	140
East Lake	2360	167
Ziyang Lake	926	65
Total	5260	372

In this two-part survey, the first part focused on socio-economic data collection, while the second part was a willingness to pay (WTP) survey. For the WTP, we employed the Contingent Valuation Method (CVM) to evaluate the non-use value of lake wetlands. The pre-survey used an open-ended (OE) questionnaire to obtain core estimates for the second survey, which utilized a payment card (PC) approach.

The successful application of CVM depends on the accurate and reasonable design of the questionnaire. The core valuation guidance techniques of CVM include iterative bidding game (IB), open-ended (OE), payment card (PC), and dichotomous choice (DC) [26]. In this study, we used OE for the first pre-survey to obtain the core estimate and its interval for the second survey; thereafter, PC was used.

The questionnaire was comprised of seven parts:

1. Description of non-use value of lake wetland resources.
2. Respondents' survey on environmental awareness of Wuhan Lake wetlands.
3. On-site photos and basic information introduction to Zhangdu Lake, East Lake, and Ziyang Lake.
4. Survey of respondents' willingness to pay.
5. Identification of non-use value components (existence value, heritage value, and option value) the respondents are willing to pay for.
6. The respondents' gender, age, occupation, education level, technical title, and annual income, among other social characteristics.
7. Guidance and suggestions for this CVM survey.

2.2.2. Secondary Data Collection

Secondary data were collected from various sources, including official documents, government reports, and data from organizations such as the Hubei Provincial Department of Water Resources Lake Division, the Hubei Environmental Monitoring Center, and the Hubei Statistics Bureau.

2.3. Analytical Approaches

2.3.1. Ecological Service Value Assessment

Using the environmental resource value theory, we divided the wetland value into use and non-use value. The use value further included direct use value (DUV) and indirect use value (IUV) [27]. We computed DUV based on the market price of ecosystem products, which included components such as aquaculture production, water conservation, and leisure tourism. Each component was assessed via different techniques: aquaculture production was evaluated by looking at the yield per unit area, the area of the lake wetland, and the market price of these products. The water conservation value was estimated based on the volumes of the urban lake wetlands at constant water levels and the construction cost of storage capacity per unit area. The leisure tourism value was gauged using the travel cost method, supplemented with questionnaire survey data.

IUV and non-use value (NUV) were estimated based on the willingness to pay (WTP) for utilizing an ecosystem service function. This involved calculating the sum of the values of carbon sequestration and oxygen release, transpiration and endothermic, water quality purification, flood control, biodiversity maintenance, and scientific research and education. Each component had its unique method for estimation: carbon sequestration and oxygen release were obtained by measuring these quantities in the aquatic vegetation of each lake and then monetizing them using the shadow price method. The transpiration value was calculated through monetization and transformation of the heat absorption function index. Water quality purification, flood control, and biodiversity maintenance values were estimated through a combination of measurement and monetization methods. The research and education value was calculated using the result reference method. Tables A4–A6 in Appendix A provide a list of indicators and methods of ecosystem services valuation used in this study.

Non-use value (NUV) was estimated as the sum of existence, heritage, and option values. These components were estimated using the Contingent Valuation Method (CVM), which involved creating an open-ended (OE) questionnaire in the pre-survey stage to capture respondents' willingness to pay, thereby indicating the perceived non-use value of the lake wetlands. A payment card (PC) approach was then used in the main survey to solidify these initial estimates.

Consequently, an array of methods, such as the market price, shadow engineering, travel cost, shadow price, efficiency alternative, pollution cost, result reference, and contingent valuation methods, were employed to evaluate various functional indicators of wetland ecosystems. The total economic value was calculated by the sum of direct, indirect, and non-use values.

Through the use of the Contingent Valuation Method (CVM), we established an index database to analyze respondents' environmental consciousness, the significance of the non-use value component, the frequency distribution of respondents' WTP, per capita WTP value, and total payment value. This method allowed us to directly measure respondents' willingness to pay (WTP) values [28]. Additionally, the survey analyzed the correlation between WTP and societal factors.

In the analysis of the degree of importance of non-use value (NUV) indicators, a 4-point scale [29] was used to grade their significance: very important (3 points), important (2 points), average (1 point), and not important (0 points). The scores were then multiplied by the frequency of respondents' evaluation of the importance of each non-use value component, and the average score was calculated to determine their order of importance. The total payoff value was calculated by multiplying the actual number of people WTP by the average total payoff value (Vwtp) of each lake [30]. The correlation analysis of respondents' social characteristics on WTP and WTP value was tested by the p value size [31].

2.3.2. Water Quality Assessment

The water quality of the three lakes in Wuhan was assessed by measuring the levels of total nitrogen (TN); total phosphorus (TP); and several heavy metals, such as copper (Cu), zinc (Zn), lead (Pb), cadmium (Cd), arsenic (As), and mercury (Hg). In March 2022, 11 surface water samples from the selected lakes were collected and sent to the laboratory of No. 6 Geological Team of the Hubei Geological Bureau for analysis. This analysis provided crucial data for estimating the water purification values of the lakes.

The determination of TN and TP was carried out according to HJ 636-2012 and GB 11893-89 standards, respectively, using an LH-25A intelligent multi-parameter digestion device and a TU-1810D UV-visible spectrophotometer [32–35]. These standards are suitable for surface water, groundwater, industrial waste water, and domestic sewage. The detection limit for TN is 0.05 mg/L, and for TP, it is 0.01 mg/L.

For the determination of TN and TP, water samples were first digested using reagents P1 and P2 and potassium persulfate (SP). After digestion, the samples were colored with specific reagents, and their absorbance was measured using a spectrophotometer at specific wavelengths: 220 nm and 275 nm for TN and 700 nm for TP.

For heavy metal analysis, the procedure varied depending on the metal. For Cu, Zn, Cd, and Pb, unacidified samples were filtered and then measured using Inductively Coupled Plasma Atomic Emission Spectroscopy (ICP-AES) [36–38]. For Hg and As, the samples were subjected to acid digestion followed by measurement using an atomic fluorescence analyzer [39–41].

Quality control for total nitrogen (TN), total phosphorus (TP), copper (Cu), zinc (Zn), lead (Pb), cadmium (Cd), arsenic (As), and mercury (Hg) was conducted meticulously. Correlation coefficients of the calibration curves for all substances met the requirements. A parallel double sample was analyzed for each batch, ensuring the relative deviation was within the acceptable range. For each batch, a standard quality control sample was also tested, with results complying with the standard values. For Cu, Zn, Pb, Cd, Hg, and

As, blanks were within standards, and certified standard samples were confirmed to fall within the uncertainty range. Intermediate point concentration was checked and met the requirements. All these steps ensured the precision and accuracy of our measurements.

3. Results

3.1. Results of Ecological Service Value Assessment

3.1.1. Direct Use Value (DUV)

Material Production Value (V1)

The material production value of the urban lake wetlands, classified as natural, semi-natural, or artificial, was assessed. The assessment involved evaluating all types of material products provided by these lakes, considering the yield per unit area, the area of the lake wetland, and the market price of these material products. The total material production value (Table 4) was determined to be 796.59 $\times$ 10^4 CNY $\times$ yr^{-1}. Among this total, natural lakes contributed 445.77 $\times$ 10^4 CNY $\times$ yr^{-1}, and semi-natural lakes contributed 350.82 $\times$ 10^4 CNY $\times$ yr^{-1}, while artificial lakes did not contribute. This indicates that natural lakes have the highest material production value, followed by semi-natural lakes, with artificial lakes not contributing to the material production value. Additional details regarding the formulae and definition of variables V1 to V9 are provided in Appendix A (Tables A4–A6).

Table 4. Direct use value of urban lake wetlands in Wuhan (in 10^4 CNY $\times$ yr^{-1}).

DUV Components	Zhangdu Lake	East Lake	Ziyang Lake
Material production value	445.77	350.82	0
Water conservation value	2784.63	6905.28	12.44
Leisure tourism value	712.80	71,940.00	1168.20
Total DUV	3943.20	79,196.10	1180.64

Water Conservation Value (V2)

Water conservation values were estimated based on the volumes of the urban lake wetlands at constant water levels and the construction cost of storage capacity per unit area. The total water conservation value (Table 4) was found to be 9702.35 $\times$ 10^4 CNY $\times$ yr^{-1}. Natural lakes contributed 2784.63 $\times$ 10^4 CNY $\times$ yr^{-1}, semi-natural lakes contributed 6905.28 $\times$ 10^4 CNY $\times$ yr^{-1}, and artificial lakes contributed 12.44 $\times$ 10^4 CNY $\times$ yr^{-1}. While semi-natural lakes had the highest overall water conservation value, a different picture emerged when looking at the value per unit area, with natural lakes surpassing both semi-natural and artificial lakes.

Leisure Tourism Value (V3)

The leisure tourism value of the three different types of lakes was estimated using the travel cost method, supplemented with questionnaire survey data. The total leisure tourism value (Table 4) came to 73,821.00 $\times$ 10^4 CNY $\times$ yr^{-1}. The breakdown of this total shows that the semi-natural East Lake contributed the majority, 71,940.00 $\times$ 10^4 CNY $\times$ yr^{-1}, followed by the artificial Ziyang Lake, 1168.20 $\times$ 10^4 CNY $\times$ yr^{-1}, and the natural Zhangdu Lake, 712.80 $\times$ 10^4 CNY $\times$ yr^{-1}. When comparing value per unit area, however, the artificial lake surpassed both the semi-natural and natural lakes.

In sum, the direct use value per unit area of urban lake wetlands is influenced by the extent of human intervention. The more significant the artificial influence, the higher the direct use value per unit area per year, while less artificial influence corresponds to a lower direct use value.

3.1.2. Indirect Use Value (IUV)

Carbon Sequestration and Oxygen Release Value (V4)

The carbon sequestration and oxygen release values were obtained by measuring these quantities in the aquatic vegetation of each lake and then monetizing them using the shadow price method. The total value of carbon sequestration and oxygen release (Table 5) for the three urban lakes amounted to 281.33 × 10^4 CNY × yr^{-1}. Zhangdu Lake, a natural lake, had the highest value, followed by the semi-natural East Lake and then the artificial Ziyang Lake.

Table 5. Indirect use value of urban lake wetlands in Wuhan (in 10^4 CNY × yr^{-1}).

IUV Components	Zhangdu Lake	East Lake	Ziyang Lake
Carbon Sequestration and Oxygen Release Value	236.37	44.43	0.53
Transpiration Value	24,758.55	120,090.39	446.04
Water Purification Value	2.31	91.58	0.11
Flood Control Value	838.28	4047.92	7.77
Biodiversity Maintenance Value	315.71	1524.32	5.85
Research and Education Value	487.68	2354.45	9.03
Total IUV	26,638.90	128,153.09	469.33

Transpiration Value (V5)

The transpiration value, a key measure of the heat absorption function of urban lake wetlands, was calculated through monetization and transformation of the heat absorption function index. The total transpiration value (Table 5) across the three urban lake wetlands was 145,294.98 × 10^4 CNY × yr^{-1}. The semi-natural East Lake contributed the most, 120,090.39 × 10^4 CNY × yr^{-1}, followed by the natural Zhangdu Lake, 24,758.55 × 10^4 CNY × yr^{-1}, and finally, the artificial Ziyang Lake, 446.04 × 10^4 CNY × yr^{-1}.

Water Purification Value (V6)

Water purification values were estimated by analyzing three different indexes of the water purification function for each type of lake and then monetizing the results. The laboratory testing results showed that the total nitrogen in the water bodies of the three studied lakes was 3.10 mg/L, 1.02 mg/L, and 0.72 mg/L, and the total phosphorus content was 0.39 mg/L, 0.05 mg/L, and 0.14 mg/L. The total value of water purification (Table 5) for the three urban lake wetlands was 94.00 × 10^4 CNY × yr^{-1}. The semi-natural East Lake had the highest water purification value, followed by the natural Zhangdu Lake and then the artificial Ziyang Lake.

Flood Control Value (V7)

The flood control values were derived by quantifying the homogenized flood function indices of the three targeted lakes and then monetizing the results. The total economic value of the flood control function (Table 5) across the three lakes was 4893.97 × 10^4 CNY × yr^{-1}. The semi-natural East Lake contributed the most to this value, followed by the natural Zhangdu Lake and then the artificial Ziyang Lake.

Biodiversity Maintenance Value (V8)

The biodiversity maintenance values of the three lakes were obtained by following Costanza [42], and Chen and Zhang [43]. The total biodiversity maintenance value (Table 5) across the three lakes was 1845.88 × 10^4 CNY × yr^{-1}. Here, too, the semi-natural East Lake contributed the most to the total value, followed by the natural Zhangdu Lake, and lastly, the artificial Ziyang Lake.

Research and Education Value (V9)

The research and education values of the three lakes were also calculated using the result reference method. The semi-natural East Lake had the highest value, followed by the natural Zhangdu Lake and, finally, the artificial Ziyang Lake.

The comparison of indirect use value per unit area of the three urban lake wetlands showed little difference, ranging between 20 and 23 CNY $\times$ m^{-2} $\times$ yr^{-1} (Table 5). This underlines the importance of urban lake wetlands and the need to protect them to ensure their sustainable existence, which plays a key role in enhancing the urban ecological environment.

3.1.3. Non-Use Value (NUV)

Respondents' Environmental Awareness

Our survey on the importance of the non-use value (NUV) of Wuhan urban lakes and wetlands revealed that 88.2% of the respondents consider the NUV of Wuhan urban lakes as highly important, 9.7% are indifferent, and 2.1% view it as not essential. In terms of the current environmental quality of urban lakes and wetlands in Wuhan, 58.8% of the respondents believe that it is deteriorating, 25.8% opine that the overall quality remains unchanged, and 15.4% assert that it is improving.

Importance of NUV Components

When asked to evaluate the importance of the three components of the NUV of urban lake wetlands—existence value, bequest value, and option value—the respondents, on average, assigned the highest importance to existence value (2.509), followed by bequest value (2.188) and, finally, option value (1.692) (as shown in Figure 2).

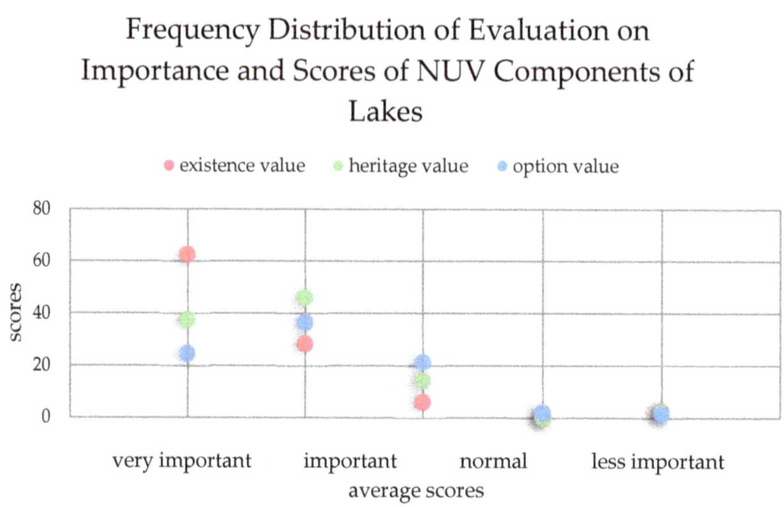

Figure 2. Frequency distribution of evaluation on importance and scores of NUV components of lakes.

Respondents' Willingness to Pay (WTP) Per Capita

The statistical analysis of the survey samples indicated that the mean WTP in Zhangdu Lake is 114.57 CNY$\times$ yr^{-1}. Applying the linear interpolation method to obtain the median value of 50%, the WTP per capita was found to be 48.4 CNY $\times$ yr^{-1}. Similarly, the mean WTP values of East Lake and Ziyang Lake were 108.71 CNY $\times$ yr^{-1} and 100.38 CNY $\times$ yr^{-1}, respectively, and the median WTP values obtained by the same method were 49.6 CNY $\times$ yr^{-1} and 44.0 CNY $\times$ yr^{-1}, respectively (Figure 3).

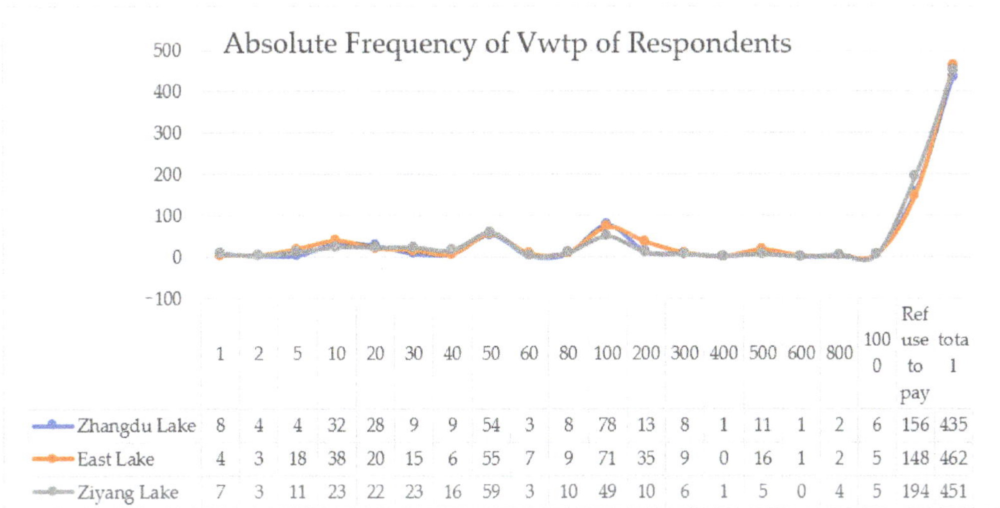

(a)

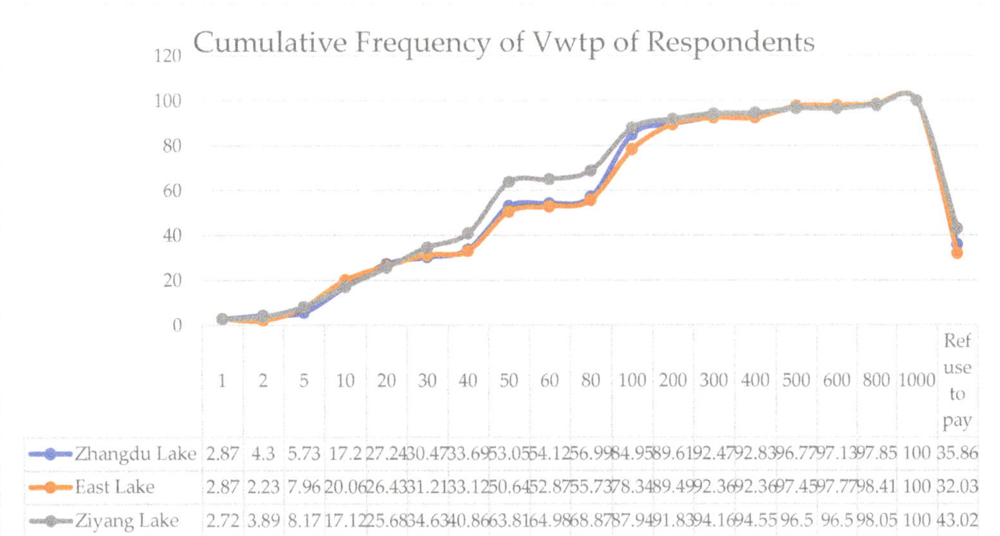

(b)

Figure 3. (**a**) Absolute frequency distribution of WTP value (Vwtp) of respondents; (**b**) cumulative frequency distribution of WTP value (Vwtp) of respondents.

Total Payment Value (Vwtp) and Its Decomposition

The median, as a positional mean, is often used to replace the arithmetic mean to calculate Vwtp to avoid large errors. According to the 2020 Wuhan Yearbook [19], the total population of Wuhan in 2018 was about 110,810,000 people. The Vwtp of each lake, as calculated with the median WTP of Zhangdu Lake, East Lake, and Ziyang Lake (70.78%, 70.09%, and 56.92%, respectively), was 2.85×10^8 CNY $\times$ yr^{-1}, 2.89×10^8 CNY $\times$ yr^{-1}, and 2.08×10^8 CNY $\times$ yr^{-1}, respectively. The respondents were willing to pay the highest proportion of existence value, which is more than 50% for each lake, followed by heritage value and option value (Figure 4).

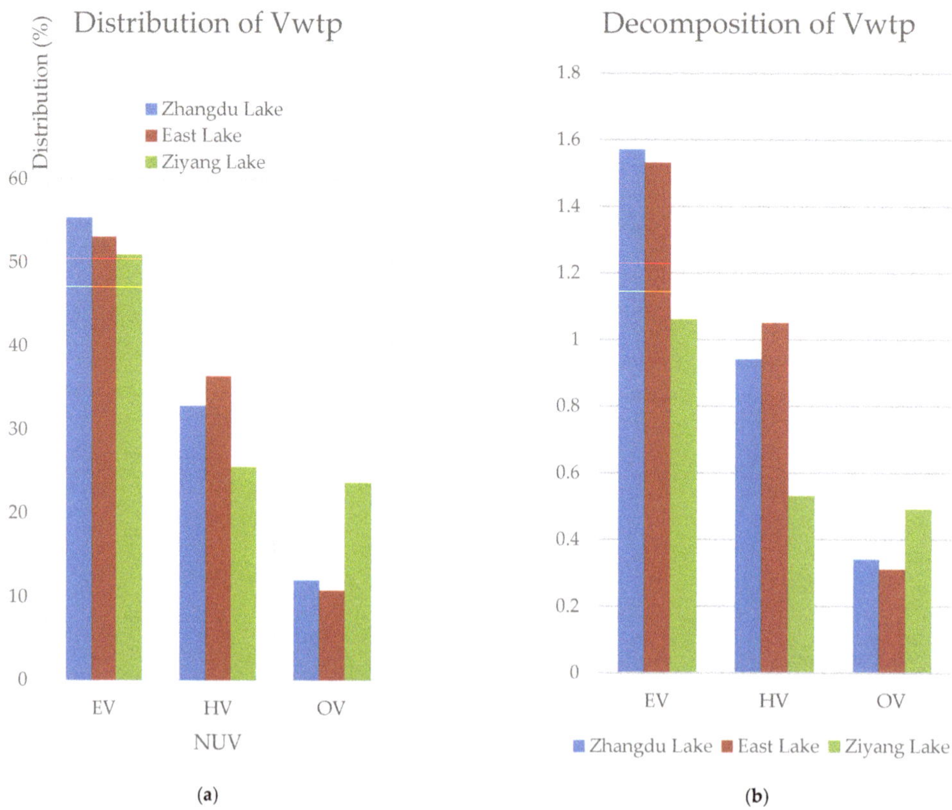

Figure 4. (**a**) Distribution of willingness to pay value (Vwtp); (**b**) decomposition of willingness to pay value (Vwtp).

Correlation Analysis between Social Characteristics of Respondents and WTP

An X^2 analysis was conducted to examine the correlation between the respondents' social characteristics and their willingness to pay (WTP). The results, summarized in Table 6, indicate that for Zhangdu Lake, the age of the respondents showed a significant influence on WTP. In contrast, factors such as gender, occupation, professional title, annual income, and education level had no notable effect.

Table 6. Correlation between WTP and social characteristics of respondents.

Social Characteristics		Gender	Age	Occupation	Professional Title	Annual Income	Education Level
	Degree of freedom (df)	1	2	8	2	4	3
X^2	Zhangdu Lake	0.238	5.234 **	7.614	1.086	3.799	2.301
	East Lake	0.899	5.070 **	6.677	1.299	5.788 *	0.093
	Ziyang Lake	0.339	0.018	4.836	2.380	13.089 **	4.023 *

Note: Significance level * $p < 0.05$, ** $p < 0.01$.

The WTP for East Lake was significantly influenced by the respondents' age and annual income, whereas gender, occupation, professional title, and education level showed no significant impact.

For Ziyang Lake, the respondents' annual income and educational level had a considerable effect on WT#$P, while gender, age, occupation, and professional title had no significant influence. Furthermore, a higher proportion of respondents of older age, higher annual income, and higher education level expressed a willingness to pay.

Correlation Analysis between Social Characteristics of Respondents and Payment Value

Further analysis demonstrated that respondents' social characteristics significantly influenced WTP values (Table 7). The Vwtp of Zhangdu Lake and East Lake were significantly influenced by the respondents' annual income, professional title, and educational level. The factors that significantly influenced the Vwtp of Ziyang Lake were the respondents' annual income and educational level. Gender, age, and occupation had no significant effect on Vwtp in the three lakes. In addition, respondents with higher annual income, higher education level, and higher professional titles expressed a higher willingness to pay.

Table 7. Correlation between Vwtp and social characteristics of respondents.

Social Characteristics		Gender	Age	Occupation	Professional Title	Annual Income	Education Level
	Degree of freedom (df)	3	6	24	6	12	9
χ^2	Zhangdu Lake	4.129	5.801	28.471	8.98 *	24.011 ***	12.929 *
	East Lake	4.211	8.063	0.029	11.729 **	36.40 **	13.578 *
	Ziyang Lake	1.852	7.719	29.530	8.371	23.541 ***	20.559 **

Note: Significance level * $p < 0.05$, ** $p < 0.01$, *** $p < 0.001$.

3.2. *Results of Water Quality Assessment*

The water quality analysis of the three targeted lakes in Wuhan (Appendix A, Tables A1–A3) reveals important insights into the respective water quality and purification values. The lakes exhibited varying nutrient levels, with Zhangdu showing the highest total nitrogen (TN) and total phosphorus (TP) concentrations, while East Lake displayed the lowest levels (Table A1). These differences in nutrient levels can significantly affect the lakes' biological productivity and overall water quality.

In terms of heavy metal concentrations, Ziyang Lake had the highest average arsenic (As) concentration (Table A2). Traces of copper (Cu) were also detectable in both Zhangdu and East Lakes. Notably, while none of the lakes exceeded any dangerous threshold levels for the metals tested (including Cu, zinc (Zn), lead (Pb), cadmium (Cd), mercury (Hg), and As), the presence of these elements underscores the importance of ongoing monitoring.

The water purification values, estimated based on nutrient levels and each lake's water volume, showed that East Lake exhibited the highest water purification value (Table A3). This is likely due to its larger water volume compared to the other two lakes. It is crucial to understand that the purification capacity of a lake is an invaluable ecosystem service, serving to naturally improve water quality.

These results shed light on the current state of these urban lakes in Wuhan, emphasizing the need for continued monitoring and management to maintain and improve their water quality and ecological health.

4. Discussion

Our findings have several implications for wetland management strategies and policy. The high non-use value (NUV) and willingness to pay (WTP) suggest that policies aimed at preserving and improving these wetlands would receive strong public support. Furthermore, this high non-use value points towards the importance of these wetlands in providing long-term, inherent benefits, such as biodiversity maintenance, carbon sequestration, and flood control. Therefore, management strategies should focus on the preservation of these inherent qualities of the wetlands. Finally, the issues related to pollution highlighted in our results suggest a need for policies and regulations that control and reduce pollution sources in and around these wetlands.

Another important finding of our study is the impact of current management issues on the ecological service function values and ecological compensation. Our research highlights that the lack of a well-established lake management system, fragmented approach to management, insufficient funding, and failure to involve beneficiary households have all contributed to the degradation of the wetlands. This degradation, in turn, reduces their ecological service function values, as the services they provide, such as water quality purification, flood control, and biodiversity maintenance, are diminished. As for ecological compensation, the aforementioned management issues make it difficult to maintain the wetlands and restore them if degraded, thus increasing the cost of ecological compensation. Therefore, improving the management of these wetlands would not only increase their ecological service function values but also reduce the cost of ecological compensation.

The water quality analysis revealed notable differences in the nutrient levels and heavy metal concentrations among the three lakes, emphasizing the need for ongoing monitoring and management. The higher concentrations of total nitrogen (TN) and total phosphorus (TP) in Zhangdu Lake and the detection of copper (Cu) in both Zhangdu and East lakes underline the importance of regulating sources of these pollutants. On the other hand, the higher arsenic (As) concentration in Ziyang Lake calls for a more thorough investigation into potential pollution sources.

The results of IUV provide valuable insights into the public's perception of the non-use value of urban lakes and wetlands in Wuhan. The considerable proportion of the non-use value, when compared to the direct and indirect use value, indicates a strong societal recognition of the inherent and long-term benefits of these wetlands, even when they are not directly used.

Interestingly, a high level of willingness to pay was observed among the respondents. This suggests strong public support for the preservation and improvement of these wetlands. However, it is also evident that there is a lack of practical understanding and concern about the current status and issues of urban lake wetlands. This calls for more extensive awareness campaigns to foster informed participation in conservation efforts.

A notable point of discussion is the role of social characteristics in shaping individuals' willingness to pay. The survey results suggest that higher income levels, advanced education, and professional titles are positively associated with a higher willingness to pay. This could potentially guide targeted educational and awareness initiatives.

The higher WTP per capita for East Lake, despite respondents' dissatisfaction with its development and degree of interference, is intriguing. It could be attributed to the lake's status as a national scenic area, which possibly raises its perceived value. Similarly, the relatively high payment values for the more natural lake, Zhangdu Lake, could be attributed to a preference for less-disturbed ecosystems.

However, it is also important to note that the size of the lakes did not significantly influence their Vwtp. This suggests that the perceived value of the lakes is not just about their size but other factors, such as their ecological health, accessibility, and amenity value. This finding underscores the importance of comprehensive lake management strategies that consider more than just the physical size of the water body.

The correlation between the paid value distribution and the importance evaluation of non-use value components of lakes underscores the consistency in respondents' perceptions of these values. This consistency could be leveraged in future communication and awareness initiatives to emphasize the importance of preserving these ecosystems for their inherent non-use values.

5. Conclusions and Recommendations

5.1. Conclusions

This study, grounded in the theories and methods of eco-economics and resource economics, evaluated the wetland resources' various functions, including resource, environment, and social functions. The water quality analysis of the three lakes in Wuhan revealed differences in nutrient levels and heavy metal concentrations, indicating the need for regular monitoring and targeted management strategies. The assessment of the value of ecosystem services also highlighted the significant role these lakes play in water purification. The findings also underscored the indirect use value (IUV) of wetlands as the most considerable, accounting for 43.47% of the total ecosystem service function value. This emphasizes the need to enhance the wetlands' ability to improve the ecological environment. Conversely, the direct use value (DUV) was the least at 23.61%, indicating that the wetlands' primary role is not merely human service.

The non-use value (NUV) of the wetlands, higher than both DUV and IUV and constituting 32.92% of the total value, demonstrated a willingness among people to invest in preserving wetlands for their environmental benefits. However, the absence of an efficient lake management system and the burden placed solely on industries have led to public pollution. This study underlines the need to consider the ecological and environmental benefits of urban lake wetlands in their preservation and management decisions. The results suggest that less human disturbance results in higher water conservation value in the wetlands.

5.2. Recommendations

Our findings underscore the need for several key strategies to ensure the sustained health and value of urban lake wetlands in Wuhan city. Chief among these is the continued monitoring and management of water quality. The variations in nutrient levels and heavy metal concentrations among the lakes highlight the necessity of regular monitoring and swift, targeted intervention when needed.

Moreover, public education and engagement are instrumental in preserving these natural resources. It is essential to foster an informed public that appreciates the importance of these wetlands and understands the role they can play in their conservation.

Regulations must be strengthened and enforced to control and prevent pollution in and around the wetlands. The introduction of stricter regulations for potential pollution sources, such as industrial activities, is of paramount importance.

Efforts to enhance the ecosystem services provided by these wetlands, such as increasing biodiversity and improving water purification capabilities, should be prioritized.

Finally, the insights derived from this study can inform policy decisions. Recognizing the economic value of these wetlands and society's willingness to contribute to their preservation can help shape policies that prioritize and secure their conservation. The integration of this knowledge into policymaking can foster an environment that supports the sustainable management of these critical urban lake wetlands.

Author Contributions: Conceptualization, J.D. and R.P.S.; Data curation, J.D.; Formal analysis, J.D.; Investigation, J.D.; Methodology, J.D. and R.P.S.; Software, J.D.; Supervision, R.P.S.; Validation, R.P.S.; Writing—original draft, J.D.; Writing—review and editing, R.P.S., V.N., T.P.L.N. and A.R. All authors have read and agreed to the published version of the manuscript.

Funding: This research was funded by the Science and Technology Project of the Hubei Geological Bureau (Grant No. KJ2023-30).

Data Availability Statement: The data presented in this study are available on request from the corresponding author.

Acknowledgments: The authors would like to thank the Institute of Geology and Mineral Resources Development of the Sixth Geological Brigade of Hubei Geological Bureau for assisting in the survey and sampling for this project and the laboratory of the Sixth Geological Brigade of Hubei Geological Bureau for assisting in the water quality testing for this project.

Conflicts of Interest: The authors declare no conflict of interest.

Appendix A

Table A1. Total nitrogen (TN) and total phosphorous (TP) in the lakes of Wuhan.

Targeted Lakes	Mean TN Value (mg/L)	Mean TP Value (mg/L)
Zhangdu	3.10	0.39
East	1.02	0.05
Ziyang	0.72	0.14

Table A2. Average heavy metal concentrations in the lakes of Wuhan.

Targeted Lakes	Cu (mg/L)	Zn (mg/L)	Pb (mg/L)	Cd (mg/L)	As (μg/L)	Hg (μg/L)
Zhangdu	0.00125	0.004475	0.025	0.0005	2.0377685	0
East	0	0.0047	0.017	0	1.134244	0
Ziyang	0.001	0.0045	0.027	0	4.99678667	0.303

Table A3. Water purification values of the lakes of Wuhan.

Targeted Lakes	Ptp	Ptn	Water Volume (10^{10} L)	Total Value (10^4 CNY yr^{-1})
Zhangdu	2.5	1.5	0.410667	2.31
East	2.5	1.5	55.335347	91.58
Ziyang	2.5	1.5	0.076923	0.11

Table A4. Indicators and methods for calculating direct use value (DUV = V1 + V2 + V3) of ecosystem services.

Value Component	Definition and Formula	Method	Data Type
V1: Material Production Value	V1 = ΣSi × Yi × P	Market price method [27]	Primary data: field survey in 2022—aquaculture product name, aquaculture area Si (m^2), yield per unit area Yi (kg); secondary data: aquaculture company—market price P (CNY)
V2: Water Conservation Value	V2 = G = ΣXi (i = 1, 2, ... n)	Shadow engineering method [44]	Primary data: field survey in 2022—the cost of storage ΣXi (CNY); secondary data: Wuhan Water Bureau—the average water level (m), the highest water level (m), the lowest water level (m), capacity per unit storage volume (m^3)
V3: Leisure Tourism Value	V3 = ΣVi	Travel cost method [45]	Primary data: questionnaire in 2022—tourism cost per capita Σvi (CNY), e.g., accommodation, food, transportation, entrance fee, etc.; secondary data: Wuhan Yearbook 2020—number of tourists (2×10^4/yr), average daily wage per person (CNY 260), opportunity wage cost (CNY 260/3 = CNY 86)

Table A5. Indirect use value (IUV = V4 + V5 + V6 + V7 + V8 + V9) indicators and methods of ecosystem services values.

Value Component (DUV)	Definition and Formula	Method	Data Type
V4: Carbon Sequestration Value	V4 = Q × P	Shadow price method [46]	Primary data: field survey in 2022—vegetation types (emergent, floating, and submerged), fresh weight (g), constant weight (g) (baked in an oven at 85 °C); secondary data: chemical equation of photosynthesis: $CO_2 + H_2O \rightarrow C_6H_{12}O_6 + O_2 \rightarrow$ polysaccharide—1 g dry matter concentrate 1.63 g CO_2 and 1.2 g O_2; price CO_2 absorption (afforestation cost: 250 CNY × t − 1); price O_2 release (afforestation cost: 352.93 CNY × t − 1) [47]

Table A5. *Cont.*

Value Component (DUV)	Definition and Formula	Method	Data Type
V5: Transpiration Value	$V5 = \Sigma (CMi\Delta Ti/3.6 \times 106) \times P$	Efficiency alternative method [44]	Secondary data: Wuhan Environmental Protection Bureau and environmental monitoring station [48]—average temperature in Jul., Aug., and Sep. (30.0 °C, 29. 5 °C, 28. 5 °C), the evaporation volume (m3), the market price of residential electricity (CNY 0.57/kilowatt hour)
V6: Water Purification Value	$V6 = M \times P$	Pollution cost method [49]	Primary data: lab analysis in 2022—total phosphorus (mg/L), total nitrogen (mg/L), heavy metal operation (Cu, Zn, Cd, Pb, and Hg); secondary data: HJ 636-2012—total nitrogen ¥1.5 /kg, GB 11893-89—total phosphorus ¥2.5 /kg [50]
V7: Homogenized Flood Value	$V7 = G = \Sigma Xi$ (i = 1,2,...n)	Shad-ow engineering method [51]	Primary data: field survey in 2022—the product of the amount of flood storage in flood season (kg); secondary data: the cost of storage capacity per unit storage volume (yuan), construction cost ¥0.67/ m^3 [52]
V8: Biodiversity Maintenance Value	Result reference method [53]		Secondary data: Costanza et al. [42], the wetland sanctuary value (\$304 hm^{-2} yr^{-1}) [18], biodiversi-ty value per unit area of global wetland ecosystem species
V9: Scientific Research & Education Value	Result reference method [54]		Secondary data: average value of China's wetland ecosystem per unit area, -the average value of Costanza et al.'s assessment [42], average value of cultural function of systematic scientific research (¥3,897. 80 /hm^2) [55]

Table A6. Non-use value (NUV = V10 + V11 + V12) indicators and methods of ecosystem services values.

Value Component (NUV)	Definition and Formula	Method	Data Type
V10: Existence Value	$Vwtp = WTP \times N$	Contingent value method [56]	Primary data: questionnaire in 2022—the Vwtp per person (CNY), the actual population willing to pay in the surveyed place (CNY)
V11: Heritage Value	$Vwtp = WTP \times N$	Contingent value method [56]	Primary data: questionnaire in 2022—the Vwtp per person (CNY), the actual population willing to pay in the surveyed place (CNY)
V12: Election Value	$Vwtp = WTP \times N$	Contingent value method [56]	Primary data: questionnaire in 2022—the Vwtp per person (CNY), the actual population willing to pay in the surveyed place (CNY)

References

1. Goodwin, E.-J. Convention on Wetlands of International Importance, especially as Waterfowl Habitat 1971 (Ramsar). Multilateral Environmental Treaties, 2017. Available online: https://en.unesco.org/about-us/legal-affairs/convention-wetlands-international-importance-especially-waterfowl-habitat (accessed on 18 February 2022).
2. Vepraskas, M.-J. Wetlands, Third Edition. *J. Environ. Qual.* **2002**, *31*, 979–1016. [CrossRef]
3. Marconi, P.; Arengo, F.; Clark, A. The arid Andean plateau waterscapes and the lithium triangle: Flamingos as flagships for conservation of high-altitude wetlands under pressure from mining development. *Wetl. Ecol. Manag.* **2022**, *30*, 827–852. [CrossRef]
4. Huang, Q.; Fang, C. Progresses of Wetland Ecosystem Field Monitoring in Nanji Wetland National Nature Reserve, South of Poyang Lake. *Wetl. Sci.* **2017**, *6*, 199–210.
5. Daily, G.-C. *Nature's Services: Societal Dependence on Natural Ecosystems*; Island Press: Washington, DC, USA, 1997.

6. United Nations. *Ecosystems and Human Well-Being: A Framework for Assessment*; United Nations: New York, NY, USA, 2003.
7. Marsh, G.P. *Man and Nature*; Charles Scribner: New York, NY, USA, 1864.
8. Warde, P.; Sörlin, S. Road to survival. *Soil Sci.* **1948**, *67*, 75.
9. King, R.-T. *Wildlife and Man*; New York State Conservation Department: New York, NY, USA, 1966.
10. Holdren, J.-P.; Ehrlich, P.-R. Human population and the global environment. *Am. Sci.* **1974**, *62*, 282.
11. Westman, W.E. How much are nature's service worth? *Science* **1977**, *197*, 960–964. [CrossRef]
12. Grand-Clement, E.; Anderson, K.; Smith, D.; Luscombe, D.; Gatis, N.; Ross, M.; Brazier, R.E. Evaluating ecosystem good and services after restoration of marginal upland peatlands in South-West England. *J. Appl. Ecol.* **2013**, *50*, 324–334. [CrossRef]
13. Jenkins, W.A.; Murray, B.C.; Kramer, R.A.; Faulkner, S.P. Valuing ecosystem services from wetlands restoration in the Mississippi Alluvial Valley. *Ecol. Econ.* **2010**, *69*, 1051–1061. [CrossRef]
14. Grossmann, M. Economic value of the nutrient retention function of restored floodplain wetlands in the Elbe River basin. *Ecol. Econ.* **2012**, *83*, 108–117. [CrossRef]
15. Han, M.; Cui, J.; Hao, Z.; Wang, Y.; Wang, R. Eco-compensation of wetlands in Yellow River Delta of Shandong Province, China. *Chin. Geogr. Sci.* **2012**, *22*, 119–126. [CrossRef]
16. Wong, C.P.; Jiang, B.; Kinzig, A.P.; Lee, K.N.; Ouyang, Z. Linking ecosystem characteristics to final ecosystem services for public policy. *Ecol. Lett.* **2015**, *18*, 108–118. [CrossRef]
17. Cuperus, R.; Canters, K.J.; Piepers, A.A. Ecological compensation of the impacts of a road. Preliminary method for the A50 road link (Eindhoven-Oss, The Netherlands). *Ecol. Eng.* **1996**, *7*, 327–349. [CrossRef]
18. Allen, A.-O.; Feddema, J.-J. Wetland loss and substitution by the Section 404 permit program in southern California, USA. *Environ. Manag.* **1996**, *20*, 263–274. [CrossRef]
19. Wang, W. *Wuhan Statistic Yearbook*; China Statistics Press: Wuhan, China, 2020.
20. Johst, K.; Drechsler, M.; Wätzold, F. An ecological-economic modelling procedure to design compensation payments for the efficient spatio-temporal allocation of species protection measures. *Ecol. Econ.* **2002**, *41*, 37–49. [CrossRef]
21. Pan, W.; Tang, T.; Deng, H.; Cai, Q. Evaluation of Lake Ecosystem Services: A Case Study of Baoan Lake in Hubei Province. *Chin. J. Appl. Ecol.* **2012**, *13*, 1315–1318.
22. Kleijn, D.; Berendse, F.; Smit, R.; Gilissen, N.; Smit, J.; Brak, B.; Groeneveld, R. Ecological Effectiveness of Agri-Environment Schemes in Different Agricultural Landscapes in The Netherlands. *Conserv. Biol.* **2004**, *18*, 775–786. [CrossRef]
23. Ning, L.M.; Wang, X.L. A Study on the Change of Ecosystem Service Value in Wuhan City Based on RS and GIS. *Ecol. Environ.* **2006**, *15*, 637–640.
24. Li, X.; Ren, L.; Liu, Y.; Craft, C.; Mander, Ü.; Yang, S. Evaluation of Wetland Ecosystem Service Value in the Yangtze River Estuary. *Resour. Environ. Yangtze Basin* **2013**, *12*, 411–416.
25. Yamane, T. *Statistics, An Introductory Analysis*, 2nd ed.; Harper and Row: New York, NY, USA, 1967.
26. Zhang, Z.Q. Evaluation of Restoration Value of Ecosystem Services in Zhangye City of the Heihe River Basin, China. *J. Nat. Resour.* **2014**, *3*, 230–238.
27. Ren, Z.Y.; Li, J. Evaluation of the Ecological Function of Vegetation in the Mountains, Southern Shanxi. *Acta Geogr. Sin.* **2013**, *58*, 503–511.
28. Rubec, C.D.A.; Hanson, A.R. Wetland mitigation and compensation: Canadian experience. *Wetl. Ecol. Manag.* **2009**, *17*, 3–14. [CrossRef]
29. Jenkins, M.; Scherr, S.J.; Inbar, M. Markets for Biodiversity Services: Potential Roles and Challenges. *Environ. Sci. Policy Sustain. Dev.* **2004**, *46*, 32–42. [CrossRef]
30. Sun, G.Y. Progress and Prospect of Wetland Science in China. *Prog. Geogr.* **2010**, *15*, 666–672.
31. Sheng, G.; Chen, H.; Ferretti-Gallon, K.; Innes, J.L.; Wang, Z.; Zhang, Y.; Wang, G. *Water Ecological Environment Value and Protection Countermeasures*; Tsinghua University Press: Beijing, China, 2014.
32. MoEP. *Water Quality—Determination of Orthophosphate and Total Phosphorus—Continuous Flow Analysis (CFA) and Ammonium Molybdate (HJ 670—2013)*; National Environmental Protection Standards of the People's Republic of China. Ministry of Environmental Protection: Beijing, China, 2013.
33. Jianwei, L. Study on the influence factors of ammonium molybdate spectrophotometric determination of total phosphorus in water. *China Resour. Compr. Util.* **2022**, *4*, 20–22.
34. Yunqi, H. Discussion on determination of total phosphorus in water by ammonium molybdate spectrophotometry. *Biol. Chem. Ind.* **2019**, *5*, 98–101. (In Chinese)
35. Ruli, G. Analysis of factors affecting the determination of total nitrogen in water by potassium persulfate method. *Water Saf.* **2023**, *5*, 85–87.
36. MoEP. *Water Quality–Determination of 65 Elements. Inductively Coupled Plasma-Mass Spectrometry (HJ 700—2014)*; National Environmental Protection Standards of the People's Republic of China. Ministry of Environmental Protection: Beijing, China, 2014.
37. Lin, Y.; Zhiwei, W. Determination of Molybdenum, cobalt, beryllium, antimony, nickel, barium, vanadium, titanium and thallium in drinking water by ICP-MS. *Spectr. Lab.* **2011**, *28*, 1852–1855.
38. Shuo, L.; Yuan, L. Determination of trace elements in Yellow River Water by inductively coupled plasma mass spectrometry. *Gansu Sci. Technol.* **2015**, *31*, 32–34. (In Chinese)

39. MoEP. *Water Quality–Determination of Mercury, Arsenic, Selenium, Bismuth and Antimony—Atomic Fluorescence Method (HJ 694–2014)*; National Environmental Protection Standards of the People's Republic of China. Ministry of Environmental Protection: Beijing, China, 2014.
40. Li, Y. Effect of potassium borohydride concentration on determination of arsenic, selenium and mercury in water by atomic fluorescence method. *Environ. Sci. Technol.* **2020**, *26*, 50–53.
41. Xiaoqing, Z.; Weiwei, Z. Experimental study on determination of Arsenic, mercury and Selenium in Water by hydrogenation-atomic fluorescence method. *J. Yunnan Norm. Univ.* **2013**, *33*, 54–58. (In Chinese)
42. Costanza, R.; d'Arge, R.; De Groot, R.; Farber, S.; Grasso, M.; Hannon, B.; Limburg, K.; Naeem, S.; O'neill, R.V.; Paruelo, J.; et al. The Value of the World's Ecosystem Services and Natural Capital. *Nature* **1997**, *387*, 253–260. [CrossRef]
43. Chen, Z.; Zhang, X. The Value of ecosystem Benefits in China. *Chin. Sci. Bull.* **2000**, *45*, 17–22.
44. Yang, L.; Kong, F.L.; Xi, M.; Li, Y.; Wang, S. Evaluation of Environmental Economic Value of Constructed Wetland and Case Study. *Res. Environ. Sci.* **2015**, *18*, 70–73.
45. Zhang, X.F.; Liu, Z.W.; Xie, Y.F.; Chen, G.R. Evaluation of Water Ecosystem Services in the Process of Urban Lake Degradation Evolution Assessment: A Case Study of the Fairy Lake in Zhaoqing, China. *Acta Ecol. Sin.* **2017**, *27*, 2349–2354.
46. Duan, X.N.; Wang, X.K.; Ouyang, Z.Y. Ecosystem Services and Value Assessment of Wuliangsuhai Wetland. *Resour. Sci.* **2015**, *27*, 110–115.
47. Sierra-Maldonado, R.; Russman, E. On the efficiency of environmental service payments: A forest conservation assessment in the Osa Peninsula, Costa Rica. *Ecol. Econ.* **2006**, *59*, 131–141. [CrossRef]
48. Wuhan Environmental Monitoring Center. Wuhan River Water Quality. *Wuhan Hydrol. Bull.* **2020**, *1*, 45–53.
49. Dietschi, S.; Holderegger, R.; Schmidt, S.G.; Linder, P. Agri-environment incentive payments and plant species richness under different management intensities in mountain meadows of Switzerland. *Acta Oecologica* **2007**, *31*, 216–222. [CrossRef]
50. Bendor, T.; Brozovic, N. Determinants of Spatial and Temporal Patterns in Compensatory Wetland Mitigation. *Environ. Manag.* **2007**, *40*, 349–364. [CrossRef]
51. Xue, D.Y.; Bao, H.S.; Li, W.H. Indirect Economic Value Assessment of Forest Ecosystem in Changbai Mountain Nature Reserve. *China Environ. Sci.* **2019**, *19*, 247–252.
52. Liu, M.C.; Li, D.Q.; Wen, Y.M. Functional Evaluation of CO_2 Emission from Vegetation Fixation in the Three-river Headwaters Region. *Ecol. Environ.* **2016**, *15*, 594–597.
53. Costanza, R.; Norton, B.G.; Haskell, B.D. *Ecosystem Health: New Goals for Environmental Management*; Island Press: Washington, DC, USA, 1992.
54. Ramsar Convention Bureau. *The Ramsar Convention: Convention on Wetlands of International Importance Especially Waterfowl Habitat; Pamphlet*; Rapport, D.J., Ed.; Dimensions of Ecosystem Health. Ecosystem Health Model; Ramsar Convention Bureau: Gland, Switzerland, 1971; pp. 34–40.
55. Cuperus, R.; Canters, K.J.; Haes, H.A.U.D.; Friedman, D.S. Guidelines for ecological compensation associated with highways. *Biol. Conserv.* **1999**, *90*, 41–51. [CrossRef]
56. He, G.Z.; Lv, Y.L.; Wang, X.L. Evaluation of Comprehensive Management of Wuli Lake in Wuxi City by Conditional Valuation Method. *Acta Ecol. Sin.* **2017**, *27*, 270–280.

economies

Article

One, Two, Three: How Many Green Patents Start Bringing Financial Benefits for Small, Medium and Large Firms?

Anastasia Semenova [1,*], Konstantin Semenov [2] and Maxim Storchevoy [3]

1 Department of Economics, St. Petersburg School of Economics and Management, HSE University (National Research University Higher School of Economics), St. Petersburg 194100, Russia
2 School of Cyber-Physical Systems and Control, Institute of Computer Science and Technology, Peter the Great St. Petersburg Polytechnic University, St. Petersburg 195251, Russia; semenov.k.k@iit.icc.spbstu.ru
3 Department of Management, St. Petersburg School of Economics and Management, HSE University (National Research University Higher School of Economics), St. Petersburg 194100, Russia; mstorchevoy@hse.ru
* Correspondence: assemenova@hse.ru

Abstract: This paper studies the relationship between environmental innovations and firms' financial performance from the perspective of environmental activism intensity. We explore how the number of green patents affects the financial performance of small, medium, and large firms and whether the growing number of green patents positively affects firms' financial performance. We employed a panel data sample of 1136 green innovative and 2395 non-green innovative firms from the USA and Europe and compared their financial results. The results show that small firms benefit financially only in the second year after the first green patent implementation. Medium-sized firms enjoy improved financial performance in the first two years after the implementation of one or two green patents; however, the third green patent does not anyhow improve the financial performance. Large firms gain financial benefits every year after issuing green patents regardless of the patents' quantity. Generally, the increase in financial performance is moderate in the first year, reaches the maximum in the second year, and becomes statistically insignificant in the third year after the last green patent's implementation.

Keywords: environmental innovations; financial performance; green patents; sustainable development

1. Introduction

Growing environmental issues have become a global challenge, engaging the corporate sector as much as other sectors of society. Growing competitive pressure, technological capabilities, and environmental policies (Horbach 2008) force firms worldwide to focus on improvements in their environmental performance. In this context, environmental innovations (EIs) and their impact on firms' performance have attracted much attention from both scholars and practitioners trying to find the right balance between economic growth and firms' environmental commitment.

Numerous studies explored the influence of EI on firms' financial performance, and many of them found a positive impact (Yi et al. 2021; Liao et al. 2021; Leyva-de la Hiz et al. 2018; Vasileiou et al. 2022). However, the theoretical framework of this influence has some space for improvement. In some cases, EI brings economic value to the firm by reducing costs or improving the product, and the firms are willing to implement them without any regulations (Horbach 2008; Andries and Stephan 2019). In other cases, the firms are required to comply with minimal environmental regulation and view further environmental investing as unprofitable (Duque-Grisales et al. 2020). In both situations, EI may bring economic benefits to the firm, and in the case of patent protection, EI may help to obtain a sustainable competitive advantage (see the resource-based view (RBV) approach to strategic management (Hart 1995)). However, there is some amount of EI that is economically optimal for the firm, and exceeding this amount will lead to a decrease in financial results. As noted by some authors (Przychodzen et al. 2020), overconcentration on EI relative to other environmental activism has negative effects on the financial performance of pioneering firms.

Therefore, it is important to collect empirical evidence to check the existence of an optimal level of EI and its factors. In this paper, we refer to EI as organizational implementations and changes focusing on the environment, with implications for firms' products, manufacturing processes, and marketing, with different degrees of novelty (Angelo et al. 2012). We may note that firms' engagement in EI differs tremendously even within the same industry. One important firm-level characteristic that heavily affects its environmental activism is the size of the firm. Small and large firms substantially differ in their strategic motivation, management practices, and access to resources (Lin et al. 2019), resulting in huge differences in their environmental activities and obtained financial results (Andries and Stephan 2019).

Firms of various sizes have different cost efficiency when it comes to EI development. Small firms tend to focus on one product or service, so it may be easier for them to develop EI related to one narrow area rather than a range of EIs of various environmental or technological natures. Larger firms are usually more diversified, both horizontally and vertically, so they have higher chances of being involved in various technological and environmental issues where conceptual ideas for EI may be naturally generated and developed. Moreover, large firms have more potential to enjoy slack resources. George (2005) defines slack as "potentially utilizable resources that can be diverted or redeployed for the achievement of organizational goals". In fact, reversed causality may take place here; slack resources proved to be positively associated with a higher number of innovations, which, in their turn, may result in better competitive advantages (Aragon-Correa and Hiz 2015) and improved financial performance. Overall, we may expect that resource constraints and narrow focus make a lower number of EI more cost-effective for small firms than for larger firms.

The mechanisms that drive the financial benefits as a result of EI introduction may also be different for firms of various sizes. Both small and large companies can enjoy cost reduction and marketing benefits caused by EI introduction, but larger firms are more likely to engage in long-lasting large-scale environmental projects entailing resource saving and increasing customer loyalty from a diversity of EI. Small firms usually focus on a particular technology or closely follow their target audience's needs to achieve financial benefits, which apparently limits the scope of the EI, so they may benefit only from EI related to a particular area. Obtaining EI outside their scope is related to additional expenditures that might never be made up and just weaken the firm (Andries and Stephan 2019).

The time period when EI may start bringing first financial benefits may also be different for small and large firms. Previous studies demonstrate that in many cases, EI introduction may have a delayed financial effect (Marín-Vinuesa et al. 2018; Qing et al. 2022). Some studies explore how different EI types impact the short-term and long-term financial performance of the firms (Ghisetti and Rennings 2014). However, the question of the timeframe when firms of various sizes may start getting financial benefits from their first, second, and consequent EI is still underexplored. This timeframe is directly associated with the mechanisms of how small and large firms may improve financial performance as a result of EI introduction. Small firms are more flexible, reactive, and, in general, have less bureaucracy (Pinget et al. 2015). Therefore, they might quicker introduce EI to solve small ad-hoc problems, i.e., emission reduction or environmentally friendly package development. Still, being a complex process, the EI introduction may imply a series of technological, production, and marketing changes (Vasileiou et al. 2022), and considering the resource constraints typical for small firms, the financial benefits would hardly outweigh the expenditures very soon after the EI introduction. Moreover, this natural time lag between the EI introduction and the first potential financial benefits provides an additional argument for small firms to introduce EI more seldom unless there is severe institutional or legal pressure. Large firms have the resources to implement successful innovations on a larger scale and get profits through better access to large markets (Aguilar-Fernández and Otegi-Olaso 2018). Their large-scale environmental projects may take quite a long time; however, having a larger influence and access to media, larger companies may start yielding the first financial results long before the whole process of EI introduction is finished, i.e., by increasing their customers' loyalty.

Unlike previous studies, this paper aims to explore the relationship between EI and financial performance from the perspective of the intensity of environmental activism of small, medium, and large firms measured by green patents. Following (Ghisetti and Rennings 2014), we contribute to the debate on when it pays to be green by differentiating how the number of EI affects the financial performance of small, medium, and large firms and, specifically, analyze the impact of the first, the second, and the third EI.

We employed a panel data sample of 1136 green innovative firms (firms that were involved in tangible green developments (Marcus and Fremeth 2009) and obtained green patents in 2013–2018) and 2395 non-green innovative firms (without green patents in 2010–2018) from the USA and Europe. We matched each green innovative firm with non-green innovative firms of the same size and checked that the matched firms had no green patents during the periods of analysis. Following Aguilera-Caracuel and Ortiz-de-Mandojana (2013), we directly compared the financial results of green innovative firms and the matched firms. Our results indicate that different numbers of green patents bring different financial results for small, medium, and large firms. We found that, on average, small firms benefit financially only in the second year after the first green patent introduction. The second and further green patents lead to a decrease in financial performance compared to non-green innovative firms. Medium-sized firms usually enjoy improved financial performance in the first two years after the introduction of one or two green patents; however, the third green patent does not anyhow improve the financial performance. Large firms gain financial benefits every year after issuing green patents regardless of the patents' quantity.

One of the major contributions of this paper is that it jointly considers the firm size, intensity of environmental activism, and the time period after the latest EI introduction as potential moderators of the relationship between EI and financial performance. Our study expands the current understanding of how EI affects firms' financial performance by providing an important context of this relationship that can be further adapted by policymakers and business owners in making decisions and planning their environmental activities.

2. Materials and Methods

Following previous studies (Leyva-de la Hiz et al. 2018; Hoang et al. 2020; Aguilera-Caracuel and Ortiz-de-Mandojana 2013), we used green patents to measure firms' environmental activism. Green patents as an EI measure have several advantages: measurability, comparability among firms from different markets, and broad usage in the relevant literature (Hoang et al. 2020). Since patents contain standardized information about new technologies and ideas, they are considered to be the most important indicator of innovation (Frietsch and Grupp 2006). Moreover, it is important that innovations are protected legally and by other mechanisms as it affects the extent to which firms profit from innovations (Frietsch and Grupp 2006).

To measure financial performance, we used return on assets (ROA) as the ratio between the annual profits and the firm's total assets value. ROA is widely used for its reliability in the literature dedicated to EI and serves as a proxy for the firm's financial performance (Duque-Grisales et al. 2020; Aguilera-Caracuel and Ortiz-de-Mandojana 2013; Wu 2017; Xie et al. 2019).

We collected data on the total assets and net income of European (87.9%) and North American (12.1%) firms in 2014–2019 from the Orbis database (sample A contains 3531 firms). Table 1 contains information on the distribution of these firms by size for each country. We used the widespread Alpha-2 codes (ISO 3166) for countries' designation.

Table 1. The firms' sample distribution by countries.

Firms' Size	Country												
	AT	*BA*	*BE*	*BG*	*BM*	*CA*	*CH*	*CZ*	*DE*	*DK*	*EE*	*ES*	*FI*
Small	0/0	0/1	6/47	5/8	0/0	0/0	0/1	11/11	9/22	0/0	0/3	66/39	11/12
Middle	0/0	0/1	3/27	0/0	0/1	0/0	0/0	23/11	73/58	0/0	0/0	18/28	6/13
Large	6/11	0/0	8/38	0/0	3/29	1/5	2/33	8/7	126/233	3/6	1/0	14/47	19/24
Total	17	2	129	13	33	6	36	71	521	9	4	212	85
Firms' Size	Country												
	FR	*GB*	*GR*	*HR*	*HU*	*IE*	*IT*	*LI*	*LT*	*LU*	*LV*	*NL*	*NO*
Small	7/17	19/65	2/1	2/1	0/17	2/1	141/100	0/0	1/0	0/0	2/4	28/26	0/0
Middle	8/10	54/82	0/0	3/0	0/14	1/2	60/42	0/0	2/3	0/0	0/0	9/17	0/0
Large	19/48	75/126	0/2	0/0	0/4	4/9	51/89	1/0	2/1	0/7	0/0	12/52	2/6
Total	109	421	5	6	35	19	483	1	9	7	6	144	8
Firms' Size	Country												
	PL	*PT*	*RO*	*RS*	*RU*	*SE*	*SI*	*SK*	*TR*	*UA*	*US*		
Small	1/3	2/6	14/9	0/0	11/192	32/56	1/16	6/2	0/0	0/4	2/3		
Middle	0/4	2/4	2/2	0/2	3/104	8/21	1/4	3/3	0/0	0/5	2/6		
Large	0/10	1/2	2/2	0/0	2/128	14/55	0/2	1/0	2/4	0/5	95/279		
Total	18	17	31	2	440	186	24	15	6	14	387		

Most of the studied firms in the sample are located in Belgium (129), Germany (521), Spain (212), France (109), Great Britain (421), Italy (483), Netherlands (144), Russia (440), Sweden (186), the USA (387). In Table 1, each cell contains two numbers divided by a slash; the first is the number of green firms, and the second is the number of firms without green patents. If a firm's size changed within the reviewed time period, then the median size was taken as its size along this time interval (Table 1).

We used the WIPO database to collect the sample of green innovative firms with at least one green patent published in 2013–2018 and non-green innovative firms with no green patents published in 2010–2018 (sample B contains 16,674 patents records). We specifically focused on the patents that were included in the WIPO category "24—environmental technology". In this paper, we use the patents' priority date as the date closest to the moment of environmental innovation introduction. In contrast, the publishing date cannot serve as a reliable indicator of the timing of ecological innovation; it can take more than a year from filing to the publication of a patent. For the patents in sample A in the article, the median time between publishing and priority date was equal to (1.79 $\pm$ 0.07) years.

We formed the intersection of these two samples as follows. First, public authorities/states/governments and firms with no recent financial data were excluded from sample B. After that, we matched the firms' names from sample A and sample B to find reliable intersections. To do that, based on (He et al. 2018; Raffo and Lhuillery 2009; Lee et al. 1999; Graham et al. 2018), we developed the following advanced procedure.

1. We turned the firms' names in samples A and B into the uppercase text;
2. We removed "the" and non-text characters such as spaces, dots, commas, dots with commas, hyphens, single and double quotes, parentheses, etc. We replaced the symbol "&" with "AND";
3. We replaced all widely used abbreviations with the common reduced form: "INCORPORATED" to "INC", "OPENED JOINT STOCK COMPANY" to "OSJC", "PUBLIC LIMITED COMPANY" to "PLC", "LIMITED LIABILITY COMPANY" to "LLC", "LIMITED" to "LTD", "COMPANY" to "CO". The order of these replacements mattered;
4. We deleted all abbreviations from the endings of firms' names in samples A and B ("PUBL", "AB", "AG", "LTD", "INC", "CO", "LLC", "PLC", "OJSC") and, after that, from their beginnings ("OJSC", "LLC", "PLC");
5. For the non-trivial errors and variations in firms' names (such as typos), we calculated the Levenstein distance between all pairs of text strings representing firms' names included in sample A and sample B. For each pair, we normalized the obtained value following (Lambert et al. 1999). We interpreted the firms' names as coincidental if the normalized distance was lower than the pre-determined threshold based on the statistic of typos occurrence in economic databases.

The threshold value was determined as follows. After completing steps 1–5, the average length of a firm's name was calculated, and it equaled 16.1 symbols. Following (Pollock and Zamora 1983) and (Damerau 1964), the typical errors in firms' names do not exceed 1–2 symbols (1-symbol typos were detected in 90–96% of cases in full-scale studies and at least 80% in small-scale studies). So, we set the threshold as 0.1.

The described procedure helped to increase the depth of the dataset's match and obtain the final accurate dataset. As a result, we obtained a sample of 1136 firms with green patents and 2395 firms without green patents.

The ROA distribution values turned out to be significantly different from the normal one (Shapiro-Francia and Shapiro–Wilk tests both reject the hypothesis on normality, p-value < 10^{-30}). To detect the outliers, we used three approaches: whiskers boxplot (Tukey 1977) to find the values lying outside the interval J covering 99% of the population; the sequential Grubbs' test based on estimating z-score (Adikaram et al. 2015) for the p-value equal to 0.01.

We also applied the DBSCAN algorithm (Ester et al. 1996) to detect the outliers (the parameters of the algorithm were equal to $n_{\min} = 0.5 \cdot n$ and $\varepsilon = 1.25 \cdot \sigma$, where $n_{\min}$ is the minimum number of points in the clustered data that are allowed to form a single cluster; n is the number of points in data to be processed; ε is the magnitude of the neighborhood of a point to connect points in one cluster; σ is a standard deviation of the population to be analyzed for outliers, such parameters correspond to the normally distributed population to detecting 1% of the data to be outliers).

If an ROA value was recognized as an outlier by at least two of the mentioned approaches, then the corresponding data were removed from further calculations.

We followed the OECD classification to identify the firms' size (Enterprises by Business Size 2021): 250 or more employers in large firms, 50 to 249 employees in medium-sized, and less than 50 in small firms.

In our study, we adopted the principles of PSM methodology and partly followed the leading papers in this research area that applied the direct comparison (Aguilera-Caracuel and Ortiz-de-Mandojana 2013; Przychodzen and Przychodzen 2015; Tugores and García 2015; Forsman 2013; Fernando et al. 2010).

We divided the initial sample into subgroups for direct comparison of green firms and firms without green patents under the same or close to other conditions. We considered the firm's size and the number of green patents introduced over the last few years as control variables. Firms' age in the sample was greater than at least 6 years.

The research question was mathematically formulated as follows. Let $P_k(n)$ be the probability that the obtainment of n green patents on average leads to an improved firm's financial performance measured by ROA, compared to the financial performance of the firms without green patents after k years since the latest patent's introduction. Let $n_{\max}(k)$ be the number of patents that maximizes $P_k(n)$: $n_{\max}(k) = \underset{n}{\operatorname{argmax}} P_k(n)$. We studied these values for small, medium, and large firms correspondingly.

To study the $n_{\max}$ value, we applied the following procedures based on the direct statistical comparison of ROA for 2014–2018 for the two groups of firms: green innovative firms and the matched ones.

1. Paired one-sided t-test for the null hypothesis stating that the ROA of the firms with n green patents on average is larger than the median ROA of firms without green patents after k years since the latest green patent has been introduced:

$$E\left[\mathrm{ROA}_k^{\mathrm{eco}}(n)\right] > med\left[\mathrm{ROA}_k^{\mathrm{conv}}\right],$$

where E is the mathematical expectation, *med* is the median value, and ROA^{eco} and ROA^{conv} are ROA values for green innovative firms and non-green innovative firms correspondingly;

2. Paired one-sided sign test for the null hypothesis stating that the ROA of the firms with n green patents is more often larger than the median ROA of the firms without green patents after k years since the latest green patent has been introduced:

$$med[\mathrm{ROA}_k^{\mathrm{eco}}(n) - med[\mathrm{ROA}_k^{\mathrm{conv}}]] > 0;$$

3. The probability estimate

$$q = \mathrm{Prob}(\mathrm{ROA}_k^{\mathrm{eco}}(n) > med[\mathrm{ROA}_k^{\mathrm{conv}}])$$

that the ROA of the firms with n green patents will be larger than the median ROA of the firms without green patents after k years since the latest patent was issued.

To test the robustness of the results, we processed the original ROA data as is and, after outliers' removal, applied the procedure mentioned above (values lying outside the interval covering 99% of the population) and found no statistically significant differences.

The descriptive statistics—in-sample distributions for firms-green patent holders and firms without green patents—are presented in Tables 2 and 3. Tables contain the values of the following sample statistical characteristics: mean value (mean), median value (med), standard deviation (s), and bounds of the confidence interval for confidence probability equal to 90% (formed by 5% and 95% quantiles). Again, if a firm's size changed within the reviewed time period, then the median size was taken as its size along this time interval in Tables 2 and 3.

Table 2. Descriptive statistics for green firms' total sample (1136 firms).

	Small Firms (381)				Medium Firms (281)				Large Firms (474)			
	Mean	Med	s	90% CI	Mean	Med	s	90% CI	Mean	Med	s	90% CI
Patents number	7.92	4	16.17	[1.00, 26.5]	9.63	4	17.47	[1.00, 35.9]	51.64	7	193.4	[1.00, 206.8]
ROA in 2012	0.061	0.036	0.253	[−0.39, 0.81]	0.048	0.048	0.087	[−0.17, 0.22]	0.045	0.043	0.061	[−0.09, 0.17]
ROA in 2013	0.044	0.028	0.297	[−0.56, 0.66]	0.048	0.053	0.083	[−0.17, 0.22]	0.052	0.046	0.058	[−0.06, 0.17]
ROA in 2014	0.021	0.023	0.271	[−0.61, 0.48]	0.046	0.041	0.081	[−0.13, 0.20]	0.052	0.048	0.054	[−0.06, 0.16]
ROA in 2015	0.049	0.024	0.339	[−0.46, 0.58]	0.048	0.039	0.084	[−0.13, 0.24]	0.046	0.044	0.057	[−0.08, 0.16]
ROA in 2016	0.022	0.027	0.300	[−0.57, 0.53]	0.050	0.040	0.085	[−0.12, 0.24]	0.047	0.045	0.060	[−0.08, 0.17]
ROA in 2017	0.024	0.018	0.206	[−0.47, 0.44]	0.047	0.038	0.081	[−0.12, 0.21]	0.047	0.042	0.054	[−0.07, 0.16]
ROA in 2018	0.020	0.013	0.235	[−0.42, 0.43]	0.048	0.039	0.088	[−0.12, 0.23]	0.046	0.044	0.060	[−0.08, 0.17]
ROA in 2019	0.059	0.013	0.327	[−0.41, 0.98]	0.071	0.037	0.136	[−0.14, 0.47]	0.067	0.048	0.120	[−0.10, 0.43]

Table 3. Descriptive statistics for the total sample of the firms without green patents (2395 firms).

	Small Firms (620)				Medium Firms (459)				Large Firms (1316)			
	Mean	Med	s	90% CI	Mean	Med	s	90% CI	Mean	Med	s	90% CI
Patents number	0	0	0	[0, 0]	0	0	0	[0, 0]	0	0	0	[0, 0]
ROA in 2012	0.045	0.027	0.174	[−0.34, 0.43]	0.035	0.030	0.095	[−0.21, 0.21]	0.035	0.030	0.052	[−0.08, 0.15]
ROA in 2013	0.036	0.025	0.177	[−0.41, 0.39]	0.037	0.030	0.102	[−0.18, 0.25]	0.039	0.035	0.051	[−0.07, 0.15]
ROA in 2014	0.053	0.030	0.154	[−0.27, 0.44]	0.038	0.032	0.094	[−0.19, 0.24]	0.041	0.036	0.047	[−0.04, 0.15]
ROA in 2015	0.043	0.026	0.143	[−0.27, 0.38]	0.038	0.034	0.098	[−0.20, 0.22]	0.038	0.033	0.049	[−0.06, 0.14]

Table 3. *Cont.*

	Small Firms (620)				Medium Firms (459)				Large Firms (1316)			
	Mean	Med	s	90% CI	Mean	Med	s	90% CI	Mean	Med	s	90% CI
ROA in 2016	0.043	0.024	0.157	[−0.32, 0.41]	0.038	0.034	0.095	[−0.20, 0.22]	0.037	0.032	0.051	[−0.07, 0.15]
ROA in 2017	0.033	0.023	0.176	[−0.38, 0.45]	0.044	0.037	0.098	[−0.21, 0.23]	0.040	0.033	0.046	[−0.04, 0.14]
ROA in 2018	0.040	0.023	0.146	[−0.31, 0.41]	0.033	0.035	0.107	[−0.23, 0.20]	0.040	0.033	0.048	[−0.05, 0.15]
ROA in 2019	0.080	0.025	0.223	[−0.22, 0.78]	0.053	0.035	0.120	[−0.19, 0.34]	0.045	0.038	0.059	[−0.07, 0.19]

3. Results

We performed the abovementioned tests for $n = 1, 2$, and 3 green patents for $k = 1, 2, 3$, and 4 years. The firms' sizes were taken exactly for the specified year (both for the green innovative firms and non-green innovative firms).

The results are presented in Tables 4–7, which contain corresponding p-values. Table 8 shows the samples' sizes in the form (x/y), where x is the sample size for firms with green patents and y—for firms without green patents. We colored p-values less than 0.05 with green, from 0.05 to 0.15 with yellow, and greater than 0.95 with red.

Table 4. The tests' p-values for $k = 1$.

Firms' Size	1 Green Patent			2 Green Patents			3 Green Patents		
	t-Test	Signtest	$1 - q$	t-Test	Signtest	$1 - q$	t-Test	Signtest	$1 - q$
Small	0.169	0.072	0.332	0.406	0.935	0.966	0.581	0.738	0.738
Medium	0.024	0.025	0.025	0.095	0.443	0.443	0.902	0.837	0.837
Large	0.204	0.008	0.008	0.197	0.076	0.076	0.004	0.010	0.010

p-values less than 0.05 are colored with green, from 0.05 to 0.15 with yellow, and greater than 0.95 with red.

Table 5. The tests' p-values for $k = 2$.

Firms' Size	1 Green Patent			2 Green Patents			3 Green Patents		
	t-Test	Signtest	$1 - q$	t-Test	Signtest	$1 - q$	t-Test	signtest	$1 - q$
Small	0.042	0.133	0.676	0.560	0.993	0.997	0.137	0.500	0.500
Middle	0.008	0.090	0.090	0.066	0.122	0.122	0.263	0.581	0.581
Large	0.236	0.124	0.124	0.037	0.111	0.111	0.030	0.039	0.039

p-values less than 0.05 are colored with green, from 0.05 to 0.15 with yellow, and greater than 0.95 with red.

Table 6. The tests' p-values for $k = 3$.

Firms' Size	1 Green Patent			2 Green Patents			3 Green Patents		
	t-Test	Signtest	$1 - q$	t-Test	Signtest	$1 - q$	t-Test	Signtest	$1 - q$
Small	0.151	0.806	0.962	0.802	0.996	0.999	0.558	0.941	0.941
Middle	0.062	0.304	0.304	0.739	0.364	0.364	0.061	0.584	0.584
Large	0.044	0.022	0.022	0.052	0.040	0.061	0.412	0.598	0.598

p-values less than 0.05 are colored with green, from 0.05 to 0.15 with yellow, and greater than 0.95 with red.

Table 7. The tests' p-values for $k = 4$.

Firms' Size	1 Green Patent			2 Green Patents			3 Green Patents		
	t-Test	Signtest	$1 - q$	t-Test	Signtest	$1 - q$	t-Test	Signtest	$1 - q$
Small	0.589	0.656	0.967	0.156	0.785	0.948	0.714	0.887	0.927
Middle	0.122	0.623	0.806	0.606	0.708	0.708	0.574	0.788	0.788
Large	0.525	0.252	0.252	0.510	0.788	0.788	0.627	0.500	0.500

p-values less than 0.05 are colored with green, from 0.05 to 0.15 with yellow, and greater than 0.95 with red.

Table 8. Samples' sizes for tests which results are in Tables 4–7.

Firms' Size	$k = 1$			$k = 2$			$k = 3$			$k = 4$		
	$n = 1$	$n = 2$	$n = 3$	$n = 1$	$n = 2$	$n = 3$	$n = 1$	$n = 2$	$n = 3$	$n = 1$	$n = 2$	$n = 3$
Small	17/508	17/508	74/489	13/485	67/485	17/485	12/502	54/502	15/502	6/474	40/474	11/474
Middle	17/428	17/428	48/415	14/426	36/426	24/426	15/426	33/426	22/426	10/411	30/411	14/411
Large	30/1236	30/1236	49/1268	27/1261	43/1261	21/1261	25/1244	33/1244	16/1244	20/1212	25/1212	13/1212

4. Discussion

Environmental problems aggravation has spurred extensive research on EI and their financial implications. An explicit understanding of how EI impacts firms' financial performance is essential as it draws more firms to EI activities and helps to design policies to ensure accurate support. However, the empirical evidence of EI financial implications shows divergent results, so this paper aimed to provide new insights into this research by adding firms' size and the length of the period after EI introduction to the context of the relationship between environmental innovations.

The design of the study also made it possible to see the difference in financial results of green firms with different numbers of green patents. For linear-case regression—a popular instrument of analysis—this dependency is usually described as $\Delta FP_t \sim \alpha \cdot GP_{t-1}$, where FP is the financial performance of a green firm, ΔFP_t is its increase for the year t, GP is the number of green patents, α is a sensitivity coefficient independent from any control and dependent variables. Such models can only show that with every subsequent green patent, firms experience permanent additional financial performance increase (if $\alpha > 0$) or decrease (if $\alpha < 0$), and no trend changing is allowed.

In our study, we constructed the dependency in a more generalized and complicated way: $\Delta FP_t \sim f(Size_{t-1}, GP_{t-1})$, where f is a non-constant function of the green patents number and firms' size. Using this approach, we separately investigated the influence of each patent instead of the averaged impact estimated in regression models and found that the larger green patents number is not generally associated with larger financial benefits for small firms, medium-sized firms, and even large enterprises. This original approach is one of the contributions of our study.

To ensure the sensitivity of our analysis, we tested the financial performance (ROA) of green firms against the median value of the ROA of firms without green patents. Using the median value is better than using mean values because of greater robustness and no sensitivity to outliers. This allowed us to indicate whether the financial results of green firms that met certain conditions exceeded the financial results of half of the firms without green patents in the selected countries. The application of this approach helped to mitigate the statistical uncertainty related to the green firms' sample sizes and shed light on the peculiarities of the relationship between the number of green patents and averaged financial benefits.

Our research demonstrated that outliers' removal did not substantially affect the final results of the study. We also see that for the firms without green patents, the difference between the mean and median of the obtained ROA values distributions does not exceed 0.15 (small firms), 0.25 (medium-sized firms), and 0.16 (large firms) of the population standard deviation for any year from the considered time period. This is much less than the statistical uncertainty caused by the relatively small size of the samples in Table 8. Therefore, our results are determined by the relation between financial performance increase and green patents number but not by the biases from the applied statistical approach.

The results presented in Table 4 were calculated for the first year after the latest green patent introduction. For small firms, the financial effect of EI introduction is quite ambiguous, even in the case of only one green patent introduction. Apparently, substantial costs on green R&D usually come as a burden for a small firm, and the expenses on green patent implementation that follow green patent obtaining are not covered by the financial benefits gained next year. Additionally, in contrast to large firms, small firms may have higher marginal costs as they cannot enjoy economies of scale (Khanna 2001). This result corroborates (Andries and Stephan 2019) in the part that "smaller firms reap fewer financial benefits from environmental innovation than their larger counterparts".

For medium-sized firms, the first green patent, on average, leads to an increased ROA in the first year after the green patent introduction compared to medium-sized firms without green patents. So, for them, the generated financial benefits exceed the costs of green R&D. Larger small and medium firms (SMEs) are more financially stable, enjoy better access to human resources (Pinget et al. 2015), and usually have a wider market presence. In the case of two green patents, we observe ambiguous results; there are no explicit financial implications in the first year after the second green patent's introduction. When three green patents are introduced, medium-sized firms experience decreased financial performance due to the high development and introduction costs.

Finally, large firms enjoy improved financial performance regardless of the number of implemented green patents. Every subsequent green patent does not make any substantial difference when it comes to large firms' financial performance.

Table 5 shows the results calculated for the second year after the latest green patent introduction. Small firms benefit financially only when one green patent is introduced. The second and further green patent introduction is associated with a decrease in financial performance due to financial constraints. For medium-sized firms, the introduction of up to two green patents in the second year after the latest green patent's introduction leads to improved financial performance, yet the introduction of the third green patent most likely causes a decrease in financial performance. The situation with large firms does not change: a green patent introduction is associated with improved financial performance regardless of the green patent quantity.

These results comply with (Bermúdez-Edo et al. 2016), who suggested that a greater number of patented EI does not increase firm performance. They are also in line with (Przychodzen et al. 2020), who showed that an increasing share of green patents in the total number of patents negatively influences current financial performance, and there are positive lagged financial performance gains from being active in the field of green innovations (one- and two-year lagged financial performance effects).

Given the financial constraints and lack of resources, small firms are more likely to obtain focused EI, i.e., innovations that invest in well-known technologies rather than unrelated technologies that require much experimentation—diversified EI (Leyva-de la Hiz et al. 2018). If SMEs obtain more than two green patents, their EI naturally becomes more diversified, provided that SMEs usually focus on one narrow area. This requires more resources for R&D but, at the same time, offers less predictability of financial success. In this case, SMEs become more financially vulnerable, and, obviously, only large firms can afford experimentations with diversified EI, not worrying about their financial performance.

Table 6 shows the results calculated for the third year after the latest green patent introduction. In the third year, the firms' financial implications of EI start shrinking compared to firms with no green patents. The results for the fourth year presented in Table 7 support this tendency. Financial implications in absolute numbers are still higher for the firms with green patents, but the probability of gaining more is close to the probability of gaining less. This fact might serve as a check for our study design; we measured the number of green patents only within the specified time period and controlled that no patents were implemented in the previous three years. If any green patents were implemented before this time, their financial implications could not affect our results.

The presented results show the following averaged estimates of $n_{\max}(k)$. For small firms, $n_{\max}(1) = 0$; $n_{\max}(2) = 1$; $n_{\max}(3) = 0$. For medium-sized firms, $n_{\max}(1) = 1$; $n_{\max}(2) = 1 \div 2$; and $n_{\max}(3) = 0 \div 1$. For large firms, $n_{\max}(1) \leq 3$; $n_{\max}(2) \leq 3$; and $n_{\max}(3) \leq 2$.

These findings empirically extend the conclusions of (Appio et al. 2019), who demonstrated that "profitability increases until a certain level of patent portfolio diversity reaches a maximum and then decreases". They are also in line with (Leyva-de la Hiz and Bolívar-Ramos 2022) in the part that overemphasis on environmental innovations may be detrimental to firm performance.

The results demonstrate that the positive effect of EI on financial performance is not only lagged from the moment of EI introduction but also limited in time as in the fourth year after green patents introduction, the difference in ROA between green innovative and non-green innovative firms ceases to be attributed to green patents existence regardless of green patents quantity. Successful innovations may be "picked up" by other players in the market and implemented by them, and thereby, the positive difference in financial performance that green innovative firms enjoy may be offset as other firms also start taking advantage of the benefits brought by this EI.

The results of this study also suggest that the assessment of how EI affects firms' financial performance should consider the size of the firms together with the length of the period after the latest green patent introduction. Moreover, it is important to consider the ratio of small, medium, and large firms in the sample to avoid disproportions and, thereby, biased results. One of the following options might be followed. The first option implies composing the sample of small, medium-sized, and large firms in the same proportion to avoid prevailing the one-size group results. The second option takes into account the industry of the firms in the sample. It implies that firms' sizes in the sample should correspond to the natural distribution that takes place in a real market or industry so that the final integral result for the whole market or industry is averaged in terms of the firm's size in the right way. For example, if the sizes of the firms in the real market are distributed as 10% for large, 20% for medium-sized, and 70% for small, the sample should be composed correspondingly. The third option implies studying the effects of each size group separately.

5. Conclusions

In an attempt to disentangle the ambiguity of the empirical results of how environmental innovations affect firms' financial performance, this research studies this relationship from the perspective of firms' size, the number of EI, and the length of the period after the latest EI introduction. Firms of different sizes are different in many aspects, and we demonstrate one additional important dimension; the financial effects of the implementation of different numbers of green patents are different for large, middle, and small firms. The application of the microeconomic methodology is an important area of improvement in the field of EI research, which quite often lacks microfoundations for theoretical comprehension of the studied phenomena.

Our findings demonstrate that when assessing the effect of EI on firms' financial performance, it is crucial to consider the size of the firm together with the number of EI and the length of the period after the latest EI introduction. Moreover, our research underscores the importance of considering the ratio of small, medium-sized, and large firms in the studies' samples to avoid disproportions and, thereby, biased results.

The results indicate that, on average, for small firms, the impact of EI on financial performance is generally negative, and only the introduction of one green patent in three years can be associated with some financial benefits in the second year after this patent's introduction.

Medium-sized firms observe an increase in financial performance in the first two years after the introduction of one or two green patents. The third green patent does not anyhow improve financial performance. Large firms, on average, gain financial benefits every year after green patents' introduction, regardless of the patents' quantity.

Generally, the increase in financial performance is normally moderate in the first year after the latest green patent introduction. It reaches the maximum in the second year and becomes statistically insignificant in the third year.

5.1. Theoretical Contributions

The theoretical contributions of this paper are as follows. First, this study has provided new insights into the relationship between EI and financial performance by showing the importance of additional contextualization when assessing this relationship. Second, our research provides nuanced empirical evidence to RBV by bringing in the quantity of EI linked with the firms' size as important conditions of the positive impact of EI on firms' financial performance. Third, for the first time, this paper empirically studied how the length of the period after the latest EI introduction affects financial performance with regard to the firm size. Overall, our work expands the boundary conditions of EI research.

Additionally, this research provides several methodological contributions related to data processing in the studies in this field. First, we developed a special algorithm to connect firms' names in databases of patents and financial data. Second, the study presents the combined procedure to detect outliers. Third, we proposed to make conclusions on the results using the set of statistical tests; it is recommended to use not only measures such as average treatment effect that deals with absolute values but also additionally apply the tests based on probabilities and ratios comparison.

5.2. Practical Implications

This study offers several practical implications. First, firms can more precisely plan their environmental activities considering their firm's size and the quantity of the upcoming EI to better balance environmental activism and economic development. Second, our research vividly demonstrates the importance of a differentiated approach toward firms of various sizes in terms of support and stimulation of EI introduction. Policymakers can pay special attention to SMEs' support as, on the one hand, smaller firms are less eager to involve in environmental activities due to financial constraints, lack of resources, and extended payback period on investment in EI, and, on the other hand, they are responsible for approximately 64% of all industrial pollution (Pinget et al. 2015). A clear understanding of how and when firms of various sizes can enjoy the positive financial effect of their EI helps to accurately design appropriate targeted policy.

5.3. Limitations

Like any other study, this study has some limitations that provide opportunities for future research. First, we measured environmental activism by green patents, so the other types of firms' EI were not considered. Future research could focus on more objective measures that would reflect the effect of overall EI.

Second, considering data availability, our firms' sample was limited to European and North American firms. The inclusion of more parts of the world in the analysis would extend theoretical development and empirical design. Our research methods could be extended to an interesting cross-country comparison. Assuming that, ceteris paribus, in the countries with advanced environmental culture and regulation, green patents might bring more marginal benefits because green activities are expected from companies and brands, and it is easier to conduct them because regulation is friendly to green investments, we may expect that the inflection point where the next green patent becomes unprofitable should be located closer to a larger number of patents. However, at the same time, we may assume that in these countries, the majority of the firms are more involved in green innovations, and some additional green patents may have a limited value for the market and for the firm. These opposing assumptions create a theoretical intrigue that may be resolved based on future research.

Third, future research could add indicators of long-term corporate financial performance, such as Tobin's Q, market value (MV), and price-to-earnings ratio (P/E).

Furthermore, future research could include the environmental status of a firm in the analysis to see if, for environmental companies, the inflection point where the next green patent becomes unprofitable is located closer to a larger number of patents.

Author Contributions: Conceptualization, A.S. and K.S.; Methodology, A.S. and K.S.; Software, K.S.; Validation, A.S., K.S. and M.S.; Formal analysis, A.S. and M.S.; Investigation, A.S. and K.S.; Resources, A.S. and M.S.; Data curation, A.S. and K.S.; Writing—original draft, A.S., K.S. and M.S.; Writing—review & editing, A.S., K.S. and M.S.; Supervision, A.S. and M.S.; Project administration, A.S. and M.S.; Funding acquisition, M.S. All authors have read and agreed to the published version of the manuscript.

Funding: The research leading to these results has received funding from the Basic Research Program at the National Research University Higher School of Economics.

Informed Consent Statement: Not applicable.

Data Availability Statement: Data is unavailable due to privacy restrictions.

Conflicts of Interest: The authors declare no conflict of interest.

References

Adikaram, Kalutara Koralalage Lasantha Britto, Marwan A. Hussein, Mathias Effenberger, and Thomas Becker. 2015. Data Transformation Technique to Improve the Outlier Detection Power of Grubbs'Test for Data Expected to Follow Linear Relation. *Journal of Applied Mathematics* 2015: 708948. [CrossRef]

Aguilar-Fernández, Maria, and José Ramon Otegi-Olaso. 2018. Firm Size and the Business Model for Sustainable Innovation. *Sustainability* 10: 4785. [CrossRef]

Aguilera-Caracuel, Javier, and Natalia Ortiz-de-Mandojana. 2013. Green Innovation and Financial Performance: An Institutional Approach. *Organization & Environment* 26: 365–85. [CrossRef]

Andries, Petra, and Ute Stephan. 2019. Environmental Innovation and Firm Performance: How Firm Size and Motives Matter. *Sustainability* 11: 3585. [CrossRef]

Angelo, Dias, Jose Chiappetta Jabbour, and Simone Vasconcellos. 2012. Environmental innovation: In search of a meaning. *World Journal of Entrepreneurship, Management and Sustainable Development* 8: 113–21. [CrossRef]

Appio, Francesco P., Luigi M. De Luca, Robert Morgan, and Antonella Martini. 2019. Patent portfolio diversity and firm profitability: A question of specialization or diversification? *Journal of Business Research* 101: 255–67. [CrossRef]

Aragon-Correa, Juan Alberto, and Dante I. Leyva-de la Hiz. 2015. The Influence of Technology Differences on Corporate Environmental Patents: A Resource-Based Versus an Institutional View of Green Innovations. *Business Strategy and the Environment* 25: 421–34. [CrossRef]

Bermúdez-Edo, Maria, Nuria E. Hurtado-Torres, and Natalia Ortiz-de-Mandojana. 2016. The Influence of International Scope on the Relationship Between Patented Environmental Innovations and Firm Performance. *Business & Society* 56: 357–87. [CrossRef]

Damerau, Fred J. 1964. A technique for computer detection and correction of spelling errors. *Communications of the ACM* 7: 171–76. [CrossRef]

Duque-Grisales, Eduardo, Javier Aguilera-Caracuel, Jaime Guerrero-Villegas, and Encarnación García-Sánchez. 2020. Does green innovation affect the financial performance of Multilatinas? The moderating role of ISO 14001 and R&D investment. *Business Strategy and the Environment* 29: 3286–302. [CrossRef]

Enterprises by Business Size. 2021. OECD. Available online: https://data.oecd.org/entrepreneur/enterprises-by-business-size.htm (accessed on 14 January 2022).

Ester, Martin, Hans-Peter Kriegel, Jörg Sander, and Xiaowei Xu. 1996. A density-based algorithm for discovering clusters in large spatial databases with noise. Paper presented at the Second International Conference on Knowledge Discovery and Data Mining, Portland, OR, USA, August 2–4; pp. 226–231.

Fernando, Chitru, Mark Sharfman, and Vahap Uysal. 2010. Does Greenness Matter? Environmental Performance, Ownership Structure and Analyst Coverage. *SSRN Paper* 1571596. [CrossRef]

Forsman, Helena. 2013. Environmental innovations as a source of competitive advantage or vice versa? *Business Strategy and the Environment* 22: 306–20. [CrossRef]

Frietsch, Rainer, and Hariolf Grupp. 2006. There's a new man in town: The paradigm shift in optical technology. *Technovation* 26: 13–29. [CrossRef]

George, Gerald. 2005. Slack Resources and the Performance of Privately Held Firms. *Academy of Management Journal* 48: 661–76. [CrossRef]

Ghisetti, Claudia, and Klaus Rennings. 2014. Environmental innovations and profitability: How does it pay to be green? An empirical analysis on the German innovation survey. *Journal of Cleaner Production* 75: 106–17. [CrossRef]

Graham, Stuart J. H., Cheryl Grim, Tariqul Islam, Alan C. Marco, and Javier Miranda. 2018. Business dynamics of innovating firms: Linking US patents with administrative data on workers and firms. *Journal of Economics & Management Strategy* 27: 372–402. [CrossRef]

Hart, Stuart. 1995. A natural-resource-based view of the firm. *Academy of Management Review* 20: 986–1014. [CrossRef]

He, Zi-Lin, Tony W. Tong, Yuchen Zhang, and Wenlong He. 2018. A database linking Chinese patents to China's census firms. *Scientific Data* 5: 180042. [CrossRef]

Hoang, Thi-Hong-Van, Wojciech Przychodzen, Justyna Przychodzen, and Elysé Segbotangni. 2020. Does it pay to be green? A disaggregated analysis of U.S. firms with green patents. *Business Strategy and the Environment* 29: 1331–61. [CrossRef]

Horbach, Jens. 2008. Determinants of environmental innovation—New evidence from German panel data sources. *Research Policy* 37: 163–73. [CrossRef]

Khanna, Madhu. 2001. Non-mandatory approaches to environmental protection. *Journal of Economic Surveys* 15: 291–324. [CrossRef]

Lambert, Bruce L., Swu-Jane Lin, Kwan-Young Chang, and Sanjay K. Gandhi. 1999. Similarity as a risk factor in drug-name confusion errors: The look-alike (orthographic) and sound-alike (phonetic) model. *Medical Care* 37: 1214–25. [CrossRef] [PubMed]

Lee, Mong Li, Hongjun Lu, Tok Wang Ling, and Yee Teng Ko. 1999. Cleansing data for mining and warehousing. Paper presented at International Conference on Database and Expert Systems Applications, Florence, Italy, August 30–September 3; pp. 751–60. [CrossRef]

Leyva-de la Hiz, Dante I., and Maria Teresa Bolívar-Ramos. 2022. The inverted U relationship between green innovative activities and firms' market-based performance: The impact of firm age. *Technovation* 110: 102372. [CrossRef]
Leyva-de la Hiz, Dante I., Vera Ferron-Vilchez, and J. Alberto Aragon-Correa. 2018. Do Firms' Slack Resources Influence the Relationship Between Focused Environmental Innovations and Financial Performance? More is Not Always Better. *Journal of Business Ethics* 159: 1215–27. [CrossRef]
Liao, Zhongju, Ping Liu, and Shuchun Liu. 2021. A meta-analysis of environmental innovation and firm performance. *Journal of Environmental Planning and Management* 64: 2047–65. [CrossRef]
Lin, Woon-Leong, Jun-Hwa Cheah, MohamedAzali, Jo Ann Ho, and Nick Yip. 2019. Does firm size matter? Evidence on the impact of the green innovation strategy on corporate financial performance in the automotive sector. *Journal of Cleaner Production* 229: 974–88. [CrossRef]
Marcus, Alfred, and Adam Fremeth. 2009. Green management matters regardless. *Academy of Management Perspectives* 23: 17–26. [CrossRef]
Marín-Vinuesa, Luz María, Sabina Scarpellini, Pilar Portillo-Tarragona, and José M. Moneva. 2018. The Impact of Eco-Innovation on Performance Through the Measurement of Financial Resources and Green Patents. *Organization & Environment* 33: 108602661881910. [CrossRef]
Pinget, Amandine, Rachel Bocquet, and Caroline Mothe. 2015. Barriers to Environmental Innovation in SMEs: Empirical Evidence from French Firms. *Management* 18: 132–55. [CrossRef]
Pollock, Joseph J., and Antonio Zamora. 1983. Collection and characterization of spelling errors in scientific and scholarly text. *Journal of the American Society for Information Science* 34: 51–58. [CrossRef]
Przychodzen, Justyna, and Wojciech Przychodzen. 2015. Relationships between ecoinnovation and financial performance—Evidence from publicly traded companies in Poland and Hungary. *Journal of Cleaner Production* 90: 253–63. [CrossRef]
Przychodzen, Wojciech, Dante I. Leyva-de la Hiz, and Justyna Przychodzen. 2020. First-mover advantages in green innovation—Opportunities and threats for financial performance: A longitudinal analysis. *Corporate Social Responsibility and Environmental Management* 27: 339–57. [CrossRef]
Qing, Lingli, Dongphil Chun, Abd Alwahed Dagestani, and Peng Li. 2022. Does Proactive Green Technology Innovation Improve Financial Performance? Evidence from Listed Companies with Semiconductor Concepts Stock in China. *Sustainability* 14: 4600. [CrossRef]
Raffo, Julio, and Stéphane Lhuillery. 2009. How to play the "Names Game": Patent retrieval comparing different heuristics. *Research Policy* 38: 1617–27. [CrossRef]
Tugores, María, and Dolores García. 2015. The Impact of Innovation on Firms' Performance: An Analysis of the Hotel Sector in Majorca. *Tourism Economics* 21: 121–40. [CrossRef]
Tukey, John W. 1977. *Exploratory Data Analysis*. Reading: Addison-Wesley, 688p, ISBN 978-0201076165.
Vasileiou, Efi, Nikolaos Georgantzis, Giuseppe Attanasi, and Patrick Llerena. 2022. Green innovation and financial performance: A study on Italian firms. *Research Policy* 51: 104530. [CrossRef]
Wu, Guo-Ciang. 2017. Environmental innovation approaches and business performance: Effects of environmental regulations and resource commitment. *Innovation* 19: 407–27. [CrossRef]
Xie, Xuemei, Jiage Huo, and Hailiang Zou. 2019. Green process innovation, green product innovation, and corporate financial performance: A content analysis method. *Journal of Business Research* 101: 697–706. [CrossRef]
Yi, Yali, Saixing Zeng, Hongquan Chen, and Jonathan Jingsheng Shi. 2021. When Does It Pay to Be Good? A Meta-Analysis of the Relationship Between Green Innovation and Financial Performance. *IEEE Transactions on Engineering Management* 1–11. [CrossRef]

International Journal of
Environmental Research and Public Health

MDPI

Article

Competitor-Weighted Centrality and Small-World Clusters in Competition Networks on Firms' Innovation Ambidexterity: Evidence from the Wind Energy Industry

Runbo Zhao [1], Huiying Zhang [1,*], Marina Yue Zhang [2], Fei Qu [3,*] and Yunlong Xu [4]

1 College of Management and Economics, Tianjin University, Tianjin 300072, China
2 Australia-China Relations Institute, University of Technology Sydney, Sydney 2007, Australia
3 Business School, Guilin University of Technology, Guilin 541004, China
4 School of Civil Engineering, Tianjin University, Tianjin 300072, China
* Correspondence: hyzhang@tju.edu.cn (H.Z.); qufeiflyfly@163.com (F.Q.)

Abstract: A firm's embedding structures in a technology competition network can influence its propensity for innovation ambidexterity. Using PCT (patent cooperation treaty) patent data of wind energy companies between 2010 and 2019, we adopted social network analysis and fixed-effects panel negative binomial regression to examine the impacts of network structural features on firm innovation ambidexterity. The results show that competitor-weighted centrality contributes to a firm's propensities for both incremental and radical green innovation. In contrast, a firm's embeddedness in small-world clusters can moderate the effect of the firm's competitor-weighted centrality positively on its incremental innovation but negatively on its radical innovation. The study makes three theoretical contributions. First, it enriches the understanding of how the competition network affects innovation ambidexterity. Second, it provides new insights into the relationship between competition network structures and technology innovation strategy. Finally, it contributes to bridging the research on the social embeddedness perspective and green innovation literature. The findings of this study have important implications for enterprises in the wind energy sector regarding how competitive relationships affect green technology innovation. The study underscores the importance of considering the competitiveness of a firm's rivals and the embedded structural features when devising green innovation strategies.

Keywords: structure embedding; technology competition network; green technology innovation; ambidexterity; wind energy; innovation performance

1. Introduction

Severe environmental challenges and energy shortages have aroused widespread concerns and led to the burgeoning of studies on green or sustainable innovation in recent years (e.g., [1–3]). Green innovation, also called sustainable innovation, is innovation applied in environmental practices, energy conservation, waste reduction, pollution prevention, etc. [4]. As a critical type of green innovation, green technology innovation can help reduce the environmental burden and drive the technological upgrading of the economy [5,6] and has become the strategic focus of enterprises. The concept of ambidexterity finds a wide range of applications across various organizational contexts. There is a growing consensus that organizational ambidexterity indicates companies' ability to simultaneously explore and exploit [7]. In this paper, the ambidexterity of green technology innovation is defined as two essential enterprise strategies: Incremental green technology innovation refers to the continuous improvement that leverages existing technology, while radical green technology innovation indicates that significant progress deviates from current technology. Enterprises gain sustainability competitiveness when they constantly engage in both exploratory and exploitative activities [8]. That is, firms must not only carefully consider the pressure on current technology regimes but also capture opportunities for new technologies [9].

How competition network structure affects firm performance has attracted considerable attention in recent years [10]. Compared with the cooperation network (e.g., [11–13]), information flow in a firm's competition network occurs at a lower frequency [14], and the information is more public rather than private. Firms cannot ask for information directly from their rivals, which urges firms to spend more effort to search for and decode information on competitors' dynamics [15], especially in more tacit technology competition networks [16]. Therefore, a favorable position in a competition network is especially crucial for a firm's innovation activities. In the study of how competitive network structure affects firm performance, relevant findings include that the size of a competition network can increase the firm's product market entry [10], the competitive density and strength affect the firm's technology competitive capability differently [16], inter-organizational linkages reduce the likelihood of competitive war [17], and the position of brokerage in the domestic technology competition network can facilitate the firm to join related international strategic alliances [15]. Although these studies have extended our understanding of the competition network structure, several essential research gaps remain to be explored further:

(1) Many studies treat competitors in a competition network equally by measuring the intensity of competition by the number of competitors. We argue that the different competitive ability of the competitors deserves academic attention. In practice, one strong competitor plays a completely different role for a local firm than many ordinary competitors combined.
(2) Previous studies rarely distinguish how structural features embedded in a competition network affect the two modes of firm-level green technology innovation, exploration vs. exploitation.

To fill these gaps, this paper examines the effect of two network structures on the ambidexterity of green technical innovation from the perspective of social embeddedness. The first structural feature is competitor-weighted centrality. (1) Centrality takes into account that nodes in different network locations have different levels of importance. A firm in a more central position can gather more information and knowledge from the network [18], which helps monitor its competitors' dynamics, reduce the uncertainty of technological direction, and seize potential opportunities for technical development [10,14,19]. (2) Competition weighting includes the centrality of competitors; that is, more weight is assigned to a focal firm's centrality if its key rival is central in the competition network. However, enterprise nodes are also affected by neighboring enterprises [20]. When facing a strong competitor, the enterprise will have greater competitive pressure, prompting it to keep searching for and learning from the network [14] and improving the firm's capabilities. Thus, focusing on a firm alone without considering its rival's competitiveness can result in a cognitive bias in understanding the strategic choice of the focal firm in technology innovation. Based on the measurement method of Qi's research (2016) in cooperative networks, this paper innovatively constructs the competitor-weighted centrality, which considers both the features of both a focal firm and its competitors, measuring the firms' technical competition in the competitive network more accurately.

The second structural feature is a firm's embeddedness in small-world clusters. This paper focuses on its moderator role in the firm's green technology innovation. Firms in small-world clusters tend to have more tensive connections with each other but relatively sparse ties with firms outside the clusters [21,22]. The tensive connections in a competition network and repeated interactions can reduce the diversity of competitors [10], make the connections more interdependent [23], and thus, promote mutual trust that aligns with the rivals' interests [24]. As a result, the more a firm is embedded with specific competitors, the more sensitive it is to changes in the rival dynamics and the quicker it can respond to the changes by conducting incremental innovation. However, the capture of homogeneous information from tight connections can spur only conventional reactions [25] on existing paths [26]. Only when firms are stimulated by new external factors will they begin to expand their innovation directions [27,28]. Furthermore, being embedded in different external environments brings firms different technical pressures, motivating them to search for new directions [29]. From this stream of argument, we deduce that embeddedness in small-world clusters can have differential impacts on the effect of the competition network structure on ambidextrous strategies for green technical innovation.

Wind energy is chosen as the empirical setting for the following reasons. In the context of a low-carbon economy, wind energy plays a vital role in reducing carbon emissions to mitigate climate change [30]. Moreover, it is the most rapidly growing and the most promising renewable energy source [30,31], which has drawn great attention from governments and companies in recent years. For example, the Chinese government has promulgated a series of subsidies and tax credits to support companies involved in wind technology [16,32]. These initiatives have greatly contributed to sustainable economic development and ensured energy security. For enterprises with increasing complexity and growing demand for advanced wind energy products, there is a constant push to innovate their technologies, which enables them to keep technological advantages in the competitive market. All of these reasons made technological innovation by wind energy enterprises an important issue worthy of further examination.

In summary, we combine the social embeddedness perspective with studying how the two structures of the competition network (the competitor-weighted centrality and embedding in small-world clusters) affect the ambidexterity of a firm's green innovations. We test the propositions in the wind energy field, which is deemed a promising source of clean energy [31], using patent data from companies between 2010 and 2019. Our contributions are as follows: (1) This study enhances our understanding of how competitive structures influence the ambidexterity of a firm's green technical innovation, including radical and incremental innovation. (2) We reveal how the weighted centrality of the technical competition network impacts a firm's innovation by counting the centrality of the focal firm's competitors. (3) We extend the social embeddedness perspective by illuminating the moderator role of embedding in small-world clusters. The small world is a suitable embodiment for understanding the competitive environment in which a firm is embedded. (4) We propose several management practices for managers of wind energy enterprises facing a competitive environment for technology.

2. Theoretical Foundation

2.1. Wind Energy Studies

According to the World Energy Outlook 2021 report, annual clean energy investments will grow to USD 4 trillion by 2030 to achieve the goal of net zero emission of CO_2 [33]. Among renewable energy sources, wind energy contributed 34% of the newly installed renewable energy capacity in 2016 [34]. In 2020, wind energy even increased at the fastest rate under the context of economic downturns during COVID-19 lockdowns [33], which shows a huge potential for wind power.

Previous studies have made great contributions by analyzing the wind energy industry from different aspects (e.g., policy, institutional logic, product design, cooperation network, etc.). Shen (2019) investigated how government regulation affected stakeholders. They uncovered that the delegation of approval authority promotes the growth of regional wind power and suggest that governments carefully consider the trade-off between different levels of approval authority [32]. Yock (2016) conducted research based on the conflict argument on wind energy (i.e., natural environment in conflict with economic prosperity) from the perspective of institutional logic and the evolution of organizational fields [30]. Yang (2021) proposed a framework to improve the success rate of the radical innovation of wind power systems. They used the QFD (quality function deployment) method to discover features of radical technologies and the TRIZ (theory of innovative problem solving) method to instruct the design process of wind energy products [35]. Liu (2021) studied the cooperation network of the wind energy industry through patent analysis. They suggested

Chinese government strengthen the construction of cooperation platforms and encourage technological innovation [13].

Although these studies have made great contributions to the wind energy industry, little attention has been paid to the perspective of innovation strategy. Our research, therefore, investigates how the embedding structures of technology competition networks affect firm performance in innovation in the wind energy sector.

2.2. The Ambidexterity of Green Technology Innovation

The ambidexterity of exploration and exploitation in green technology innovation is a vital strategy for firm performance: incremental innovation, based on exploiting existing technologies, contributes to firms' current growth; while radical innovation, based on exploring breakthrough technologies, facilitates firms' long-term performance [36]. As an enterprise's resources are limited, a trade-off between these two types of innovation strategies needs to be carefully considered for firms to optimize their performance.

We follow the classification of green innovation in Cui's (2022) research and divide green technology innovation into two types: incremental and radical green technology innovation [1]. Incremental green technology innovation refers to green technology development with continuous changes based on exploiting existing technology. Radical green technology innovation indicates that technological advancements have significantly progressed by exploring new technologies away from the present ones. These two types of green technology innovation are essential for enterprises' growth. On the one hand, radical innovation can lead to valuable technical and financial improvement [37], and it also provides enterprises with considerable technical advantages, innovation competitiveness, and reputation [38]; however, it also carries intrinsic uncertainties and risks (March, 1991). On the other hand, incremental innovation is more likely to succeed [39,40], but it risks missing breakthroughs in the industry and being interrupted by newcomers (Christensen, 1997).

Patent data is an important indicator of technological innovation and is widely used to measure the technological innovativeness of firms and nations (e.g., [41,42]). Compared with other intellectual properties, patents are transferable assets with both economic and technological value, making patents better reflections of the reality of technological development [43]. We collected the green patents according to the IPC (International Patent Classification) Green Inventory, developed by experts in the World Intellectual Property Office, to reflect enterprise green technology innovation activities. In this paper, green patents are divided into two categories according to their novelty. Radical green patents are patents that contain new green technological sub-classes, and incremental green patents are patents that contain only existing green technological sub-classes [41]. The detailed measurement is introduced in Part 4.3.1. Based on these green patents, we can understand the input and output of enterprises in green technology research and development, competitive relations, and other information, all of which help provide useful suggestions to improve enterprises' green technology innovation and improve energy efficiency.

2.3. Antecedents of Firms' Green Technology Innovation

Previous studies have investigated various antecedents of a firm's green technology innovation, including the following aspects: (1) External factors that mainly influence the policy environment. Scholars have explored the effect of carbon emission trading policy on a firm's technology innovation [3], the impact of environmental regulation on technological innovation efficiency [44], how different political competition influences the enterprise's green technology innovation [45], the effect of China's R&D investment on green innovation performance [46], and how the green credit policy impacts a firm's green technology innovation [47]. (2) Internal factors that mainly influence the organizational learning process. The research includes, for example, how firms change their open innovation strategies to develop green competence-destroying technologies [48], and the exploration process of green technology innovation from a learning perspective is based on a case study [49].

These studies have contributed significantly to the green technology innovation literature. In addition, a firm's performance can also be affected by its surrounding enterprises [20]. A unique position in a technology competition network can be a lever that creates useful differentiation for a focal firm, particularly regarding its innovation, and may help explain why seemingly similar firms differ in their innovation outcomes. Therefore, how structural embedding features influence a firm's innovation strategy in a technology network merits more academic investigation.

2.4. Definition of Two Structural Features

Competitor-weighted centrality refers to the degree to which a focal enterprise competes with other firms in a technology competition network. This feature considers not only the number of competitors but also the quality of the competitors in the network. That is, the focal firm with more central competitors in the competition network is given more 'weight', and a firm with more peripheral rivals is given less 'weight', when calculating its value of competitor-weighted centrality. This is because the more central a firm's position is in the network the more easily it can attract resources in the network [50,51]. Figure 1 illustrates three common forms of network centrality. Unlike degree centrality and betweenness centrality, which treat other nodes as having equal importance, eigenvector centrality is more sophisticated [11], as it can capture the heterogeneities of other nodes. Therefore, we employ eigenvector centrality to measure the 'competitor-weighted centrality' of focal firms in this paper. Given that competitor-weighted centrality measures not only how many competitors the focal firm competes with but also the difference in the competitiveness of the competitors, we believe this measure is closer to the real-life situation of competition.

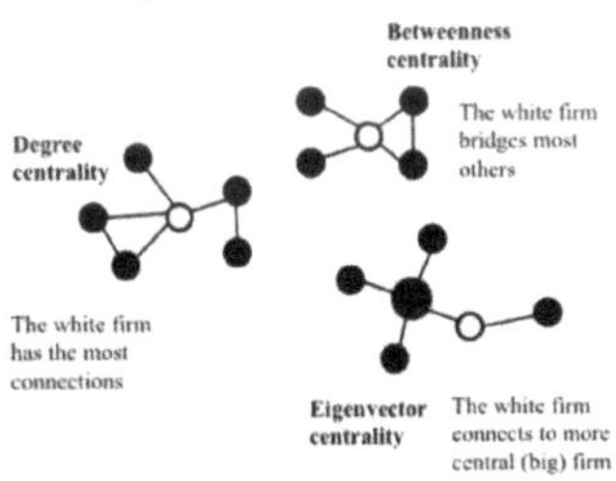

Figure 1. Three common forms of network centrality (Reprinted with permission from Ref. [11]. 2017, John Wiley & Sons—Books).

Small-world clusters can reflect the external competitive environment in which an enterprise is embedded. Figure 2 shows the distinguishing features of whether an enterprise is embedded in small-world clusters. Firms embedded in small-world clusters tend to build more connections with each other, while they have few connections outside of the clusters [12,21,22]. Since the environment in which a firm is embedded has a significant influence on its competitive capacity [16] and search strategies [52], exploring the differences in enterprises' embeddedness in small-world clusters may help explain why similar firms, at face value, differ in their innovation performance.

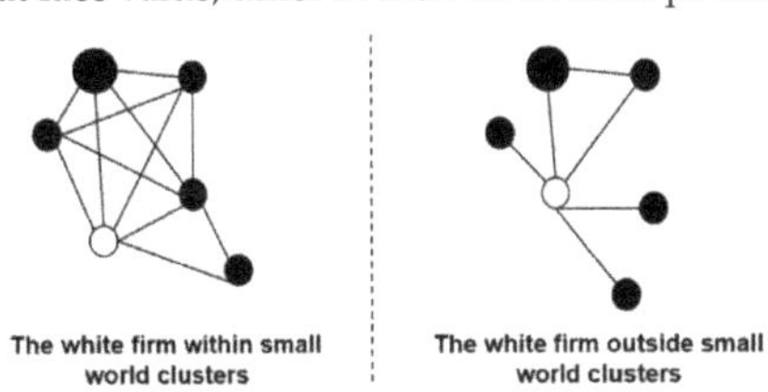

Figure 2. The embeddedness within small-world clusters.

3. Hypotheses

This section analyzes the effects of the two significant embeddedness structures of a competition network. Hypotheses 1a and 1b examine the main effect of competitor-weighted centrality on network firms' radical and incremental green innovation, respectively. Hypotheses 2a and 2b investigate the moderator effect of embeddedness in small-world clusters on network firms' ambidexterity of green technology innovation, respectively.

3.1. The Main Effect of Competitor-Weighted Centrality

First, enterprises with higher competitor-weighted centrality suffer from higher competitive pressure, which is one of the strongest drivers of corporate innovation. Generally, if a firm has higher centrality in a technology competition network, it means that it has formed a competitive relationship with many enterprises in the network and can, therefore, receive more competitive pressure. This type of pressure from the competitors can stimulate a focal firm to search for and learn from the competitors [14], which can be translated into the firm's technology strategy [53]. Hence, we argue that competitor-weighted centrality can better capture the pressure of competitive relations surrounding the focal firm and will benefit their green innovations.

Second, competitor-weighted centrality can bring diverse and timely information to the firms within a network, which is a valuable resource for firms' innovation. The competitive relationships within a competition network can be an important channel for information flows among rivals. Since information can be conveyed through the competitors' actions [54], enterprises can, thus, gather information by monitoring the dynamics of their competitors. A high frequency of interaction with various competitors can help capture the timing and diverse information, which can benefit the central firms in the network [15]. This kind of information can be helpful in competitive analysis and strategical formulation [55], and it can bring new opportunities to firms occupying such positions [54,56]. Furthermore, information is crucial to innovation, as it helps reduce the uncertainty of technological competition; prepare, in advance, for potential technological threats; and predict the future directions of technology development [14]. In addition, if a focal firm is more likely to acquire technical information that others cannot access, it will obtain the first-mover advantage, leaving its rivals in a position that is hard to catch up from in a short time [14].

Altogether, competitor-weighted centrality enables focal firms to perceive higher competitive pressure, which can motivate them to actively engage in searching and learning activities and, hence, obtain useful information. As a result, such centrality can lead to both exploratory and exploitative innovations in green technology. Hence, we posited the following hypotheses between a firm's competitor-weighted centrality in a competition network and its green technology innovation.

H1a: *A firm's competitor-weighted centrality positively affects its radical green technology innovation in a competition network.*

H1b: *A firm's competitor-weighted centrality positively affects its incremental green technology innovation in a competition network.*

3.2. Moderating Effect of Small-World Clusters

First, firms can experience different competitive pressures when they are inside and/or outside small-world clusters. The pressure that a firm experiences within the same small world tends to be homogeneous, while the pressure is heterogeneous when they are not in the same small world. Different types of competitive pressure can lead to different motivations for innovation [29]. This is because a dense competition network can reduce the diversity of competitors [10] and enable the competitors to act more interdependently and in a similar fashion [23]. Repeated interactions can facilitate knowledge exchange and resource sharing [17], leading to the promotion of mutual trust and alignment of interests among competitors [24,57]. Consequently, focal firms in small-world clusters are more sensitive to their competitors' technological dynamics and act more rapidly in response to others' actions at a lower cost. Hence, embeddedness in small-world clusters can lead focal firms to have a higher tendency toward incremental innovation. However, a homogenous technological foundation can deter the firms' radical innovation. Potential opportunities, new trends, and radical innovation, thus, become difficult if firms are embedded too deeply in specific clusters [24]. By contrast, firms outside of small-world clusters face competitors from more diverse backgrounds. The surrounding environment makes it impossible for such firms to respond as quickly as the ones inside small-world clusters, and, thus, drives firms outside of the small-world cluster to explore new innovation directions that match their environment.

Second, firms embedded in small-world clusters obtain different types of information from those outside the clusters, which can instigate different reactions (search orientation) within firms. Since information within a small world is shared in an intensified information flow [58], firms inside the small-world cluster are more likely to capture familiar information on a common technical foundation. When capturing conventional information, firms tend to respond in a conventional way and engage in less exploratory innovation [25], whereas novel or unfamiliar information has the ability to attract focal firms' attention and encourage their search for solutions away from their technological trajectories [14]. The unknown result brought by novel or unfamiliar information can spur such firms to engage in outward searching and exploratory innovation. We argue that embeddedness in a small world contributes to a firm's incremental innovation but not to its radical innovation. Over the long term, internal embeddedness can reinforce firms' cognitive barriers and lock them in knowledge isolation, resulting in a narrower scope of innovation [59]. On the contrary, information with connections to outside firms can stimulate firms to search for novel technologies [22,60], and technical knowledge from distant fields can encourage creativity and foster new ideas that are more likely to become radical innovations [25,61,62].

Based on the above arguments, enterprises embedded within or outside of small-world clusters tend to experience different competitive pressures and receive different information, which spurs their different responses and leads to them adopting innovation strategies. Hence, we posited the following hypothesis:

H2a: *Embeddedness in small-world clusters negatively moderates a firm's relationship between its competitor-weighted centrality and radical green technology innovation.*

H2b: *Embeddedness in small-world clusters positively moderates a firm's relationship between its competitor-weighted centrality and incremental green technology innovation.*

Based on the preceding discussion, our research framework is shown in Figure 3.

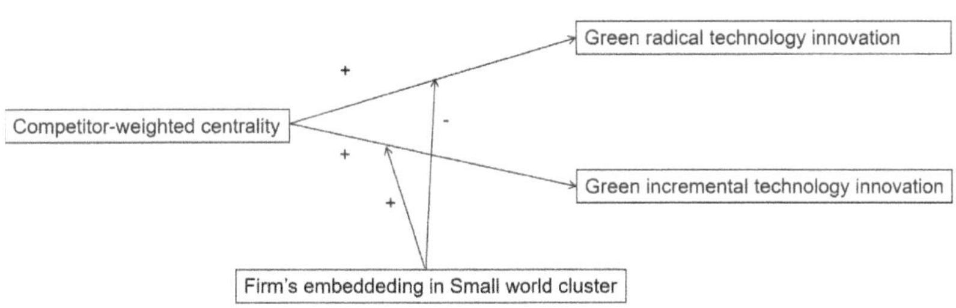

Figure 3. Research framework. "+" indicates the positive effect; "−" indicates the negative effect.

4. Methodology

4.1. Data and Collection

The research panel data mainly included patent data and enterprise attribute information. First, we followed a series of steps to ensure the accuracy of the patent data. (1) The formulation of a search string. Our collection of wind energy green patents was built on the integrated use of keywords and green technical categories. We referred to previous studies [31,63], to obtain the relevant keywords for wind energy. We learned of the green technical category of F03D for wind energy from the IPC Green Inventory, developed by experts in the World Intellectual Property Office (https://www.wipo.int/classifications/ipc/green-inventory/home (accessed on 1 January 2023)). (2) The choice of patent application type. Since a patent application is a perfect agent for a firm's innovation output [41], we employed application data from the United States Patent Office (USPTO) during 2010–2019. The reason for choosing patents applied for within the USPTO was because it is a good representation of a firm's global innovation performance [64], and USPTO contains plenty of high-quality patents within the largest renewable energy market of wind energy. (3) The selection of the patent period. The review of patent applications sometimes takes two or three years, depending on the complexity of the patent text and other factors. It was likely that not all patents applied for after 2020 would have completed their examination by 30 June 2022 (the date we collected these data). This prompted us to choose the years 2010–2019 as our observation period and not include the very recent years. In this way, we could ensure that all patents applied for during this period were collected, and thus, reflected the technical innovation performance of the company in the best possible way.

Altogether, our search string was CTB = ("wind power" OR "wind energy" OR "wind turbine" OR "wind generator" OR "wind electricity" OR "wind farm" OR "windmill" OR "energy of wind" OR "energy from wind" OR "wind rotor" OR "wind axis" OR "wind blade") AND ICR = (F03D) AND AY $\geq$ (2010) AND AY $\leq$ (2019) AND AC = (US). We conducted data collection on 30 June 2022, within the Derwent Innovation (DI) database—the largest commercial patent database in the world, composed of many authoritative organizations worldwide [65]. Eventually, 5034 patents were obtained.

Second, we collected enterprise attribute information. Since the 5034 collected patents covered more than 1000 enterprises, we found that the number of patents owned by the top 120 enterprises was 3752, accounting for about 74.5% of the total. According to the Pareto principle, the top enterprises create the vast majority of patent resources. We finally collected the attribute information of the top 120 representative enterprises. The attribute information, including the number of employees, firm age, and cash flow, was then collected from the Compustat database within Wharton Research Data Services (WRDS), the EBSCO database, and some corporate annual reports.

4.2. Construction of Firm's Technical Competition Network

Figure 4 shows the construction process of the firms' green technology competitive network. First, the IPC codes, representing the wind energy technologies in a firm's patent, were extracted. Second, enterprises with the same IPC codes indicated that they had technically competitive relationships. Thus, the greater the number of common IPC codes, the more intense their technical competition was. Third, according to the competitive relationship obtained in the previous step, we summarized their competitive relationship within the firm co-occurrence matrix. Finally, the firm competition network was constructed based on the firm co-occurrence matrix.

Further, although the firm competition network is the main focus of this paper, we would like to briefly mention another interesting idea: the technology convergence network. As Figure 4 illustrates, we also constructed a technology convergence network. Different firms with the same IPC code tended to have a competitive relationship; however, at the technology level, the higher frequency of different IPC codes within the same enterprise indicated a closer relationship between these IPC codes. That is, there was a higher convergent tendency for these different technologies. Similarly, a technology convergence network could be constructed based on the technology co-occurrence matrix [66]. We speculate that there is a specific relationship between technology convergence and firm competition networks, and their interaction will be an interesting topic in future research to provide insights into firms' technical competition activities.

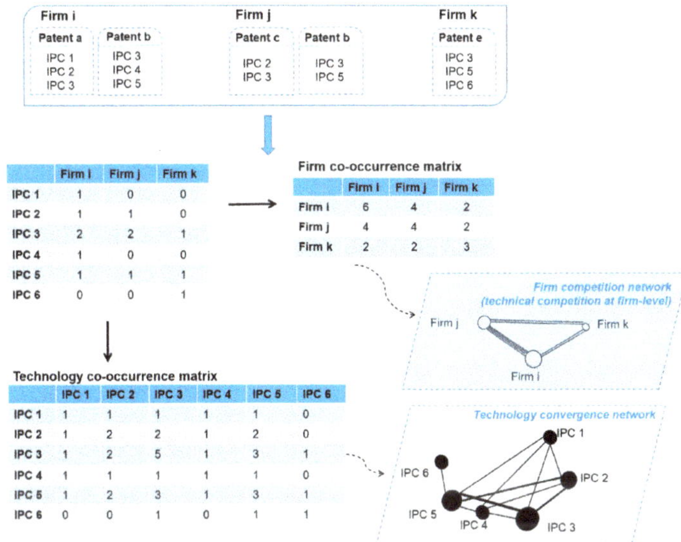

Figure 4. Construction of a firm's competition network.

Figure 5 shows an example of a competition network of wind energy companies in 2010. Each node within the network is a firm with wind energy patents. The size of a node represents its competition-weighted centrality. The link between them indicates that they had common IPC codes. Therefore, the thicker the link, the more overlapping the two firms' technologies areas were, and the stronger their competitive relationship.

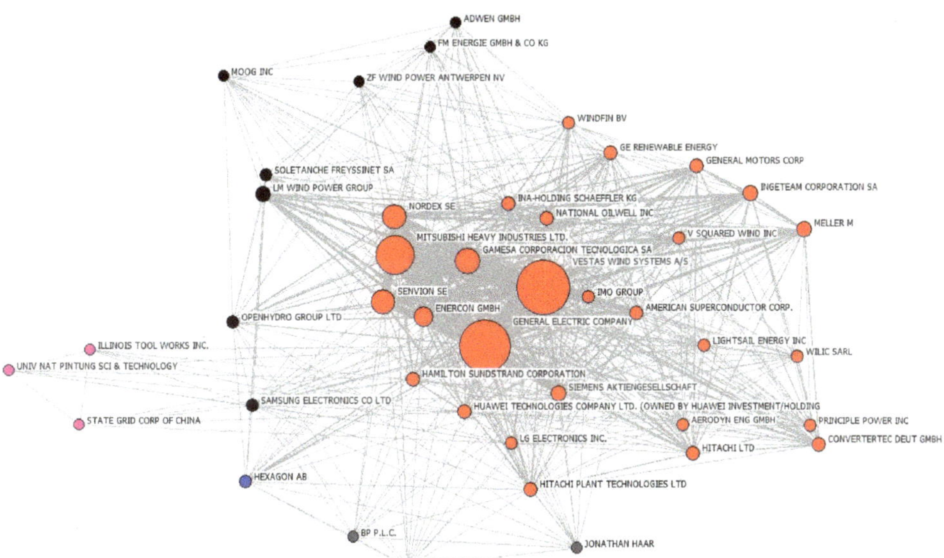

Figure 5. Firm competition network in 2010.

4.3. Measurement

4.3.1. Dependent Variables

Based on existing literature on the measurement of exploration and exploitation innovation [41], this paper measured the ambidexterity of green technical innovation in the following ways. Green technical radical innovation indicated the degree of that firm's utilization of new green technology, which was measured by the total number of patents containing new green technological classes. A new technological class referred to a technical class that a firm had not used in patents filed within the previous five years. The five-year window is appropriate for estimating innovation and is widely used [64,65,67,68]. Therefore, a patent that utilized a new technological class was categorized as a technical exploration outcome. Similarly, green technical incremental innovation captured the firm's reuse of existing green technologies, which were measured by the total number of utilizations of familiar technical classes (which had appeared at least once) within the previous five years. Patents that used only these familiar technical classes were considered incremental outcomes.

4.3.2. Independent Variables

We measured the competitor-weighted centrality by the eigenvector centrality degree. As stated earlier, eigenvector centrality is a more advanced index that can capture the differences between adjacent nodes. If its adjacent nodes also have higher centrality, then the focal node will have more influential power and capability than the others [11,69,70]. The formula is as follows:

$$\mathrm{eig}(firm_i) = \frac{1}{\lambda}\sum_{k\in M} \mathrm{eig}(firm_k) \tag{1}$$

Among these, $\mathrm{eig}(firm_i)$ indicates the eigenvector centrality of $firm_i$, $\frac{1}{\lambda}$ indicates the eigenvalue, and k indicates the direct competitor of firm i (i.e., having a direct competition with the firm i). M is the set of all direct competitors of firm i. These parameters and each node's eigenvector centrality were calculated using the Ucinet 6 tool.

4.3.3. Moderator

The average clustering coefficient is widely used to measure the degree of node clusters with neighboring nodes within a closely related group [12,71,72]. A higher value meant that a firm had dense connections with neighboring knowledge within a small-world cluster, and all connections in the network were scattered in the distribution [12,72]. The formula for the average clustering coefficient is as follows:

$$ACC_i = \frac{n_k}{n_i(n_i - 1)/2} \tag{2}$$

ACC_i represents the average clustering coefficient of firm i; that is, the small-world value of firm i. n_k stands for the number of connections among all n_i direct competitors, k, of the focal firm, i.

4.3.4. Control Variables

To avoid the endogeneity problem, we selected control variables based on firms, unlike the dependent and independent variables built on patents. The detailed control variables included the firm's size, cash flow from operating activities (CFOA), R&D expenditures, age, and technology elements. All these variables impacted the firm's technology innovation activities, and their respective measurements, references, and data sources are represented in Table 1.

Table 1. Summary of variables.

Variables	Measures	Literature Support	Data Source
Radical green innovation (RGI)	The number of patents that contain new technical categories. The new technology category refers to the technical class that a firm has never used in patents filed in the past five years.	[41,64,73,74]	Derwent Innovation database
Incremental green innovation (IGI)	The number of patents that contain only existing technical categories. The existing technical category refers to a familiar technical class (which has appeared at least once) in the past five years.		
Competitor-weighted centrality (CWC)	Eigenvector centrality	[11,16,69]	Firm competition network
Small-world clusters (Smallworld)	The degree to which the focal firm clusters with neighboring firms into a group (the average clustering coefficient)	[12,71,72]	Firm competition network
Firm size	The number of employees (thousands)	[12,75]	Standard and Poor's Compustat Xpressfeed data; Compustat North America/Global database; Corporate Financial annual report; Company websites;EBSCOhost company information
Cash flow from operating activities (CFOA)	The annual cash flow of operating activities of firms (million dollars (log)).	[2,67]	
R&D expenditure	The annual investment in firms' research and development activities (million dollars (log)).	[75]	
Firm age	The number of years since the established date	[12]	
Firm's technology elements	The total number of IPC classes contained in firm patents	[16,76]	Derwent Innovation database

4.4. Analysis Strategy

Fixed-effects models can help control for time-invariant factors [14]. Moreover, since our dependent variables were non-negative integer values with a skewed distribution, linear regression models could lead to inconsistent, biased, and inefficient estimates [77]. Therefore, the fixed-effects panel negative binomial model was the most appropriate for this study. Further, we clustered by firm and used robust standard errors in all analyses to correct the correlation of observations from the same firm over the year.

5. Results

5.1. Descriptive Statistical Analysis

This section illustrates the results of our procedure. Table 2 reports the descriptive statistics and correlations. The variance inflation factors (VIFs) for all the variables were below the recommended cut-off of 5.0, indicating that collinearity was unlikely to affect our results. Generally, firms produce an average of 2.7 radical and 3.7 incremental patents, yearly. In addition, competitor-weighted centrality has a positive effect on both the firm's radical and incremental innovation ($\beta_1 = 0.917$, $p_1 < 0.01$; $\beta_2 = 0.956$, $p_2 < 0.01$). Notably, we observed that the Smallworld variable negatively affected radical innovation to a greater extent than incremental innovation ($\beta_1 = -0.259$, $p_1 < 0.01$; $\beta_2 = -0.213$, $p_2 < 0.01$), which indicates that a firm's truck in a small world is not beneficial to their innovation outcome, especially the radical outcome. The impact of the interaction between the small world and competitor-weighted centrality on innovation will be discussed in the next section.

Table 2. Descriptive statistics and correlations.

	Mean	SD	RGI	IGI	CWC	Smallworld	Size	lnCFOA	lnRD	Age
RGI	2.747	4.668								
IGI	3.711	11.33	0.828 ***							
CWC	0.0598	0.115	0.917 ***	0.956 ***						
Smallworld	1.745	1.177	−0.259 ***	−0.213 ***	−0.285 ***					
Size	78.62	116.7	0.128 ***	0.156 ***	0.144 ***	−0.03				
lnCFOA	53.02	49.12	0.029	0.055	0.029	−0.017	0.243 ***			
lnRD	10.09	15.03	0.022	0.038	0.015	−0.052	0.284 ***	0.559 ***		
Age	6.910	2.907	0.078 *	0.102 **	0.100 **	0.007	0.445 ***	0.058	0.113 **	
Tcs	5.943	2.576	0.936 ***	0.879 ***	0.937 ***	−0.279 ***	0.138 ***	0.055	0.028	0.075 *

*** $p < 0.01$, ** $p < 0.05$, * $p < 0.1$.

5.2. Regression Analysis

We first ran the fixed-effects negative binomial panel regression, and the results of the analyses are presented in Table 3. Models 1.1 to 1.4 provide the results of radical green innovation, and Models 2.1 to 2.4 show the results of incremental green innovation.

Table 3. Results of the fixed-effects negative binomial panel analysis.

VARIABLES	Radical Green Innovation				Incremental Green Innovation			
	Model 1.1	Model 1.2	Model 1.3	Model 1.4	Model 2.1	Model 2.2	Model 2.3	Model 2.4
CWC		2.344 ***	2.014 **	3.722 ***		4.785 ***	4.273 ***	2.744 ***
Smallworld			−0.219 ***	−0.191 **			−0.340 **	−0.456 ***
CWC * Smallworld				−2.629 *				2.897 **
Size	−0.003 *	−0.003	−0.003	−0.003	0.004 **	0.001	0.001	0.002
lnCFOA	−0.026	−0.019	−0.022	−0.018	0.013	0.068 *	0.054	0.051
lnRD	0.029	−0.013	−0.01	−0.014	0.024	−0.064	−0.048	−0.05
Age	−0.057 ***	−0.037 **	−0.028	−0.022	−0.015 ***	−0.001	0.003	0.005
Tcs	0.030 ***	0.020 ***	0.020 ***	0.026 ***	0.011 ***	−0.008 **	−0.006 *	−0.014 ***
Constant	14.881	12.077	8.896	8.171	1.734 ***	1.096 *	1.530 **	1.757 ***
Observations	252	252	252	252	207	207	207	207
Number of firms	53	53	53	53	38	38	38	38
Firm fixed effects	Include	Include	Include	Include	Include	Include	Include	Include
Wald chi-square	156.5	158.97	165.89	169.57	27.88	98.35	102.28	111.53
Log-likelihood	−301.3	−296.8	−291.1	−289.4	−266	−250.3	−246.4	−244.1

*** $p < 0.01$, ** $p < 0.05$, * $p < 0.1$.

Hypotheses 1a and 1b predicted that competitor-weighted centrality would be positively related to radical and incremental green innovation, respectively. Their coefficients in the results were significant and positive across the models, supporting Hypotheses 1a and 1b.

Hypothesis 2a predicted that being embedded in a small world would weaken the effect of competitor-weighted centrality on a firm's radical innovation. We observed that the coefficient of Smallworld ($\beta = -0.219$, $p < 0.01$) was negative and significant (Model 1.3), which means that enterprises being structured in the small world was not beneficial for their radical innovation. In a further step, we examined the interaction between small worlds and technical competition pressure in Model 1.4. The results showed that their interaction ($\beta = -2.629$, $p < 0.1$) intensified the tendency against the radical innovation of enterprise, supporting Hypothesis 2a.

Hypothesis 2b predicted that being embedded in a small world would strengthen the effect of competitor-weighted centrality on a firm's incremental innovation. Similarly, we observed that the coefficient of Smallworld ($\beta = -0.340$, $p < 0.05$) was negative and significant (Model 2.3), which means that enterprises being structured in the small world was also not beneficial for their innovation. Furthermore, we examined the interaction between firms in a small world and competitor-weighted centrality in Model 2.4. The results showed that their interaction ($\beta = 2.897$, $p < 0.05$) significantly enhanced the enterprises' exploitative innovation, supporting Hypothesis 2b.

5.3. Robustness Test

To enhance the reliability of the results, we conducted three robustness tests. The results are presented in Table 4. First, the fixed-effects Poisson (FEP) panel regression was used as an alternative to the abovementioned models. Second, since the dependent variable contained many 0 values, we also conducted a robustness test with a zero-inflation Poisson (ZIP) regression. Finally, we changed the measurement window of the dependent variable. The 5-year moving window was replaced with 4 years. Radical and incremental patents were judged by whether the patent IPC code had occurred within the previous four years. The robustness test results once again proved the reliability of our results.

Table 4. Robustness test.

VARIABLES	Radical Innovation			Incremental Innovation		
	FEP	ZIP	4-Year	FEP	ZIP	4-Year
CWC	3.732 ***	2.972 ***	3.901 ***	3.318 ***	3.024 ***	2.684 ***
Smallworld	−0.191 **	−0.415 ***	−0.183 **	−0.678 ***	−1.487 ***	−0.527 ***
CWC * Smallworld	−2.626 *	−2.942 **	−2.264 *	2.470 ***	4.130 ***	2.955 **
Size	−0.003	−0.001 *	−0.003	0.002 *	−0.001 *	0.002
lnCFOA	−0.018	0.006	−0.02	0.078 ***	0.123 ***	0.051
lnRD	−0.014	−0.019	−0.016	−0.069	−0.134 ***	−0.053
Age	−0.023	0.001	−0.013	0.043 ***	0.003 ***	0.006
Tcs	0.026 ***	0.028 ***	0.022 ***	−0.016 ***	−0.013 ***	−0.014 ***
Constant		1.221 ***	6.781		2.324 ***	1.714 **
Observations	252	259	252	207	259	199
Number of firms	53		53	38		36
Wald chi-square	175.5		169.77	169.29		109.86
LR chi		2657.99			2657.99	
Log-likelihood	−289.4	−443.6	−292.3	−244.8	−420.1	−239.3

*** $p < 0.01$, ** $p < 0.05$, * $p < 0.1$.

6. Conclusions and Implications

Based on the constructed technical competition network of wind energy enterprises, this study examined how two significant network structures (competitor-weighted centrality and embeddedness in small-world clusters) affect network firms' ambidexterity of green technology innovation from a social embeddedness perspective. Our results verified our hypotheses. (1) Competitor-weighted centrality contributed to an enterprise's incremental and radical green innovation. This is because stronger competitors put the focal firm under higher technical pressure, which motivates the firm to keep exploring and exploiting to improve its innovation. Moreover, powerful rivals from the network can provide the focal firm with more useful information that can stimulate its green innovation. (2) The interaction between competitor-weighted centrality and being embedded in small-world clusters can promote the focal firm's incremental green innovation but can hinder its radical green innovation. Our findings elucidate the critical roles of competitors and the embedded structural features, providing enterprises with a benchmark to balance their incremental and radical innovative activities.

6.1. Theoretical Contributions

First, our research enhances the understanding of how competition affects innovation by exploring the ambidexterity of green innovation. Our investigation of innovation ambidexterity provides an integrated picture of how competitive relationships affect firm innovation. Firms embedded in small-world clusters face competitors that are more stable and less diverse [10], and the repeated interactions between those firms facilitate information transfer [17]. These factors make firms more sensitive to the dynamics of their rivals and, thus, take faster actions to respond. Therefore, we argue that stable competition promotes enterprise innovation, but only incremental innovation. From a long-term perspective, this stable competitive relationship would reinforce cognitive barriers and information insularity, which is not beneficial to a firm's radical innovation.

Second, our study provides new insights into the relationship between competition structures and technological innovation management. We focus on two significant structures of a competition network. One is the competitor-weighted centrality of a firm in the network. Although previous studies have illuminated the importance of the centrality of a firm's position in a collaboration network [11,18], there are some differences between cooperation networks and competition networks in terms of information exchange [14,15]. To the best of our knowledge, this is the first study that investigates the competitiveness of a focal firm's competitors and how its competitor-weighted centrality affects the firm's green technology innovation. Our findings suggest that the more central a firm is within a competition network, the higher its performance in green technology innovation. This is partly because a central competitor can impose higher competitive pressure on the focal firm and provide it with more valuable information, which instigates the focal firm to constantly search for and learn from its competitors. These findings support the argument from previous studies that competitors can be a valuable resource rather than a threat, as most previous studies have suggested [19].

Third, the findings related to a firm's embeddedness in small-world clusters contribute to research on the social embeddedness perspective and green innovation literature. On the one hand, this paper complements previous studies on the social embeddedness perspective, which focuses their attention merely on other network structures [10,15,16,19]. Small-world clusters are important but often overlooked network structural features. We discovered that embeddedness in a small world is a suitable embodiment for understanding the competitive relationship between firms within a competition network. Therefore, this study enriches the research on the antecedent factors of green innovation. Similar to environmental policy, the small-world characteristic is an important dimension for measuring the external environment at the firm level. Our results show that firms embedded in different positions within a small-world cluster can receive different kinds of competitive pressure and obtain different technical information, which can lead the firms to employ different reactions and adopt different technological innovation strategies. These findings also support the idea that the environment in which a firm is embedded has a significant influence on its competitive capacity [16] and search strategies [52].

6.2. Managerial Implications

Our research has several important managerial implications for practitioners involved in green technology innovation. (1) Managers should treat their competitors as vital resources rather than threats to their technology innovation. Since competitive relationships can be an important channel for information flow between rivals, competitors with high technical capabilities can inspire the focal firm to engage in more innovation activities, and thus, improve their performance through such information flows and identify promising technological directions. (2) Managers can leverage competitive pressure to enhance their innovation performance. Managers need to be aware that an appropriate level of competitive pressure can become a driving force for firms to improve their abilities in innovation. (3) A familiar competition environment can be conducive to incremental innovation. Stable and repeated linkages in a small world can provide regular information and share technical improvements on slim trajectories, which can help firms perceive the insignificant dynamics of the competitors and respond to those changes quickly, thus improving incremental green innovation. (4) Finding new competitors is an efficient way to achieve radical green innovation. Enterprises should appropriately extend their sights to competitors outside their clique, such as firms in different markets or utilizing different technologies. These competitors can provide fresh ideas to stimulate the firm's explorative capabilities and contribute to its radical green innovation.

6.3. Limitations and Future Directions

This study has some limitations. First, the measurement of green technology innovation outcomes can be more diverse. Although patents are an important indicator of technological innovativeness, other measurements, such as tacit knowledge, technology or trade secrets, and technical standards, can also be used to measure technology innovation outcomes. Second, this study selected only the top 120 leading enterprises as the empirical sample. Future research may apply more advanced data pre-processing and analysis methods to increase the sample size. For example, methods such as natural language processing and big data analytics can be used as efficient complementary methods in data collection and text pre-processing [78,79]. Finally, future studies can explore the features of the technology convergence network mentioned in Section 4.2. This study only investigated the competition network constructed based on the collected patent data, whereas the technology convergence network that laid the foundations for a firm's competition has not been explored. The interactions of the two networks may provide new insight into the competitive behaviors of the focal firms.

Author Contributions: Data curation, Y.X.; formal analysis, R.Z.; funding acquisition, H.Z. and F.Q.; methodology, R.Z.; supervision, H.Z.; visualization, R.Z.; writing—original draft, R.Z.; writing—review and editing, H.Z. and M.Y.Z. All authors have read and agreed to the published version of the manuscript.

Funding: This work was supported by the Project of the National Natural Science Foundation of China (No. 72062011). Authors Huiying Zhang and Fei Qu have received research support from this project.

Data Availability Statement: The data that support the findings of this study are available from the corresponding author upon reasonable request.

Acknowledgments: The authors would like to thank the editor and anonymous referees for their helpful suggestions. Thanks to the support of the National Natural Science Foundation of China (Project No. 72062011).

Conflicts of Interest: The authors have no relevant financial or non-financial interest to disclose.

References

1. Cui, J.; Dai, J.; Wang, Z.; Zhao, X. Does Environmental Regulation Induce Green Innovation? A Panel Study of Chinese Listed Firms. *Technol. Forecast. Soc. Chang.* **2022**, *176*, 12378. [CrossRef]
2. Huang, J.-W.; Li, Y.-H. Green Innovation and Performance: The View of Organizational Capability and Social Reciprocity. *J. Bus. Ethics* **2015**, *145*, 309–324. [CrossRef]
3. Zhang, Y.J.; Shi, W.; Jiang, L. Does China's carbon emissions trading policy improve the technology innovation of relevant enterprises? *Bus. Strategy Environ.* **2019**, *29*, 872–885. [CrossRef]
4. Chen, Y.-S.; Lai, S.-B.; Wen, C.-T. The Influence of Green Innovation Performance on Corporate Advantage in Taiwan. *J. Bus. Ethics* **2006**, *67*, 331–339. [CrossRef]
5. Rennings, K.; Ziegler, A.; Ankele, K.; Hoffmann, E. The influence of different characteristics of the EU environmental management and auditing scheme on technical environmental innovations and economic performance. *Ecol. Econ.* **2006**, *57*, 45–59. [CrossRef]
6. Wang, X.; Zhou, J.; Maitlo, Q. Effect of green technology innovation on the upgrading of the manufacturing value chain: Evidence from China. *Front. Environ. Sci.* **2022**, *10*, 1–16. [CrossRef]
7. O'Reilly III, C.A.; Tushman, M.L. Ambidexterity as a dynamic capability: Resolving the innovator's dilemma. *Res. Organ. Behav.* **2008**, *28*, 185–206. [CrossRef]
8. Katila, R.; Ahuja, G. Something old, something new: A longitudinal study of search behavior and new product introduction. *Acad. Manag. J.* **2002**, *45*, 1183–1194. [CrossRef]
9. Schiederig, T.; Tietze, F.; Herstatt, C. Green innovation in technology and innovation management–an exploratory literature review. *RD Manag.* **2012**, *42*, 180–192. [CrossRef]
10. Skilton, P.F.; Bernardes, E. Competition network structure and product market entry. *Strateg. Manag. J.* **2015**, *36*, 1688–1696. [CrossRef]
11. Qi Dong, J.; McCarthy, K.J.; Schoenmakers, W.W.M.E. How Central Is Too Central? Organizing Interorganizational Collaboration Networks for Breakthrough Innovation. *J. Prod. Innov. Manage.* **2017**, *34*, 526–542. [CrossRef]
12. Potter, A.; Wilhelm, M. Exploring supplier–supplier innovations within the Toyota supply network: A supply network perspective. *J. Oper. Manag.* **2020**, *66*, 797–819. [CrossRef]

13. Liu, W.W.; Song, Y.F.; Bi, K.X. Exploring the patent collaboration network of China's wind energy industry: A study based on patent data from CNIPA. *Renew. Sust. Energ. Rev.* **2021**, *144*, 110989. [CrossRef]
14. Thatchenkery, S.; Katila, R. Seeing What Others Miss: A Competition Network Lens on Product Innovation. *Organ. Sci.* **2021**, *32*, 1346–1370. [CrossRef]
15. Choi, S.; Kim, W.; Kim, N. International alliance formations: The role of brokerage in technology competition networks. *J. Bus. Res.* **2022**, *144*, 440–449. [CrossRef]
16. Zhang, J.J.; Guan, J.C. The impact of competition strength and density on performance: The technological competition networks in the wind energy industry. *Ind. Mark. Manag.* **2019**, *82*, 213–225. [CrossRef]
17. Yu, T.; Guo, W.; Gimeno, J.; Zhang, Y. Glue or Gasoline? The Role of Interorganizational Linkages in the Occurrence and Spillover of Competitive Wars. *Acad. Manag. J.* **2022**, *65*, 304–330. [CrossRef]
18. Kilduff, M.; Brass, D.J. Organizational social network research: Core ideas and key debates. *Acad. Manag. Ann.* **2010**, *4*, 317–357. [CrossRef]
19. Wang, L.L.; Gao, Y. Competition network as a source of competitive advantage: The dynamic capability perspective and evidence from China. *Long Range Plan.* **2021**, *54*, 102052. [CrossRef]
20. Choi, T.Y.; Kim, Y. Structural Embeddedness and Supplier Management: A Network Perspective. *J. Supply Chain Manag.* **2008**, *44*, 5–13. [CrossRef]
21. Watts, D.J.; Strogatz, S.H. Collective dynamics of 'smallworld' networks. *Nature* **1998**, *393*, 440–442. [CrossRef]
22. Sharma, A.; Kumar, V.; Yan, J.; Borah, S.B.; Adhikary, A. Understanding the structural characteristics of a firm's whole buyer–supplier network and its impact on international business performance. *J. Int. Bus. Stud.* **2019**, *50*, 365–392. [CrossRef]
23. Salancik, G.R.; Burt, R.S. Structural Holes: The Social Structure of Competition. *Adm. Sci. Q.* **1995**, *40*, 345. [CrossRef]
24. Uzzi, B. Social structure and competition in interfirm networks. *Adm. Sci. Q.* **1997**, *42*, 37–69.
25. March, J.G. Exploration and Exploitation in Organizational Learning. *Organ. Sci.* **1991**, *2*, 71–87. [CrossRef]
26. Helfat, C.E. Evolutionary trajectories in petroleum firm R&D. *Manag. Sci.* **1994**, *40*, 1720–1747.
27. Ahuja, G.; Katila, R. Where do resources come from? The role of idiosyncratic situations. *Strateg. Manag. J.* **2004**, *25*, 887–907. [CrossRef]
28. Nelson, R.R. *An Evolutionary Theory of Economic Change*; Harvard University Press: Cambridge, MA, USA, 1985.
29. Schniederjans, D.G.; Khalajhedayati, M. Competitive sustainability and stakeholder engagement: Exploring awareness, motivation, and capability. *Bus. Strategy Environ.* **2021**, *30*, 808–824. [CrossRef]
30. York, J.G.; Hargrave, T.J.; Pacheco, D.F. Converging Winds: Logic Hybridization in the Colorado Wind Energy Field. *Acad. Manag. J.* **2016**, *59*, 579–610. [CrossRef]
31. Zhu, H.M.; Du, Z.F.; Wu, J.L.; Sun, Z.H. Innovation environment and opportunities of offshore wind turbine foundations: Insights from a new patent analysis approach. *World Pat. Inf.* **2022**, *68*, 102092. [CrossRef]
32. Shen, X.; Lyu, S. Wind power development, government regulation structure, and vested interest groups: Analysis based on panel data of Province of China. *Energy Policy* **2019**, *128*, 487–494. [CrossRef]
33. IEA. *World Energy Outlook 2021*; IEA: Paris, France, 2021.
34. Wang, X.F.; Zeng, X.W.; Li, J.L.; Yang, X.; Wang, H.J. A review on recent advancements of substructures for offshore wind turbines. *Energy Convers. Manag.* **2018**, *158*, 103–119. [CrossRef]
35. Yang, W.; Cao, G.; Peng, Q.; Sun, Y. Effective radical innovations using integrated QFD and TRIZ. *Comput. Ind. Eng.* **2021**, *162*, 107716. [CrossRef]
36. Li, Z.; Zhang, M.Y.; Zhang, H. Firm Growth Performance and Relative Innovation Orientation of Exploration vs Exploitation: Moderating Effects of Cluster Relationships. *Manag. Organ. Rev.* **2020**, *17*, 143–172. [CrossRef]
37. Fleming, L. Finding the organizational sources of technological breakthroughs: The story of Hewlet" Packard's thermal in‹ jet. *Ind. Corp. Chang.* **2002**, *11*, 1059–1084. [CrossRef]
38. Gittelman, M.; Kogut, B. Does Good Science Lead to Valuable Knowledge? Biotechnology Firms and the Evolutionary Logic of Citation Patterns. *Manag. Sci.* **2003**, *49*, 366–382. [CrossRef]
39. Yamakawa, Y.; Yang, H.B.; Lin, Z.A. Exploration versus exploitation in alliance portfolio: Performance implications of organizational, strategic, and environmental fit. *Res. Policy* **2011**, *40*, 287–296. [CrossRef]
40. Oerlemans, L.A.G.; Knoben, J.; Pretorius, M.W. Alliance portfolio diversity, radical and incremental innovation: The moderating role of technology management. *Technovation* **2013**, *33*, 234–246. [CrossRef]
41. Guan, J.C.; Yan, Y. Technological proximity and recombinative innovation in the alternative energy field. *Res. Policy* **2016**, *45*, 1460–1473. [CrossRef]
42. Wang, W.; Chen, H. Extrusion Effect or Promotion Effect? The Effect of Environmental Regulation on Enterprise Green Innovation. *Int. J. Environ. Res. Public Health* **2023**, *20*, 1748. [CrossRef]
43. Li, S.; Garces, E.; Daim, T. Technology forecasting by analogy-based on social network analysis: The case of autonomous vehicles. *Technol. Forecast. Soc. Chang.* **2019**, *148*, 119731. [CrossRef]
44. Li, G.; Li, X.; Wang, N. Research on the influence of environmental regulation on technological innovation efficiency of manufacturing industry in China. *Int. J. Environ. Sci. Technol.* **2022**, *19*, 5239–5252. [CrossRef]
45. Deng, Y.L.; You, D.M.; Wang, J.J. Optimal strategy for enterprises' green technology innovation from the perspective of political competition. *J. Clean. Prod.* **2019**, *235*, 930–942. [CrossRef]

46. Fan, J.; Teo, T. Will China's R&D investment improve green innovation performance? An empirical study. *Environ. Sci. Pollut. Res. Int.* **2022**, *29*, 39331–39344. [CrossRef]
47. Wang, H.; Wang, S.; Zheng, Y. China green credit policy and corporate green technology innovation: From the perspective of performance gap. *Environ. Sci. Pollut. Res. Int.* **2022**, *18*, 1682. [CrossRef]
48. Cabigiosu, A. Sustainable development and incumbents' open innovation strategies for a greener competence-destroying technology: The case of electric vehicles. *Bus. Strategy Environ.* **2022**, *31*, 2315–2336. [CrossRef]
49. Wicki, S.; Hansen, E.G. Green technology innovation: Anatomy of exploration processes from a learning perspective. *Bus Strategy Environ.* **2019**, *28*, 970–988. [CrossRef] [PubMed]
50. Cross, R.; Cummings, J.N. Tie and Network Correlates of Individual Performance in Knowledge-Intensive Work. *Acad. Manag. J.* **2004**, *47*, 928–937. [CrossRef]
51. Durmuşoğlu, S.S. Merits of Task Advice during New Product Development: Network Centrality Antecedents and New Product Outcomes of Knowledge Richness and Knowledge Quality. *J. Prod. Innov. Manag.* **2013**, *30*, 487–499. [CrossRef]
52. Shi, X.X.; Zhang, Q.P.; Zheng, Z.L. The double-edged sword of external search in collaboration networks: Embeddedness in knowledge networks as moderators. *J. Knowl. Manag.* **2019**, *23*, 2135–2160. [CrossRef]
53. Bowen, H.P.; Wiersema, M.F. Foreign-based competition and corporate diversification strategy. *Strateg. Manag. J.* **2005**, *26*, 1153–1171. [CrossRef]
54. Huber, G. Organizational information systems: Determinants of their performance and behavior. *Manag. Sci.* **1982**, *28*, 138–155. [CrossRef]
55. Rashid, Z.A.; Al-Tabakha, M.M.; Alomar, M.J. Proper Counseling and Dispensing of Isotretinoin Capsule Products by Community Pharmacists in UAE: A Simulated Patient Study. *Clin. Cosmet. Investig. Derm.* **2020**, *13*, 405–414. [CrossRef] [PubMed]
56. Huber, G.P.; Daft, R.L. *The Information Environments of Organizations*; Sage Publication: Newbury Park, CA, USA, 1987; pp. 130–164.
57. Zaheer, A.; Venkatraman, N. Relational governance as an interorganizational strategy: An empirical test of the role of trust in economic exchange. *Strateg. Manag. J.* **1995**, *16*, 373–392. [CrossRef]
58. Vestal, A.; Danneels, E. Technological Distance and Breakthrough Inventions in Multi-Cluster Teams: How Intra- and Inter-Location Ties Bridge the Gap. *Adm. Sci. Q.* **2022**, *67*, 167–206. [CrossRef]
59. George, G.; Kotha, R.; Zheng, Y.F. Entry into Insular Domains: A Longitudinal Study of Knowledge Structuration and Innovation in Biotechnology Firms. *J. Manag. Stud.* **2008**, *45*, 1448–1474. [CrossRef]
60. Galaskiewicz, J. Studying Supply Chains from a Social Network Perspective. *J. Supply Chain Manag.* **2011**, *47*, 4–8. [CrossRef]
61. Audia, P.G.; Goncalo, J.A. Past success and creativity over time: A study of inventors in the hard disk drive industry. *Manag. Sci.* **2007**, *53*, 1–15. [CrossRef]
62. Heil, S.; Enkel, E. Exercising Opportunities for Cross-Industry Innovation: How To Support Absorptive Capacity in Distant Knowledge Processing. *Int. J. Innov. Manag.* **2015**, *19*, 1550048. [CrossRef]
63. Tsai, Y.-C.; Huang, Y.-F.; Yang, J.-T. Strategies for the development of offshore wind technology for far-east countries—A point of view from patent analysis. *Renew. Sustain. Energy Rev.* **2016**, *60*, 182–194. [CrossRef]
64. Gilsing, V.; Nooteboom, B.; Vanhaverbeke, W.; Duysters, G.; van den Oord, A. Network embeddedness and the exploration of novel technologies: Technological distance, betweenness centrality and density. *Res. Policy* **2008**, *37*, 1717–1731. [CrossRef]
65. Guan, J.C.; Liu, N. Exploitative and exploratory innovations in knowledge network and collaboration network: A patent analysis in the technological field of nano-energy. *Res. Policy* **2016**, *45*, 97–112. [CrossRef]
66. Miao, H.; Zhao, R.; Huang, H. Research on technology convergence measurement based on LDA-SVM classification algorithm. *Sci. Sci. Manag. S. T.* **2018**, *39*, 13–29.
67. Morandi Stagni, R.; Fosfuri, A.; Santaló, J. A bird in the hand is worth two in the bush: Technology search strategies and competition due to import penetration. *Strateg. Manag. J.* **2021**, *42*, 1516–1544. [CrossRef]
68. Gambeta, E.; Koka, B.R.; Hoskisson, R.E. Being too good for your own good: A stakeholder perspective on the differential effect of firm-employee relationships on innovation search. *Strateg. Manag. J.* **2018**, *40*, 108–126. [CrossRef]
69. Guo, K.; Zhang, T.; Liang, Y.; Zhao, J.; Zhang, X. Research on the promotion path of green technology innovation of an enterprise from the perspective of technology convergence: Configuration analysis using new energy vehicles as an example. *Environ. Dev. Sustain.* **2022**, 1–20. [CrossRef]
70. Dong, J.Q.; Yang, C.-H. Being central is a double-edged sword: Knowledge network centrality and new product development in U.S. pharmaceutical industry. *Technol. Forecast. Soc. Chang.* **2016**, *113*, 379–385. [CrossRef]
71. Basole, R.C. Topological analysis and visualization of interfirm collaboration networks in the electronics industry. *Decis. Support Syst.* **2016**, *83*, 22–31. [CrossRef]
72. Sharma, A.; Pathak, S.; Borah, S.B.; Adhikary, A. Is it too complex? The curious case of supply network complexity and focal firm innovation. *J. Oper. Manag.* **2019**, *66*, 839–865. [CrossRef]
73. Vanhaverbeke, W.I.M.; Beerkens, B.; Duysters, G. Explorative and Exploitative Learning Strategies in Technology-Based Alliance Networks. *Acad. Manag. Proc.* **2004**, *2004*, J1–J6. [CrossRef]
74. Manso, G.; Balsmeier, B.; Fleming, L. Heterogeneous innovation and the antifragile economy. 2019; *Unpublished working paper*.
75. Kim, K.; Jung, S.; Hwang, J. Technology convergence capability and firm innovation in the manufacturing sector: An approach based on patent network analysis. *R D Manag.* **2019**, *49*, 595–606. [CrossRef]

76. Yayavaram, S.; Ahuja, G. Decomposability in knowledge structures and its impact on the usefulness of inventions and knowledge-base malleability. *Adm. Sci. Q.* **2008**, *53*, 333–362. [CrossRef]
77. Greene, W.H. *Economic Analysis*; Prentice Hall: Upper Saddle River, NJ, USA, 2003.
78. Teng, F.; Sun, Y.L.; Chen, F.; Qin, A.N.; Zhang, Q. Technology opportunity discovery of proton exchange membrane fuel cells based on generative topographic mapping. *Technol. Forecast. Soc. Chang.* **2021**, *169*, 120859. [CrossRef]
79. Minatogawa, V.L.F.; Franco, M.M.V.; Rampasso, I.S.; Anholon, R.; Quadros, R.; Durán, O.; Batocchio, A. Operationalizing Business Model Innovation through Big Data Analytics for Sustainable Organizations. *Sustainability* **2020**, *12*, 277. [CrossRef]

energies

MDPI

Article

Evolutionary Game and Simulation Analysis of Power Plant and Government Behavior Strategies in the Coupled Power Generation Industry of Agricultural and Forestry Biomass and Coal

Dan Yu [1], Caihong Zhang [1,*], Siyi Wang [2,*] and Lan Zhang [1]

[1] School of Economics and Management, Beijing Forestry University, Beijing 100083, China
[2] Chinese Research Academy of Environmental Sciences, Beijing 100012, China
* Correspondence: zhangcaihong@263.net (C.Z.); wang_si_yi@126.com (S.W.)

Abstract: Under the background of "dual carbon", the coupled power generation of agricultural and forestry biomass (AFB) and coal, as a new path of coal-power transformation, is key to achieving energy conservation and reducing emissions in the power sector. Timely and effective government subsidies as well as regulation policies will play important roles in the development of the coupled power generation industry. Previous studies usually assumed government policy as singular and static, rarely considering the dynamic changes in government policies. In this study, evolutionary game theory and systematic dynamics research methods were combined. The game relationship and the dynamic evolution process of the behavioral strategies of both sides are analyzed through the construction of a mixed-strategies game model of the government and power plants. A system dynamics model is built for simulations based on the results of the dynamic game evolution, and the influence paths of key factors on the behavioral strategies of the government and power plants were further demonstrated. The results indicated the following: (1) The behavioral strategies of the government and power plants were not stable for a long period of time, but fluctuated during their mutual influence. The dynamic policies and measures formulated by the government according to changes in the behavioral strategies of power plants will promote industrial development more effectively. (2) Increasing subsidization and the strengthening of supervision caused by government policy can increase the enthusiasm of power plants to choose the coupled power generation of AFB and coal. (3) If the government improves the benefits or reduces the transformation costs caused by coupled power generation the industry will be fundamentally improved. The results clearly show the interactions as well as adjustment processes of the behavioral strategies of power plants and the government in the coupled power generation industry of AFB and coal, and the specific effects of key factors on the behavioral strategies of power plants and the government were investigated. This study can provide a theoretical basis for the government to formulate reasonable industrial policies and measures for the coupled power generation of AFB and coal, in addition to being a valuable reference for other countries to develop a coupled power generation industry.

Keywords: agricultural and forestry biomass; coupled power generation; evolutionary game; system dynamics; behavioral strategy

1. Introduction

China has been committed to reducing international carbon emissions, and has actively promoted as well as fulfilled its climate change commitments. At the UN General Assembly in 2020, China proposed the goal of "carbon peaking and carbon neutrality", that is, carbon emissions will peak before 2030 and achieve carbon neutrality around 2060 [1]. As a major contributor to global energy consumption and CO_2 emissions, and the largest industrial source of greenhouse gas emissions in China, the transformation of the energy utilization structure in the power sector is key to achieving the "dual carbon" goal [2]. The Intergovernmental Panel on Climate Change (IPCC) has noted that energy use is closely related to increased greenhouse gas emissions; the use of renewable energy can help to break this correlation and thus contribute to sustainable development [3]. As a kind of renewable energy, the CO_2 absorbed by biomass energy in the growth process can offset the CO_2 released in combustion, and it has the characteristic of "zero carbon emissions". It can replace fossil energy in the field of power supply and heating, and has great potential for carbon emission reduction [4]. Biomass energy includes multiple forms, such as forest biomass energy, agricultural biomass energy, animal feces, food residue, and municipal solid waste. Among them, agricultural and forestry biomass energy usually refers to agricultural residues (cereal straw, maize stover, rapeseed straw, etc.) and forestry residues (shrub residues, branches, sawdust, etc.) [5].

Biomass energy technologies have been widely used in the field of power generation, including direct combustion, gasification, and coupled power generation technologies [6], among which the coupled power generation of biomass and coal is regarded as one of the economic, effective, and executable ways of reducing carbon emissions in thermal power plants [7]. The coupled power generation of agricultural and forestry biomass (AFB) and coal uses the mixed combustion of AFB raw materials and coal to generate electricity, which is a comprehensive utilization method of traditional and renewable energy [8]. On the one hand, this power generation technology reduces the usage costs of AFB raw materials, enabling them to be fully utilized and avoiding the air pollution caused by direct combustion (such as straw incineration returning to the field) [9]. On the other hand, by replacing a certain proportion of coal for power generation, AFB can not only ensure better combustion calorific value and efficiency in the power generation process but also greatly reduce CO_2 emissions [10]; however, the policies and measures for the agricultural and forestry biomass and coal coupling power generation industry are not consistent in China. In 2016, the National Energy Administration proposed promoting the coupled power generation technology of biomass and coal during the 13th Five-Year Plan period [11]. In 2017, the Ministry of Ecology and Environmental issued the Notice on Carrying out the Technical Transformation Pilot Work of Coal-Biomass Co-firing Power generation, aiming at developing the technical transformation project of the coupled power generation of AFB and coal [12]; however, in 2018 this technology project was officially removed from the national subsidy catalog [13]. In 2019, the Guidance Catalogue of Industrial Structure Adjustment 2019, issued by the National Development and Reform Commission, once again listed coal–biomass co-firing power generation as an industry to be encouraged [14]; however, in 2020 the National Development and Reform Commission issued a supplementary notice on matters related to "Several Opinions on Promoting the Healthy Development of Non-Water Renewable Energy Power Generation", which made it clear that biomass-related power generation projects will not be eligible for subsidies if they run for 15 years or if the number of reasonable utilization hours in the whole life cycle reaches 82,500 h [15]. The instability of these industrial policies leads to the failure of the steady development of the coupled power generation industry of AFB and coal; therefore, it is very important for the formulation of industrial policies to systematically analyze the behavioral strategies of the government and the mutual influence paths between the behavioral strategies of the government and power plants.

In the existing literature, the research on China's coupled power generation industry of AFB and coal mainly focuses on technological research as well as development [16–21] and the evaluation of economic and environmental benefits [22–28], while the research on industrial policy [29–32] is relatively lacking. In some of the literature scholars focused on the macro-level to study the effect of industrial policies on the development of the coupled power generation industry of AFB and coal. It is generally believed that reasonable policies can promote industrial development, but there are few studies on how to formulate the rationality of policies and how policies affect the development of enterprises. Some studies have pointed out that the behavioral strategies between government and enterprises are mutually driven and influenced [33]. Studies have also confirmed that enterprises will change their strategies according to the policies and measures of the government, and the design as well as choice of policies by the government play important roles in guiding and demonstrating the behavioral strategies of enterprises [34,35]; sustainable and stable industrial policies can better promote industrial development [36].

Many scholars have also studied the relevant subjects in the field of biomass energy utilization, which aimed to figure out the influence of the changes in the behavioral strategies of various stakeholders to better improve the government's policies and measures. The specific research can be divided into two aspects: (1) As for the study of the influence on behavior strategies of enterprises and inter-enterprises, Nasiri and Zaccour [37] built a static game model among power plants, brokers, and farmers to study the behavior strategies of stakeholders and pointed out that the quantity of biomass raw materials supplied by farmers was affected by the purchase price of brokers and power plants. Tan et al. [38] and Zhang et al. [39] argued that the behavioral strategy of farmers to supply quantities of biomass raw materials would be affected by the type and quantity of participants in the supply chain. (2) As for the influence of behavioral strategies between the government and enterprises, Zhai et al. [40] constructed a game model among the government, power plants, and farmers, and concluded that government subsidy strategies would improve the income as well as participation enthusiasm of all of participants, and the level of a subsidy policy was positively correlated with the income as well as enthusiasm of the participants. Luo and Miller [41] analyzed the influence of government policies on the development of the AFB market with game theory, and believed that appropriate incentive policies could improve the supply quantity of biomass raw materials in addition to the benefits of relevant participants.

By reviewing and summarizing the literature, we found that the existing literature has enriched our understanding of the research on the influence of game theory on industrial policy formulation and the strategies of players, in addition to laying the foundation for this study; however, there are still some shortcomings. For instance, the aforementioned research on industrial policies usually assumed that government policies are singular or static, but policies are not eternal; however, research on how dynamic changes in policies affect enterprise behavior and strategies are rare. In order to remedy these defects, the research methods of evolutionary game theory and system dynamics are used in this study. Evolutionary game theory is often used to study the existence of players' behavior strategies and dynamic evolutionary stability [42], while system dynamics (SD) can explore the causal feedback relationship between various factors in the model [43]. Based on this, the specific objectives of this study are as follows: (1) to establish an evolutionary game model between the government and power plants under the condition of fully considering the externalities of the coupled power generation of AFB and coal, as well as to clarify the dynamic evolution process of the game relationship and behavioral strategies of both sides; (2) based on the results of the evolutionary game model, an SD model of the government and power plants is constructed to analyze the influence path of key factors on the behavior strategies of the government and power plants. This study will provide a theoretical basis and scientific reference for the government to formulate reasonable policy measures, promoting more enterprises to choose the coupled power generation of AFB and coal.

2. Evolutionary Game Model

2.1. Applicability of Evolutionary Game Theory

In the coupled power generation industry of AFB and coal, the government is the policy maker in regard to subsidies and regulation, whilst enterprise is the practitioner of policy measures; the government mainly considers subsidies and regulatory costs, while enterprises pay attention to the actual gains and losses in subsidies and regulatory policies. In the process of strategy selection, two players will constantly adjust their strategies according to the behaviors of the other side, eventually evolving into a stable strategy, which is in line with the basic characteristics of an evolutionary game [44].

Evolutionary game theory is a combination of biological evolutionary theory and game theory. It believes that game players are bounded, rational, and will constantly adjust their strategies according to changes in a situation through trial and error to finally reach a state of dynamic equilibrium [45]. This method abandons the assumption of complete rationality and focuses on the analysis of dynamic adjustment processes, which makes up for the deficiencies in rational as well as static perspectives and can better deal with some management problems; it is suitable for the study of repeated game processes between two groups in dynamic processes [46,47]. Therefore, it is reasonable and feasible to use evolutionary game theory to study the behavior policies of the government and power plants in the coupled power generation industry of AFB and coal.

2.2. Game Players

In this study, power plants and government are the game players in the dynamic evolutionary game model.

A power plant refers to a factory using traditional coal burning technology to generate electricity that can be transformed into a power plant using AFB and coal for coupled power generation. Power plant behavior strategies can be divided into "choose the coupled power generation of AFB and coal" and "do not choose the coupled power generation of AFB and coal".

For the consideration of environmental benefits, the government has taken corresponding policy measures to encourage more power plants to reduce carbon emissions and develop the coupled power generation of AFB and coal. The government's behavioral strategies are divided into "regulate and subsidize" and "not regulate and not subsidize". To "regulate and subsidize" the power plants that choose coupled power generation technology can be regarded as the government's investment in and management of the environment. The purpose is to reduce the carbon emissions of the power generation industry and achieve the coordinated development of economic and environmental benefits.

2.3. Model Hypotheses

Hypothesis 1. *Due to the information asymmetry between the two sides of the game, both the government and power plants adopt bounded rational behaviors and play repeated games under the condition of asymmetric information* [45].

Hypothesis 2. $x \in (0,1)$ *represents the behavior strategies of power plants.* $x = 0$ *means that a power plant does not choose the "coupled power generation" strategy and* $x = 1$ *means that a power plant chooses the "coupled power generation" strategy. When* $y \in (0,1)$ *represents the government's behavior strategies,* $y = 0$ *means that the government chooses the "not regulation and not subsidy" strategy for power plants whilst* $y = 1$ *means that the government chooses the "regulation and subsidy" strategy for power plants.*

Hypothesis 3. *When the government chooses the subsidy and regulation strategy it will subsidize power plants that choose coupled power generation technology and take punitive measures for power plants that do not choose the technology. Power plants need to pay certain carbon emission fees to the government.*

Hypothesis 4. *When the government chooses the strategy of non-subsidy and non-regulation, it will not subsidize power plants that choose coupled power generation technology, nor will it punish power plants that do not choose the technology.*

Hypothesis 5. *When power plants choose the coupled power generation strategy the government will obtain certain environmental benefits regardless of whether the government chooses the subsidy and regulation strategy* [48].

2.4. Model Construction

Based on the above analysis of the applicability of the evolutionary game model, the game players, and the hypothesis of the model, an evolutionary game model with power plants and the government as the main players in the coupled power generation industry is constructed. The payoff matrix of the government and power plants is shown in Table 1.

Table 1. The payoff matrix for power plants and the government.

Strategies of Power Plants	Strategies of the Government	
	Regulation and Subsidy (y)	No Regulation and No Subsidy (1−y)
Coupled power generation is chosen (x)	$EC_b + S + B - C$, $EN_b - S - M$	$EC_b + B - C$, EN_b
Coupled power generation is not chosen (1 − x)	$EC_c - F$, $F - M - G$	EC_c, $-G$

EC_b represents the economic benefits generated when power plants choose coupled power generation. EC_c represents the economic benefits generated by using traditional coal-fired power generation. Due to the high collection cost of AFB, the economic benefits generated by coupled power generation are smaller than those generated by coal-fired power generation [49], that is, $EC_b < EC_c$. EN_b represents the environmental benefits obtained by the government when power plants choose coupled power generation [48]. S represents the policy subsidy amount given by the government when power plants choose coupled power generation. B represents the indirect benefits obtained when power plants choose coupled power generation, such as preferential power generation, full integration into power grid policy, sales of by-products, etc. [41,50]. C represents the transformation cost required for power plants to choose coupled power generation [51]. M is the regulatory cost that the government has to pay to regulate power plants. F is the carbon emission charge that power plants need to pay by burning coal when the government regulates power plants. G is the amount of money the government has to pay to clean up the environment when the power plant only uses coal to generate electricity.

According to the payoff matrix, the expected benefits when the power plant chooses the coupled power generation strategy are as follows:

$$U_{11} = y(EC_b + S + B - C) + (1 - y)(EC_b + B - C) \quad (1)$$

When the power plant does not choose the coupled power generation strategy the expected benefit is as follows:

$$U_{12} = y(EC_c - F) + (1 - y)(EC_c) \quad (2)$$

The average benefit of the power plant is as follows:

$$\overline{U_1} = xU_{11} + (1 - x)U_{12} \quad (3)$$

Similarly, the expected benefits of the government choosing the strategies of "regulation and subsidy" and "not regulation and not subsidy", as well as the average benefit of the government, are shown as follows, respectively:

$$U_{21} = x(EN_b - S - M) + (1 - x)(F - M - G) \quad (4)$$

$$U_{22} = xEN_b + (1 - x)(-G) \quad (5)$$

$$\overline{U_2} = yU_{21} + (1 - y)U_{22} \quad (6)$$

According to the definition of the replicator dynamics equation in evolutionary game theory [42], if the probability of a decision maker choosing a strategy at time t is x, then the rate of change of x at the next time will be related to the probability, x, at the previous time and the difference between the game player's corresponding strategy benefits and the expected average benefits at this time, which can be expressed as follows:

$$F(x) = \frac{dx}{dt} = x\left(U_{11} - \overline{U_1}\right) = x(1-x)[y(S+F) + EC_b + B - C - EC_c] \quad (7)$$

In the same way:

$$F(y) = \frac{dy}{dt} = y\left(U_{21} - \overline{U_2}\right) = y(1-y)[-(S+F)x + F - M] \quad (8)$$

2.5. Analysis of the Evolutionary Game Results

2.5.1. Unilateral Strategy Analysis of Power Plants

According to the replicator dynamics, Equation (7), of power plant strategies, when $y = \frac{EC_c+C-EC_b-B}{S+F}$, $F(x) = 0$, then the game is stable at any value of x;

when $y \neq \frac{EC_c+C-EC_b-B}{S+F}$, if $F(x) = 0$, it can be obtained that $x = 0$, $x = 1$ are two stable points of the replicator dynamics equation of power plant strategies. According to evolutionary game theory, when $F'(x) < 0$ or $F'(y) < 0$ the system will have an equilibrium point, such that it can be divided into two cases for discussion:

(a) When $0 < y < \frac{EC_c+C-EC_b-B}{S+F}$, then $F'(x) = (1-2x)[y(S+F) + EC_b + B - C - EC_c]$, $F'(0) < 0$, $F'(1) > 0$, so '$x = 0$' is the equilibrium point. At this time, the power plant strategy is not to choose coupled power generation technology;

(b) When $\frac{EC_c+C-EC_b-B}{S+F} < y < 1$, $F'(0) > 0$, $F'(1) < 0$, so $x = 1$ is the equilibrium point. At this time, the power plant strategy is to choose coupled power generation technology.

Therefore, the unilateral strategies of power plants depend on the probability value of the government's strategies, and the dynamic evolution trend is shown in Figure 1. When the probability that the government adopts the regulation and subsidy strategy is greater than a certain value, $(\frac{EC_c+C-EC_b-B}{S+F})$, power plants will adopt the coupled power generation strategy. When the probability that the government will adopt the subsidy and supervision strategy is less than a certain value, $(\frac{EC_c+C-EC_b-B}{S+F})$, power plants will adopt the strategy of not using coupled power generation technology.

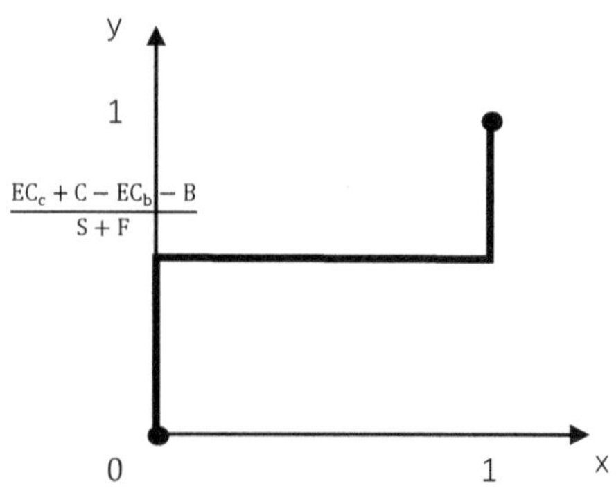

Figure 1. Dynamic evolution trend in the unilateral strategies of power plants.

2.5.2. Unilateral Strategy Analysis of the Government

According to the replicator dynamics, Equation (8), of the government's strategies, when $x = \frac{F-M}{S+F}$, $F(y) = 0$, the game is stable at any value of y;

when $x \neq \frac{F-M}{S+F}$, if $F(y) = 0$, it can be obtained that $y = 0$, $y = 1$ are two stable points of the replicator dynamics equation of the government's strategies. It can also be divided into two cases for discussion:

(c) When $0 < x < \frac{F-M}{S+F}$, $F\prime(y) = (1-2y)[-(S+F)x+F-M]$, $F\prime(0) > 0$, $F\prime(1) < 0$, so $y = 1$ is the equilibrium point, the government will regulate and subsidize power plants;

(d) When $\frac{F-M}{S+F} < x < 1$, $F\prime(0)(0, F'(1))0$, so $y = 0$ is the equilibrium point. At this time, the government's strategy is not to regulate and subsidize power plants.

Similarly, the dynamic evolution trend in the government's unilateral strategies is shown in Figure 2. The choice of strategy depends on the probability value of power plants to adopt a strategy. When the probability of power plants choosing the coupled power generation strategy is less than a certain value, ($\frac{F-M}{S+F}$), the government will adopt the "subsidize and regulate" strategy. When the probability of power plants choosing the coupled power generation strategy is greater than a certain value, ($\frac{F-M}{S+F}$), the government will no longer regulate and subsidize power plants.

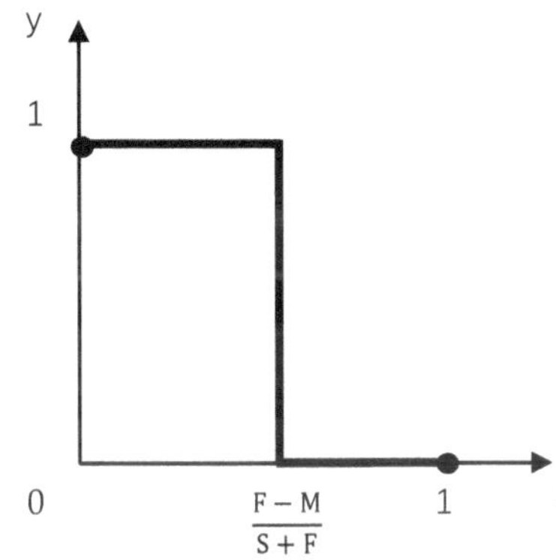

Figure 2. Dynamic evolution trend in the unilateral strategies of the government.

2.5.3. Equilibrium Strategy Analysis of Mixed Evolutionary Game Model of Power Plants and the Government

According to the replicator dynamics equation of power plants and the government, $F(x) = 0$, $F(y) = 0$, five equilibrium points of the system can be obtained, namely $(0, 0)$, $(1, 0)$, $(0, 1)$, $(1, 1)$, and (x^*, y^*), where $x^* = \frac{F-M}{S+F}$, $y^* = \frac{EC_c+C-EC_b-B}{S+F}$.

According to the analytical method proposed by Friedman [42], the stability of equilibrium points in evolutionary games can be judged by the local stability of the Jacobian matrix of the system. The Jacobian matrix, J, of the game system of power plants and the government is as follows:

$$J = \begin{bmatrix} \frac{\partial F(x)}{\partial x} & \frac{\partial F(x)}{\partial y} \\ \frac{\partial F(y)}{\partial x} & \frac{\partial F(y)}{\partial y} \end{bmatrix} = \begin{bmatrix} (1-2x)[y(S+F)+EC_b+B-C-EC_c] & x(1-x)(S+F) \\ y(1-y)(-S-F) & (1-2y)[-(S+F)x+F-M] \end{bmatrix} \tag{9}$$

The determinant, det(J), of the Jacobian matrix, J, is as follows:

$$\begin{aligned} \det(J) &= \begin{vmatrix} (1-2x)[y(S+F)+EC_b+B-C-EC_c] & x(1-x)(S+F) \\ y(1-y)(-S-F) & (1-2y)[-(S+F)x+F-M] \end{vmatrix} \\ &= (1-2x)[y(S+F)+EC_b+B-C-EC_c] \\ &\quad \cdot(1-2y)[-(S+F)x+F-M] - x(1-x)(S+F)\cdot y(1-y)(-S-F) \end{aligned} \tag{10}$$

The trace, tr(J), of the Jacobian matrix, J, is as follows:

$$tr(J) = (1-2x)[y(S+F)+EC_b+B-C-EC_c] + (1-2y)[-(S+F)x+F-M] \tag{11}$$

When $\det(J) > 0$ and $\mathrm{tr}(J) < 0$, the system equilibrium for the ESS can be determined, and the system is stable. The stability analysis of the above five equilibrium points is shown in Table 2.

Table 2. Stability analysis of five equilibrium points in the system.

Local Equilibrium Point	det(J)	tr(J)	Nature
(0, 0)	−	±	Instability
(0, 0)	−	±	Instability
(0, 0)	−	±	Instability
(0, 0)	−	±	Instability
(x^*, y^*)	−	0	Saddle point

According to the above discussion on the unilateral strategies of power plants and the government (Sections 2.5.1 and 2.5.2), it can be seen that the characteristic roots corresponding to the point $(x^*, y^*) = \left(\frac{F-M}{S+F}, \frac{EC_c+C-EC_b-B}{S+F}\right)$ are a pair of pure virtual roots. According to the relevant literature [52], (x^*, y^*) is the saddle point. That is to say, the mixed evolutionary game matrix of power plants and the government has no stable equilibrium point, and the game strategies of power plants and the government cannot be stable, always changing according to the strategies of the other side.

3. System Dynamics Model Construction and Simulation Analysis Based on an Evolutionary Game

System dynamics is a scientific method used to analyze and study the dynamic and dialectical relationships among information feedback systems, the core of which is to explore the causal feedback relationships among various factors in a model [53]. In consideration of the fact that the evolutionary game is limited to the explanatory description of the process and equilibrium results, it cannot describe the decision-making evolution paths of the government and power plants. In order to more profoundly explain and verify the model structure and changes in the players' behavior strategies, as well as explore the influence paths of different factors in the model on the behavior strategies of power plants and the government, the method of combining game theory and system dynamics is used to model and simulate the strategies of power plants and the government based on the results of the evolutionary game model.

3.1. System Dynamics Model Construction

In this study, the system dynamics simulation software VENSIM PLE 7.3.5 is used to model and simulate the game behavior strategies of power plants and the government, the dynamic processes of the decision changes of both sides, and to build the stock and flow diagram of a system dynamics model of the behavior strategies, as shown in Figure 3. The model includes two level variables, two rate variables, four auxiliary variables, and nine constants. The two level variables are the probability that power plants choose coupled power generation and the probability that the government subsidizes and regulates, respectively. The two rate variables are the rate of change in the probability that power plants choose coupled power generation and the rate of change in the probability that the government subsidizes and regulates, respectively. The four auxiliary variables are the expected benefits of power plants choosing coupled power generation, the expected benefits of power plants not choosing coupled power generation, the expected benefits of the government subsidizing and regulating, and the expected benefits of the government not subsidizing and regulating. The nine constants are the economic benefit of coupled power generation technology, the environmental benefits of coupled power generation, the indirect benefits of coupled power generation, the transformation cost of coupled power generation, the government subsidy amount, the government regulation cost, the government environmental treatment cost, the economic benefits of coal-fired power generation, and the carbon emission charge of coal-fired power generation.

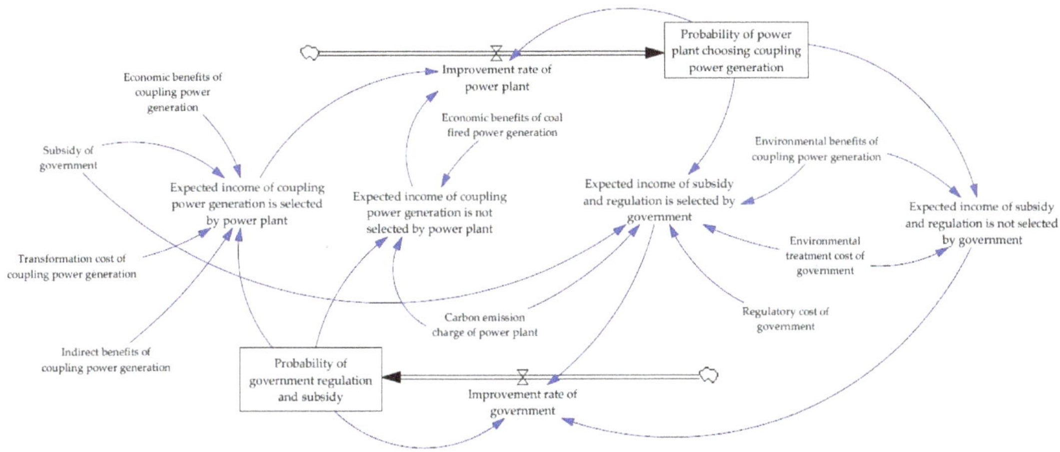

Figure 3. The stock and flow diagram of a system dynamics model of the government and power plants.

3.2. *Analysis of Simulation Results*

The initial parameters of the model are set as INITIAL TIME = 0, FINAL TIME = 100, and TIME STEP = 1. Based on existing research [26,28,54,55] and field survey data, the following values were assigned to relevant variables: $EC_c = 10$, $EC_b = 8$, $EN_b = 2$, $C = 4$, $B = 2$, $F = 5$, $S = 2$, $M = 3$, and $G = 2$.

3.2.1. Initial Simulation Results

The initial value of the probability, x, of power plants choosing the coupled power generation strategy is set as 0.5, and the initial value of the probability, y, for the government to choose the regulation and subsidy strategy is also set as 0.5. The simulation results are shown in Figure 4.

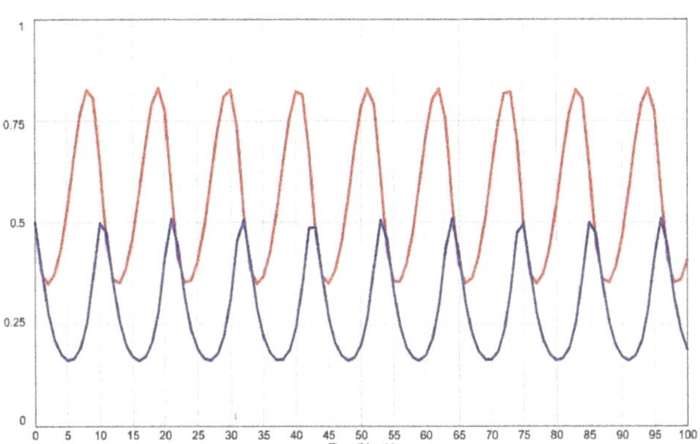

Figure 4. Initial simulation results of an evolutionary game model of power plants and the government.

As shown in Figure 4, the probability value, x, of power plants choosing the coupled power generation strategy fluctuates between [0.15, 0.55], while the probability value, y, of the government choosing the regulation and subsidy strategy fluctuates between [0.35, 0.85]. The strategies of power plants and the government cannot tend to a certain stable strategy, but change their own strategies according to each other in the long run. When the probability of power plants choosing coupled power generation is lower than a certain value, the government will increase the probability of regulating and subsidizing power plant. With the increase in the probability of government regulation and subsidization, power plants will increase their probability of choosing coupled power generation; however, when the probability of power plants choosing coupled power generation increases to a certain extent the government will reduce the probability of regulating and subsidizing power plants, after which the probability of power plants choosing coupled power generation will decrease, and the change trend in the strategies of power plants and the government is recurrent fluctuation in the long term.

3.2.2. The Influence of a Government Subsidy, S, on the Behavioral Strategies of the Players

The initial values of the system simulation remain unchanged and only the values of S are changed, which are set as S = 1 and S = 3. The influences of a reduction (S = 1) and increase (S = 3) in a subsidy on the behavior strategies of power plants and the government are analyzed successively. The simulation results are shown in Figure 5. When the subsidy amount is reduced, the probability value, x, of power plants choosing the coupled power generation strategy fluctuates between [0.2, 0.6], and the probability value, y, of the government choosing the regulation and subsidy strategy fluctuates between [0.4, 0.9]. When the subsidy amount increases, the probability value, x, of power plants choosing the coupled power generation strategy fluctuates between [0.1, 0.5], while the probability value, y, of the government choosing the regulation and subsidy strategy fluctuates between [0.25, 0.8].

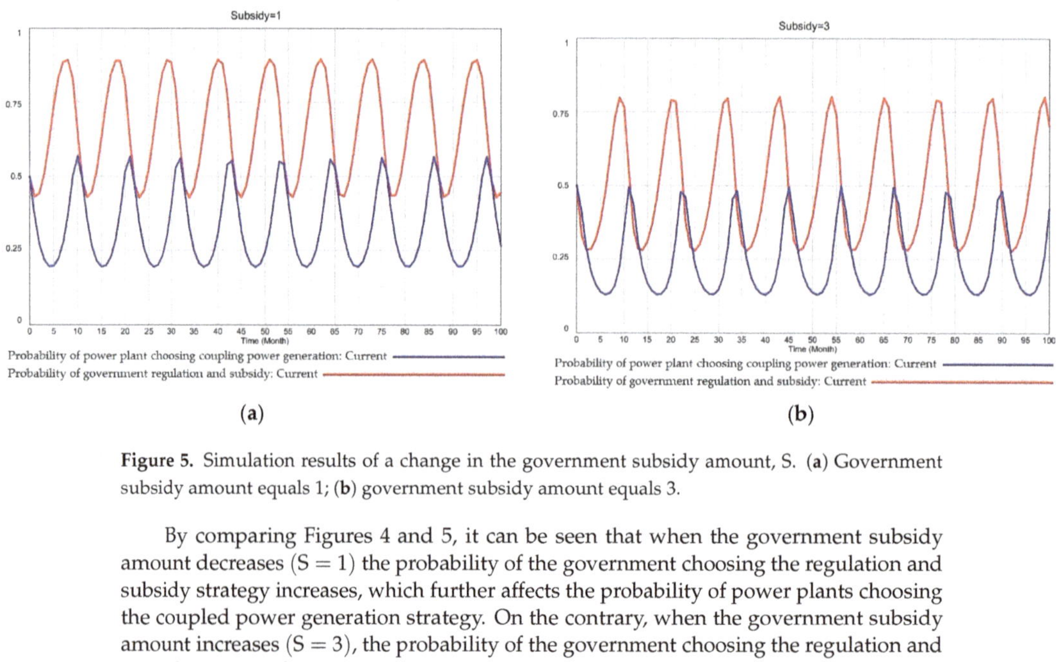

Figure 5. Simulation results of a change in the government subsidy amount, S. (**a**) Government subsidy amount equals 1; (**b**) government subsidy amount equals 3.

By comparing Figures 4 and 5, it can be seen that when the government subsidy amount decreases (S = 1) the probability of the government choosing the regulation and subsidy strategy increases, which further affects the probability of power plants choosing the coupled power generation strategy. On the contrary, when the government subsidy amount increases (S = 3), the probability of the government choosing the regulation and subsidy strategy for power plant decreases, resulting in a decrease in the probability of power plants choosing the coupled power generation strategy. Meanwhile, as can be seen from the change trend in power plants' strategies in Figure 5, when the government subsidy amount, S, decreases, the change in the power plant strategy curve slows down, while when the subsidy amount, S, increases, the change in the power plant strategy curve intensifies. This indicates that a change in the government subsidy amount can not only change the probability of power plants' and the government's strategy selection, but also improve or reduce the degree of influence of the government's strategies on the strategies of power plants. Therefore, if the government wants to promote the development of the coupled power generation industry in a relatively short period of time and improve the probability of power plants choosing coupled power generation technology, it should increase the probability and amounts of subsidies to power plants.

3.2.3. The Influence of the Carbon Emission Charge, F, Paid by Power Plants on the Behavioral Strategies of the Players

The initial value of the system simulation remains unchanged, and only the value of F is changed, that is, the carbon emission charge (i.e., penalty amount) that power plants need to pay for coal-fired power generation under government regulation. The F is set as F = 3.5 and F = 10. The influence of reducing (F = 3.5) and increasing (F = 10) the penalty amount on the behavior strategies of power plants and the government is analyzed successively. The simulation results are shown in Figure 6. When the carbon emission charge to be paid by the power plant is reduced, the probability value, x, of power plants choosing the coupled power generation strategy fluctuates between [0.02, 0.55], while the probability value, y, of the government choosing the regulation and subsidy strategy fluctuates between [0.2, 1]. When the carbon emission charge to be paid by power plants increases the probability value, x, of power plants choosing the coupled power generation strategy fluctuates between [0.4, 0.8], while the probability value, y, of the government choosing the regulation and subsidy strategy fluctuates between [0.2, 0.55].

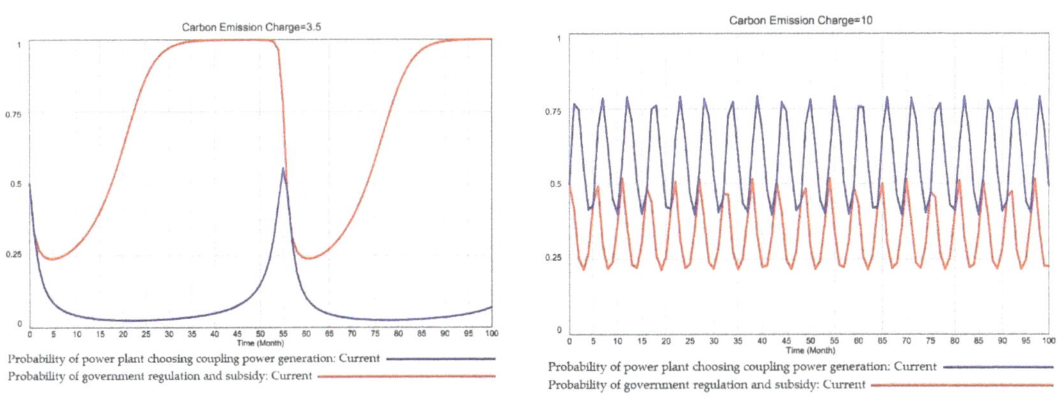

(a) (b)

Figure 6. Simulation results of a change in the carbon emission charge, F. (**a**) Carbon emission charge equals 3.5; (**b**) carbon emission charge equals 10.

By comparing Figures 4 and 6, it can be seen that when the carbon emission charge of power plants is reduced under government regulation (F = 3.5) power plants are very reluctant to choose coupled power generation due to the lax government supervision. Only when power plants notice that the government will definitely choose the strategy of supervision and subsidy (i.e., the probability of government supervision and subsidy is 1), will they improve the probability of choosing the coupled power generation strategy. When the carbon emission charge to be paid by power plants increases (F = 10), since the fee is very high once it is regulated by the government, power plants are reluctant to take risks and prefer to choose the coupled power generation strategy. As long as the government slightly increases the probability of choosing regulation and subsidies, the probability of choosing coupled power generation for the power plant will increase. Therefore, the government will increase the carbon emission charge paid by coal-fired power plants, which will encourage more power plants to choose coupled power generation.

3.2.4. The Influence of the Economic Benefits, EC_b, of Coupled Power Generation on the Behavior Strategies of the Players

The initial simulation values of the system remained unchanged, and only the values of the economic benefits, EC_b, of the coupled power generation of AFB and coal are changed, which are set as $EC_b = 6$ and $EC_b = 12$, respectively, so as to analyze the influence of a reduction ($EC_b = 6$) and increase ($EC_b = 12$) in economic benefits on the behavior strategies of power plants and the government. The simulation results are shown in Figure 7. When the economic benefits of coupled power generation decrease, the probability value, x, of power plants choosing the coupled power generation strategy fluctuates between [0.1, 0.7], while the probability value, y, of the government choosing the regulation and subsidy strategy fluctuates between [0.4, 1]. When the economic benefits of coupled power generation increase, the initial value of the probability, x, of power plants choosing the coupled power generation strategy is between [0.85, 1] and soon becomes stable at 1, while the initial value of the probability, y, of the government choosing the regulation and subsidy strategy is between [0, 0.3] and soon becomes stable at 0.

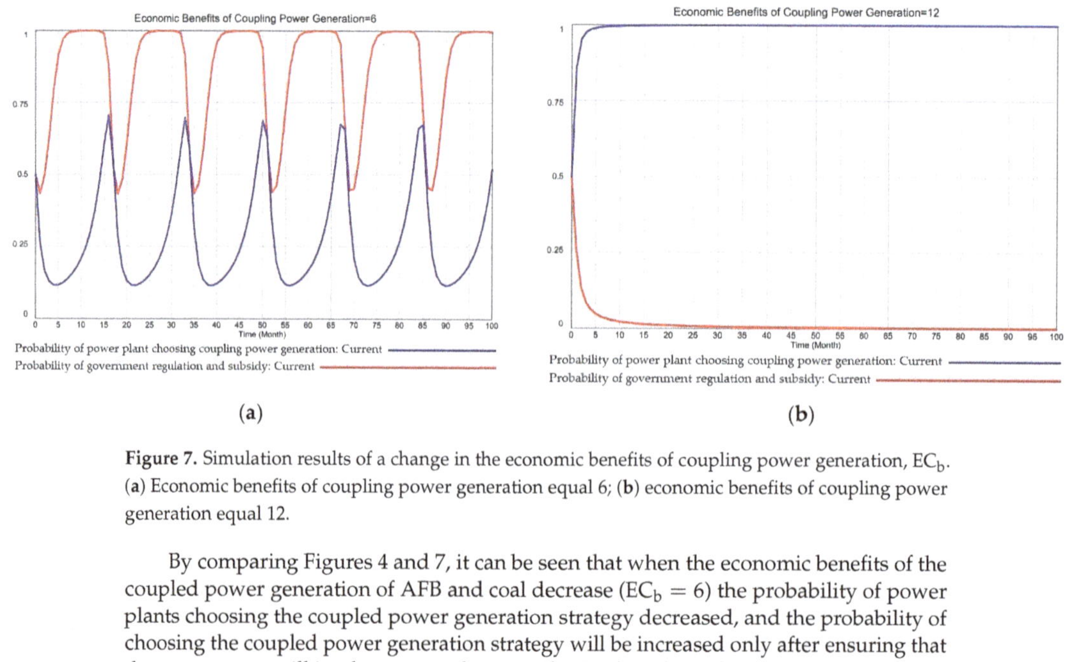

Figure 7. Simulation results of a change in the economic benefits of coupling power generation, EC_b. (**a**) Economic benefits of coupling power generation equal 6; (**b**) economic benefits of coupling power generation equal 12.

By comparing Figures 4 and 7, it can be seen that when the economic benefits of the coupled power generation of AFB and coal decrease ($EC_b = 6$) the probability of power plants choosing the coupled power generation strategy decreased, and the probability of choosing the coupled power generation strategy will be increased only after ensuring that the government will implement regulatory and subsidy policies for power plants (i.e., the probability of government regulation and subsidies is 1). By comparing Figures 4 and 7, it can be seen that when the economic benefits of coupled power generation decrease ($EC_b = 6$) the main reason for this is that the economic benefits generated by coal-fired power generation are far greater than those obtained by coupled power generation. Without strong government regulatory and subsidy policies, power plants will not actively choose coupled power generation. On the contrary, when the economic benefits of coupled power generation increase ($EC_b = 12$) the economic benefits of power plants choosing the coupled power generation of AFB and coal are greater than those of coal-fired power generation. Power plants will actively choose the coupled power generation strategy to obtain larger economic benefits, and the government will no longer supervise and subsidize power plants.

3.2.5. The Influence of the Indirect Benefits, B, of Coupled Power Generation on the Behavior Strategies of the Players

The initial numerical value of the system simulation remains unchanged, and only the indirect benefits, B, generated by the coupled power generation of AFB and coal are changed, which are set as B = 0 and B = 8, respectively, so as to analyze the influence of a reduction (B = 0) and increase (B = 8) in the indirect benefits of coupled power generation on the behavior strategies of power plants and the government. The simulation results are shown in Figure 8. When the indirect benefits of coupled power generation decrease, the probability value, x, of power plants choosing the coupled power generation strategy fluctuates between [0.1, 0.7], and the probability value, y, of the government choosing the regulation and subsidy strategy fluctuates between [0.4, 1]. When the indirect benefits of coupled power generation increase, the initial value of the probability, x, of power plants choosing the coupled power generation strategy is between [0.95, 1] and soon becomes stable at 1, while the initial value of the probability, y, of the government choosing the subsidy and regulation strategy is between [0, 0.25] and soon becomes stable at 0.

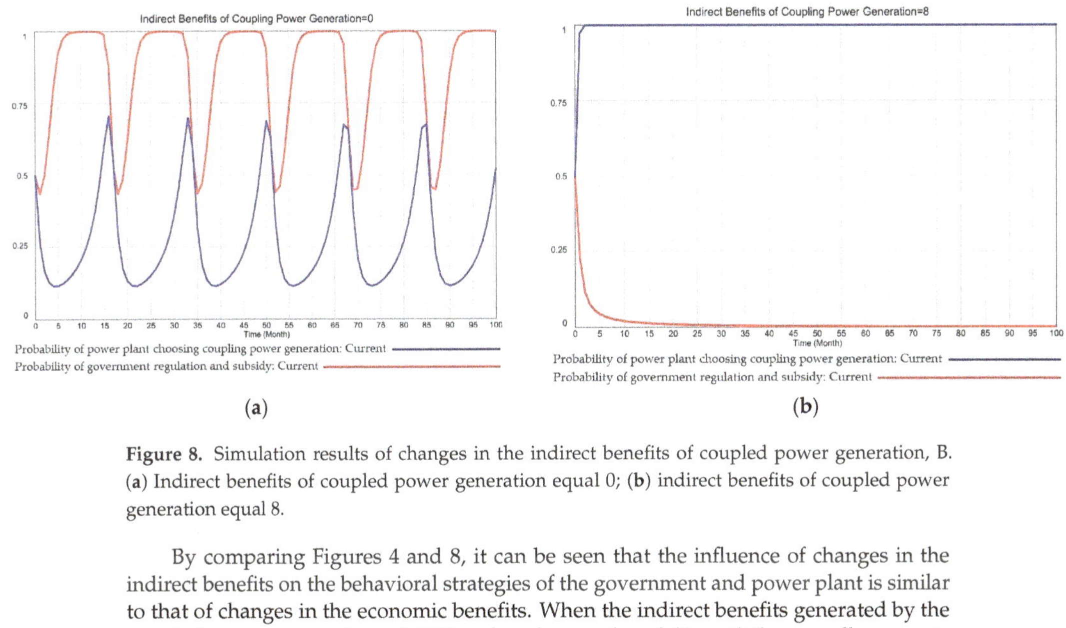

(a) (b)

Figure 8. Simulation results of changes in the indirect benefits of coupled power generation, B. (**a**) Indirect benefits of coupled power generation equal 0; (**b**) indirect benefits of coupled power generation equal 8.

By comparing Figures 4 and 8, it can be seen that the influence of changes in the indirect benefits on the behavioral strategies of the government and power plant is similar to that of changes in the economic benefits. When the indirect benefits generated by the coupled power generation of AFB and coal are reduced (B = 0) the overall economic benefits will also be reduced, and the probability of power plants choosing the coupled power generation strategy will be reduced. Similarly, the probability of choosing the coupled power generation strategy is increased only after ensuring that the government will implement regulatory and subsidy policies for power plants (i.e., the probability of the government adopting regulation and subsidies is 1); however, when the indirect benefits of the coupled power generation of AFB and coal increase (B = 8), if the total economic benefits generated by the coupled power generation of power plants exceed coal-fired power generation, power plants will actively choose coupled power generation and the government will no longer regulate and subsidize power plants.

3.2.6. The Influence of the Transformation Cost, C, of Coupled Power Generation on the Behavior Strategies of the Players

The initial values of system simulation remain unchanged, and only the value of the coupled power generation transformation cost, C, is changed, which is set as C = 0 and C = 6, so as to analyze the influence of a reduction (C = 0) and increase (C = 6) in the transformation cost on the behavior strategies of power plants and the government. The simulation results are shown in Figure 9. When the coupled power generation transformation cost is reduced, the initial value of the probability, x, of power plants choosing the coupled power generation strategy is between [0.85, 1] and soon becomes stable at 1, while the initial value of the probability, y, of the government choosing the subsidy and supervision strategy is between [0, 0.3] and soon becomes stable at 0. When the coupled power generation transformation cost increases, the probability value, x, of power plants choosing the coupled power generation strategy fluctuates between [0.1 and 0.7], and the probability value, y, of the government choosing the regulation and subsidy strategy fluctuates between [0.4, 1].

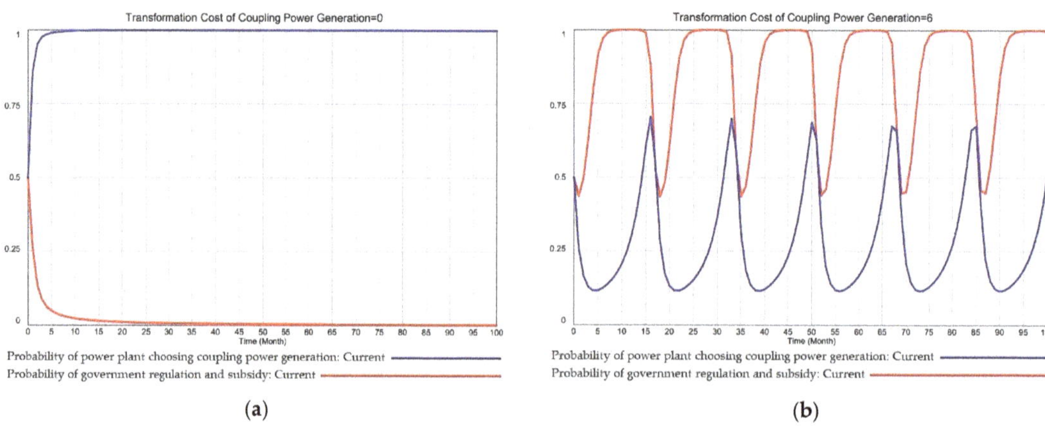

Figure 9. Simulation results of a change in the coupled power generation transformation cost, C. (**a**) Transformation cost of coupled power generation equals 0; (**b**) transformation cost of coupled power generation equals 6.

By comparing Figures 4 and 9, it can be seen that the influence of the coupled power generation transformation cost on the behavior strategies of the government and power plants is exactly the opposite of the influence of the coupled power generation economic benefits and indirect benefits. When there is no transformation cost (C = 0) power plants are very willing to choose the coupled power generation strategy, and the government will no longer regulate and subsidize power plants. When the transformation cost increases (C = 6), power plants need to pay a higher cost to choose the coupled power generation strategy, and the probability of choosing this strategy will decrease, while the government's regulatory and subsidy policies are needed to improve the probability of power plants choosing coupled power generation.

4. Discussion

In order to make up for the lack of consideration of the dynamic changes in government policies in previous research, this study explores the dynamic evolution process of the government and power plants' behavior strategies in the long term by constructing an evolutionary game model and system dynamics game. The influence paths of key factors on the government and power plants' behavior strategies are demonstrated to better formulate policies and measures to promote the development of the coupled power generation of AFB and coal industry. At the same time, this study can provide some references for other countries in the development of the coupled power generation of AFB and coal.

With regard to the results in this study, they can be discussed from three perspectives:

The first is a unilateral behavior strategy analysis of power plants and the government. When the influence of the other side's behavior strategies is not considered, the unilateral behavior strategies of power plants and the government are very stable. For power plants, as long as the government provides sufficient financial subsidies and implements regulatory measures for the coupled power generation of AFB and coal, they will choose coupled power generation. For the government, the essential purpose of subsidies for and the regulation of power plants is to reduce carbon emissions and improve environmental benefits. When many power plants do not adopt coupled power generation technology the government will increase subsidies and regulation to encourage more power plants to choose it.

The second is to study the dynamic evolution processes of power plants and the government's behavior strategies. According to the results of the evolutionary game model, the behavioral strategies of power plants and the government interact with and influence each other from a long-term development perspective, and the behavioral strategies of both sides cannot be stabilized but fluctuate. This is consistent with the research results of Li et al. [56] as well as Gao and Xi [57], that is, it is difficult for fixed policies and measures to effectively promote industrial development, and policies as well as measures should be constantly adjusted to adapt to the changes in enterprise behavior and strategy. When the probability of power plants choosing coupled power generation is lower than a certain value the government will increase the probability of subsidies for and the regulation of power plants. With the increase in the probability of government subsidies and regulation, power plants will increase the probability of choosing coupled power generation; however, when the probability of power plants choosing coupled power generation increases to a certain extent the government will reduce the probability of subsidies and regulation, after which the probability of power plants choosing coupled power generation decreases.

The third is an analysis of the influence paths of the key factors on the government and power plants' behavior strategies. This study selects key factors, such as the government subsidy amount, the carbon emissions charge paid by power plants, the economic benefits of coupled power generation, the indirect benefits of coupled power generation, and the transformation cost of coupled power generation, and analyzes how changes in these factors affect the behavior strategies of the government and power plants. When the government increases the subsidy amount for selected coupled power plants, power plants will be motivated to choose the coupled power generation strategy, which is also in line with the research expectations of Zhai et al. [40] as well as Luo and Miller [41], that is, an increase in the government subsidy amount improves an enterprise's participation enthusiasm, but an increase in the subsidy amount only increases the probability of choosing coupled power generation. The reason for this is that the subsidy amount is an external factors for power plants to choose coupled power generation. Once the external factors disappear, the probability of power plants choosing coupled power generation will be greatly reduced. At present, China has not formulated specific subsidy policies in the field of the coupled power generation industry of AFB and coal; however, there are related subsidy policies for biomass power generation, such as an electricity price subsidy policy. In 2010, the notice issued by the National Energy Administration of China was that the electricity generated by biomass energy was all incorporated into the national grid at a price of RMB 0.75. The price was the sum of the benchmark price and the subsidy price for renewable energy. The benchmark price was RMB 0.3068 and the subsidy price was RMB 0.4432 [58]. This electricity price subsidy has greatly promoted the development of the biomass power generation industry. This is consistent with the conclusion of this study, that subsidy policies can promote the development of the coupled power generation industry of AFB and coal. Similarly, when the carbon emission charge paid by power plants increases, enterprises face a very large risk of being punished with an increase in the probability of government regulation. In order to reduce the risk and cost, enterprises are more willing to choose the coupled power generation strategy. The carbon emission charge is also an external factor. When the carbon emission charge is reduced, enterprises are reluctant to choose coupled power generation because coal-fired power generation can bring more benefits.

However, changes in the economic benefits, indirect benefits, and transformation cost of the coupled power generation of AFB and coal will essentially change the behavioral strategies of power plants and the government. When the economic and indirect benefits generated by coupled power generation exceed the benefits obtained by coal-fired power generation, power plants will independently choose coupled power generation and the government will no longer need to formulate corresponding subsidy policies and regulatory measures. Similarly, when the transformation cost of coupled power generation is zero power plants are very willing to choose the coupled power generation strategy, and the government will no longer subsidize and regulate power plants; however, when the transformation cost is very high, far exceeding the benefits brought by coupled power generation, few power plants are willing to choose it. In this case, the government needs to formulate very high policy subsidies and strict regulatory measures. It can be seen that improving the economic and indirect benefits of coupled power generation in addition to reducing the transformation cost will encourage more power plants to choose it.

Despite the in-depth research in this study, some shortcomings must be admitted. First of all, the relevant subjects involved in this study are the government and power plants, and the subjects in the coupled power generation of AFB and coal also include raw material suppliers, brokers, logistics, and so on. It is hoped that more research subjects can be added in future research. Secondly, the selection of key factors affecting the subjects' strategies is limited in this study, most of which focus on benefits and costs. It is hoped that the selection scope of key factors can be expanded in future studies to bring research closer to reality.

5. Conclusions and Policy Recommendations

This study analyzes and simulates the behavior strategies of power plants and the government in the coupled power generation of AFB and coal, aiming at developing effective policies and measures to promote the development of the coupled power generation industry. In summary, we figured out the interactions and adjustment processes of the behavioral strategies of the government and power plants, and the specific effects of key factors on the behavioral strategies of the government and power plants were clarified. The conclusion is consistent with the hypothesis. The main research conclusions are as follows:

(1) In the process of the development of the coupled power generation of AFB and coal, the behavioral strategies of power plants and the government dynamically influence each other. The government's subsidy strategies and regulatory measures for the coupled power generation industry will drive more power plants to choose it.

(2) The government will increase the amount of policy subsidies and strengthen the punishment of coal-fired power plants, which will encourage more power plants to choose the coupled power generation of AFB and coal. Meanwhile, improving the economic benefits, increasing the indirect benefits, and reducing the transformation cost of coupled power generation can encourage more power plants to choose coupled coal power generation.

(3) The government's corresponding policy regulatory and subsidy measures can only change the enthusiasm of power plants to choose the coupled power generation of AFB and coal. In order to fundamentally enable power plants to independently choose it, it is necessary to increase the income obtained by the choice or reduce the transformation cost of coupled power generation.

According to the above research conclusions, corresponding suggestions are further proposed as follows: (1) Improving the timeliness of industrial policy measures. Since the behavioral strategies of power plants and the government can influence each other, the government should adjust and feed back the behavioral strategies of power plants quickly and accurately when formulating industrial policies, so as to ensure the timeliness of policy measures. (2) Improve the standards of industrial subsidy and punishment mechanisms. The government should subsidize and reward power plants that adopt the coupled power generation of AFB and coal, punish coal-fired power plants, and formulate as well as improve the levels and standards of the corresponding subsidy and punishment mechanisms, which will effectively encourage more power plants to choose coupled power generation. (3) Deepen technological research as well as development and improve the raw material supply system. Increasing the research on as well as innovation of the coupled power generation technology of AFB and coal, in addition to improving the supply systems of AFB raw materials, can reduce the economic cost of coupled power generation and further promote the development of this industry.

Author Contributions: Conceptualization, D.Y., C.Z. and S.W.; methodology, D.Y. and C.Z.; software, D.Y.; validation, D.Y., C.Z. and S.W.; formal analysis, D.Y. and C.Z.; investigation, S.W. and L.Z.; resources, C.Z.; data curation, D.Y. and L.Z.; writing—original draft preparation, D.Y., C.Z. and S.W.; writing—review and editing, D.Y., C.Z. and L.Z.; visualization, C.Z.; supervision, C.Z. and S.W.; project administration, S.W. and C.Z.; funding acquisition, S.W. and C.Z. All authors have read and agreed to the published version of the manuscript.

Funding: This research was funded by National Social Science Fund of China (Grant No. 21BJY247).

Institutional Review Board Statement: Not applicable.

Informed Consent Statement: Not applicable.

Data Availability Statement: Not applicable.

Conflicts of Interest: The authors declare no conflict of interest.

References

1. Mallapaty, S. How China could be carbon neutral by mid-century. *Nature* **2020**, *586*, 482–483. [CrossRef] [PubMed]
2. Franke, K.; Sensfuss, F.; Bernath, C.; Lux, B. Carbon-neutral energy systems and the importance of flexibility options: A case study in China. *Comput. Ind. Eng.* **2021**, *162*, 107712. [CrossRef]
3. Bongaarts, J. Intergovernmental panel on climate change special report on global warming of 1.5 °C Switzerland: IPCC, 2018. *Popul. Dev. Rev.* **2019**, *45*, 251–252. [CrossRef]
4. Ma, X.; Wu, L.; Zhu, Y.; Wu, J.; Qin, Y. Simulation of vegetation carbon sink of arbor forest and carbon mitigation of forestry bioenergy in China. *Int. J. Environ. Res. Public Health* **2022**, *19*, 13507. [CrossRef]
5. Wieruszewski, M.; Mydlarz, K. The potential of the bioenergy market in the European union—An overview of energy biomass resources. *Energies* **2022**, *15*, 9601. [CrossRef]
6. Roni, M.S.; Chowdhury, S.; Mamun, S.; Marufuzzaman, M.; Lein, W.; Johnson, S. Biomass co-firing technology with policies, challenges, and opportunities: A global review. *Renew. Sust. Energy Rev.* **2017**, *78*, 1089–1101. [CrossRef]
7. Aberilla, J.M.; Gallego-Schmid, A.; Azapagic, A. Environmental sustainability of small-scale biomass power technologies for agricultural communities in developing countries. *Renew. Energy* **2019**, *141*, 493–506. [CrossRef]
8. Agbor, E.; Oyedun, A.O.; Zhang, X.; Kumar, A. Integrated techno-economic and environmental assessments of sixty scenarios for co-firing biomass with coal and natural gas. *Appl. Energy* **2016**, *169*, 433–449. [CrossRef]
9. Dundar, B.; Mcgarvey, R.G.; Aguilar, F.X. Optimal multi-state partnerships for woody biomass co-firing incorporating a demand-response function for biomass procurement. *Optim. Lett.* **2022**, *16*, 821–844. [CrossRef]
10. Askarova, A.; Zamorano, M.; Martín-Pascual, J.; Nugymanova, A.; Bolegenova, S. A review of the energy potential of residual biomass for coincineration in Kazakhstan. *Energies* **2022**, *15*, 6482. [CrossRef]
11. National Energy Administration, PRC. The 13th Five-Year Plan for Biomass Energy Development. Available online: http://www.gov.cn/xinwen/2016-12/06/content_5143612.htm (accessed on 28 October 2016).
12. National Energy Administration, PRC. Notice of the Ministry of Ecology and Environment of the People's Republic of China on Carrying out the Technical Transformation Pilot Work of Coal-Fired Coupled Biomass Power Generation. Available online: http://zfxxgk.nea.gov.cn/auto84/201712/t20171204_3065.htm (accessed on 27 November 2017).
13. National Energy Administration, PRC. Notice on the Publication of the Catalogue of Additional Fund Subsidies for Electricity Price of Renewable Energy (The Seventh Batch). Available online: http://www.gov.cn/xinwen/2018-06/17/content_5299251.htm (accessed on 17 June 2018).
14. National Development and Reform Commission, PRC. Guiding Catalogue of Industrial Structure Adjustment (2019 Edition). Available online: https://www.ndrc.gov.cn/fgsj/tjsj/cyfz/zzyfz/201911/t20191106_1201659.html (accessed on 6 November 2019).
15. National Development and Reform Commission, PRC. Supplementary Notice on Relevant Issues of Several Opinions on Promoting the Healthy Development of Power Generation from Non-Water Renewable Energy Sources. Available online: http://jjs.mof.gov.cn/zhengcefagui/202010/t20201015_3604104.htm (accessed on 29 September 2020).
16. Liu, H.; Ye, C.; Zhao, Y.; Li, G.; Xu, Y.; Tang, Y.; Luo, G.; Wang, Q. Performance analysis of biomass gasification coupled with ultra-supercritical power generation system. *Chem. Eng. Process. Process Intensif.* **2022**, *179*, 109093. [CrossRef]

17. Ouyang, D.; Han, Y.; Wang, F.; Zhao, X. All-iron ions mediated electron transfer for biomass pretreatment coupling with direct generation of electricity from lignocellulose. *Bioresour. Technol.* **2021**, *344*, 126189. [CrossRef]
18. Kamari, M.L.; Maleki, A.; Nazari, M.A.; Sadeghi, M.; Rosen, M.A.; Pourfayaz, F. Assessment of a biomass-based polygeneration plant for combined power, heat, bioethanol and biogas. *Appl. Therm. Eng.* **2021**, *198*, 117425. [CrossRef]
19. Cai, Y.; Tay, K.; Zheng, Z.; Yang, W.; Wang, H.; Zeng, G.; Li, Z.; Siah, K.B.; Prabakaran, S. Modeling of ash formation and deposition processes in coal and biomass fired boilers: A comprehensive review. *Appl. Energy* **2018**, *230*, 1447–1544. [CrossRef]
20. Wang, X.; Rahman, Z.U.; Lv, Z.; Zhu, Y.; Ruan, R.; Deng, S.; Zhang, L.; Tan, H. Experimental study and design of biomass co-firing in a full-scale coal-fired furnace with storage pulverizing system. *Agronomy* **2021**, *11*, 810. [CrossRef]
21. Krzywanski, J.; Czakiert, T.; Zylka, A.; Nowak, W.; Sosnowski, M.; Grabowska, K.; Skrobek, D.; Sztekler, K.; Kulakowska, A.; Ashraf, W.M.; et al. Modelling of SO_2 and NO_x emissions from coal and biomass combustion in Air-Firing, Oxyfuel, iG-CLC, and CLOU conditions by fuzzy logic approach. *Energies* **2022**, *15*, 8095. [CrossRef]
22. Agbor, E.; Zhang, X.; Kumar, A. A review of biomass co-firing in North America. *Renew. Sust. Energy Rev.* **2014**, *40*, 930–943. [CrossRef]
23. Truong, A.H.; Ha-Duong, M.; Tran, H.A. Economics of co-firing rice straw in coal power plants in Vietnam. *Renew. Sust. Energy Rev.* **2022**, *154*, 111742. [CrossRef]
24. Yang, B.; Wei, Y.M.; Liu, L.C.; Hou, Y.B.; Zhang, K.; Yang, L.; Feng, Y. Life cycle cost assessment of biomass co-firing power plants with CO_2 capture and storage considering multiple incentives. *Energy Econ.* **2021**, *96*, 105173. [CrossRef]
25. Sugiyono, A.; Febijanto, I.; Hilmawan, E.; Adiarso. Potential of biomass and coal co-firing power plants in Indonesia: A PESTEL analysis. *IOP Conf. Ser. Earth Environ. Sci.* **2022**, *963*, 012007. [CrossRef]
26. Nawaz, Z.; Ali, U. Techno-economic evaluation of different operating scenarios for indigenous and imported coal blends and biomass co-firing on supercritical coal fired power plant performance. *Energy* **2020**, *212*, 118721. [CrossRef]
27. Ashizawa, M.; Otaka, M.; Yamamoto, H.; Akisawa, A. CO_2 emissions and economy of co-firing carbonized wood pellets at coal-fired power plants: The case of overseas production of pellets and use in Japan. *Energies* **2022**, *15*, 1770. [CrossRef]
28. Picciano, P.; Aguilar, F.X.; Burtraw, D.; Mirzaee, A. Environmental and socio-economic implications of woody biomass co-firing at coal-fired power plants. *Resour. Energy Econ.* **2022**, *68*, 101296. [CrossRef]
29. Xu, Y.; Yang, K.; Zhou, J.; Zhao, G. Coal-biomass co-firing power generation technology: Current status, challenges and policy implications. *Sustainability* **2020**, *12*, 3692. [CrossRef]
30. Lintunen, J.; Kangas, H.L. The case of co-firing: The market level effects of subsidizing biomass co-combustion. *Energy Econ.* **2010**, *32*, 694–701. [CrossRef]
31. Nurariffudin, M.; Hashim, H.; Tin, L.C. Spatial biomass resource planning framework for co-firing under carbon policy scheme. *Chem. Eng. Trans.* **2018**, *63*, 445–450. [CrossRef]
32. Xu, J.; Huang, Q.; Lv, C.; Feng, Q.; Wang, F. Carbon emissions reductions oriented dynamic equilibrium strategy using biomass-coal co-firing. *Energy Policy* **2018**, *123*, 184–197. [CrossRef]
33. Tevapitak, K.; Helmsing, A.H.J.B. The interaction between local governments and stakeholders in environmental management: The case of water pollution by SMEs in Thailand. *J. Environ. Manag.* **2019**, *247*, 840–848. [CrossRef]
34. Moledina, A.A.; Coggins, J.S.; Polasky, S.; Costello, C. Dynamic environmental policy with strategic firms: Prices versus quantities. *J. Environ. Econ. Manag.* **2003**, *45*, 356–376. [CrossRef]
35. Li, B.; Geng, Y.; Xia, X.; Qiao, D. The impact of government subsidies on the low-carbon supply chain based on carbon emission reduction level. *Int. J. Environ. Res. Public Health* **2021**, *18*, 7603. [CrossRef]
36. Galik, C.S.; Benedum, M.E.; Kauffman, M.; Becker, D.R. Opportunities and barriers to forest biomass energy: A case study of four U.S. states. *Biomass Bioenergy* **2021**, *148*, 106035. [CrossRef]
37. Nasiri, F.; Zaccour, G. An exploratory game-theoretic analysis of biomass electricity generation supply chain. *Energy Policy* **2009**, *37*, 4514–4522. [CrossRef]
38. Tan, Q.; Wang, T.; Zhang, Y.; Miao, X.; Zhu, J. Nonlinear multi-objective optimization model for a biomass direct-fired power generation supply chain using a case study in China. *Energy* **2017**, *139*, 1066–1079. [CrossRef]
39. Zhang, X.; Luo, K.; Tan, Q. A game theory analysis of China's agri-biomass-based power generation supply chain: A co-opetition strategy. *Energy Procedia* **2017**, *105*, 168–173. [CrossRef]
40. Zhai, M.; Zhang, X.; Cheng, F.; Zhou, X.; Su, X. A game-theoretic analysis of the government's role on the biomass supply chain construction. *Int. J. Ambient Energy* **2017**, *38*, 444–458. [CrossRef]
41. Luo, Y.; Miller, S.A. Using game theory to resolve the "chicken and egg" situation in promoting cellulosic bioenergy development. *Ecol. Econ.* **2017**, *135*, 29–41. [CrossRef]
42. Friedman, D. Evolutionary games in economics. *J. Econom.* **1991**, *59*, 637–666. [CrossRef]
43. Alghamdi, F.; Tatari, O.; Alghamdi, L. Enhancing the decision-making process for public-private partnerships infrastructure projects: A socio-economic system dynamic approach. *J. Eng. Appl. Sci.* **2022**, *69*, 67. [CrossRef]
44. Xing, Y.; Deng, X. Evolutionary game model study of construction green supply chain management under the government intervention. *IOP Conf. Ser. Earth Environ. Sci.* **2017**, *94*, 012059. [CrossRef]
45. Mahmoudi, R.; Rasti-Barzoki, M. Sustainable supply chains under government intervention with a real-world case study: An evolutionary game theoretic approach. *Comput. Ind. Eng.* **2018**, *116*, 130–143. [CrossRef]

46. Jamali, M.B.; Rasti-barzoki, M.; Khosroshahi, H.; Altmann, J. An evolutionary game-theoretic approach to study the technological transformation of the industrial sector toward renewable electricity procurement: A case study of Iran. *Appl. Energy* **2022**, *318*, 119083. [CrossRef]
47. Chen, W.; Zeng, Y.; Xu, C. Energy storage subsidy estimation for microgrid: A real option game-theoretic approach. *Appl. Energy* **2019**, *239*, 373–382. [CrossRef]
48. Kanwal, F.; Ahmed, A.; Jamil, F.; Rafiq, S.; Ayub, H.M.U.; Ghauri, M.; Khurram, M.S.; Munir, S.; Inayat, A.; Abu Bakar, M.S.; et al. Co-combustion of blends of coal and underutilised biomass residues for environmental friendly electrical energy production. *Sustainability* **2021**, *13*, 4881. [CrossRef]
49. Juan, J.S.; Sy, C. Multi-objective target-oriented robust optimization of biomass co-firing networks under quality uncertainty. *J. Sustain. Dev. Energy Water Environ. Syst.* **2021**, *9*, 1080364. [CrossRef]
50. Khademi, A.; Eksioglu, S. Optimal governmental incentives for biomass cofiring to reduce emissions in the short-term. *IISE Trans.* **2021**, *53*, 883–896. [CrossRef]
51. San Juan, J.L.G.; Aviso, K.B.; Tan, R.R.; Sy, C.L. A multi-objective optimization model for the design of biomass co-firing networks integrating feedstock quality considerations. *Energies* **2019**, *12*, 2252. [CrossRef]
52. Taylor, P.D.; Jonker, L.B. Evolutionary stable strategies and game dynamics. *Math. Biosci.* **1978**, *40*, 145–156. [CrossRef]
53. Garces, E.; Franco, C.J.; Tomei, J.; Dyner, I. Sustainable electricity supply for small off-grid communities in Colombia: A system dynamics approach. *Energy Policy* **2023**, *172*, 113314. [CrossRef]
54. Cebrucean, D.; Cebrucean, V.; Ionel, I. Modeling and performance analysis of subcritical and supercritical coal-fired power plants with biomass co-firing and CO_2 capture. *Clean Technol. Environ. Policy* **2020**, *22*, 153–169. [CrossRef]
55. Ko, S.; Lautala, P. Advanced woody biomass logistics for co-firing in existing coal power plant: A case study of the great lakes states. *Transp. Res. Rec.* **2018**, *2672*, 93–106. [CrossRef]
56. Li, Y.; Yang, T.; Zhang, Y. Evolutionary game theory-based system dynamics modeling for community solid waste classification in China. *Util. Policy* **2022**, *79*, 101451. [CrossRef]
57. Gao, X.; Xi, Z. Evolutionary game of government and enterprise pollution discharge behavior under combined measures. *China Environ. Sci.* **2020**, *40*, 5484–5492. (In Chinese) [CrossRef]
58. National Energy Administration of China. Notice on Improving the Price Policy of Agricultural and Forest Biomass Power Generation. Available online: http://www.nea.gov.cn/2010-07/28/c_131097727.htm (accessed on 28 July 2010).

water

MDPI

Article

The Impact of Water Resources Tax Policy on Water Saving Behavior

Lei Wang [1], Muniba [2,*], Zoltán Lakner [3,4] and József Popp [5,6]

1 YueTai Medical Treatment (Shandong) Limited Company, Taian 253000, China
2 School of Insurance and Economics, University of International Business and Economics, Beijing 100029, China
3 Department of Agricultural Business and Economics, Institute of Agricultural and Food Economics, Hungarian University of Agriculture and Life Sciences, 2100 Budapest, Hungary
4 Faculty of Agriculture, University of Science and Technology, Aliero 863102, Nigeria
5 Hungarian National Bank–Research Center, John von Neumann University, 6000 Kecskemét, Hungary
6 College of Business and Economics, University of Johannesburg, Johannesburg 2006, South Africa
* Correspondence: munibauibe@hotmail.com

Abstract: The Chinese water administration department has continuously explored and formulated regulatory and market-oriented water control policies to alleviate the contradiction between water shortage and economic and social development and promote the new idea of 'water-saving first' water control. Among them, implementing a water resources tax policy as a price means has achieved initial success. The water-saving effect of water resources tax collection is one of the important bases for determining whether the tax reform will be promoted nationwide in the next stage. Based on this, taking Hubei Province, the first tax reform pilot in China, as an example, water resource elements are integrated into the economic system and a dynamic stochastic general equilibrium model (DSGE) is constructed, embedded in water resources tax to simulate the persistent impact of such a tax on water saving objectives. The research shows that: (1) A water resources tax can effectively achieve the goal of water-saving and improve the utilization efficiency of water resources. (2) Levying a water resources tax helps to improve the water-saving awareness of enterprises and residents and promotes enterprises to optimize their production structure. (3) Rational and efficient use of special water resources protection funds is the basis for ensuring the effective implementation of a water resources tax. It can also improve the recycling capacity of water resources. This means that the government should speed up the exploration of the relationship between supply and demand for comprehensive water resources, to establish a reasonable range of water resources tax rates to guarantee people's livelihoods, and to accelerate the construction of water resources tax guarantee measures, in order to achieve a relatively steady-state of water resources utilization and protection, realizing the dual goal of sustainable economic development and sustainable use of water resources.

Keywords: water resources tax; water saving behavior; DSGE model; policy effect; Hubei province

1. Introduction

The uneven spatial and temporal distribution of water resources, frequent water drought disasters, human activities interfering with the water cycle, and low water use efficiency are gradually becoming major factors hindering China's green and sustainable development [1]. To solve this problem, the government is exploring the development of new water resources management policy initiatives, attempting to shift from levying water resources fees to water resources taxes and establishing a water resources protection tax system in the country [2]. Based on the coexistence of water scarcity and sustainable development requirements in China, water administration departments and scholars generally believe that the current design of water resources tax should not consider only its fiscal revenue but should focus more on its important significance in resource conservation, ecological protection, and green development in the field of water resources [3,4]. On the one hand, the water resources tax is an effective compensation for the use of natural resources, which concentrates on the labor value, service value, and ecological value of water resources [5]; on the other hand, a water resources tax effectively curbs the unreasonable demand for water in the region [6], and tax collection and management is more reasonable and transparent [7]. It improves the efficiency of water use and reduces the exploitation of groundwater, which plays a role in the protection of water resources [8–10], thus promoting environmental protection to a certain extent. In addition, from the various practices of water resources tax in foreign countries, the water-saving effect is also significant [11,12], but it also has a severe negative impact on agricultural production [13]. Therefore, the water-saving effect must be combined with the country's actual situation. If there is a lack of experience and inefficient water resources management, the policy effect of a water resources tax may be unsatisfactory [14].

Since China is in the critical period of transforming from a water resources fee to a water resources tax, Chinese scholars have researched the applicability and consensuality of the water resources tax system. Ref. [15] proposed that the water resources tax be administered to match the country's water conditions and geography. A water resources tax administration model with Chinese characteristics should be established. As the water resources tax pilot work found, the water resources tax burden standard should be adjusted simultaneously with change in the local economic level. The improvement of the ecological environment [16] and the earmarking of tax revenue should also be clarified to emphasize enhancement of the regulating function of the water resources tax [17]. However, implementing a water resources tax policy in the short term also harms socio-economic development. Ref. [18] point out that the water resources tax has increased the tax burden of urban water supply enterprises. Ref. [19] argued that the current water resources tax levy approach had not been applied to agricultural water use and the efficiency of agricultural water use is still low, and proposed an agricultural water resources tax levy and management model that is tailored to local conditions and is simple, rather than difficult. Ref. [20] found that although the introduction of a water resources tax has raised the awareness of water conservation, reduced the water demand of enterprises, and improved the efficiency of water use in various sectors, it has not played a significant role in reducing the total amount of water used in multiple sectors. There are still specific problems in tax collection and administration.

As a new tax system, the impact of the water resources tax on residents' life, enterprise production, and social development still need to be studied in depth. Regarding its characteristics, the water resources tax is a fiscal policy tool. Its regulating effect is not only for current economic benefit and water-saving effects but also should be based on protecting water resources. The DSGE model can meet the current demand for a more comprehensive study of the water resources tax. Since [21] first used the DSGE model to study the time-series characteristics of the U.S. macro economy, the effects of energy price shocks [22], the impact of stochastic technology shocks [23], and coal and carbon resources [24] have been continuously introduced to optimize this. Subsequently, [25] used a DSGE model to simulate the dynamic responses of total output and environmental quality before and after the carbon tax, comparing and analyzing the effects of imposing a carbon tax, and increasing the carbon tax rate under different carbon emission intensities. Ref. [26] used a DSGE model to find that increasing the tax rate could reduce carbon emissions in the power sector. Still, at the same time, the price of electricity would also increase, making carbon emissions reductions and economic growth difficult to achieve simultaneously. Ref. [27] introduced the effects of environmental technologies and energy prices in DSGE to simulate the response changes of the ecological–economic system. Ref. [28] used a DSGE model to conclude that financing constraints can amplify the impact of fuel tax shocks, and the stronger the constraints, the more pronounced the stimulating effect on the economy.

Reviewing the existing literature, there is a lack of research on the overall and lasting effects of water resources tax policies, and most of the research analyses are short-term or static, such as using the CGE model to study the optimal tax rate for water resources [29], which has not yet formed an internal logical structure for the impact of water resources tax policies on economic growth. It is difficult to analyze further the transmission path of a water resources tax on economic impact. On the other hand, although the theoretical basis of the DSGE model can solve the above problems, there is a lack of a DSGE model for extended water resources, and the available references come from the research results on the DSGE model on carbon taxation. Based on this, this paper constructs a DSGE model incorporating water resources based on the characteristics and economic value of water resources and takes the first pilot tax reform in Hubei Province as an example to simulate the long-term dynamic response mechanism of the water resources tax in order to explore how the behavior of micro-entities influences decisions and to evaluate the effect of water resources tax on social water conservation and water resources protection from the perspective of long-term development. The DSGE model was used to simulate the long-term dynamic response mechanism of water taxation in Hubei Province, the first pilot project.

2. Theoretical Background

As a kind of resource tax, the analysis of the water resources tax policy effect is in line with the paradigm of fiscal policy analysis [30]. In this paper, the policy effect of the water resources tax refers to its economic impact and the combined effect of water users' change in water consumption behavior. Therefore, the analysis of the behavior of micro-actors directly affected by the implementation of the water resources tax policy can help to understand its transmission mechanism and help to understand why each actor chooses to conserve water resources and reduce unreasonable water demand through such a tax; moreover, it can help to reflect the way to achieve the structural change in water consumption and the restructuring of water abstraction by the differential tax rate. The analysis of the effect of the water resources tax policy can explain how the water resources tax can alleviate the contradiction between the supply and demand of water resources, change the water supply structure and promote industrial upgrading, as well as to infer and argue for its impact on economic development.

2.1. Mechanisms of the Impact of Water Resources Tax on People's Lives

As per the current policy document of the pilot tax reform, the water resources tax follows the principle of 'Tax and fee for translation'. Therefore, the reform measures implemented in the current tax reform pilot will have almost no direct impact on residents. However, in the long term, the existing taxation principle makes it difficult to adjust the price of water in a short time. Still, the price of water is bound to increase in the future because the water resource fee that residents should bear is temporarily transferred to the urban public water supply sector. In addition, with the improvement of the water resource tax system and the promotion of tax reform, residents will gradually realize the importance and urgency of water conservation and water resource protection, thus reducing the unreasonable demand for water and achieving the goal of protecting water resources.

As the water resources tax reform is at its early stage, each pilot area's overall tax rate is still low. Still, the demand for water resources in social development and the scarcity of water resources increase the value of water resources. Therefore, it is reasonable to consider the increase in the water resources tax rate in this paper, and this will eventually reach a reasonable range that integrates the supply and demand of water resources and the protection of people's livelihood.

2.2. The Mechanism of the Impact of Water Resources Tax on Enterprise Production

The reverse regulation of a water resources tax reduces the production scale of high water-consuming enterprises, promotes the introduction of water-saving technology, and optimizes the industrial structure. As current high water-consuming enterprises generally do not use water resources efficiently, the levy of a water resources tax will significantly impact high water-consuming enterprises. From the perspective of the production chain, as the production cost of high water-consuming enterprises increases, to keep their profits unharmed they transfer the tax burden to downstream enterprises or consumers by raising the price of their products in the short term [31], and the resulting impact can be divided into the following two cases.

Case 1: The high water-consuming enterprises produce products that are inelastic in demand, and the transfer of the tax burden ensures that their interests are not damaged because the products made are in immediate demand, and consumers who buy such products bear the pressure of the tax burden, which reduces the consumption level of consumers and damages their interests.

Case 2: The high water-consuming firm produces an elastic product, and the price increase makes consumers choose substitutable products—the higher costs caused by the tax need to be borne by themselves.

Since the current water resources tax is based on the principle of 'Tax and fee for translation' and is an in-price tax, Case 1 is not in line with the actual situation, while the comprehensive result of Case 2 is that a high water-consuming enterprise will reduce the scale of production, improve the efficiency of water use, and promote the optimization and upgrading of its industrial structure. The positive regulation of a water resources tax enables enterprises to innovate water conservation techniques and guides them to switch from wasteful to economic production and use of water. Since the water resources tax is a general, special tax, the tax revenue is used for the construction of water conservation facilities and for the incentive of water conservation effectiveness. On the other hand, in a fully competitive market, innovative water-saving enterprises are rewarded for their low production costs and water-saving production, which increases the market competitiveness of their products and motivates their competitors to transform to water-saving production, further spreading the impact to the whole production chain. The effect will be transmitted to the entire production chain, leading to the optimization and upgrading of industrial structure and improving the efficiency of water use.

2.3. Comprehensive Impact Mechanism of Water Resources Tax on Social Development

The water resources tax levy will have a negative impact on economic development in the short term. At the same time, the economy tends to stabilize in the long term, the efficiency of water use increases, and the value of water use increases. By the national income accounting expenditure method, the main driving factors of a country's economy include consumption, investment, and government spending, if only closed economic conditions are considered. In the short term, the reduction in consumption occurs instantaneously. It is directly reflected in economic indicators, while the promotion effect of investment on the economy takes a longer time to respond. With the gradual emergence of investment effects, the widespread popularity of water-saving technologies, the improvement of water resources utilization efficiency, and the decline of production water costs, the level of wages and benefits and consumption levels will return to a steady state. At that time, the supply of water resources meets the demand for water resources. The water ecological environment, the natural cycle of water, and the ecological functions of water bodies are protected. Therefore, although the water resources tax reform needs to pay the price of damaged economic development in the short term, from the perspective of long-term sustainable development the water resources tax is the most effective means to solve the contradiction between the current water shortage and the growing water demand in China.

3. Material and Methods

3.1. Data

The water resources data used in this study are from the Hubei Water Resources Bulletin from 2001 to 2020. The data on social and economic indicators such as capital volume and GDP are from the China Statistical Yearbook from 1990 to 2020. Data that cannot be obtained directly from statistical data are estimated according to previous research methods.

3.2. Assumptions

Dynamic stochastic general equilibrium is a kind of equilibrium state that various economic entities achieve in pursuing their respective goals according to their behavior rules and habits. Therefore, to simplify the complex reality, the DSGE model constructed in this paper is a closed model, including households, firms, and government. The operation of the whole economic system is simulated by constructing the behavior equation of each micro subject. To establish the model equation, it is necessary to put forward corresponding assumptions from the four aspects of production, consumption, factor capital, and government departments before constructing the model. The following four model assumptions are proposed regarding the previous research experience [32] and combined with the research needs of the water resources tax issue.

Assumption 1: On the production side, the market structure is assumed to be a perfectly competitive market in which manufacturers follow the principles of cost minimization and profit maximization when producing, and the production function has the characteristic of constant returns to scale. Assumption 2: On the consumption side, residents consume various goods according to the principle of utility maximization, and their utility is modeled using the constant relative risk aversion (CRRA) function. Assumption 3: Capital, labor, and water resources are freely mobilizable, while wages conform to the supply and demand theorem and can change instantaneously. Assumption 4: Only the government department levies the water resources tax, and it is used only for water resources protection management and infrastructure construction expenditure.

3.3. Hypothesis

To simplify the complex reality, the DSGE model with embedded water resources tax is a closed model that includes three sectors, households, firms, and government, and simulates the operation of the whole economic system by constructing the behavior equations of each micro-actor. To make the model equations hold, the corresponding assumptions are made before constructing the model. Regarding the previous research experience [33–35] and combined with the research needs of the water resources tax issue, the following four model assumptions are proposed.

(1) **Hypothesis 1**: On the production side, the market structure is assumed to be a perfectly competitive market in which manufacturers follow the principles of cost minimization and profit maximization when producing, and the production function has the characteristic of constant returns to scale.
(2) **Hypothesis 2**: On the consumption side, residents consume various goods according to the principle of utility maximization, and their utility is modeled using the constant relative risk aversion (CRRA) function.
(3) **Hypothesis 3**: Capital, labor, and water resources are freely mobilizable, while wages conform to the supply and demand theorem and are free to change instantaneously.
(4) **Hypothesis 4**: only the government department levies the water resources tax, and it is used only for water resources protection management and infrastructure construction expenditure.

3.4. Model

The basic framework of this paper is as an extension of the research results of [36]. Since there is no research on applying the DSGE model to the analysis of the implementation effect of water resources tax reform policy, this paper draws on the research method of [25] on carbon tax. It incorporates water resources as a critical element into the model. Based on the research of [33,34] a water resources embedded DSGE model including households, firms and the government is constructed.

3.4.1. Households

Based on the principle of tax equalization in the water resources "fee-to-tax" reform, theoretically, the water resources tax will not directly impact the households sector, so the construction of the households sector follows the classical RBC construction. Assuming that there is an infinite representative household, the utility of consumption and labor is in the form of CES, and the utility of money balance is in the form of a logarithm to simplify the model. Thus the utility function of the representative household is:

$$\max_{C_t,N_t,B_{t+1},M_t,K_{t+1}} E_0\sum_{t=0}^{\infty}\beta^t\left(\frac{C_t{}^{1-\sigma}}{1-\sigma}-\frac{N_t{}^{1+\eta}}{1+\eta}+\ln\frac{M_t}{P_t}\right) \quad (1)$$

where C denotes consumption, N denotes labor, B denotes bond holdings, I denotes investment, M denotes the quantity of money, P denotes the price level, M/P denotes real money holdings, β denotes the discount factor, δ denotes the inverse of the intertemporal elasticity of substitution of consumption, and η denotes the inverse of the Frisch elasticity of labor supply.

The utility maximization of the residential sector is subject to the constraint that the income in each period is higher than or equal to the expenditure

$$C_t+(K_{t+1}-(1-\delta)K_t)+\frac{B_{t+1}}{P_t}+\frac{M_t-M_{t-1}}{P_t}\le w_tN_t+R_tK_t+(1+i_{t-1})\frac{B_t}{P_t} \quad (2)$$

where $w_t=\frac{W_t}{P_t}$, R^K denotes the return on capital per period and i denotes the bond rate. In $I_t=K_{t+1}-(1-\delta)K_t$, the investment I that the households sector can decide in the utility function can also be replaced by the capital K.

The Lagrange equation can be constructed from the utility function and constraints

$$L=E_0\sum_{t=0}^{\infty}\beta^t\{\frac{C_t{}^{1-\sigma}}{1-\sigma}-\frac{N_t{}^{1+\eta}}{1+\eta}+\ln\frac{M_t}{P_t}+\lambda_t(w_tN_t+R_tK_t+(1+i_{t-1})\frac{B_t}{P_t}-C_t-(K_{t+1}-(1-\delta)K_t)-\frac{B_{t+1}}{P_t}-\frac{M_t-M_{t-1}}{P_t})\} \quad (3)$$

Derivatives for C_t, N_t, K_t, B_t and M_t, respectively, give the following first-order conditions

$$\lambda_t=C_t^{-\sigma} \quad (4)$$

$$N_t^{\eta}=C_t^{-\sigma}w_t \quad (5)$$

$$\lambda_t=\beta E_t(\lambda_{t+1}(R_{t+1}+1-\delta)) \quad (6)$$

$$\lambda_t=\beta E_t(\lambda_{t+1}(1+r_t)) \quad (7)$$

$$\frac{M_t}{P_t}=C_t^{\sigma}\left(\frac{i_t}{1+i_t}\right)^{-1}\# \quad (8)$$

Since money is introduced in the utility function, this paper assumes that the money supply is a non-stationary time series and therefore considers the (logarithmic) growth rate of the money supply M_t as a policy instrument with the following money growth rate equation

$$g_t^m=(1-\rho_m)\log\pi-\log\pi_t+\rho_m g_{t-1}^m+\rho_m\log\pi_{t-1}+\varepsilon_t^m \quad (9)$$

where π denotes the steady-state nominal money supply growth rate. If we use the real money balance m_t CPI inflation rate is π_t, i.e., $m_t=\frac{M_t}{P_t}$, $\pi_t=\frac{P_t}{P_{t-1}}$.

3.4.2. Firms

(1) Introducing the water resources factor into the production function.

Since the levy of a water resources tax will directly affect the production of enterprises, it is necessary to include water resources as a production factor in the production function. In this paper, with reference to the research results of [37,38], the extended form of the CD function is used to introduce water resources into the production function at the same level of production factors as capital and labor supply. The equation of the production function is as follows

$$Y_t = A_t {K_t}^{\alpha} {N_t}^{\lambda} {Z_t}^{1-\alpha-\lambda} \tag{10}$$

where Y_t denotes t period output, α is the output elasticity of capital, and λ is the output elasticity of labor; K_t denotes capital input in t period; N_t denotes labor supply in period t; Z_t denotes water use in t period; and A_t denotes the technology level in t period, assuming that technological progress obeys the AR (1) process, yielding

$$\log A_t = \rho_A \log A_{t-1} + \varepsilon_t^A, \varepsilon_t^A \sim N(0, \sigma_A^2) \tag{11}$$

where ρ_A denotes the duration parameter of the technology shock, ε_t^A denotes the random error under the technology shock, and σ_A denotes the standard deviation.

(2) Introduction of water resources tax into the production cost of enterprises.

In levying the water resource tax, the taxation department levies according to the number of water resources used, so the amount of water resource tax payable by the enterprise in the period t is $T_t Z_t$, where T_t denotes the water resource tax rate in the period t. To simulate the dynamic effect of the water resource tax, it is assumed to obey the AR (1) process

$$T_t = \rho_T T_{t-1} + \varepsilon_t^T, \varepsilon_t^T \sim \left(0, \sigma_T^2\right) \tag{12}$$

where ρ_T denotes the persistence parameter of the water tax shock, ε_t^T denotes the random error of the water tax shock, and σ_T denotes the standard deviation.

The maximization of corporate profits can be expressed as

$$\max \Pi_t = A_t {K_t}^{\alpha} {N_t}^{\lambda} {Z_t}^{1-\alpha-\lambda} - (R_t^K K_t + W_t N_t + T_t Z_t). \tag{13}$$

Because the firm seeks to maximize profit, the optimal first-order condition can be obtained by taking derivatives of K_t N_t and Z_t, respectively.

$$R_t = \alpha A_t {K_t}^{\alpha-1} {N_t}^{\lambda} {Z_t}^{1-\alpha-\lambda} = \alpha \frac{A_t {K_t}^{\alpha} {N_t}^{\lambda} {Z_t}^{1-\alpha-\lambda}}{K_t} = \alpha \frac{Y_t}{K_t} \tag{14}$$

$$W_t = \lambda A_t {K_t}^{\alpha} {N_t}^{\lambda-1} {Z_t}^{1-\alpha-\lambda} = \lambda \frac{A_t {K_t}^{\alpha} {N_t}^{\lambda} {Z_t}^{1-\alpha-\lambda}}{N_t} = \lambda \frac{Y_t}{N_t} \tag{15}$$

$$T_t = (1-\alpha-\lambda) A_t {K_t}^{\alpha} {N_t}^{\lambda} {Z_t}^{-\alpha-\lambda} = \alpha \frac{A_t {K_t}^{\alpha} {N_t}^{\lambda} {Z_t}^{1-\alpha-\lambda}}{Z_t} = (1-\alpha-\lambda) \frac{Y_t}{Z_t} \tag{16}$$

3.4.3. Government

In addition to the effectiveness of water savings, the water resources tax policy effect should also focus on what impact its collection may have on socio-economic development. The water resources tax policy effect is the combined effect on economic development and the water-saving effect of water-saving behavior after optimal choices by a series of actors guided by the government's water-saving objectives. To simplify the model, it is assumed that the source of government tax revenue is the water resources tax. The income tax revenue is set up as a special fund for water resources protection, entirely for water resources protection. The government revenue equation is as follows

$$G_t = T_t Z_t \tag{17}$$

where G_t denotes the t period of government expenditure.

Since water resources are incorporated into the economic system cycle as a factor of production, according to the general equilibrium model idea, water resources need to be resource-constrained to converge to the steady state; based on this, this paper assumes the water cycle equation

$$Z_t = (1-\psi)Z_{t-1} + G_t \cdot Z_t / Y_t \tag{18}$$

where ψ denotes the rate of water depletion per period and Z_t/Y_t represents the amount of water used per unit of output. In the steady state, this equation indicates that the value transfer function of the water resources tax in the next period will compensate for the converted value of water resources depletion in the current period.

3.4.4. Market Equilibrium

When the market reaches clearing equilibrium, the resource constraint is

$$Y_t = C_t + I_t + G_t \tag{19}$$

4. Results

4.1. Parameter Estimation

4.1.1. Calibration of Structural Parameters

The calibration of capital elasticity parameter α and labor elasticity λ can be used to replace capital income indicator by total capital formation, and labor compensation by labor income indicator, and the long-term average value of α can be calculated as 0.49 and λ as 0.5; the subjective discount parameter β can be calculated as 0.976 based on the average value of the historical one-year national bond yield of 2.6% of Ying for Finance; the capital depreciation parameter β is calculated as 0.976 based on the 1990–2020 China Statistical Yearbook; the ratio of long-term average value of depreciation of fixed assets to long-term average value of capital stock in Hubei Province is calibrated to obtain δ as 0.08; the parameter ψ of water resources depletion rate is obtained as a proxy variable based on the water consumption rate of water resources in the Hubei Province Water Resources Bulletin from 2001–2020 as ψ = 0.27; parameter σ is set to 1 by referring to the study of [38]; parameter η is set by referring to [39], calibrated η to 3.

4.1.2. Dynamic Parameter Estimation

In this paper, H.P. filtering is used to obtain the fluctuation components of the observed data. Then the regression is performed by least-squares according to the first-order autoregressive assumption of the shock, the coefficient of the first-order lag term can be used as the persistence parameter, and the standard deviation of the regression can be used as the estimate of the standard deviation of the exogenous shock, taking the natural logarithm of both sides of the output function and making the first-order difference, obtained as follows

$$\log A_{t+1} - \log A_t = \log Y_{t+1} - \log Y_t - \alpha(\log K_{t+1} - \log K_t) - \lambda(\log N_{t+1} - \log N_t) - (1-\alpha-\lambda)(\log Y_{t+1} - \log Y_t) \tag{20}$$

The time series data of GDP Y, capital K, and labor N are substituted into the above equation, and the volatility components are obtained through H.P. filtering. The AR (1) of the tax rate variable T is used as the regression equation to obtain the persistence parameter of the water tax rate.

The regression equation for the monetary variable M with AR (1) of first-order difference is

$$\log M_t - \log M_{t-1} = (1-\rho_m)\log \pi + \rho_m(\log M_{t-1} - \log M_{t-2}) + \varepsilon_t^m \tag{21}$$

The persistence parameter of the monetary shock is obtained by the same method.

The results were obtained as follows.

As shown in Table 1, the regression results of dynamic parameters are all significant and plausible. All parameters are shown in Table 2.

Table 1. Regression results.

Variable	First-Order Lag Coefficient	Standard Error of the Coefficient	*t*-Test	*p* v.	Regression Standard Deviation
Tax rate T	0.518	0.162	3.196	0.004	0.095
Currency M	0.546	0.135	4.045	0.000	0.043

Table 2. Summary of parameter setting values.

Parameter	Representative Meaning	Calibration Value
β	Subjective discount factor	0.976
σ	Reciprocal of the intertemporal elasticity of substitution of consumption	1
η	Inverted Frisch elasticity of labor supply	3
δ	Capital depreciation rate	0.08
α	The elasticity of capital output	0.49
λ	The elasticity of labor output	0.51
ψ	The loss rate of water resources	0.27
ρ_m	Persistent parameters of monetary policy shocks	0.56
ρ_T	Persistent parameters of water resources tax shocks	0.53

4.2. Calculation of Steady-State Values

The steady-state value is generally calculated by setting A to 1 and then manually solving some of the steady-state values based on this. In addition, the steady-state value can be set first for some variables, such as labor; assuming a person works 8 h a day, the steady-state of n can be set first to $1/3$. The long-term per capita water use in Hubei province is 275 m^3/person, and the logarithmic value of 2.44 can be used as the steady-state value of water use z. After setting the steady-state value, Dynare is used to calculate the steady-state value of the whole system.

4.3. Impulse Response Analysis of Water Resources Tax Shocks

4.3.1. Responses of Enterprises to Water Resources Tax Shocks

Compared with the imposition of water resources fees, the overall price of water in Hubei Province increases after the imposition of a water resources tax. In this context, the scenario of a 1% increase in the water resources tax rate is simulated. As seen in Figure S1, the water resources tax shock causes a short-term decline in water consumption and a gradual return to a steady state in the long term. In periods 1–7, the water resources tax rate gradually returns to a steady state, while the effect of water use suppression gradually diminishes. In periods 8–20, the water resources tax rate is steady, while water use is still suppressed; this suggests that the water resources tax has effectively reduced water use and suppressed water demand for a longer time. In periods 1–7, both the effect of raising water prices and reducing water use is gradually diminishing, with the change in water prices being greater than the change in water use and an increase in tax revenues, all of which are used to invest in water conservation and other areas, enhancing the recycling capacity of water resources. In periods 8–20, water price has returned to a stable state, while water consumption is still in gradual recovery. Compared with the initial state, tax revenue also shrink, but at the same time the enhanced recycling capacity of water resources is restored, and the required input also shrinks. Tax revenue is in a relatively stable state. In contrast, water consumption gradually returns to a steady state; compared with the initial state, water use does not shrink, but the water ecosystem's continued restoration of water ecosystems can meet the increasing water demand. In the long run, the increased recycling capacity of water resources meets the growing water demand and ensures sustainable economic and social development.

The impact of the water resources tax reduces labor and lowers the wage level. On the one hand, due to the effects of the water resources tax, the industrial structure is optimized and adjusted, resulting in structural unemployment and reduction of labor quantity; on the other hand, production enterprises, facing the increase in tax rate, will take measures such as introducing water-saving equipment, upgrading water-saving technology, improving the production process, etc., which brings about the improvement of efficiency making labor time decrease. In periods 1–5, there is a vertical recovery of labor, some unemployed people find jobs again, and labor time gradually recovers. After period 6, the recovery of labor slows, and some unemployed people need longer to search for matching positions, while the labor time of employed workers recovers. With the gradual employment of the unemployed, labor returns to a steady state by period 45. A decline in wage levels accompanies a decrease in labor. On the one hand, the impact of the water resources tax makes it more expensive for firms to produce water, with lower output obtained per unit of labor and lower welfare pay; on the other hand, the reduction in the amount of labor also reduces the wages required to be paid. In periods 1–5, there is a vertical decline in wages. Still, due to the gradual recovery of labor time and the increase in the value of unit labor, the wage level gradually increases after period 6. It returns to a steady state in period 65.

The impact of the water resources tax reduces both capital and capital gains. With higher costs, lower output value, and lower profits, the return per unit of capital also decreases, and the nature of capital for profit reduces capital investment. However, with the general implementation of the water resources tax policy, the demand for water-saving technology and water-saving equipment has increased, and the prospect of research and development in water-saving technology and manufacturing of water-saving equipment is promising. The investment in such development will gradually increase. In addition, the government supports enterprises to change their production methods and improve their water-saving capacity, encouraging them to purchase water-saving equipment and develop water-saving technology. They will also invest a lot of capital in water-saving. Although the water resources tax affects the capital investment in the short term, in the long time, due to government support and enterprise demand, the capital factor investment will return to a steady state. In periods 1–3, capital and capital gains continue to decrease simultaneously. In periods 4–65, capital gains gradually recover due to the positive market outlook for water-saving technologies and equipment, and capital investment also recovers, returning to steady-state levels in period 65.

4.3.2. Resident Response to the Water Resources Tax Shock

The shock of water resources tax leads to a decrease in wage levels, a reduction of disposable income of residents, and, consequently, a decrease in consumption. Higher production costs, lower profits, and lower-wage levels for businesses are caused by a higher water resources tax. There is a decline in short-term income of residents as workers directly reduce the level of consumption in the short term, and consumption decreases. From Figure S2, in the period of 1–5, the wage level continues to decline; industrial restructuring, continuous decline in short-term profits, reduction in capital investment, and increase in the unemployed population all aggravate the decline in wage level; in contrast, the decrease in consumption level is alleviated; as the constraint of residents' consumption is tightened, enterprises have to recover funds in time by lowering product prices, the price level can fall, the residents' real money balance rises, and the real purchasing power is elevated, which promotes the recovery of consumption. In periods 6–60, the reduction in unemployed workers, the improvement of the labor efficiency of employed workers, the popularization and application of water-saving technologies, the reduction of water costs, and the recovery of corporate profits promoted the gradual return of the wage level to the steady state. With the recovery of wages, disposable income increased, and the consumption level was further restored, returning to a steady state in 60.

The impact of the water resources tax makes residents save more and promotes the increase of long-term investment. Under the influence of the water resources tax, the decline in corporate profits leads to the reduction of residents' wages, the income of residents in the current period decreases, the opportunity cost of consumption increases, and the propensity of residents to save increases. Under the lower-income level, people tend to worry about future life security, curtail unnecessary current consumption and save to ensure that basic consumption in the future is satisfied. In periods 1–7, the increase in savings gradually slows down. On the one hand, the continuous decline in wage level makes the funds available for saving also gradually shrink; on the other hand, the gradual recovery of consumption makes more funds available for consumption, which weakens the incentive to save. In the 8th period, savings return to a steady state. With the recovery of wages, the increase in income and consumption reaches a relative balance, and the share of savings remains unchanged. Since saving funds equals investable social funds, saving is the supply of funds, and investment is the demand for funds. Both are two sides of the same coin, so the dynamics of long-term investment should be the same as saving.

The impact of the water resources tax has reduced the real money balance held by residents. On the one hand, the decline in wages reduces residents' disposable income; on the other hand, with the continued impact of the water resources tax, there is a strong demand for water-saving equipment and technologies, which gives companies related to the development of water-saving technologies an incentive to obtain R&D financing by rising bond rates. The increase in bond returns makes the opportunity cost of holding money higher for residents, who will allocate more of their wealth to bond investments. The water resources tax only raises the inflation rate in period 1. In contrast, from period 2 until the steady-state, the implementation of the water resources tax, in turn, lowers the inflation rate, raises real purchasing power, and promotes residents to hold money, but even though the purchasing power of money increases, in the face of future consumption concerns and reduced income, residential consumers still allocate more wealth to savings and bonds with higher short-term investment returns. From Figure S2, the impact of savings is short-lived, just as in periods 1–8, so the most important reason for residents to reduce their money holdings is the higher opportunity cost of holding money due to lower incomes. In periods 1–60, the real money balance remains steady as the bond rate gradually decreases. In the long run, residents' wealth distribution is still determined by their expected profits.

4.3.3. Comprehensive Impact of Water Resources Tax on the Economy and Society

In general, the level of aggregate output measures the overall state of economic development. From Figure S3, the shock of the water resources tax causes the economy's total output to fall in the short run and return to a steady state in the long run. In the case of a closed economy, total output is composed of consumption, investment, and government spending. Total output briefly rises in period 1 due to the increase in tax expenditures and investment outweighing the decrease in consumption. Still, from period 2 onward, total output enters a phase of sustained decline. Combining the changes in total output fluctuations in the previous periods, it can be assumed that total output has a downward trend in the short run. In periods 2–5, total output continues to decline, the pulling effect of tax expenditures and investment on the economy weakens, and the impact of reduced consumption on total output increases. In periods 6–60, tax expenditures and investments are steady. Changes in total output are mainly influenced by changes in consumption, with both total output and consumption showing an upward trend. As water-saving technologies become widespread, corporate profits return to normal, and wage levels recover, consumption levels rise and total output gradually recovers. In the 60th period, total output and consumption return to a steady state.

Overall, under the impact of a water resources tax, water consumption drops significantly in the short term. Still, the increase in water consumption cost makes enterprises' profits drop, resulting in lower wage and benefit levels and less labor quantity and time. At the same time, enterprises introduce water-saving technologies and equipment to improve water use efficiency and reduce production costs to ensure long-term growth and respond to the government's water resources tax policy. In contrast, as workers, residents see their savings and money holdings fall, their bond investments rise as they receive less income, and their consumption shrinks. The increase in residents' savings and the rise in bond investment boosts business investment, accelerating the development of water-saving technologies and the diffusion of water-saving equipment. In the long run, due to the enhanced water conservation technology, the cost of water use decreases, enterprises regain average profits, wages return to normal levels, residents' consumption capacity continues to recover, and socio-economic development returns to a steady state.

5. Discussion

(1) The collection of a water resource tax can effectively achieve the water-saving goal, and help promote water-saving production in enterprises, change the industrial water consumption structure, and improve the utilization efficiency of water resources.

The implementation of the water resource tax reduces the proportion of industrial and agricultural water use. It increases the proportion of ecological and domestic water use while the overall water use remains relatively stable. Since the water resource tax adopts differential tax rates based on the method used and the amount of water drawn by enterprises, the implementation of the water resource tax firstly affects the production of enterprises and increases the overall production cost of enterprises, while the cost of the living water of residents is not directly affected for the time being, but indirectly affected by changes in income level and commodity consumption. The increase in water cost reduces the amount of water resources used by enterprises. Meanwhile, it reduces enterprises' current output and unit labor value, thus reducing the wage level. The decline of residents' income as laborers in the current period directly affects their willingness to distribute wealth: they reduce consumption, hold money, and increase savings to guarantee future consumption demand. The decrease in current consumption and abandonment of high-water consumption products further promote enterprises to introduce water-saving technologies and equipment to reduce the cost of water use. The increase in current investment and the decrease in income are the transition problems enterprises face. The rise in savings provides a stable supply of funds to ensure that enterprises can pass through the difficult period of transition in the short term through financing. At the same time, tax funds used for water resources protection, on the one hand, can increase the capacity of sustainable water supply.

On the other hand, enterprises should be encouraged to develop water-saving technologies, adopt water-saving equipment for production, improve production processes, and promote the upgrading of industrial structures. Through the investment of special tax funds, the production cost of enterprises decreases, profit returns to the average level, the wage level gradually recovers, consumption activities increase, and economic development returns to stability. Overall, the water resources tax reduces industrial water demand and improves water efficiency. Although it harms economic growth in the short term, the economy returns to stability after long-term recovery. The sustainable supply capacity of water resources is improved to achieve the dual goals of economic development and water environment protection.

(2) The collection of the water resources tax is conducive to improving the water-saving consciousness of enterprises and residents, and the price mechanism adjustment helps encourage enterprises to change the water intake mode and optimize the production structure.

The mandatory water resources tax effectively promotes the necessity of water-saving production in enterprises; especially, if high water consumption enterprises do not improve the awareness of water-saving, change the method of water intake and introduce water-saving technology, the increase of production costs will lead to the decrease of enterprise profits, the decline of market competitiveness and the reduction of production scale. From the view of long-term development, the transformation and upgrading of enterprise production is the inevitable trend. When the price of water is at a certain level, the cost of water should be paid more attention to. For example, the water price of the Beijing special industry is 160 yuan/m^3. The high cost of water inhibits the extensive water use behavior of special industry enterprises, the awareness of water-saving is enhanced, and each water account is actuarially calculated. The implementation and publicity for the water resources tax policy have also improved residents' awareness of water-saving and enhanced the awareness of water scarcity and the urgency of water resources protection. At the same time, the income of enterprises affects the income level of residents, resulting in reduced consumption, especially of high water consumption products. For example, due to the increase in car washing costs, the extent of car washing is reduced. The choice of consumers forces enterprises to introduce water-saving technology and equipment for water-saving production and improve the production process to promote industrial structure towards an intensive production transformation. With the improvement of water-saving awareness of residents and enterprises, water consumption gradually decreases, the utilization efficiency of water resources steadily increases, and the carrying capacity of the water environment consolidates and strengthens. Water resources' supply and demand capacity reach equilibrium, promoting the economy's sustainable development and that of the water resources environment.

(3) The collection of the water resources tax effectively protects the water ecological environment and improves the recycling capacity of water resources and the water supply capacity in meeting water demand.

The water resources tax restrains the desire of enterprises to develop and utilize water resources endlessly, ensures the intergenerational equity of water resources utilization, and provides the necessary means for water resources protection and sustainable utilization. Although the impact of the water resources tax will have a negative effect on the economy in the short term, the promotion of the water resources tax is still worthy and must be implemented. Tax funds provide a financial guarantee for local governments to control environmental water pollution, restore water ecological functions, and use the funds obtained from production enterprises to repair the environmental damage caused by them, which also reflects the effect of fiscal and tax policies to achieve social wealth redistribution and social equity. Therefore, the rational and efficient use of water resources tax funds is an important guarantee for an effective water resources tax policy. Currently, the expenditure items of special funds for water resources tax need to be further improved, the implementation of funds needs to be additionally supervised, and the corresponding water-saving incentive measures need to be further supplemented. Water resources tax policy requires a perfect means of rigid and more comprehensive incentive coverage. The correct guidance and water-saving incentives from the government enhances the confidence of enterprises in water-saving priority and innovative development, accelerates the process of enterprises in completing the introduction of water-saving technology facilities, promotes the optimization and upgrading of industrial structure, and improves the utilization efficiency and use-value of water resources. The water resources tax absorbs the profits of enterprises and supplements the ecological environment, which ensures the recovery of water resources' recycling capacity and the improvement of water supply capacity under the condition of increasing water demand and realizes the dual goals of sustainable economic development and sustainable use of water resources [40].

6. Conclusions

This paper analyzes the theoretical mechanism of the water resources tax, establishes a DSGE model with embedded water resources tax, and simulates the comprehensive effect of the water resources tax on water conservation and social development in Hubei Province through impulse response analysis. It was found that, on the one hand, the water resource tax can effectively achieve the goal of water conservation and help promote water-saving production, change the industrial water use structure, and improve water use efficiency; on the other hand, the water resource tax can help raise the awareness of water conservation among enterprises and residents, and help prompt enterprises to change their water extraction methods and optimize their production structure through price mechanism adjustment. In addition, the effective implementation of a water resources tax is guaranteed by the reasonable and efficient use of special water resources protection funds to effectively protect the water ecological environment and improve the recycling capacity of water resources, and the ability to supply water to meet water demand.

Therefore, the government should speed up the exploration of a reasonable range of water resources tax rates that integrates water resources supply and demand and protects people's livelihood so that the water resources tax levied within this tax range can meet the social demand for water and protect people's livelihood at the same time, and achieve a relatively steady state of water resources utilization and protection. The water resources tax should be earmarked for specific purposes, and should be used for water resources protection reasonably and efficiently, which can increase the sustainable supply capacity of water resources on the one hand and motivate enterprises to develop water-saving technologies, improve production processes and promote industrial structure upgrading on the other. At the same time, correct guidance from the government to enterprises can enhance their confidence in prioritizing water conservation and innovative development, accelerate the process of completing the introduction of water conservation technology facilities, and achieve the dual goals of sustainable economic growth and sustainable use of water resources.

Supplementary Materials: The following supporting information can be downloaded at: https://www.mdpi.com/article/10.3390/w15050916/s1, Figure S1: Responses of enterprises to water resources tax shocks; Figure S2: Resident response to the water resources tax shock; Figure S3: Overall impact of water resources tax on economy and society.

Author Contributions: Writing—review & editing, Validation, Supervision, Funding acquisition, L.W.; Writing—original draft, Conceptualization, Data curation, Software, M.; Writing—original draft, Formal analysis, Investigation, Visualization, Z.L.; Writing—original draft, Project administration, Resources, J.P. All authors have read and agreed to the published version of the manuscript.

Funding: Project no. 132805 has been implemented with the support provided from the National Research, Development and Innovation Fund of Hungary, financed under the K_19 funding scheme.

Data Availability Statement: The datasets used and/or analyzed during the current study are available from the corresponding author upon reasonable request.

Conflicts of Interest: The authors declare that they have no competing interest.

References

1. Wang, Y.; Yang, J.; Chang, J. Development of a coupled quantity-quality-environment water allocation model applying the optimization-simulation method. *J. Clean. Prod.* **2019**, *213*, 944–955. [CrossRef]
2. Zhai, J.L.; Feng, R.G.; Xia, J. Constraining factors to sustainable utilization of water resources and their countermeasures in China. *Chinese Geogr. Sci.* **2003**, *13*, 310–316. [CrossRef]
3. Majeed, A.; Wang, Y.; Muniba; Islam, M.A. The Impact of Social Preferences on Supply Chain Performance: An Application of the Game Theory Model. *Complexity* **2023**, *2023*, 1–12. [CrossRef]
4. Thomas, A.; Zaporozhets, V. Bargaining Over Environmental Budgets: A Political Economy Model with Application to French Water Policy. *Environ. Resour. Econ.* **2017**, *68*, 227–248. [CrossRef]
5. Lago, M.; Mysiak, J.; Gómez, C.M.; Delacámara, G.; Maziotis, A. Defining and assessing economic policy instruments for sustainable water management. In *Global Issues in Water Policy*; Springer: Berlin/Heidelberg, Germany, 2015; Volume 14, pp. 1–13. ISBN 9783319182872.
6. Xu, Q. Water Resource Tax Reform to Build Resource-saving Society. *J. Coast. Res.* **2020**, *115*, 506–509. [CrossRef]
7. Ma, Z.; Zhao, J.; Ni, J. Green tax legislation for sustainable development in china. *Singapore Econ. Rev.* **2018**, *63*, 1059–1083. [CrossRef]
8. Sabohi, M.; Soltani, G.R.; Zibaie, M. Evaluation of the strategies for groundwater resources management: A case study in Narimani plain, Khorasan Province. *J. Sci. Technol. Agric. Nat. Resour.* **2007**, *11*, 475–484.
9. Llop, M.; Ponce-Alifonso, X. Identifying the role of final consumption in structural path analysis: An application to water uses. *Ecol. Econ.* **2015**, *109*, 203–210. [CrossRef]
10. Berrittella, M.; Rehdanz, K.; Roson, R.; Tol, R.S.J. The economic impact of water taxes: A computable general equilibrium analysis with an international data set. *Water Policy* **2008**, *10*, 259–271. [CrossRef]

11. Berck, P.; Moe-Lange, J.; Stevens, A.; Villas-Boas, S. Measuring Consumer Responses to a Bottled Water Tax Policy. *Am. J. Agric. Econ.* **2016**, *98*, 981–996. [CrossRef]
12. Qin, C.; Jia, Y.; Su, Z.; Bressers, H.T.A.; Wang, H. The economic impact of water tax charges in China: A static computable general equilibrium analysis. *Water Int.* **2012**, *37*, 279–292. [CrossRef]
13. Mushtaq, S.; Khan, S.; Dawe, D.; Hanjra, M.A.; Hafeez, M.; Asghar, M.N. Evaluating the impact of Tax-for-Fee reform (Fei Gai Shui) on water resources and agriculture production in the Zhanghe Irrigation System, China. *Food Policy* **2008**, *33*, 576–586. [CrossRef]
14. Kilimani, N.; van Heerden, J.; Bohlmann, H. Water taxation and the double dividend hypothesis. *Water Resour. Econ.* **2015**, *10*, 68–91. [CrossRef]
15. Shen, J.; Luo, C. Overall review of renewable energy subsidy policies in China - Contradictions of intentions and effects. *Renew. Sustain. Energy Rev.* **2015**, *41*, 1478–1488. [CrossRef]
16. Xin, C.; Guo, F.; Wang, A. Exploring the impacts of China's water resource tax policies: A trade-off between economic development and ecological protection. *Front. Environ. Sci.* **2022**, *10*, 2163. [CrossRef]
17. Ouyang, R.; Mu, E.; Yu, Y.; Chen, Y.; Hu, J.; Tong, H.; Cheng, Z. Assessing the effectiveness and function of the water resources tax policy pilot in China. *Environ. Dev. Sustain.* **2022**, 1–17. [CrossRef]
18. Mu, L.; Zhang, X.; Cheng, S.; Song, P. The effectiveness of a water resource tax policy in improving water-use efficiency: A quasi-natural experiment-based approach. *Water Policy* **2022**, *24*, 899–922. [CrossRef]
19. Mu, L.; Mou, M.; Tang, H. Does the water resource 'fee to tax' policy alleviate water poverty? Evidence from a quasi-natural experiment. *Water Supply* **2022**, *22*, 8465–8482. [CrossRef]
20. Liu, Y.; Zhang, Z.; Zhang, F. Challenges for water security and sustainable socio-economic development: A case study of industrial, domestic water use and pollution management in Shandong, China. *Water* **2019**, *11*, 1630. [CrossRef]
21. Kydland, F.E.; Prescott, E.C. Time to Build and Aggregate Fluctuations. *Econometrica* **1982**, *50*, 1345. [CrossRef]
22. Dhawan, R.; Jeske, K.; Silos, P. Productivity, energy prices and the great moderation: A new link. *Rev. Econ. Dyn.* **2010**, *13*, 715–724. [CrossRef]
23. Angelopoulos, K.; Economides, G.; Philippopoulos, A. First-and second-best allocations under economic and environmental uncertainty. *Int. Tax Public Financ.* **2013**, *20*, 360–380. [CrossRef]
24. Filardo, A.J. Monetary Policy and Asset Price Bubbles: Calibrating the Monetary Policy Trade-Offs. *SSRN Electron. J.* **2011**. [CrossRef]
25. Sun, C.; Xu, Z.; Zheng, H. Green transformation of the building industry and the government policy effects: Policy simulation based on the DSGE model. *Energy* **2023**, *268*, 126721. [CrossRef]
26. Benavides, C.; Gonzales, L.; Diaz, M.; Fuentes, R.; García, G.; Palma-Behnke, R.; Ravizza, C. The impact of a carbon tax on the chilean electricity generation sector. *Energies* **2015**, *8*, 2674–2700. [CrossRef]
27. Majeed, A.; Ye, C.; Chenyun, Y.; Wei, X. Muniba Roles of natural resources, globalization, and technological innovations in mitigation of environmental degradation in BRI economies. *PLoS ONE* **2022**, *17*, e0265755. [CrossRef]
28. Argentiero, A.; Bollino, C.A.; Micheli, S.; Zopounidis, C. Renewable energy sources policies in a Bayesian DSGE model. *Renew. Energy* **2018**, *120*, 60–68. [CrossRef]
29. Llop, M.; Ponce-Alifonso, X. A never-ending debate: Demand versus supply water policies. A CGE analysis for Catalonia. *Water Policy* **2012**, *14*, 694–708. [CrossRef]
30. Xiong, Z.; Li, H. Ecological deficit tax: A tax design and simulation of compensation for ecosystem service value based on ecological footprint in China. *J. Clean. Prod.* **2019**, *230*, 1128–1137. [CrossRef]
31. Chen, Z.R.; Yuan, Y.; Xiao, X. Analysis of the Fee-to-Tax Reform on Water Resources in China. *Front. Energy Res.* **2021**, *9*, 641. [CrossRef]
32. Pan, D. The Economic and Environmental Effects of Green Financial Policy in China: A DSGE Approach. *SSRN Electron. J.* **2019**. [CrossRef]
33. Li, H.; Peng, W. Carbon Tax, Subsidy, and Emission Reduction: Analysis Based on DSGE Model. *Complexity* **2020**, *2020*. [CrossRef]
34. Sun, D.; Wang, F.; Chen, N.; Chen, J. The impacts of technology shocks on sustainable development from the perspective of energy structure—a dsge model approach. *Sustainability* **2021**, *13*, 8665. [CrossRef]
35. Zhang, J.; Zhang, Y. Examining the economic and environmental effects of emissions policies in China: A Bayesian DSGE model. *J. Clean. Prod.* **2020**, *266*, 122026. [CrossRef]
36. Acemoglu, D.; Aghion, P.; Bursztyn, L.; Hemous, D. The environment and directed technical change. *Am. Econ. Rev.* **2012**, *102*, 131–166. [CrossRef] [PubMed]
37. Shao, S.; Yang, L.; Huang, T. Theoretical Model and Experience from China of Energy Rebound Effect. *Econ. Res. J.* **2013**, *48*, 96–109.
38. Niu, T.; Yao, X.; Shao, S.; Li, D.; Wang, W. Environmental tax shocks and carbon emissions: An estimated DSGE model. *Struct. Chang. Econ. Dyn.* **2018**, *47*, 9–17. [CrossRef]

39. Yang, B.; Chen, X. Improvement on the congested government productive spending in stochastic endogenous growth model. In Proceedings of the 2012 International Conference on Modelling, Identification and Control, ICMIC 2012, Wuhan, China, 24–26 June 2012; pp. 426–428.
40. Majeed, A.; Wang, L.; Zhang, X.; Muniba; Kirikkaleli, D. Modeling the dynamic links among natural resources, economic globalization, disaggregated energy consumption, and environmental quality: Fresh evidence from GCC economies. *Resour. Policy* **2021**, *73*, 102204. [CrossRef]

water

MDPI

Article

Effect of Potassium Dihydrogen Phosphate Combined with Thermally Activated Nano Serpentine and Thermally Activated Nano Zeolite on Cadmium in Soil

Xiuli Wang [1,2,3,4,†], Hongdou Liu [1,2,3,†] and Hongtao Zou [1,2,3,*]

1 College of Land and Environment, Shenyang Agricultural University, Shenyang 110866, China
2 National Engineering Research Centre for Efficient Utilization of Soil and Fertilizer Resources, Shenyang 110866, China
3 Key Laboratory of Arable Land Conservation in Northeast China, Ministry of Agriculture and Rural Affairs, Shenyang 110866, China
4 Liaoning Agricultural Technical College, Yingkou 115009, China
* Correspondence: hongtaozou208@163.com
† These authors contributed equally to this work.

Abstract: The combined application of potassium dihydrogen phosphate (KH_2PO_4) and thermally activated nano serpentine and KH_2PO_4 and thermally activated nano zeolite could immobilize cadmium (Cd) in contaminated soils by increasing soil pH value. The results showed that adding nPS_{700}-2.0 (KH_2PO_4 and thermally activated nano serpentine activated at 700 °C, 2% addition) exhibited better performance under the same treatment condition; it reduced DTPA-Cd by 57.8% and exchangeable Cd by 48.76%. Adding nPF_{700} (KH_2PO_4 and thermally activated nano zeolite activated at 700 °C) reduced DTPA-Cd by 35.49–44.17% and exchangeable Cd by 35.89–42.57%, respectively. The increase of active adsorption points and the surface area of thermally activated nano serpentine reduced the bioavailability of Cd in soil, indicating that the combined application of phosphate and thermally activated nano serpentine has great potential for the immobilization of Cd in soil.

Keywords: phosphate; thermal activation; nano material; cadmium pollution; serpentine

1. Introduction

Cadmium (Cd) is recognized as the first among 12 dangerous chemical substances of global significance by the United Nations Program [1]. It can cause damage to red blood cells and lead to anemia [2], and it can cause periodontal disease or induce local bone, cartilage [3], liver, and kidney destruction [4], etc. Cd was also declared as the first carcinogen by the International Agency for Research on cancer [5]. When it circulates into the environment, about 2% of Cd enters the atmosphere, 4% enters water, and 94% enters soil [6]. In the United States, 63% of a total of 1200 soil samples were investigated that were contaminated by heavy metals, in which 8% were contaminated by Cd, according to the National Priority List (NPL) [7]. In Japan, about 0.47 million hm^2 of farmland was polluted by Cd, accounting for 82% of the total area of farmland polluted by heavy metals [8]. In China, cultivated land polluted by heavy metals was 2.67 million hm^2 in 1980 and 6.7 million hm^2 in 1988, and it actually reached 20 million hm^2 in 2012 [9], accounting for 1/5 of the national farmland area [10,11]. Cd pollution in soil is a global thorny problem. It is difficult to be degraded, transformed, and enriched with the characteristics of crypticity, being long-lasting, and being irreversible [12–15], and it influences human health via the food chain. The topic of how to reduce Cd pollution has become a hot spot in recent years.

At present, there are generally two categories of remediation in soil heavy metal pollution: One is ectopic remediation technology, which directly removes heavy metals from the soil but has a high cost, is time- and labor-consuming, and damages original soil structure. The second is situ remediation technology, which can reduce mobility and bioavailability by changing the form of heavy metals from the activated state to a stable state. It has the advantages of small investment, quick effect, complete treatment, and preservation of original soil structure [16,17]. Phosphate, clay minerals, red mud, lime, biochar, chitosan, arbuscular mycorrhizal fungi, etc., of heavy metals are all found as common remediation substances [18–21]. Among them, clay minerals, as one component of soil, can immobilize the heavy metals effectively by coprecipitation, coordination, and adsorption with specific surface and crystal layers, and they do not destroy the soil structure [22]. Zeolite [23–25] and serpentine [26,27] have been investigated more efficiently to be used for immobilization remediation in heavy metal-contaminated soils with outstanding resource advantages. Thermal-activated serpentine obtained by heating natural serpentine exhibited good adsorption performance to Cd [28]. A previous study reported that phosphate can precipitate with heavy metals under certain conditions [29], and high-temperature thermal activation treatment can be used to increase its adsorption capacity. As we all know, nanomaterial technology is one of the most promising technologies of the 21st century, and it can be used to increase the specific surface area of materials and accelerate the suction rate of heavy metals. Although the use of potassium dihy-drogen phosphate (KH_2PO_4), serpentine, and zeolite alone could stabilize heavy metal Cd in different degrees, there is rarely reports about phosphate combined with zeolite and serpentine. Therefore, an efficient and environmentally friendly technology is still urgently needed.

In this study, (1) we created two methods for reducing soil-available Cd: KH_2PO_4 combined with thermal-activated nano-serpentine, and KH_2PO_4 combined with thermal-activated nano-zeolite. (2) The best thermal activation temperature of the two methods was explored, respectively, by simulated incubation. (3) We analyzed DTPA Cd content, the exchange of Cd and pH in the soil, and the combined application of KH_2PO_4 with thermally activated nano serpentine and thermally activated nano zeolite. We also provided a theoretical and application basis for Cd removal from soil.

2. Materials and Methods

2.1. Experimental Soils and Material

Uncontaminated soil was collected at a depth of 0–20 cm from the Test Base of Shenyang Agricultural University in Shenyang, Liaoning Province (123°56′ E, 41°82′ N, 43 m above sea level), China, and the soil type was brown soil. The basic physical and chemical properties of the soil are shown in Table 1, which were determined according to the method of the literature [30]. The soil composition and content are shown in Table 1.

Table 1. Physical and chemical properties of the tested soil.

Soil Type	pH	Organic Matter ($g \cdot kg^{-1}$)	Alkali-Hydrolyzable Nitrogen ($mg \cdot kg^{-1}$)	Available Phosphorus ($mg \cdot kg^{-1}$)	Available Potassium ($mg \cdot kg^{-1}$)	Total Phosphorus ($g \cdot kg^{-1}$)	Total Cd ($mg \cdot kg^{-1}$)
Brown soil	6.73	26.93	50.02	71.27	183.59	0.63	0.18

Natural nano zeolite and natural nano serpentine were obtained from Liaoning Province, China, and were crushed into 600 nm powder by high-energy nano impact grinding. The main composition and content of the natural nano zeolite and serpentine were tested by a Philips MagiX X-ray fluorescence spectrometer (PANalytical B.V, Almelo, The Netherlands) and are shown in Tables 2 and 3, respectively.

Table 2. The main components and content of zeolite.

Composition	SiO_2	Al_2O_3	MgO	K_2O	CaO	Na_2O	Fe_2O_3
Content in zeolite (wt. %)	76.32	12.49	3.78	3.24	2.30	1.29	0.58

Table 3. The main components and content of serpentine.

Composition	SiO_2	Al_2O_3	MgO	CaO	Na_2O	Fe_2O_3
Content in serpentine (wt. %)	57.24	0.11	37.97	3.20	0.18	1.30

The 600 nm natural nano serpentine powder was put into a crucible and calcined in a muffle furnace at 350, 550, 700, and 850 °C for 2 h at constant temperature [31]. After the serpentine cooled to room temperature, it was put in zip-lock bags and stored in a desiccator. The resulting thermally activated nano serpentine was briefly recorded as nS_T, namely nS_{350}, nS_{550}, nS_{700}, and nS_{850}, where n represented nano-treated, S represented serpentine, T represented activation temperature, and natural nano serpentine was recorded as nS_0. In the same way, thermally activated nano zeolite was produced. The resulting thermally activated nano zeolite was simply noted as nZ_T, namely nZ_{350}, nZ_{550}, nZ_{700}, and nZ_{850}, where n represented nano-treated, Z represented zeolite, T represented activation temperature, and natural nano zeolite was recorded as nF_0. The phosphate for testing was KH_2PO_4 (Analytical Reagent), denoted as P.

2.2. Experimental Design

The simulated cadmium-contaminated soil without added phosphate and silicate was recorded as the reference (CK) treatment. The simulated cadmium-contaminated soil was supplemented with 0.5%, 1.0%, 2.0% low, medium, and high soil weight doses of KH_2PO_4, and the mixture of thermal-activated nano serpentine (KH_2PO_4: different nano serpentine = 1:2), denoted as nPS_T-0.5, nPS_T-1.0, and nPS_T-2.0, respectively, was mixed thoroughly. In the same way, 0.5%, 1.0%, 2.0% low, medium, and high soil weight doses of KH_2PO_4 and mixtures of thermal-activated nano zeolite (KH_2PO_4: different nano zeolite = 1:2) were denoted as nPZ_T-0.5, nPZ_T-1.0, and nPZ_T-2.0, respectively. There was a total of 31 samples, each repeated 3 times.

2.3. Sample Analysis

The air-dried soil had $CdCl_2 \cdot 2.5H_2O$ (AR) added in the form of a solution, and the exogenous Cd content in the soil reached 10 $mg{\cdot}kg^{-1}$. An accurate weight of 50.0 g of each of the above-mentioned soils with different treatments was put into jars. The jars were placed in a thermotank and cultivated at 25 ± 2 °C, and the soil water content was maintained at 70.0% of the field water-holding capacity by using deionized water every other day, according to the weighing method. At 0, 7, 14, 28, and 56 days (d) of culture, an appropriate amount of naturally air-dried and ground soil samples were weighed, and the pH and available Cd content of the soil were determined.

The soil pH was measured by a pH meter (SCHOTT Group, Mainz, Germany) with a 1:2.5 water/soil ratio [32]. Soil available Cd was extracted by the DTPA method [33] and measured by an atomic absorption spectrophotometer (HITACH, Tokyo, Japan). The exchangeable Cd content in soil used the Tessier continuous extraction method [34] and was measured by an atomic absorption spectrophotometer (HITACH, Japan).

2.4. Data Processing and Analysis

Microsoft Excel 2003, SPSS 19.0 (Duncan test) and origin 8.0 software were used for analysis and to graph data.

3. Results and Analysis

3.1. Effect on Soil pH

Soil pH affected the ionic composition and various chemical reactions, changes in the occurrence forms of heavy metals, and the bioavailability of heavy metals [35–37]. In Figure 1, compared to CK, with the increase of the dose level, after 56 d, for the culture with the combined KH_2PO_4 and thermally activated nano zeolite, the soil pH presented an uptrend, and that with the thermally activated temperature showed first a rising and then a falling trend. Compared to CK, the soil pH increased by 0.82–1.30, 0.84–1.35, and 0.90–1.39 units after the application of nPS0-0.5, nPS0-1.0, and nPS0-2.0, respectively; it increased by 0.87–1.34, 0.91–1.42, and 1.08–1.62 units after applying nPS350-0.5, nPS350-1.0, and nPS350-2.0, respectively; it increased by 1.55–2.03, 2.11–2.57, and 1.61–1.83 units after applying nPS550-0.5, nPS550-1.0, and nPS550-2.0, respectively; and it increased by 1.84–1.91, 1.86–2.12, and 1.89–2.15 units after applying nPS850-0.5, nPS850-1.0, and nPS850-2.0, respectively. After applying nPS_{700}-0.5, nPS_{700}-1.0, and nPS_{700}-2.0, the soil pH increased by 1.94–2.34, 1.98–2.39, and 2.03–2.42 units, respectively, which obtained the highest growth rate. Natural serpentine contains a large amount of OH^- with strong chemical activity. After high-temperature thermal activation, much of the OH^- was lost, and the maximum loss happened at 700 °C; the alkalinity is enhanced when entering the environment. Another reason is that the Mg-OH bond in the soil solution undergoes proton migration to form new Mg-OH and MgO [38], and MgO hydrolyzes and enhances the pH value.

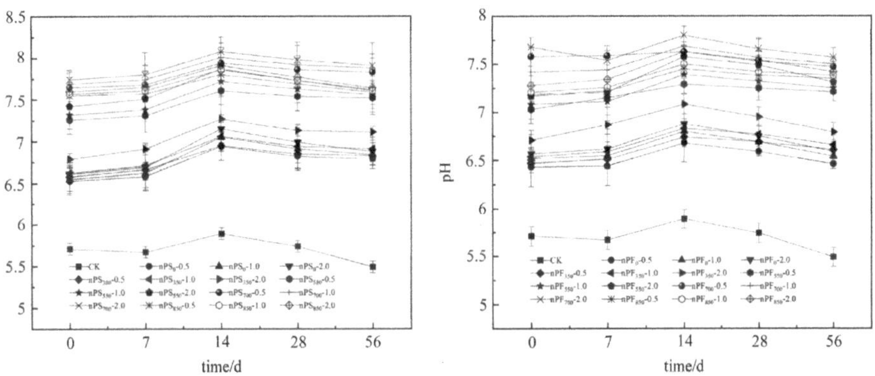

Figure 1. Changes of pH in Cd-contaminated soil with different treatments. Error bars (n = 3).

With the increase of dose level and thermally activated temperature, pH showed an uptrend during the combined application of KH_2PO_4 and thermally activated nano zeolite. After 56 d of culturing, the maximum increase of pH value was nPF_{700}; however, there was no significant difference with nPF_{850}. The soil pH increased by 1.74–1.98, 1.71–2.01, and 1.87–2.08 units, respectively, after applying nPF_{700}-0.5, nPF_{700}-1.0, and nPF_{700}-2.0. Compared to CK, the soil pH value increased by 0.72–0.97, 0.81–1.05, and 0.86–1.10 units after applying nPF_0-0.5, nPF_0-1.0, and nPF_0-2.0, respectively; it increased by 0.76–1.12, 0.83–1.17, and 0.90–1.20 units after applying nPF_{350}-0.5, nPF_{350}-1.0, and nPF_{350}-2.0, respectively; it increased by 1.32–1.72, 1.37–1.76, and 1.46–1.82 units after applying nPF_{550}-0.5, nPF_{550}-1.0, and nPF_{550}-2.0, respectively; and it increased by 1.47–1.86, 1.50–1.89, and 1.57–1.93 units after applying nPF_{850}-0.5, nPF_{850}-1.0, and nPF_{850}-2.0, respectively. At the same temperature, the pH value increased with the increase of zeolite dosage.

In general, under the same conditions, the combined application of KH_2PO_4 with thermally activated nano serpentine was better than KH_2PO_4 with thermally activated nano zeolite in improving the soil pH value.

3.2. *Effect on Soil DTPA-Cd and Exchangeable Cd by Different Modifiers*

DTPA-Cd is the easiest to transform and migrate, and it is absorbed and utilized by plants to enter the food chain, which causes harm to the environment, animals, and plants, as well as human beings. In the nPS treatments in Figure 2, the content of DTPA-Cd gradually decreased with the increase of thermal activation temperature, except over 700 °C, and the highest reduction of DTPA-Cd was shown in nPS_{700} treatments (nPS_{700}-0.5, nPS_{700}-1.0, nPS_{700}-2.0). After 56 d, compared to nPS_0-0.5, the DTPA-Cd decreased by 14.06%, 14.06%, 30.16%, and 17.19%, respectively, in nPS$_{350}$-0.5, nPS$_{550}$-0.5, nPS$_{700}$-0.5, and nPS$_{850}$-0.5. Compared to nPS$_0$-1.0, DPTA-Cd decreased by 5.31%, 22.67%, 21.51%, and 22.53%, respectively, in nPS$_{350}$-1.0, nPS$_{550}$-1.0, nPS$_{700}$-1.0, and nPS$_{850}$-1.0. Compared to nPS$_0$-2.0, DPTA-Cd decreased by 5.86%, 27.43%, 46.07%, and 27.58%, respectively, in nPS$_{350}$-2.0, nPS$_{550}$-2.0, nPS$_{700}$-2.0, and nPS$_{850}$-2.0. Compared to CK, the DTPA-Cd contents of nPS$_0$-0.5, nPS$_0$-1.0, and nPS$_0$-2.0 treatments were reduced by 0.81–20.33%, 0.62–13.22%, and 0.24–19.24%, respectively; nPS$_{350}$-0.5, nPS$_{350}$-1.0, and nPS$_{350}$-2.0 were reduced by 1.19–17.81%, 1.19–23.91%, and 1.78–31.51%, respectively; the contents of nPS$_{550}$-0.5, nPS$_{550}$-1.0, and nPS$_{550}$-2.0 were reduced by 2.73–31.51%, 0.12–32.88%, and 1.31–41.34%, respectively; and the contents of nPS$_{850}$-0.5, nPS$_{850}$-1.0, and nPS$_{850}$-2.0 were reduced by 3.32–33.40%, 0.24–32.75%, and 0.59–41.47%, respectively. In the same treatment, with the increase of cultivation time, the reduction rate of soil DPTA-Cd was higher. With the higher doses of KH_2PO_4 and the mixture of thermal-activated nano serpentine, the removal rates rose steadily, such as nPS$_{700}$-2.0 becoming 25.05% higher than nPS$_{700}$-0.5. The final DPTA-Cd removal of nPS$_{700}$-2.0 was 57.80%, well above the rate of single serpentine [28].

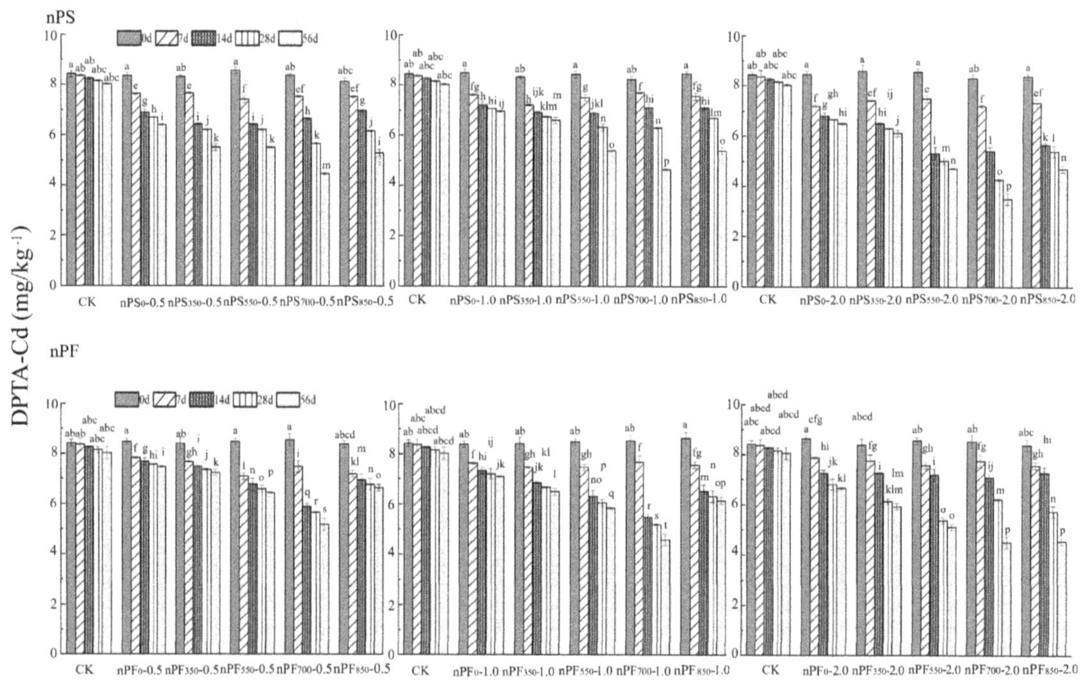

Figure 2. Effect of different treatments on the soil's DTPA-Cd content. $n = 3$, Different lower-case letters marked in Figure 2 indicate a significant difference across different treatments ($p < 0.05$). Error bars ($n = 3$, Mean ± SD).

In the nPF treatments, after 56 d of culturing, we found 700 °C was also the best thermal activation temperature, regardless of the concentration of mixed additives. It had the highest removal rate of DPTA-Cd. Compared to CK, the DPTA-Cd of nPF$_0$-0.5, nPF$_{350}$-0.5, nPF$_{550}$-0.5, nPF$_{700}$-0.5, and nPF$_{850}$-0.5 treatment decreased by 6.85%, 9.84%, 19.93%, 35.49%, and 17.43%, respectively. DPTA-Cd decreased by 11.58%, 19.18%, 27.15%, 40.97%, and 23.54%, respectively, in nPF$_0$-1.0, nPF$_{350}$-1.0, nPF$_{550}$-1.0, nPF$_{700}$-1.0, and nPF$_{850}$-1.0, and it decreased by 6.65%, 5.94%, 5.11%, 44.17%, and 43.42%, respectively, in nPF$_0$-2.0, nPF$_{350}$-2.0, nPF$_{550}$-2.0, nPF$_{700}$-2.0, and nPF$_{850}$-2.0. The removal rate became higher significantly as the added dose of KH_2PO_4 and the mixture of thermal-activated nano zeolite increased. Compared to nPF_0-0.5, DPTA-Cd decreased by 3.21%, 14.04%, 30.88%, and 11.36%, respectively, in nPF_{350}-0.5, nPF_{550}-0.5, nPF_{700}-0.5, and nPF_{850}-0.5. Compared to nPF_0-1.0, DPTA-Cd decreased by 8.59%, 17.61%, 38.50%, and 13.52%, respectively, in nPF_{350}-1.0, nPF_{550}-1.0, nPF_{700}-1.0, and nPF_{850}-1.0. Compared to nPF_0-2.0, DPTA-Cd decreased by 10.68%, 23.16%, 40.28%, and 31.43%, respectively, in nPF_{350}-2.0, nPF_{550}-2.0, nPF_{700}-2.0, and nPF_{850}-2.0.

The content changes of exchangeable Cd via different treatments are shown in Figure 3. The heavy metal Cd forms used consisted of five forms: exchangeable, carbonate bound, organic bound, iron-manganese oxide bound, and residue. Among them, the higher the content of exchangeable Cd, the greater its instability, activity, and harm to the soil [39,40]. Compared to CK, the contents of exchangeable Cd decreased significantly after adding different mixtures of KH_2PO_4 and thermal-activated nano serpentine or nano zeolite, the amounts of which were dependent on the activation temperature. Moreover, the decreasing tendency became prominent as the incubation time and the dosage of the mixture increased. After 56 d, compared to nPS_0-0.5, the exchangeable Cd decreased by 28.01%, 25.03%, 33.69%, and 19.35%, respectively, in nPS_{350}-0.5, nPS_{550}-0.5, nPS_{700}-0.5, and nPS_{850}-0.5. Compared to nPS_0-1.0, exchangeable Cd decreased by 23.61%, 25.81%, 29.33%, and 26.17%, respectively, in nPS_{350}-1.0, nPS_{550}-1.0, nPS_{700}-1.0, and nPS_{850}-1.0. Compared to nPS_0-2.0, exchangeable Cd decreased by 39.32%, 37.61%, 41.03%, and 39.36%, respectively, in nPS_{350}-2.0, nPS_{550}-2.0, nPS_{700}-2.0, and nPS_{850}-2.0. Compared to CK, the exchangeable Cd content of nPS_0-0.5, nPS_0-1.0, and nPS_0-2.0 treatments reduced by 0.78–7.86%, 2.35–12.47%, and 7.43–14.96%, respectively; the conten nPS_{350}-0.5, nPS_{350}-1.0, and nPS_{350}-2.0 reduced by 0.25–33.67%, 1.24–35.04%, and 0.25–46.88%, respectively; the content of nPS_{550}-0.5, nPS_{550}-1.0, and nPS_{550}-2.0 reduced by 2.72–30.92%, 0.62–36.91%, and 0.25–45.39%, respectively; and the content of nPS_{850}-0.5, nPS_{850}-1.0, and nPS_{850}-2.0 reduced by 1.61–25.69%, 0.62–39.90%, and 1.98–48.38%, respectively.

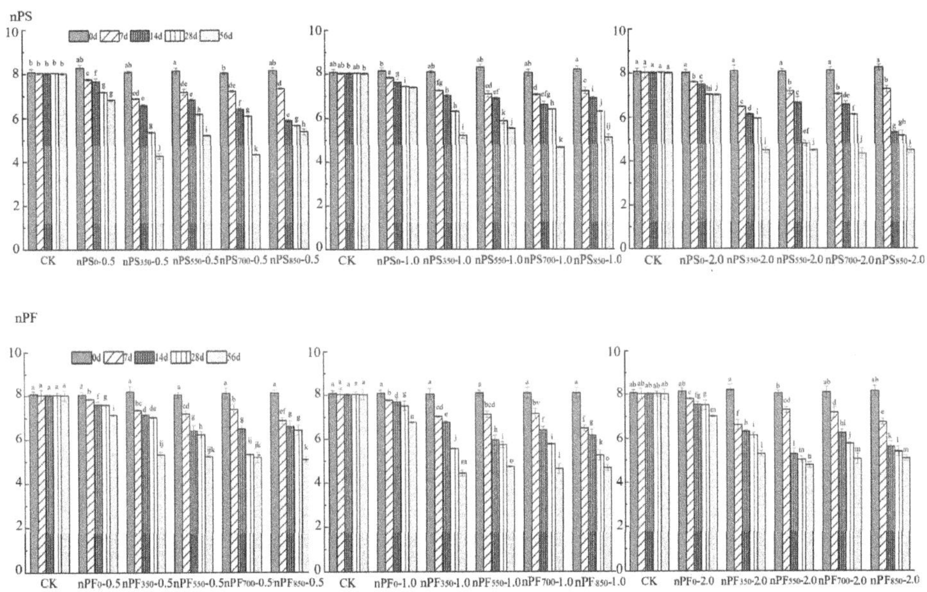

Figure 3. Effect of different treatments on the soil's exchangeable Cd content. Different lower-case letters marked in Figure 3 indicate a significant difference across different treatments ($p < 0.05$). Error bars ($n = 3$, Mean ± SD). In the nPF treatments, compared to nPF_0-0.5, the exchangeable Cd decreased by 25.53%, 26.65%, 28.75%, and 28.61%, respectively, in nPF_{350}-0.5, nPF_{550}-0.5, nPF_{700}-0.5, and nPF_{850}-0.5. Compared to nPF_0-1.0, exchangeable Cd decreased by 24.54%, 27.96%, 31.81%, and

27.53%, respectively, in nPF_{350}-1.0, nPF_{550}-1.0, nPF_{700}-1.0, and nPF_{850}-1.0. Compared to nPF_0-2.0, exchangeable Cd decreased by 34.47%, 31.36%, 42.01%, and 41.72%, respectively, in nPF_{350}-2.0, nPF_{550}-2.0, nPF_{700}-2.0, and nPF_{850}-2.0. Compared to CK, the exchangeable Cd content of nPF_0-0.5, nPF_0-1.0, and nPF_0-2.0 treatments reduced by 0.25–11.10%, 0–12.59%, and 0.74–15.71%, respectively; the content of nPF_{350}-0.5, nPF_{350}-1.0, and nPF_{350}-2.0 reduced by 1.36–33.79%, 0.50–34.04%, and 1.49–44.76%, respectively; the content of nPF_{550}-0.5, nPF_{550}-1.0, and nPF_{550}-2.0 reduced by 0.37–34.79%, 0.12–40.40%, and 0.37–44.39%, respectively; and the content of nPF_{850}-0.5, nPF_{850}-1.0, and nPF_{850}-2.0 reduced by 0.50–36.53%, 0.12–36.66%, and 0.99–46.47%, respectively.

The nPS700 additions reduced exchangeable Cd by 39.36–48.76% compared to CK, and nPF reduced it by 35.89–42.57%. The nPS was higher than previous results, such as a palygorskite addition that reduced exchangeable Cd by 11–32% [41] and single serpentine that reduced it by around 39% [28].

4. Discussion

4.1. pH

The pH value was found to have a negative correlation with DPTA-Cd [42]; thus, in our study, the increasing soil pH (Figure 1) led to the increase of the negative charge of colloid, and it increased the electrical adsorption of Cd^{2+}. Cd^{2+} can combine with $CO_3{}^{2-}$, OH^-, and $PO_4{}^{3-}$ to form insoluble $CdCO_3$, $Cd(OH)_2$, and $Cd_3(PO_4)_2$ precipitation, with the increase of OH^- concentration in soil solution. This reduces the availability of heavy metals [43,44]. On the one hand, the pH values of serpentine and zeolite themselves are higher; on the other hand, both of them contain K^+, Ca^{2+}, and Mg^{2+}, which increase the soil pH [45]. Furthermore, natural serpentine contains a large amount of OH^- with strong chemical activities. The Mg-OH^- bond in the soil solution undergoes proton migration to form new Mg-OH and MgO [38], and MgO hydrolyzes and improves pH value.

4.2. Cd Content

Reducing the content of exchangeable Cd and DTPA-Cd can reduce the pollution of heavy metals to the soil. The exchangeable Cd is bioavailable, it has great mobility, and it is most easily absorbed and utilized by organisms. Studies have found that natural zeolites can absorb heavy metals due to its porous structure [46]. In our study, zeolite was still effective in the reduction of Cd bioavailability. The serpentine unit was composed of a six square mesh of the silicon-oxygen tetrahedron, with an eight-surface layer of magnesium hydroxide [47]. Cd ions can be adsorbed in the unit layer and combined with hydroxyl to enrich the mineral surface. Therefore, after adding serpentine, the content of DTPA-Cd in soil was significantly decreased (Figure 2). The O-Si-O bond of serpentine is broken in thermal activation, which leads to the increase of the active adsorption point and the increase of the surface area [31]. When serpentine was activated by 700 °C, the surface increased by two times due to the original layered structure collapsing to form a slit-shaped pore-like structure [31], which further increased its adsorption capacity of heavy metal in soil. This is why we added thermal-activated serpentine compared to natural serpentine. Phosphate is a cheap and effective chemical fixative used for the remediation of Cd-contaminated soil [29]. We found the effect of decreasing exchangeable Cd using phosphate (KH_2PO_4) combined with thermal-activated serpentine was better than only using thermal-activated serpentine. The removal rate of the latter was only 23.76–36.49% [28], less than what we obtained in Section 3.2. Therefore, the combined application of KH_2PO_4 and thermally activated nano serpentine is a better method to remove cadmium pollution in soil.

5. Conclusions

Adding KH_2PO_4 + thermally activated nano serpentine and KH_2PO_4 + thermally activated nano zeolite could significantly increase soil pH, and the improvement of pH depends on the thermal activation temperature, dosage, and soil incubation time. The removal rate of DTPA-Cd and exchangeable Cd became higher significantly with the added dose of the mixture. The best removal performance dosage was 2.0%, and the best thermally activated temperature was 700 °C in both serpentine and zeolite treatment. In our study, zeolite and serpentine exhibited different immobilization capabilities for Cd soils. The best treatment was nPS_{700}-2.0; the removal rates of DTPA-Cd and exchangeable Cd were 57.80% and 48.76%, respectively, 13.63% and 6.19% higher than nPF_{700}-2.0. The mixture of KH_2PO_4 + thermally activated nano serpentine could effectively convert the bioavailable Cd speciation to have less bioavailable speciation, accounting for the reduction of Cd bioaccessibility.

Author Contributions: X.W.: conceptualization, data curation, formal analysis, investigation, methodology, writing—original draft, funding acquisition. H.L.: investigation, data curation, methodology, formal analysis, writing—original draft, visualization, supervision. H.Z.: conceptualization, supervision, funding acquisition, project administration, writing—review and editing. All authors have read and agreed to the published version of the manuscript.

Funding: General Items of National Natural Science Foundation of China (32072677); Liaoning Province Applied Basic Research Program (2022JH2/101300173); Liaoning Xingliao Talents Project Funding: (XLYC2006014).

Data Availability Statement: Data available on request due to restrictions eg privacy or ethical. The data presented in this study are available on request from the corresponding author. The data are not publicly available due to the 2 first authors are still under their PhD researches.

Conflicts of Interest: The authors declare no conflict of interest.

References

1. Ding, Y.W.; Liu, H.J.; Jiang, J.J.; Chen, C.K.; Liang, T.X.; Li, H.B. Study on determination of cadmium in rice by absorption spectrometry. *Food Ferment. Sci. Technol.* **2019**, *54*, 111–114.
2. Demchenkov, E.L.; Nagdalian, A.A.; Budkevich, R.O.; Oboturova, N.P.; Okolelova, A.I. Usage of atomic force microscopy for detection of the damaging effect of CdCl2 on red blood cells membrane. *Ecotoxicol. Environ. Saf.* **2021**, *208*, 111683. [CrossRef] [PubMed]
3. Browar, A.W.; Leavitt, L.L.; Prozialeck, W.C.; Edwards, J.R. Levels of Cadmium in Human Mandibular Bone. *Toxics* **2019**, *7*, 31. [CrossRef]
4. Andjelkovic, M.; Buha Djordjevic, A.; Antonijevic, E.; Antonijevic, B.; Stanic, M.; Kotur-Stevuljevic, J.; Spasojevic-Kalimanovska, V.; Jovanovic, M.; Boricic, N.; Wallace, D.; et al. Toxic Effect of Acute Cadmium and Lead Exposure in Rat Blood, Liver, and Kidney. *Int. J. Environ. Res. Public Health* **2019**, *16*, 274. [CrossRef] [PubMed]
5. Kazuo, N.; Hiroko, N. Cadmium-induced renal dysfunction: New mechanism, treatment and prevention. *J. Trace Elem. Exp. Med.* **1998**, *11*, 275–288. [CrossRef]
6. Lu, R.K.; Xiong, L.M.; Shi, Z.Y. Study on cadmium in soil-crop system. *Soils* **1992**, *24*, 129–132.
7. Huang, Y.Z.; Hu, Y.; Liu, Y.X. Combined toxicity of copper and cadmium to six rice genotypes (*Oryza sativa* L.). *J. Environ. Sci.* **2009**, *21*, 647–653. [CrossRef]
8. Liao, Z. *Environmental Chemistry and Biological Effects of Trace Elements*; Environmental Science Publishing House: Beijing, China, 1993; pp. 301–303.
9. Hu, N.J.; Li, Z.Q.; Huang, P.; Cheng, W.Y. The pollution, prevention and remediation of heavy metals in infield land in some suburb areas, China. *Bull. Mineral. Petrol. Geochem.* **2003**, *22*, 251–254.
10. Cui, L.T.; Geng, S.G.; Li, Z.W. Current situation and Prevention Countermeasures of cadmium pollution in farmland soil in China. *Mod. Agric. Sci. Technol.* **2006**, *21*, 184–186. [CrossRef]
11. Yang, K.B. Heavy metal pollution and its phytoremediation in farmland soils in China. *World Agric.* **2007**, *8*, 58–61. [CrossRef]
12. Zhao, M.L.; Wang, H.J.; Sun, J.X.; Tang, R.; Cai, B.Y.; Song, X.Y.; Huang, X.M.; Huang, J.; Fan, Z.Q. Spatio-temporal characteristics of soil Cd pollution and its influencing factors: A Geographically and temporally weighted regression (GTWR) method. *J. Hazard. Mater.* **2023**, *446*, 130613. [CrossRef] [PubMed]
13. Huang, Y.Z.; Zhu, Y.G. A review on cadmium contamination in forest ecosystem. *Acta Ecol. Sin.* **2004**, *24*, 101–108. [CrossRef]
14. Hashem, A.; Adam, E.; Hussein, H.A.; Sanousy, M.A.; Ayoub, A. Bioadsorption of Cd (II) from contaminated water on treated sawdust: Adsorption mechanism and optimization. *J. Water Resour. Prot.* **2013**, *5*, 82–90. [CrossRef]
15. Sun, Y.B.; Xu, Y.; Xu, Y.M.; Wang, L.; Liang, X.F.; Li, Y. Reliability and stability of immobilization remediation of Cd polluted soils using sepiolite under pot and field trials. *Environ. Pollut.* **2016**, *208*, 739–746. [CrossRef]

16. Komrek, M.; Vaněk, A.; Ettler, V. Chemical stabilization of metals and arsenic in contaminated soils using oxides—A review. *Environ. Pollut.* **2013**, *172*, 9–22. [CrossRef] [PubMed]
17. Sun, P.C.; Huang, Z.B.; Tang, K.; Zhang, Y. Research progress of chemical solidification on administering soil heavy metal pollution. *Environ. Eng.* **2014**, *32*, 158–161. [CrossRef]
18. Veysel, T. Confident performance of chitosan and pistachio shell biochar on reducing Ni bioavailability in soil and plant plus improved the soil enzymatic activities, antioxidant defense system and nutritional quality of lettuce. Ecotoxicology and Environmental. *Safety* **2019**, *183*, 109594. [CrossRef]
19. Veysel, T. Arbuscular mycorrhizal fungi and pistachio husk biochar combination reduces Ni distribution in mungbean plant and improves plant antioxidants and soil enzymes. *Physiol. Plant.* **2021**, *173*, 418–429. [CrossRef]
20. Liang, X.; Xu, Y.; Wang, L.; Sun, G.; Sun, Y. In-situ immobilization of cadmium and lead in a contaminated agricultural fields by adding natural clays combined with phosphate fertilizer. *Acta Sci. Circumstantiae* **2011**, *31*, 1011–1018.
21. Chang, Y.T.; His, H.C.; Hseu, Z.Y.; Jheng, S.L. Chemical stabilization of cadmium in acidic soil using alkaline agronomic and industrial by-products. *J. Environ. Sci. Health* **2013**, *48*, 1748–1756. [CrossRef]
22. Homa, G.; Mehrnoosh, A.; Tazkieh, G.; Behzad, A.; Saeed, M. Rapid and effective removal of heavy metal ions from aqueous solution using nanostructured clay particles. *Results Surf. Interfaces* **2023**, *10*, 100097. [CrossRef]
23. Hamidpour, M.; Afyuni, M.; Kalbasi, M.; Khoshgoftarmanes, A.H.; Inglezakis, V.J. Mobility and plant-availability of Cd(II) and Pb(II) adsorbed on zeolite and bentonite. *Appl. Clay Sci.* **2010**, *48*, 342–348. [CrossRef]
24. Taparcevska, J.; Markovska, L.; Koumanova, B.; Meshko, V. Diffusion models for adsorption kinetics of Zn($^{2+}$), Cd($^{2+}$) and Pb($^{2+}$) onto natural zeolite. *Water Sci. Technol.* **2010**, *62*, 1136–1142. [CrossRef] [PubMed]
25. Li, M.Y.; Zhang, Y.; Du, L.Y.; Su, X.N.; Dong, S.P.; Lan, X.P. Effect of biocarbon and zeolite mixing on morphological conversion of soil Cd. *J. Soil Water Conserv.* **2014**, *28*, 248–252. [CrossRef]
26. Guo, J.X.; Yuan, C.G. Experimental study on the treatment of heavy metals in sewage by serpentine adsorption. *Fine Chem.* **2000**, *17*, 586–589. [CrossRef]
27. Luo, W.H. Research on Mineralogy Characteristics and Adsorption Capability of Serpentine from Jianchaling. Master's Thesis, Chang'an University, Xi'an, China, 2008. [CrossRef]
28. Cao, C.Y.; Yu, B.; Wang, M.; Zhao, Y.Y.; Wan, X.; Zhao, S. Immobilization of cadmium in simulated contaminated soils using thermal-activated serpentine. *Soil Sci. Plant Nutr.* **2020**, *66*, 499–505. [CrossRef]
29. Raicevic, S.; Kaludjerovic-Radoicic, T.; Zouboulis, A.I. In situ stabilization of toxic metals in polluted soils using phosphates: Theoretical prediction and experimental verification. *J. Hazard. Mater.* **2005**, *117*, 41–53. [CrossRef]
30. Bao, S.D. *Agrochemical Analysis of Soil*, 3rd ed.; China Agriculture Press: Beijing, China, 2000; pp. 25–200.
31. Cao, C.Y.; Liang, C.H.; Yin, Y.; Du, L.Y. Thermal Activation of Serpentine for Adsorption of Cadmium. *J. Hazard. Mater.* **2017**, *329*, 222–229. [CrossRef]
32. Lu, R.K. *Analytical Methods of Soil Agrochemistry*; China Agricultural Science and Technology Press: Beijing, China, 1999; pp. 12–139.
33. Liu, M.; Liu, F.Z.; Liu, B.F. Determination of available lead and cadmium in Soil. *J. Agro-Environ. Sci.* **2007**, *26*, 300–302. [CrossRef]
34. Tessier, A.; Campbell, P.G.C.; Bisson, M. Sequential extraction procedure for the speciation of particulate trace metals. *Anal. Chem.* **1979**, *51*, 844–851. [CrossRef]
35. Majdan, M.; Pikus, S.; Kowalska-Ternes, M.; Gładysz-Płaska, A.; Staszczuk, P.; Fuks, L.; Skrzypek, H. Equilibrium study of selected divalent d-electron metals adsorption on A-type zeolite. *J. Colloid Interface Sci.* **2003**, *262*, 321–330. [CrossRef] [PubMed]
36. Hui, K.S.; Chao, C.Y.H.; Kot, S.C. Removal of mixed heavy metal ions in wastewater by zeolite 4A and residual products from recycled coal fly ash. *J. Hazard. Mater.* **2005**, *127*, 89–101. [CrossRef] [PubMed]
37. Sheikhhosseini, A.; Shirvani, M.; Shariatmadari, H. Competitive sorption of nickel, cadmium, zinc and copper on palygorskite and sepiolite silicate clay minerals. *Geoderma* **2013**, *192*, 249–253. [CrossRef]
38. Li, Y.M.; Wang, K.Q.; Liu, Z.Q.; Wang, J.Y.; Zhou, X. Effect of measure of engineering preparation to soil water in Yunnan Dry-Hot River Valley. *J. Soil Water Conserv.* **2006**, *20*, 15–19. [CrossRef]
39. Ren, L.X. Soil heavy metal morphology and solubility organic matter environmental behavior. *Environ. Sci. Technol.* **2008**, *31*, 69–73. [CrossRef]
40. Li, X.J.; Qiu, S.; Zhao, J.; Zhang, C.P.; Zeng, D.J.; Liu, S.H. Studies on manganese absorption and soil remediation of four ornamental plants. *J. South. Agric.* **2010**, *41*, 951–954. [CrossRef]
41. Liang, X.J.; Han, Y.; Xu, Y.; Sun, L.; Wang, X. In Situ Field-scale Remediation of Cd Polluted Paddy Soil Using Sepiolite and Palygorskite. *Geoderma* **2014**, *235–236*, 9–18. [CrossRef]
42. Jafarnejadi, A.R.; Sayyad, G.; Homaee, M.; Davamei, A.H. Spatial Variability of Soil Total and DTPA-extractable Cadmium Caused by Long-term Application of Phosphate Fertilizers, Crop Rotation, and Soil Characteristics. *Environ. Monit. Assess.* **2013**, *185*, 4087–4096. [CrossRef]
43. Hooda, P.S.; Alloway, B.J. Cadmium and lead sorption behavior of selected English and Indian soils. *Geoderma* **1998**, *84*, 121–134. [CrossRef]
44. Ding, J.H.; Wen, Y.M.; Shu, Q. Fraction Transformation of cadmium and zinc in soils. *Urban Environ. Urban Ecol.* **2001**, *14*, 47–49.
45. Chen, M.; Chen, Y.H.; Shen, Z.G.; Shen, Q.R. Amelioration of aluminum toxicity on wheat plants grown in acid red soil by pig manure. *J. Plant Nutr. Fertil.* **2002**, *8*, 173–176. [CrossRef]

46. Zhang, J.; Sui, Q.; Zhong, H.; Meng, X.; Wang, Z.; Wang, Y.; Wei, Y. Impacts of zero valent iron, natural zeolite and Dnase on the fate of antibiotic resistance genes during thermophilic and mesophilic anaerobic digestion of swine manure. *Bioresour. Technol.* **2018**, *258*, 135–141. [CrossRef] [PubMed]
47. Li, X.J.; Wang, L.J.; Lu, A.H.; Wang, C.Q. A Discussion on Activation Mechanism of Atom Groups in Serpentine. *Acta Petrol. Mineral.* **2003**, *22*, 386–390. [CrossRef]

International Journal of
Environmental Research and Public Health

MDPI

Article

Research on the Cooperative Governance Path of Multiple Stakeholders in Doctor–Patient Disputes under the Environment of Information Asymmetry

Mao-Min Jiang [1,†], Zheng-Yu Wu [1,†] and Ai-Xian Tu [2,*]

1 School of Public Affairs, Xiamen University, Xiamen 361005, China
2 School of Management, Hainan Medical University, Haikou 571199, China
* Correspondence: tuaixian@hainmc.edu.cn
† These authors contributed equally to this work.

Abstract: The number of doctor–patient conflicts and disputes in China has been increasing recently. In order to solve the current social problems of the tense doctor–patient relationship and frequent medical disputes, this article, based on grounded theory, uses qualitative analysis software to conduct grounded coding on 622 cases of doctor–patient disputes randomly selected by stratification. After successively adopting open, axial, and selective coding, the relationship structure between the causes and development of medical disputes is summarized. Furthermore, this relationship structure can be used to analyze further and discuss the causes of doctor–patient disputes from the perspective of multiple co-governance and the governance framework of doctor–patient disputes. Finally, it reminds us of the need to standardize government's regulatory responsibilities, promote the equal distribution of medical resources, strengthen the communication awareness between doctors and patients, reduce the information asymmetry between doctors and patients, and build a preventive mechanism-oriented multi-subject collaborative governance path to promote the relationship between doctors and patients in China. We provide relevant countermeasures and suggestions for harmonious development and the smooth resolution of doctor–patient conflicts.

Keywords: doctor–patient disputes; dispute governance; grounded theory; multiple co-governance

1. Introduction

The Chinese government, which has permanently attached great importance to building a harmonious relationship between doctors and patients, put forward the core value concept of "patient-centered" care, advocating the public welfare and humanization of medical services, and has promoted the construction of a new type of doctor–patient relationship; the government has also implemented a people's mediation mechanism, dispute prevention mechanism and other measures to create a positive and healthy relationship between doctors and patients. However, the relationship between doctors and patients in China has not improved fundamentally, and the number of judicial cases between doctors and patients is also increasing yearly [1]. The contradiction between doctors and patients has become a hot and challenging issue in constructing a harmonious society. Doctor–patient disputes result from the evolution of tense relationships. Usually, they refer to disputes and conflicts between doctors and patients due to medical services or other factors other than medical services. Doctor–patient disputes will not only directly lead to the outbreak of conflicts between doctors and patients but also lead to the two sides confronting each other head-on; it will also reduce the level of trust between doctors and patients, bring the diagnosis and treatment process to a deadlock, and affect the quality and level of diagnosis and treatment [2]. In a new era and stage for the implementation of an epidemic prevention and control strategy and the implementation of the Healthy China strategy, the governance of doctor–patient disputes reflects the distinctive characteristics of the times, a clear problem orientation, and vital innovation needs [3]. In the post-epidemic era, how to scientifically and effectively govern doctor–patient disputes has aroused extensive discussions in all walks of life [4]. The research on doctor–patient dispute governance is not only a direct response to building a harmonious doctor–patient relationship but also a requirement of the times to promote the implementation of the "Healthy China" strategy and the construction of a harmonious society [5]. Based on the theoretical framework of collaborative governance and from the perspective of collaborative participation of multiple subjects, this study uses grounded theory to study and discuss the path of medical dispute governance, providing a theoretical reference for government departments or medical institutions to prevent and resolve medical disputes correctly.

Scholars at home and abroad have conducted much research on doctor–patient disputes. After sorting, it is not difficult to find that the current research mainly focuses on the cause analysis and countermeasures of doctor–patient disputes. In terms of the causes of doctor–patient disputes, Parsons put forward the theory of the patient's role, in 1951, pointing out that there is a significant asymmetry between doctors and patients, and misunderstandings are prone to lead to opposition between doctors and patients [6]; in 1956, Hollender and Szasz also proposed a doctor–patient relationship consisting of a "mutual participation model, guided cooperation model, and active-passive model," pointing out that because doctors fail to use different diagnosis and treatment models according to their conditions during diagnosis and treatment, it has caused tension between doctors and patients [7,8]. However, Allman believes that communication between doctors and patients plays a crucial role. The deep cause of doctor–patient disputes stems from the lack of in-depth communication and communication [9]. When Chinese scholars discuss the causes of doctor–patient disputes, they attribute it to the lack of trust between doctors and patients, information asymmetry, and uneven distribution of medical resources. For example, Li pointed out that medical staff believed that patients who came to seek medical treatment perceived doctors as having a "condescending" mentality, and that they ignored listening to patient's complaints and questions [10]; Qi said that with the professionalization of doctors and the industrialization of medical institutions, doctors and the trust between patients gradually dissipated, which eventually led to the occurrence of doctor–patient disputes [3]; Shao and Hu believed that information asymmetry between doctors and patients led to wrong choices, moral hazard and one-sided judgments, which made patients feel the dissatisfaction of doctors accumulates and promotes the transformation of the doctor–patient relationship from dissatisfaction to disputes [11]; Wu and Meng believed that the imbalance of medical resources made the medical needs of patients unable to be met from the perspective of social and economic development, and this caused doctor–patient disputes [12].

Scholars at home and abroad have also conducted in-depth discussions on the path and countermeasures of doctor–patient disputes. For example, Chen Zheng and Peng Hua proposed that hospitals should make full use of the information system's key monitoring role in medical safety indicators and establish a risk sentinel early warning mechanism to deal with potential disputes [13]; Gerds and Lerner pointed out that artificial intelligence should be used to reduce medical risks The advantages of prediction and management are to quantify medical risk early warning indicators and establish a risk early warning system. Hospitals should closely prevent and control potential risk factors with higher risk levels to make effective interventions as soon as possible when doctor–patient disputes occur [14,15]; Rubin believes that hospitals should focus on the quality of medical care, play an active role in the risk management of evidence-based medicine in the process of diagnosis and treatment, and curb the emergence of doctor–patient conflicts from the source [16]; while Zhang, Hu, Tao, etc. believe that the government should play a role in governance. They pointed out that the government should scientifically design the framework of laws and regulations for doctor–patient disputes in China, clarify critical points such as risk prevention and patient-centeredness [17–19], and should give full play to the role of third-party mediation mechanisms in resolving doctor–patient disputes [20,21].

The existing research literature shows that although domestic and foreign scholars have conducted rich research on the causes and solutions of doctor–patient disputes, the current research still presents fragmentation and local characteristics and fails to respond from a systematic and global perspective. The governance theory emphasizes "multiple subject participation, clear responsibilities, and fair and reasonable goals". Grounded theory is a qualitative research method that uses systematic procedures to summarize and guide grounded theories according to the development direction of social problems. Therefore, from the perspective of multi-subject collaborative governance, this paper uses grounded theory to explore the path framework of doctor–patient dispute governance under multi-stakeholders.

2. Methodology

We used grounded theory to conduct coding analysis on the research text data. We used NVIVO software to perform open, axial, and selective coding during the coding process. During the coding process, the Delphi method was used to obtain the opinions of external experts for some objectionable content. Open coding is a process of breaking up data, assigning concepts to the data, and then recombining the data in a new way through the careful study of phenomena, that is, the process of analyzing, inspecting, conceptualizing, summarizing, and comparing data. The purpose of open coding is to discover conceptual categories from the data and name the categories to determine the attributes and dimensions of the categories. The code is saturated. In the process of open coding, a dozen or even hundreds of concepts will be generated, so researchers must classify similar concepts, but the concept of categories in this stage is temporary and may be changed at any time due to situations where a discovery necessitates modification of the genus [22]. Axial coding is a complex process of linking similar codings together through continuous comparison, and its main task is to select and construct the content of the main categories [23] and connect the main concept categories with the secondary concept categories to find the underlying logical connections between the various genera eventually form tighter categories [24]. The purpose of selective coding is to further systematically deal with the relationships between categories. The main work is to summarize the "Core category" through integration and to condense and explain all phenomena in the form of a "Story Line" [25]. The core category is the keywords obtained after condensing all the analysis results, and the condensed keywords are enough to explain the connotation of the fundamental research.

The data used in the analysis of this article come from the 2020 medical dispute cases of China Judgment Documents Network, and the judicial cases of doctor–patient disputes are selected as the analysis data of this article. The cause, mediation, and result of the event are representative, fair, and traceable, so we can better understand the evolution and development of doctor–patient disputes; Secondly, judicial cases have recorded the causes and consequences of doctor–patient disputes in detail, and the records are completely authentic and reliable, and the quality of the materials is relatively high. In the China Judgment Documents Network, we searched by setting the case type as a civil case, the cause of the case as medical liability dispute, the trial procedure as a civil first instance, and the date of judgment as after 2020. A total of 15,210 case texts were retrieved, including 13,533 cases of doctor–patient conflicts in public hospitals. After eliminating 7316 invalid texts, such as duplicate case names, undisclosed content, and civil rulings, 6217 samples remained. The sample size of the analysis is between 30–500 [26,27], so we used IEAD software to conduct stratified sampling according to the ratio of 1/10, to try to include every province in China in the research scope, and, finally, screen out 622 case texts. The total number and sampling quantity of doctor–patient disputes in various provinces of China are shown in Table 1.

Table 1. Summary of doctor–patient disputes data in various provinces in 2020.

District	Quantity	Number of Public Hospital Cases	Sample Quantity	District	Quantity	Number of Public Hospital Cases	Sample Quantity
Jiangsu	1871	1665	75	Heilongjiang	372	331	15
Shandong	1634	1454	66	Fujian	366	326	15
Henan	1507	1341	61	Guangxi	363	323	15
Beijing	1080	961	43	Shanxi	360	320	14
Jilin	807	718	33	Shaanxi	252	224	10
Sichuan	746	664	30	Guizhou	251	223	10
Hebei	672	598	27	Shanghai	246	219	10
Hubei	615	547	25	Inner Mongolia	241	214	10
Zhejiang	524	466	21	Xinjiang	138	123	6
Hunan	499	444	20	Tianjin	134	119	5
Anhui	493	439	20	Ningxia	46	41	5
Liaoning	396	352	16	Gansu	28	25	3
Yunnan	394	351	16	Hainan	8	7	2
Guangdong	391	348	16	Qinghai	6	5	2
Chongqing	387	344	16	Tibet	1	1	1
Jiangxi	382	340	15				

3. Results

3.1. Open Coding

In this paper, after reading and analyzing the text data word by word and carrying out initial conceptualization, 1306 original sentences and corresponding initial concepts were finally obtained, the initial concept of frequency three and above. Table 2 shows the obtained initial concepts and categories. In order to save space, three to four original sentences for each category are excerpted in Table 2.

Table 2. Open coding summary table.

Generic	Original Statement
Delay in diagnosis and treatment 288/622 (46.30%)	1. The defendant was at fault during the treatment process, and the rescue was not timely, causing the plaintiff to be disabled. 2. The doctor did not treat in a time when the patient's fetus was in distress. 3. Before the cesarean section, all the tests were not timely. 4. There are delays in the hospital's diagnosis and treatment of patients with craniocerebral trauma, and the surgical treatment is not timely.
Missed diagnosis and misdiagnosed disease 52/622 (8.36%)	1. The doctor blindly concluded that the diagnosis is: asthma, chronic bronchitis, which is a misdiagnosis that violates common sense. 2. Diagnosed as acute pharyngitis in the case of abnormal test results, used antibiotics for treatment. 3. The hospital's misdiagnosis and hysterectomy caused the plaintiff's cervical cancer. 4. A misdiagnosis of delayed encephalopathy caused by carbon monoxide poisoning delayed the best treatment period, and the doctor had apparent faults in the diagnosis and treatment process.
Improper use of drugs 73/622 (11.74%)	1. A drug skin test was not given to the patient before intravenous administration, resulting in drug anaphylactic shock and death. 2. The medical practice of overdose in the use of antipyretics by the doctor; the overdose of the drug aggravated the inflammatory response of the child's disease and aggravated the condition. 3. Due to the doctor's wrong prescribing of medication this directly led to the complete damage of the left facial nerve and the hypofunction of the bilevel semicircular canal.

Table 2. *Cont.*

Generic	Original Statement
Insufficient diagnostic examination 41/622 (6.59%)	1. Without taking any ECG monitoring and perfecting relevant examinations, the doctor blindly concluded and diagnosed asthma and chronic bronchitis. 2. In the process of diagnosis and treatment, the hospital failed to perform the required inspection to rule out the possible risk of disease. 3. The hospital did not check the patient's right eye injury in place, underestimated the severity of the patient's condition, and delayed the diagnosis and treatment of the disease.
Improper nursing operation 41/622 (6.59%)	1. Postoperative medication and nursing violated the norms of diagnosis, treatment, and nursing, and there were significant mistakes. 2. When the nurse on duty was nursing the patient, she accidentally stuck tape on her glove and directly took the intubation tube out of the patient's mouth. 3. The nursing operation violated the "Clinical Nursing Practice Guidelines (2011 Edition)".
Improper surgical operation 145/622 (23.31%)	1. There are problems such as non-compliance with the regulations in the doctor's operation records. 2. The doctor did not grasp the surgical indications rigorously and performed improper operations during the operation. 3. The hospital's operation was not rigorous enough, and the patient was required to be transferred to other hospitals for treatment after the operation.
Insufficient communication between doctors and patients 62/622 (9.97%)	1. Insufficient communication between doctors and patients on the significance of cerebral angiography examination, insufficient communication between doctors and patients. 2. Insufficient assessment and treatment of the severity of the children's condition, insufficient communication between doctors and patients. 3. The doctor arranged for the patient to have an ECT examination at an inappropriate time, and there was no communication record between the doctor and the patient
Lack of professionalism 16/622 (2.57%)	1. The doctor concealed the patient's condition and did not inform the patient's actual condition to their guardian in time. 2. The doctor was highly irresponsible and lost the test sheet, resulting in severe hypoxia after birth. 3. The doctor was irresponsible, yelling at the patient and ignoring the patient's pain.
Insufficient risk notification 67/622 (10.77%)	1. The doctor did not explain the medical alternatives to the patient, nor did he inform the pros and cons of the plan. 2. Failure to inform patients and their families about the risk of brain damage during surgery. 3. Failure to inform patients of postoperative risks and preventive measures after discharge.
Violation of medical record writing 161/622 (25.88%)	1. Failing to make, write and save outpatient medical records. 2. The writing of medical records was not standardized, and there were apparent differences between the medical records and the actual situation. 3. The outpatient medical records attached to the hematology department upon admission were handwritten and did not meet the norms.
Inferior equipment and consumables 36/622 (5.79%)	1. The use of unqualified internal fixation products by the doctor directly caused the bone destruction of the patient's proximal right clavicle and sternal manubrium. 2. The chemical composition of the same batch of broken steel nails did not meet the requirements of national standards, and the microstructure and hardness did not meet the requirements of relevant standards, resulting in the final breakage of steel nails. 3. The disposable sterile package expired, causing infection incidents.
Practicing medicine beyond the scope 26/622 (4.18%)	1. The scope of practice of the treating physician did not meet the requirements. 2. Doctors who examine, diagnose and operate on patients do not have a license to practice. 3. The doctor does not have the qualifications to perform tertiary surgery, and the medical staff do not have the qualifications.

Table 2. *Cont.*

Generic	Original Statement
Lack of trust in diagnosis and treatment 88/622 (14.15%)	1. A patient believed that the doctor's delivery method was improper and directly led to the death of their newborn. 2. The patient provided the precautions collected on the Internet to prove the doctor's mistake, but the evidence belongs to the network knowledge, and there is no corresponding factual and legal basis for whether it meets the medical regulations. 3. During the hospitalization period, privately contacted acquaintances for barium dialysis of the digestive tract, which aggravated the condition of the patient.
Failure to cooperate with diagnosis and treatment 36/622 (5.79%)	1. The doctor suggested deep vein catheterization for parenteral nutrition, but the patient was hesitant and did not cooperate with the treatment in time. 2. The patient was sent for rescue treatment in the ICU, because the relatives of the patient refused to cooperate with the treatment. 3. The doctor advised her to give birth by cesarean section, but the puerpera and her family insisted on continuing vaginal trial delivery, which could lead to severe consequences such as obstructive dystocia, uterine rupture, vesicovaginal fistula, neonatal intracranial hemorrhage, and intrauterine death.
Disrupting the order of diagnosis and treatment 16/622 (2.57%)	1. During the doctor's examination, the patient thought the doctor's examination caused his child's constant crying, so he maliciously pushed and shoved and caused the two parties to fight each other. 2. Because the doctor was silent during the diagnosis and treatment, the patient thought that his attitude was horrible, and the doctor only cared about prescribing medicine to make money, so he violently attacked the doctor. 3. Because the patient waited a long time for a prenatal checkup, she believed that the hospital service was not appropriate, so she insulted and beat the nurse on duty.
Dissatisfied with medical expenses 10/622 (1.61%)	1. Expend hundreds of thousands of yuan in medical expenses, unable to pay for the cost of surgery to remove the steel plate. 2. In addition to the medical expenses incurred by normal childbirth, the remaining expenses are unreasonable or too high. 3. The patient claims that the fee is unreasonable, and the subject of the claim should be the social security institution.
The patient's condition is poor 57/622 (9.16%)	1. Elderly patients, who also have a variety of underlying diseases in the past and have immune dysfunction. 2. The patient is older and has suffered from asthma for 30 years, and the threshold for resisting the influence of various internal or external factors on the body is lower than that of ordinary people. 3. The diagnosis and treatment process is correct, and the surgical measures taken are appropriate. During the operation, due to laparoscopic surgery, the operation is minimally invasive, but the technology required is relatively high, and the patient's disease is the reason for ureteral injury.
Resource and technical limitations 21/622 (3.38%)	1. It is a rare disease that is a medical problem, and the medical technology level of the county-level medical institutions could not diagnose the genetic disease in the fetal stage at that time. 2. Factors such as local medical technology conditions. 3. There is no necessary rescue equipment and no rescue technology, resulting in the death of the patient in the hospital due to lack of proper treatment after suffocation.

3.2. *Axial Coding*

This paper summarizes six principal categories according to the central axis coding through our repeated derivation and induction. The meaning of each main category and its corresponding sub-categories is shown in Table 3.

Table 3. Summary of main axis coding.

Main Generic	Sub-Generic	Number of Cases	Interpretation
Medical care diagnosis and treatment fault	Delay in diagnosis and treatment	228	Disputes caused by doctors' mistakes in the process of diagnosis and treatment resulting in physical or psychological losses to patients
	Misdiagnosed and misdiagnosed	52	
	Improper use of drugs	73	
	Failed to perfect inspection	41	
	Improper nursing practice	41	
	Improper operation	145	
Lack of medical literacy	Insufficient doctor–patient communication	62	Conflicts caused by patients' dissatisfaction with medical treatment due to insufficient service awareness and service ability of medical staff
	lack of professionalism	16	
Lack of hospital management	Insufficient risk communication	67	Conflicts caused by patients' dissatisfaction with medical treatment due to inadequate implementation of relevant hospital management measures
	Medical record writing violations	161	
Lack of government regulation	Inferior equipment consumables	36	Due to the lack of government supervision, the illegal behavior of the hospital is not regulated, which leads to the harm of patients' medical treatment
	Practicing medicine beyond the scope	26	
Patient dissatisfaction	Lack of trust in diagnosis and treatment	88	Conflicts with doctors due to differences in patients' cognition of diseases and partial dissatisfaction with medical treatment
	Not cooperating with medical treatment	36	
	disrupt the medical order	16	
	Dissatisfied with medical expenses	10	
objective factors	The patient's condition is poor	57	Refers to the contradictions and conflicts caused by objective factors that exist and are difficult to change
	Resource Technology Constraints	21	

3.3. Selective Coding

Through the analytical step of selective coding, researchers developed faction types on which to build a theory. In this paper, 17 sub-categories were identified, such as "improper use of drugs, improper operation, insufficient risk notification, and lack of trust in diagnosis and treatment," as well as "doctor's fault in diagnosis and treatment, improper medical behavior, lack of hospital management, lack of government supervision, subjective dissatisfaction of patients, and objective factors. The investigation of six main categories, such as "factor constraints", determined the story line of multi-stakeholders' collaborative participation in governance: the government, hospitals, doctors, patients, and other stakeholders have some improper behaviors in medical activities, and these behaviors directly or indirectly affect doctor–patient disputes. The emergence and development of doctor–patient disputes have an impact, and to avoid doctor–patient disputes requires multiple subjects to manage these direct or indirect influencing factors jointly.

4. Discussions

According to the typical relationship structure between the main categories in Table 4, the internal formation mechanism of doctor–patient disputes from the perspective of multiple subjects can be drawn, that is, lack of government supervision, inadequate hospital management, improper medical and nursing behavior, and doctor's fault in diagnosis and treatment. The mutual influence and effect of other factors directly or indirectly lead to the occurrence and development of doctor–patient disputes.

Table 4. Typical relational structure of main categories.

Typical Relational Structure	The Connotation of Relational Structure	Theoretical Theme
Lack of government supervision → lack of hospital management → doctor–patient disputes	The lack of supervision of hospitals by competent government departments has led to the failure to detect and stop the illegal activities of some medical institutions in a timely manner. For example, hospitals use substandard medical consumables and equipment, and carry out related diagnosis and treatment activities beyond their scope of practice. Suffering from conflict and burying hidden dangers	Controlling Prisoners Affects Policy Implementation
Insufficient hospital management → Improper behavior of doctors and nurses → Disputes between doctors and patients	As a relatively special service organization, medical care has the responsibility and obligation to strengthen the service level and service awareness of medical staff in the organization, and regulate the behavior of medical staff, otherwise it will inevitably lead to "customers" dissatisfaction with the service.	
Insufficient hospital management → doctor's fault in diagnosis and treatment → doctor–patient dispute	As the main place of medical activities, hospitals have the responsibility and obligation to manage and standardize the diagnosis and treatment activities of the hospital and require medical staff to standardize the performance of diagnosis and treatment procedures. doctor–patient conflict	
Doctor's fault in diagnosis and treatment → doctor–patient dispute	In the diagnosis and treatment activities, the doctor's fault diagnosis and treatment behavior caused by misdiagnosis; misdiagnosi and improper clinical operation directly affects the physical and mental health of patients and the treatment of diseases, and is the primary factor that induces doctor–patient conflict.	Damage to rights will inevitably lead to conflicts
Improper medical behavior → doctor–patient disputes	As medical service personnel, medical staff should have a good sense of service for patients. When receiving patients, the lack of service level and service awareness will lead to patients' dissatisfaction with medical treatment, which will lead to conflicts.	
Subjective dissatisfaction of patients → doctor–patient disputes	Due to the patient's cognitive asymmetry between the disease itself and the disease diagnosis and treatment information, the understanding of the disease treatment is biased, the failure to cooperate with the diagnosis and treatment or the wrong understanding of some diagnosis and treatment behaviors, eventually lead to conflicts and dissatisfaction between the doctor and the patient.	Information asymmetry leads to misunderstanding between doctors and patients
Restricted by objective factors → doctor's fault in diagnosis and treatment → doctor–patient dispute	Due to the poor health of the patient, or the limitation of current medical technology, equipment, talents and other resources, this can lead to unavoidable non-fault diagnosis and treatment errors, which lead to contradictions and conflicts between doctors and patients.	The dilemma of technology and resources will inevitably lead to conflicts

4.1. In Order to Obtain the Most Incredible Economic Benefits, Medical Institutions Lay Hidden Dangers of Disputes

Under China's current medical and health system, government departments generally adopt the method of formulating rules to regulate independently operated medical institutions in the supervision of medical institutions. Usually, government departments do not directly supervise the daily operation of medical institutions [28]. Due to the large number of medical institutions in China, the operation of the regulatory mechanism requires a large workforce, material resources, and financial resources [29], and there are no low regulatory costs. The income and development of medical institutions bring "income generation" to government departments. Therefore, some government departments tend to relax or even downplay their regulatory responsibilities for medical institutions out of self-interest considerations. As a unique service organization, medical institutions ensure their essential operation and development needs. Taking measures to obtain profits has become the primary goal that medical institution managers must implement. Therefore, they will adopt non-practicing medical services and use inferior medical equipment. These behaviors will adversely affect the service level of medical institutions and will also lay hidden dangers for future doctor–patient disputes. At the same time, to obtain the most significant economic benefits, doctors will also prescribe more medicines, more expensive medicines, and more checklists, which not only aggravates the dissatisfaction of patients with medical staff but also catalyzes the emergence of doctor–patient disputes and development.

4.2. Lack of Medical Resources Increases the Probability of Doctor–Patient Disputes

Disease diagnosis and treatment depend on high-quality medical technology and advanced medical equipment [30]. When seeing a doctor, medical equipment and staff, as critical medical resources, directly affect the patient's diagnosis and treatment. In an example of a dispute case, the hospital "did not have the necessary rescue equipment and rescue technology, resulting in the death of the patient in the hospital because he could not receive proper treatment after suffocation," and "the level of medical technology at that time could not be adjusted to the fetal stage. Diagnosis of hereditary diseases", leading to patients' high dissatisfaction with the results and quality of diagnosis and treatment, which eventually led to disputes and conflicts between the two parties. Currently, the total medical resources in China are insufficient and unevenly distributed between regions. There are differences in medical resources and technologies between urban and rural hospitals and between public and private hospitals. The medical service system and the people's medical treatment needs differ. There will inevitably be an increase in the probability of doctor–patient disputes.

4.3. Insufficient Communication Leads to Disputes

Communication is a bridge of trust between people and a way of interaction and collision between thoughts and emotions. Good communication is conducive to the construction of trust and friendly interpersonal relationships [31]. In the process of seeing a doctor, the patient's anxiety and expectation of disease treatment make him subjectively eager to get communication and feedback from the doctor to relieve his inner anxiety and panic. However, in the communication process between doctors and patients, doctors cannot fully understand the emotional needs of patients, and they also lack psychological empathy for patients. In most cases, doctors only pay attention to understanding the patient's physiological condition and specific symptoms in the process of diagnosis and treatment, ignoring the information communication and emotional exchange with the patient. Just as in the case, the medical staff "did not notice the abnormal behavior of the patient Li, and the mood swings were obvious," "the communication with the patient about the significance of cerebral angiography was insufficient, and the communication between the doctor and the patient was insufficient," there was no communication between the doctor and the patient. With direct, effective, and sufficient communication and exchanges, patients can understand the doctor's diagnosis, treatment ideas, and treatment plans. Doctors also cannot feel the patient's psychological changes and inner dissatisfaction. The lack of communication between the two parties has resulted in a lack of information exchange and emotional connection. Obstacles catalyze the emergence of conflicts between doctors and patients and eventually lead to disputes.

4.4. Moral Hazard Promotes the Development of Disputes

In medical and health services, hospitals and medical staff have absolute information advantages, which are not only reflected in the grasp of medical knowledge but also in the selection of treatment plans and drugs. It is because doctors have absolute information advantages. Doctors are prone to moral risks such as insufficient respect for the right to informed consent and inducing demands during the treatment process, which leads to misunderstandings between doctors and patients and aggravates the development of disputes [32]. For example, "doctors did not inform patients and their families about the risks and alternatives of brain damage from surgery" before surgery, and "the doctor communicated insufficiently and inadequately with patients on the significance of cerebral angiography". However, these measures will not directly affect the diagnosis and treatment results. However, it violates favorable principles from the medical ethics perspective, which will inevitably deepen the misunderstanding between doctors and patients and then quickly escalate disputes between doctors and patients. In addition, due to the interference of factors such as economic benefits and imperfect supervision mechanisms, some doctors use the advantages of information to induce demand for patients and provide patients with medical services that do not meet the needs of diagnosis and treatment. Trust will plummet, and the doctor–patient relationship will undoubtedly deteriorate faster.

4.5. Medical Errors Stimulate the Outbreak of Disputes

In the disease diagnosis and treatment process, patients and their families often have higher requirements and expectations for the results of diagnosis and treatment. However, some medical staff may have misdiagnoses, improper clinical operation, and other faulty diagnosis and treatment behaviors due to a lack of professional skills [33]. For example, in one of the cases, the medical staff "diagnosed acute pharyngitis as acute pharyngitis and treated it with antibiotics in a hanging bottle when the test results were abnormal," and "the wrong implementation of hysterectomy caused the plaintiff's cervical cancer gland"; it may also be due to objective technical resource conditions. Restrictions lead to mistakes in diagnosis and treatment. For example, in this case, "the hospital did not need rescue equipment, and the patient died in the hospital because he could not get the proper treatment after suffocation". However, medical errors, no matter the cause, will seriously lower the expectations of patients and their families and lead to dissatisfaction among patients. In addition, patients have an insufficient accumulation of medical professional knowledge or lack of communication, and the dissatisfaction and grievance of patients quickly accumulate. The doctor–patient relationship deteriorates rapidly, resulting in doctor–patient disputes. In addition, some patients and their families are emotionally unable to accept all negative treatment results and have little tolerance for medical errors within the normal range. Psychologically provoked incidents make the contradictions and conflicts between doctors and patients more intense.

5. Recommendations

Participation in the whole process is the basic premise of the collaborative governance of multiple subjects [34]. The governance of doctor–patient disputes involves multiple stages and is long-term. Only by participating in the process can we ensure a comprehensive grasp of doctor–patient dispute information and avoid further conflicts caused by information asymmetry [35]. In addition, communication is the information link for the collaborative governance of multiple subjects. In the process of doctor–patient dispute governance, the communication and exchange of multiple subjects on dispute governance measures can not only enrich the information and requirements related to dispute governance, improve the rationality and scientificity of dispute governance measures, but also better help each participant. Understand and accept governance measures [36]. Moreover, the governance goal is a highly concentrated governance direction and concept, representing doctor–patient dispute governance's value orientation. Because of the particularity of doctor–patient disputes and the bottom-line thinking of people first and safeguarding the people's fundamental interests, protecting the legitimate rights and interests of both doctors and patients is undoubtedly the fundamental goal of doctor–patient dispute governance [37]. Therefore, multi-subject collaborative governance should not only realize the disadvantaged position and low risk-tolerance of patients in the medical business but also pay attention to the interests of patients in the governance process and ensure the participation and voice of patients. In most cases, doctors have no intention to create doctor–patient disputes subjectively and intentionally. Finally, promoting the healthy development of the medical industry is the inherent goal of managing doctor–patient disputes. It is necessary to follow the natural law of disease outcomes and to accommodate seemingly disadvantaged groups only a little. The governance measures need to consider realistic factors objectively and fairly.

5.1. Standardize the Responsibilities of Government Supervision and Improve the Internal Management of Hospitals

Clarifying the government's regulatory responsibilities and standardizing the government's regulatory behavior is a crucial way to urge government departments to improve the service efficiency of medical institutions. First of all, it is necessary to promptly stop hospitals' improper behavior through the government's accountability mechanism. At the same time, it is necessary to promote further the formation of the "separation of management and management" pattern from the practical level so that the regulators and the regulated have no common interests and promote the supervision of medical institutions by the regulatory authorities to normalization, standardization, and institutionalization, development in the direction of refinement. Regarding the internal management problems of the hospital, public hospitals should further promote the reform of the hospital's internal personnel management, salary system, performance appraisal, etc., so that the hospital will gradually change from "pursuing benefits" to "returning to public welfare," and by solving the "income worries" of medical staff, "To avoid their intrinsic motivation to pursue economic interests and guide them to have the correct value orientation. As for private hospitals, they should improve their internal supervision and control mechanism, give full play to the supervisory role of third-party organizations such as news media and industry associations, improve the self-discipline of medical staff in private hospitals, strengthen publicity and education for medical staff, and cultivate a sense of identity, belonging, and honor to one's profession, correcting career motivations, and truly "saving the dying and healing the wounded".

5.2. Balance the Distribution of Medical Resources and Deepen the Reform of the Diagnosis and Treatment System

Given the current situation of uneven distribution of medical resources, the government should give full play to the function of macro-control, starting from the aspects of hardware facilities, medical staff, science and technology, and encouraging learning to promote the continuous reduction of medical disparities between urban and rural areas and between regions. For example, for areas with relatively weak medical resources, the government should make full use of financial transfer, policy assistance, etc., increase capital investment, and improve the hardware facilities of local medical institutions, so that they will not be unable to meet the needs of patients due to backward equipment and defective equipment. With respect to the demand for medical treatment, medical staff can be selected from the tertiary hospitals with relatively concentrated medical resources to assist in areas with relatively weak medical resources, or excellent medical staff can be selected from the areas with relatively weak medical resources to go to the tertiary hospitals for further training, improving the overall level of local medical staff. In the face of insufficient resources, the development and improvement of the hierarchical diagnosis and treatment system should be promoted. By accelerating the improvement of the medical pattern of "first diagnosis at the grassroots level, two-way referral, separation of acute and chronic diseases, and linkage between upper and lower levels," the situation of tight medical resources can be effectively solved. For example, the government should concentrate on building grassroots medical and health institutions, provide preferential policies to attract investment, and accelerate the development of grassroots health services; optimize and redistribute existing medical and health resources, merge medical institutions with overlapping functions and close geographical locations, and large hospitals' partnerships with grassroots hospitals can be established to realize the sharing of medical resources. The medical service system can be further optimized through models such as medical alliances and alliances. The effect of valuable medical resources can be amplified so that people can seek medical treatment in time at their doorsteps.

5.3. Strengthen the Awareness of Doctor–Patient Communication and Improve the Art of Communication between Diagnosis and Treatment

Doctor–patient communication is an art, a required professional course for medical staff, and a meaningful way to ease disputes between doctors and patients. For the current domestic situation, strengthening the communication between doctors and patients must start from two aspects: changing the service concept of medical staff and improving communication skills. For example, in the doctor's inpatient training stage, the concept of doctor–patient communication is taken as one of the critical points of the training. The training content can cover the service concept, the standard of words and deeds of the service process, the importance of doctor–patient communication, the emotional intelligence education of medical staff, and the development of doctor–patient communication. The form and content of the system, etc., guide medical staff to form the concept of friendly communication from the intrinsic value orientation. At the same time, the communication skills of medical staff need to be improved. When communicating with patients, medical staff should pay attention to empathy, respect, and patience, maintain a good habit of listening, and communicate with patients after fully grasping the patient's condition, treatment and examination results, and medical expenses. It is also necessary to pay attention to the patient's emotional response to the disease treatment and avoid forcing the patient to accept the fact of the disease, avoid stimulating the other party's emotions, or excessively exploiting the challenges faced by the other party when doctors use professional vocabulary for communication.

5.4. Narrowing the Knowledge and Information Gap and Enhancing the Equal Status of Information

The knowledge gap and information asymmetry between doctors and patients are unavoidable. However, it is possible to change. Due to the absolute advantage of medical knowledge and information, it is necessary to focus on patients to reduce the knowledge and information asymmetry between doctors and patients. Therefore, the channels for obtaining medical information can be broadened for patients to increase their understanding of their diseases, such as other authoritative information channels to understand the development history of the disease, the current status of diagnosis and treatment, and treatment methods to improve their medical quality. For the government, it can cooperate with the media to promote national health education and popularize medical knowledge; for example, in the community, by holding free lectures, distributing publicity materials, placing information display boards, playing publicity videos, etc., conducting health knowledge lectures on common diseases, and conduct health knowledge lectures. For the current topics of concern or diseases that need vigilance, targeted publicity and education programs are formulated, and relevant special editions or feature films are produced to popularize health knowledge.

5.5. Build an Error Prevention Mechanism and Improve the Error Early Warning Mechanism

Establishing a medical error prevention mechanism is one of the important ways to prevent medical errors. Improving medical technology improves communication, and training platforms should be established to avoid unnecessary diagnosis and treatment errors. For example, hospitals can conduct training and exchange activities on medical technology skills on a regular and irregular basis to improve the medical skills and level of medical staff. On the other hand, the hospital should give full play to the role and function of the medical quality committee, supervise, control, evaluate and punish errors in the hospital, ensure the safety of patients, and exhort medical staff to summarize experience and lessons promptly to avoid secondary errors. Finally, the information early warning system should be improved, the safe use of drugs and medical equipment should be automatically prompted, and early warning through modern information technology and the role of information technology in improving and preventing the quality of medical care should be brought into play. For example, by developing a scientific and practical prescription review system, an evidence-based medical system plays the role of automatic error warning.

6. Conclusions

Doctor–patient conflicts and disputes should be paid constant attention. The internal causes of doctor-patient disputes mainly include the government, hospitals, doctors, and patients. The lack of government supervision, inadequate hospital management, insufficient medical resources, improper behavior of doctors and nurses, faults of doctors in diagnosis and treatment, and subjective dissatisfaction of patients are important factors leading to doctor-patient disputes.

7. Limitation

This study is a qualitative analysis, and 622 case texts from the China Judgment Documents Network were selected for analysis. The sample has certain limitations. Later studies could include more case texts as well as actual cases. At the same time, there was little information about the doctors and patients involved in each case, and there was a lack of analysis of doctors' data. Through interviews, subsequent research can explore the correlation between doctors' personal characteristics and medical errors. In addition, subsequent research can also use quantitative methods and analyze the correlation between medical malpractice and doctors and hospitals.

Author Contributions: Conceptualization, M.-M.J. and A.-X.T.; methodology, Z.-Y.W.; software, Z.-Y.W.; validation, M.-M.J. and A.-X.T.; formal analysis, M.-M.J.; resources, A.-X.T.; data curation, Z.-Y.W.; writing—original draft preparation, M.-M.J.; writing—review and editing, M.-M.J.; visualization, A.-X.T.; supervision, A.-X.T.; project administration, A.-X.T.; funding acquisition, A.-X.T. All authors have read and agreed to the published version of the manuscript.

Funding: This work was supported by the High-level Talents Program of Hainan Natural Science Foundation (Grant Numbers 720RC621). This work was supported by the Hainan Medical College Talent Introduction Research Project (Grant Numbers XRC202127), Hainan Health Industry Research Project (Grant Numbers 20A200485). Project supported by Hainan Province Clinical Medical Center. The project sponsors had an essential role in the study design, data collection, analysis, interpretation, and manuscript writing.

Institutional Review Board Statement: Not applicable.

Informed Consent Statement: Not applicable.

Data Availability Statement: The data sets used and/or analyzed during the current study are available from the corresponding author on reasonable request.

Acknowledgments: The authors thank all the participants, assistants, and researchers for their contribution to this study. Thanks especially to Geriatric Hospital of Hainan Jin-sheng Lu for his guidance and revision of the article, as well as the financial support from the Geriatric Hospital of Hainan. Meanwhile, I would like to thank Hui Li from The Second Affiliated Hospital of Hainan Medical University for his guidance on the article.

Conflicts of Interest: The authors declare that the research was conducted in the absence of any commercial or financial relationship that could be construed as potential conflict of interest.

References

1. Du, L.; Xu, J.; Chen, X. Rebuild doctor–patient trust in medical service delivery in China. *Sci. Rep.* **2020**, *10*, 21956. [CrossRef] [PubMed]
2. Wang, F.; Song, Z.; Zhang, W.; Xiao, Y. Medical Humanities play an important role in improving the doctor-patient relationship. *BioSci. Trends* **2017**, *11*, 134–137. [CrossRef] [PubMed]
3. Qi, X.X. Cause Analysis and Legal Response to the "Intensification" of Doctor-Patient Disputes—Taking Three Cases of Violent Attacks on Doctors as Concerns. *Qiushi Acad. J.* **2020**, *47*, 12–18. [CrossRef]
4. Xu, B. The impact of covid-19 on the doctor-patient relationship in China. *Front. Public Health* **2022**, *10*, 1–15. [CrossRef]
5. Chang, P.C.; Wu, T.; Du, J. Psychological contract violation and patient's antisocial behaviour. *Int. J. Confl. Manag.* **2020**, *31*, 647–664. [CrossRef]
6. Liang, J.L. Development of Western Health Sociology Research. *Foreign Soc. Sci.* **2010**, *6*, 93–99.
7. Schauer, A.J. *Ethics in Medicine*; Vandenhoeck & Ruprecht: Göttingen, Germany, 2001.
8. Robinson, G. A thought-provoking exchange. *Nurs. Manag.* **1989**, *20*, 80J. [CrossRef]
9. Allman, R.L. The relationship between physicians and the pharmaceutical industry: Ethical problems with the every-day conflict of interest. *HEC Forum* **2003**, *15*, 155–170. [CrossRef]
10. Yang, T.W.; Lu, W.T. A review of the research on conflict between doctors and patients at home and abroad. *Chin. Med. Ethics* **2006**, *1*, 46–48.
11. Shao, H.; Hu, X.J. On the Refined Path of Doctor-Patient Dispute Prevention. *Jinyang Acad. J.* **2017**, *3*, 82–89. [CrossRef]
12. Wu, L.H.; Meng, F.Y. Using Marxist philosophy to analyze the doctor-patient relationship and solutions in the new era. *Contin. Med. Educ.* **2020**, *34*, 48–49.
13. He, W.; Wang, X.; Zhou, X.; Xu, L. Negative expectations and bad relationships: Effects of negative metastereotypes on doctor–patient relationships. *Asian J. Soc. Psychol.* **2019**, *23*, 103–108. [CrossRef]
14. Gerds, T.A.; Kattan, M.W. *Medical Risk Prediction*; Chapman and Hall/CRC: Boca Raton, FL, USA, 2021; Volume 10, pp. 108–109. [CrossRef]
15. Lerner, I.; Veil, R.; Nguyen, D.P.; Luu, V.P.; Jantzen, R. Revolution in health care: How will data science impact doctor–patient relationships? *Front. Public Health* **2018**, *6*, 99. [CrossRef] [PubMed]
16. Rubin, G.L.; Frommer, M.S. Evidence-based medicine—Time for a reality check. *Med. J. Aust.* **2001**, *174*, 214–215. [CrossRef] [PubMed]
17. Zhang, X.; Ma, L.; Ma, Y.; Yang, X. Mobile information systems usage and doctor-patient relationships: An empirical study in China. *Mob. Inf. Syst.* **2021**, *2021*, 6684448. [CrossRef]
18. Hu, Y.; Zhou, H.; Chen, Y.; Yao, J.; Su, J. The influence of patient-generated reviews and doctor-patient relationship on online consultations in China. *Electron. Commer. Res.* **2021**, *10*, 1–8. [CrossRef]
19. Tao, S.; Liu, C.; Wu, Q. Developing a scale measuring the doctor–patient relationship in China from the perspective of doctors. *Fam. Pract.* **2021**, *39*, 527–536. [CrossRef]
20. Ren, W.; Sun, L.; Tarimo, C.S.; Li, Q.; Wu, J. The situation and influencing factors of outpatient satisfaction in large hospitals: Evidence from Henan Province, China. *BMC Health Serv. Res.* **2021**, *21*, 500. [CrossRef]
21. Zeng, W.; Ma, S.; Callan, V.J.; Wu, L. Exploring the doctor-patient relationship as a challenge job demand: Application of the job demands–resources model in a chinese public hospital. *Psychol. Health Med.* **2021**, *27*, 1661–1671. [CrossRef]
22. Woolf, N.H.; Silver, C. The architecture of NVivo. *Qual. Anal. Using NVivo* **2017**, *10*, 66–110. [CrossRef]
23. Woolf, N.H.; Silver, C. Orientation to nvivo. *Qual. Anal. Using NVivo* **2017**, *12*, 61–65. [CrossRef]
24. Kendall, J. Axial coding and the grounded theory controversy. *West. J. Nurs. Res.* **1999**, *21*, 743–757. [CrossRef] [PubMed]
25. Strübing, J. Grounded theory und situationsanalyse: Zur Weiterentwicklung der grounded theory. *Grounded Theory* **2014**, *10*, 97–111. [CrossRef]
26. Zhang, L. On the construction of the characteristic indicators of effective education and scientific research in primary and secondary schools—Based on the NVivo coding analysis of 54 texts. *Curric. Textb. Teach Methods* **2021**, *41*, 136–143. [CrossRef]
27. Mason, M. Sample Size and Saturation in PhD Studies Using Qualitative Interviews. *Forum Qual. Soz./Forum Qual. Soc. Res.* **2010**, *11*, 1–19. [CrossRef]
28. Lai, Y.; Chen, S.; Li, M.; Ung, C.O.; Hu, H. Policy interventions, development trends, and service innovations of internet hospitals in China: Documentary analysis and qualitative interview study. *J. Med. Internet Res.* **2021**, *23*, e22330. [CrossRef]

29. Xu, C.; Luo, L.; Zeng, S.; He, X.; Li, J.; Zhu, G. What promotes medical overuse: Perspective on evolutionary game between administration and Medical Institutions. *Comput. Math. Methods Med.* **2022**, *2022*, 1–15. [CrossRef]
30. Cui, C.; Zuo, X.; Wang, Y.; Song, H.; Shi, J.; Meng, K. A comparative study of patients' satisfaction with different levels of hospitals in Beijing: Why do patients prefer high-level hospitals? *BMC Health Serv. Res.* **2020**, *20*, 1–10. [CrossRef]
31. Lawton, A.J.; Rosenberg, L.B. Carpe communication: Seizing the small moments to teach interpersonal and communication skills on inpatient services. *J. Grad. Med. Educ.* **2022**, *14*, 386–389. [CrossRef]
32. Liu, J.; Yu, C.; Li, C.; Han, J. Cooperation or conflict in doctor-patient relationship? an analysis from the perspective of evolutionary game. *IEEE Access* **2020**, *8*, 42898–42908. [CrossRef]
33. Brüggemann, A.J.; Wijma, B.; Swahnberg, K. Abuse in health care: A concept analysis. *Scand. J. Caring Sci.* **2011**, *26*, 123–132. [CrossRef] [PubMed]
34. Rongjuan, W. How multiple interactions between policy instruments and the policy environment affect environmental governance efficiency. *Energy Environ.* **2022**, *10*, 0958305X221107346. [CrossRef]
35. Butalid, L.; Verhaak, P.F.; Boeije, H.R.; Bensing, J.M. Patients' views on changes in doctor-patient communication between 1982 and 2001: A mixed-methods study. *BMC Fam. Pract.* **2012**, *13*, 80. [CrossRef] [PubMed]
36. Scheppokat, K.D. Lehren aus den Erfahrungen einer schlichtungsstelle für arzthaftpflichtfragen. *Ther. Umsch.* **2005**, *62*, 185–190. [CrossRef]
37. Jiang, F.; Zhou, H.; Hu, L. Psychiatry residents in China: Socio-demographic characteristics, career satisfaction, and related factors. *Front. Psychiatry* **2019**, *10*, 177. [CrossRef]

International Journal of
Environmental Research and Public Health

MDPI

Article

Research on the Impact of Green Innovation Network Embeddedness on Corporate Environmental Responsibility

Junli Wang and Wendong Lv *

Business School, University of International Business and Economics, Beijing 100029, China
* Correspondence: lwd@uibe.edu.cn; Tel.: +86-13021165755

Abstract: In the process of China's economic transformation, enterprises urgently need to use green innovation networks to realize corporate sustainability. Based on resource-based theory, this study explores the internal mechanism and boundary conditions of green innovation network embeddedness that affect corporate environmental responsibility. This paper conducts an empirical study based on panel data of listed companies engaged in green innovation in China from 2010 to 2020. Drawing on network embeddedness theory and resource-based theory, we found that relational and structural embeddedness influenced green reputation, which affected corporate environmental responsibility. We also identified the importance of ethical leadership and examined its role in moderating the effect of green innovation network embeddedness. A further investigation revealed that the impact of network embeddedness on corporate environmental responsibility was particularly pronounced in the samples of enterprises with high-level political ties, loose financing restrictions, and nonstate ownership. Our findings highlight the advantages of embedded green innovation networks and offer theoretical references and recommendations for enterprises considering network participation. Enterprises should attach great importance to the network embedding strategy of green innovation for corporate environmental responsibility and actively integrate the concept of green development into network relationship embedding and network structure embedding. Moreover, the relevant government department should provide necessary environment incentive policies according to the enterprise's development needs, especially for the enterprises with low-level political ties, high financing restrictions, and state ownership.

Keywords: innovation network embeddedness; corporate environmental responsibility; green reputation

1. Introduction

Corporate environmental responsibility is integral to corporate social responsibility, closely related to corporate sustainability [1]. Especially with the rapid growth of China's economy, its environmental problems are becoming increasingly prominent. The scholarly community has turned its attention toward figuring out how to fulfill the objective of sustainable growth of man and nature. Enterprises play a crucial role in nations' economic, social, and technological development as significant contributors to economic activity, significant employers, and significant promoters of technological advancement, which is a major force for environmental responsibility [2]. Recently, corporate environmental responsibility has changed from a simple "business responsibility" to a strategic competitive resource. The performance of corporate environmental responsibility will be used as the basis for purchasing or selling when institutional investors in the industry, who have access to increasing amounts of capital, make investment decisions. However, what has plagued companies as they take on environmental responsibilities is balancing their particular interests and broader social goals. At present, it is widely recognized by the government, investors, consumers, and other stakeholders that green innovation activities can help enterprises balance economic and social benefits [3]. An increasing number of businesses are incorporating the concept of corporate social responsibility with "green innovation" consciousness. Green innovation activities involve many participants, such as governments, service providers, banks, scientific research institutions, and supply chain enterprises [4]. How to coordinate the relationship between all parties and conduct practical cooperation reflects the complexity of the green innovation process [5–7]. In other words, the more frequent network activities among enterprises, the more likely they are to achieve green innovation and benefit from it [8–10]. Some studies have shown that by incorporating such a vast open system as the green innovation network, all innovative subjects in the network collaborate, promote the vitality and potential of green innovation within firms, and assist enterprises in reducing environmental pollution [3,11], which improves resource and energy utilization efficiency [12,13], reduces ecological degradation, and increases social welfare [14]. In addition, these enterprises also establish an excellent green reputation and increase stakeholder trust in the enterprise [12,15], which enhances the enterprise's commitment to corporate environmental responsibility. Thus, green innovation network embeddedness plays a vital role in improving their corporate environmental responsibility performance. It has become crucial to thoroughly investigate how Chinese businesses might increase their corporate environmental responsibility through the embeddedness of green innovation networks.

First of all, the existing research generally believes that innovation network embeddedness will significantly impact enterprise economic performance. For example, some researchers suggest that innovation network embeddedness can promote the enterprises' financial performance (such as their net profit margin, return on equity, and return on assets) [16,17] and innovation performance (such as the enterprises' new product development, innovation efficiency, patent number, and other indicators) [18,19]. Additionally, although the study has shown that networks could affect corporate social responsibility [20], the way that green innovation networks' embeddedness affects the corporate environment remains to be studied. Additionally, the methods used in the extant literature on network embeddedness are still theoretical analyses and questionnaire surveys. Extensive sample empirical analysis is rarely used in research to give this kind of proof. Therefore, this paper first intends to explore the impact of green network embeddedness on corporate green environmental responsibility based on the data of listed companies, which is the first research question.

Secondly, most current research on network embeddedness follows the traditional theoretical logic of "network embeddedness-resources/capabilities-enterprise performance". However, few studies take green reputation as the mediating variable to discuss the impact of network embeddedness on corporate environmental responsibility. The corporate social network is an informal mechanism for enterprises to obtain resources. Networks' embeddedness can provide valuable, rare, hard-to-imitate, and hard-to-replace resources. These resources have the potential to boost businesses' profits and give them a competitive edge [21–23]. In the era of the green economy, a green reputation is regarded as a valuable and intangible asset that cannot be duplicated. The green innovation network embeddedness will influence enterprises' green economic operations, which will have a variable impact on the green reputation by stakeholders [24,25]. In addition, Friedman and Miles [26] proposed that corporate reputation may be one of the main drivers of corporate social responsibility. A good reputation will affect the long-term viability of enterprise value, organizational performance, and financial status [27–29]. However, few studies have incorporated green reputation into green innovation network resources to explore how network embeddedness can enhance corporate environmental responsibility by obtaining a green reputation.

Finally, green innovation network embeddedness has a situational dependence on the mechanism of corporate environmental responsibility. However, few studies have explored the moderating effect of leader characteristics on network embeddedness. Therefore, the third research question explores the boundary conditions between the embeddedness of green innovation networks and corporate environmental responsibility. However, previous literature has explored the moderating effect of network embeddedness. For example, internal factors include absorptive capacity and internal value creation capacity [30,31], and external factors include environmental dynamics and network models [32,33]. However, there is still a lack of further research based on leader characteristics. Undertaking corporate environmental responsibility requires enterprises to pay a specific cost. Leaders' ethics can complement environmental regulation when it is challenging to implement it. As Fu et al. [34] believed, in the sustainable development strategy, the individual differences of leaders can affect the extent to which enterprises make beneficial activities.

Based on this, we aim to (1) reveal the impact of green innovation network embeddedness on corporate environmental responsibility; (2) reveal the mediating effect of green reputation between green innovation network embeddedness and corporate environmental responsibility based on resource-based theory; (3) Explore the boundary mechanism of ethical leadership in the logic chain of "green innovation network embeddedness, green reputation, corporate environmental responsibility".

In the CPC patent classification system issued by the European Patent Office and the United States Patent Office, this paper takes Y02 and Y04 classification as essential indicators of green patents and builds a green innovation network based on the green patents issued by Chinese A-share listed companies from 2010 to 2020. Moreover, this paper introduces Chinese enterprises' green innovation network embeddedness through a large sample empirical study and explains whether the green innovation network embeddedness improves corporate environmental responsibility. It was found that both relational embeddedness and structural embeddedness can significantly promote corporate environmental responsibility. In terms of a mediating mechanism, green innovation network embeddedness aids in enhancing enterprises' green reputation, and green reputation supports firms' commitment to corporate environmental responsibility. Further analysis shows that ethical leadership positively regulates the intermediary effect of green reputation between green innovation network embeddedness and corporate environmental responsibility. In addition, embeddedness in green innovation networks in fostering corporate environmental responsibility is especially apparent in enterprises with high-level political ties, no financial constraints, and nonstate ownership. A variety of robustness and endogenous testing strategies are used to improve the unbiased and consistent estimation of the model.

We intend to make the following contributions: (1) By integrating "green innovation" and "network embeddedness", we explore the social consequences of green innovation network embeddedness and expand the application of network embeddedness theory in the field of green innovation. (2) This study expands on previous research on the relationship between green innovation network embeddedness and corporate social performance. It explores the implications of green innovation network embeddedness on corporate environmental responsibility. (3) It provides a theoretical foundation for enterprises to play the beneficial function of network embeddedness in corporate environmental responsibility. It increases their awareness of the driving forces behind corporate environmental responsibility. (4) By introducing resource-based theory, this study examines the intermediary role of "green reputation" between green innovation network embeddedness and corporate environmental responsibility. Furthermore, it reveals the impact of network embeddedness on green resources and their economic consequences. (5) The scope of existing network embeddedness theory research is increased by looking at the moderating role of ethical leadership in the green innovation network embeddedness affecting corporate environmental responsibility, and the theoretical development of the green innovation network embeddedness process affecting corporate performance is enhanced.

The rest of this paper is organized as follows: The second part is the theoretical grounding and hypothesis development. The third part is the research design. The fourth part is the empirical test results and analysis of the research hypothesis. Finally, we summarize the theoretical contributions, put forward the corresponding practical suggestions, and point out the shortcomings of the research and future research directions.

2. Theoretical Grounding and Hypothesis Development

2.1. Green Innovation Network Embeddedness and Corporate Environmental Responsibility

Granovetter [35] first classified network embeddedness into relational and structural embeddedness, the most widely used classification in network embeddedness research. "Relational embeddedness" means that the behavior subject is embedded in the relationship network in which it is located, and its behavior is affected by its social relationship network, emphasizing the relationship characteristics between the enterprise and other members of the network. Relational embeddedness is a close and special cooperative relationship that mainly focuses on the mutual relationship under the trust mechanism. The strong tie is the primary indicator of "relational embeddedness" [36]. In contrast to "relational embeddedness", "structural embeddedness" refers to the position of node enterprises within a sizeable social relationship network and encompasses a broader range of actors. The centrality metric is used in this paper to represent the "structural embeddedness". A higher degree of network centrality gives enterprises more network power, making it easier to obtain and control new information related to innovation in the network [37].

Relational embeddedness has two critical effects on corporate environmental responsibility. Maintaining a close network connection between enterprises can give them flexibility and speed, so they can change their plans in response to shifting market conditions [38,39]. Therefore, information exchange through solid ties is essential for enterprises to fully understand the social demand for green technology. Taking customers' preference for environmental protection products as an example, this external information helps enterprises better meet these needs in the process of green innovation to assume their environmental and social responsibilities better. For another, the basis of interest in social networks is the value generated by mutual trust between the two sides [40], particularly the connection state of frequent interaction, close emotion, close ties, and mutual benefit exhibited by strong relationships [35], which can offer productive value [41]. This cooperative relationship has a solid exclusive feature: the resource demands between the two parties are relatively single minded. In this situation, enterprises may further solidify mutual trust and build the groundwork for long-term cooperation if they actively uphold their social obligations and defend the rights and interests of stakeholders [42]. Therefore, in the green innovation network, enterprises with solid ties are often conducive to promoting enterprises to fulfill their environmental and social responsibilities.

Regarding the effect of structural embeddedness on corporate environmental responsibility, it is essential to note that by taking on environmental and social duties, central enterprises can improve their competitive advantages in green resources. This is because the higher the centrality of an enterprise, the more information and resource channels it can access and the more potent its ability to access and control resources [21,33]. However, it is not always possible to fully manage the timeliness and accuracy of the data and resources that significant firms can access. Instead, it mostly depends on the willingness of marginal enterprises [43]. Therefore, to obtain more critical environmental information and resources from the green innovation network, central enterprises need to be supported by marginal enterprises to establish a long-term cooperation model. A lack of legitimacy and trust will reduce the willingness of edge enterprises to support information and resources [44]. Enterprises can achieve consistency in their own economic and social goals and legalize their economic operation by actively taking on environmental and social responsibilities. They can also gain the trust of marginal enterprises by doing this, which is beneficial for forming long-term cooperative relationships and lowering risks [45–47].

On the contrary, it is simple to be exposed to dire repercussions or even be imitated if the firm does not actively meet its environmental and social duties. This concern enables stakeholders to have governance effects on enterprises with high centrality. Based on this, enterprises usually respond positively to pressure and supervision from stakeholders [48]. Therefore, enterprises in strategic locations are more likely to be prepared to take on corporate environmental responsibility and have higher green ratings, strengthening their advantages in accessing green resources. To sum up, we propose the following:

Hypothesis 1a. *Relational embeddedness is positively related to corporate environmental responsibility.*

Hypothesis 1b. *Structural embeddedness is positively related to corporate environmental responsibility.*

2.2. Network Embeddedness, Green Reputation, and Corporate Environmental Responsibility

The impact of green innovation network embeddedness on green reputation mainly includes three points. Granovetter [35] believed that the so-called "embeddedness" means that economic activities will be limited by the social structure and social relations in which they are located, which determine the form and results of economic activities. "Embeddedness" is a process in which social structure and relations affect economic activities. That is, the embeddedness of green innovation networks will affect the green economic activities of enterprises. However, a comprehensive assessment of an enterprise's green reputation is made when stakeholders take into account its historical environmental protection practices and future potential. That is, the past green economic behavior of enterprises will inevitably lead to stakeholders' evaluation of corporate reputation [24,25]. Therefore, the embeddedness of the green innovation network will influence enterprises' green economic operations, which will have a variable impact on stakeholders when assessing a green reputation. That is, green innovation network embeddedness will affect the green reputation of enterprises.

1. Relational embeddedness and Green Reputation

First of all, if network members can forge close bonds of cooperation, they place a higher value and regard on one another [49]. That is, if two individuals are closely connected, it will affect their positive evaluation of each other [50]. Strong ties indicate that network members have high levels of trust, frequent interactions, closeness, and reciprocity [51]. This also implies that network members have rich emotional connections, which will strengthen the positive evaluation between the two sides and subsequently positively affect corporate reputation. Furthermore, reputation is the result of signal diffusion [52]. Through embedding green innovation network embeddedness, enterprises can show positive green behavior in their communication and interaction with other network members, thus releasing positive signals. Therefore, the more supportive the network relationship is of its green conduct, the more supportive it is of developing a superb green image among stakeholders, and the more supportive it is of increasing the green reputation of enterprises.

Secondly, information asymmetry is the main reason for opportunistic behavior. However, strong relationships entail a high level of trust between businesses and ongoing collaboration and contact. The degree of information asymmetry between the two parties is significantly reduced, and the higher the cost of enterprises breaking the network contract or engaging in opportunity-costing behavior [33,53,54]. Therefore, with the deepening of cooperation, enterprises will pay more attention to managing their own environmentally friendly conduct, signal that they are actively doing so, and enhance their environmental reputation.

Finally, effective green management must be the foundation for developing a green reputation. In the green innovation network, partners with strong ties often communicate frequently, which helps enterprises learn rich green management knowledge, including establishing good green management awareness, establishing green management organizations, improving green reputation management systems, cultivating green reputation management talents, and focusing on green reputation management communication. Therefore, strong ties can influence corporate green reputation by influencing enterprises' learning of green management knowledge.

2. Structural embeddedness and Green Reputation

First, according to resource-based theory, the ability of an enterprise to create value depends on the resources it has. Corporate reputation is a comprehensive evaluation of the past behavior of enterprises by stakeholders, which reflects the ability of enterprises to provide valuable output to stakeholders [52]. Enterprises in a central location can connect with other users in the network and widen the channels for resource acquisition thanks to the benefits of their network placement [21]. Additionally, centrality shows the network's control over and reliance on the green resource. Due to their advantageous location, more external organizations may request information from them, and central enterprises' perceived impact will also grow [41]. Green information and resources are shared among enterprises and can spread in the network, thus changing the reputation evaluation of enterprises by stakeholders in the network [55]. Therefore, this ability to access and control resources by using the network's central location will inevitably affect stakeholders' evaluation of the enterprise's green reputation [56,57].

Secondly, maintaining tight ties with trustworthy businesses can help them build their reputation because a good reputation can have ripple effects. This is especially true when partners have a robust technological base and a competitive advantage [58]. While obtaining diversified resources, the central enterprises can contact diversified partners, allowing them to cooperate with enterprises with high green reputations. Therefore, through the careful selection of partners, the businesses in the center decide to work with high-quality businesses with a solid reputation for being environmentally friendly and a wealth of green resources [59]. This helps them build their good reputations for being environmentally friendly.

Finally, many studies have shown that network embeddedness prevents companies from simply engaging in opportunistic activity since the repercussions are severe and others can quickly discover their behavior [54]. The network's central location is a symbolic location, which has a significant symbolic effect. Once the reputation is damaged, the loss will be more significant [60]. Therefore, enterprises in the central position will cherish their reputation more, restrain their behavior, and release good green behavior signals. In addition, a network system's reputation is crucial for regulating and constraining behavior. A good reputation is crucial for choosing whom to collaborate with and whom to avoid, further enhancing reputation [55].

3. Green reputation and corporate environmental responsibility

According to stakeholders' references to particular standards, enterprise green reputation is a comprehensive assessment of the past green environmental protection behavior and future possibilities of firms [47]. In the era of a green economy, a green reputation is regarded as a valuable and intangible asset that cannot be imitated and an important symbol of the soft power of enterprises [61]. First of all, social responsibility requires a specific cost. An excellent green reputation can send a positive signal, which is an essential basis for financial institutions to provide loans. Therefore, the better the green reputation, the easier it is for enterprises to obtain credit financing and show good performance levels [62]. Therefore, enterprises with an excellent green reputation are more capable of assuming corporate environmental responsibility.

Secondly, a green reputation can significantly enhance investors' confidence in the future profitability of enterprises. Previous studies have shown that institutions with high reputations may have greater legitimacy, thus enhancing their ability to attract funds, customers, information, and other resources [55,63]. Therefore, enterprises with high green reputations tend to show higher expected stock returns [64]. An excellent green reputation is believed to enable the company to obtain a premium and attract investors in the capital market more quickly. Because of this, enterprises with strong green reputations will be more driven to uphold their environmental and social obligations and send out positive environmental signals by disclosing information about their corporate environmental responsibility, which is a crucial foundation for investors to use when making investing decisions. Finally, the characteristics of a green reputation are long term and vulnerable [65]. Long term refers to cultivating and accumulating a green reputation, which requires a lot of human, financial, and material resources. It is a long-term accumulation process. Vulnerability means losing a green reputation is a short-term and fragile process. Once an adverse event occurs, the long-term accumulated reputation of an enterprise may be destroyed. In order to protect their green reputation from being disparaged at will, avoid ethical issues, and gain the competitive advantage provided by corporate environmental responsibility, enterprises will pay more attention to controlling their behavior and actively fulfilling their environmental and social responsibilities as their green reputation continues to grow. Promoting a green reputation will therefore encourage enterprises to fulfill their social obligations actively.

To sum up, we propose the following:

Hypothesis 2a. *Relational embeddedness indirectly affects corporate environmental responsibility via green reputation.*

Hypothesis 2b. *Structural embeddedness indirectly affects corporate environmental responsibility via green reputation.*

2.3. The Moderating Effect of Ethical Leadership

First of all, by encouraging internal employees to imitate and learn from the leaders' "ethical models" so that more employees agree with the necessity of green production, ethical leaders can more accurately identify the needs and expectations of the social system on environmental norms and raise the enterprise's overall environmental awareness [66]. At the same time, ethical leaders are more inclined to care for their subordinates and create a leadership image of integrity, honesty, and clear rewards and punishments [67]. This leadership image helps form a resource-saving corporate culture, promotes employee communication and knowledge sharing, and improves resource-use efficiency [68]. A solid green culture can motivate businesses to innovate more sustainably through network connections, to operate sustainably by environmental protection standards, to promote green products more quickly, and to develop positive environmental perceptions of their brands. Secondly, ethical leaders can use their right to coordinate essential resources, provide the resources needed by employees, develop their green creativity, take into account staff suggestions for green innovation [69], and master the ability to build and maintain relationships with various stakeholders during business operations.

When making business decisions, the ecological environment will be considered by creating an incentive system to promote cooperation and strengthen the commitment and motivation of sustainable development [70], such as offering customers green products and services. Therefore, as enterprises are deeply embedded in the green innovation network, stakeholders will perceive the importance that enterprises attach to environmental protection to enhance the evaluation of enterprises' green reputation. Finally, ethical leaders tend to care for their subordinates by creating a leadership image of integrity, fairness, and clear rules of reward and punishment in their employees' minds [67]. Therefore, ethical leaders are likelier to obtain "trust rewards" from employees. Employees will thus feel more compelled to demonstrate behaviors at work that are advantageous to the company and endeavor to raise their level of performance, which will aid the company in exhibiting commendable green behavior and enhancing its green reputation. To sum up, ethical leadership can enhance the positive effect of green innovation network embeddedness on corporate green reputation and ultimately enhance corporate environmental responsibility. To sum up, we propose the following:

Hypothesis 3a. *Ethical leadership positively moderates the effect of relational embeddedness on green reputation.*

Hypothesis 3b. *Ethical leadership positively moderates the effect of structural embeddedness on green reputation.*

The conceptual framework of this study is shown in Figure 1:

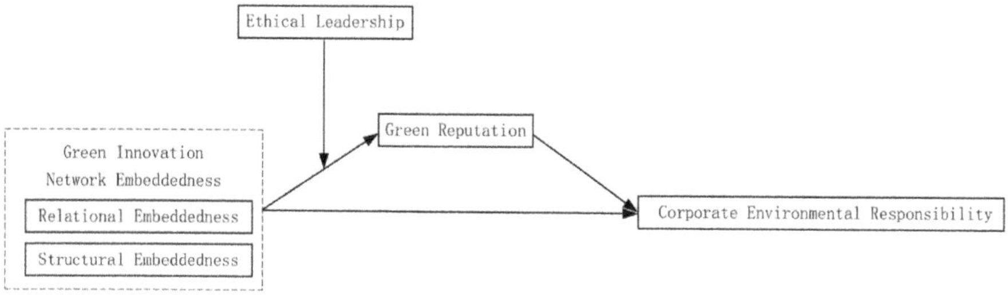

Figure 1. Hypothesized research mode.

3. Research Design

3.1. Sample Selection and Data Collection

This study takes Chinese A-share listed enterprises engaged in green innovation as an example to conduct empirical research. The data were downloaded from the IncoPat patent database and the WIND database of Clarification Analytics. The specific data processing steps are listed: First, search the IncoPat commercial patent database for all the green patent data applied by each A-share listed enterprise from 2010 to 2020. The second step is to use Python to split the applicant's obtained patent data and match the enterprise's listed stock code to facilitate the subsequent indicator matching and fusion processing. The third step is data cleaning, excluding enterprises that do not match the stock code and those that do not submit patent applications during the observation period. Finally, we obtained a total of 534 Chinese A-share listed enterprises that have conducted green innovation from 2010 to 2021. The fourth step is index calculation, which uses pandas, networks, and other packages in Python to build interconnected networks or models. The year and the five years serve as the foundation for calculating indicators. The annual solid ties and centrality of businesses in the corporate green innovation network are calculated using each year as the time unit in the robustness test.

In order to obtain more reliable data, we followed the following steps: (1) exclude ST and PT company samples; (2) remove the company samples with missing data values of relevant indicators; (3) remove the company samples with abnormal data values of relevant indicators; (4) exclude the sample of financial industry companies; (5) winsorize continuous variables.

3.2. Model Design

We designed the following models to test the hypothesis of this paper.

First, in order to consider the impact of network embeddedness on corporate environmental responsibility (Hypothesis 1a and 1b), we designed the following models:

$$CER_{i,t} = \beta_0 + \beta_1 NR_{i,t} + \gamma Controls + \gamma Year + \gamma Industry + \varepsilon_{i,t} \quad (1)$$

$$CER_{i,t} = \beta_0 + \beta_1 NS_{i,t} + \gamma Controls + \gamma Year + \gamma Industry + \varepsilon_{i,t} \quad (2)$$

Among them, CER is the proxy variable of corporate environmental responsibility. NR is the proxy variable of relational embeddedness, while NS is the proxy variable of structural embeddedness. The control variables include firm age (Age), capital structure (Lev), operating profit ratio (Profit), enterprise growth (Growth), corporate ownership (Soe), corporate value (CV), cash holdings (Cash), board size (Board) and proportion of independent directors (Indep), i and t represent enterprises and years, respectively, and $\varepsilon_{i,t}$ represents the residual items.

Secondly, in order to discuss whether green reputation plays a mediating effect (Hypothesis 2a and 2b), we used a three-step method and designed the following models:

$$CER_{i,t} = \beta_0 + \beta_1 NR_{i,t} + \gamma Controls + \gamma Year + \gamma Industry + \varepsilon_{i,t} \quad (3)$$

$$GR_{i,t} = \beta_0 + \beta_1 NR_{i,t} + \gamma Controls + \gamma Year + \gamma Industry + \varepsilon_{i,t} \quad (4)$$

$$CER_{i,t} = \beta_0 + \beta_1 NR_{i,t} + \beta_2 GR_{i,t} + \gamma Controls + \gamma Year + \gamma Industry + \varepsilon_{i,t} \quad (5)$$

$$CER_{i,t} = \beta_0 + \beta_1 NS_{i,t} + \gamma Controls + \gamma Year + \gamma Industry + \varepsilon_{i,t} \quad (6)$$

$$GR_{i,t} = \beta_0 + \beta_1 NS_{i,t} + \gamma Controls + \gamma Year + \gamma Industry + \varepsilon_{i,t} \quad (7)$$

$$CER_{i,t} = \beta_0 + \beta_1 NS_{i,t} + \beta_2 GR_{i,t} + \gamma Controls + \gamma Year + \gamma Industry + \varepsilon_{i,t} \quad (8)$$

Among them, the mediating variable GR is the proxy variable of green reputation, and other variables are the same as above. Suppose the β_1 in the model (3) and model (6), β_1 in the model (4) and model (7), and β_2 in the model (5) and model (8) are significant. In that case, the mediating effect—the function of green reputation in mediating the relationship between network embeddedness and corporate environmental responsibility—is also significant.

Finally, in order to check whether ethical leadership has a moderating effect on the relationship between green innovation network embeddedness and green reputation (Hypothesis 3a and 3b), we designed the following models:

$$CER_{i,t} = \beta_0 + \beta_1 NS_{i,t} + \beta_2 NR{*}Eth_{i,t} + \gamma Controls + \gamma Year + \gamma Industry + \varepsilon_{i,t} \quad (9)$$

$$CER_{i,t} = \beta_0 + \beta 1 NR_{i,t} + \beta 2 NS{*}Eth_{i,t} + \gamma Controls + \gamma Year + \gamma Industry + \varepsilon_{i,t} \quad (10)$$

Among them, the moderating variable Eth is the proxy variable of ethical leadership, and the other variables are the same as above. If the coefficient of the interaction term (β_2) in model (9) and model (10) is significantly positive, Hypotheses 3a and 3b are supported. This means that Ethical leadership positively moderates the effect of network embeddedness on green reputation.

3.3. Measurements

1. Corporate environmental responsibility

CSR is a kind of international private business self-regulation [71]. With increasingly serious environmental problems, environmental protection has become an essential part of corporate social responsibility. Corporate environmental responsibility mainly refers to the responsibility of enterprises for environmental pollution control and ecological environment protection [2,72]. This study chose the environmental responsibility ratings in the social responsibility report evaluation method of Hexun-listed firms to measure corporate environmental responsibility. The original data came from the social responsibility reports and annual reports released by the listed companies of the Shanghai Stock Exchange and the Shenzhen Stock Exchange through official websites.

2. Network embeddedness

In the green innovation network, relational embeddedness (NR), typically assessed by solid ties, primarily assesses the level of trust, time investment, emotional engagement, and reciprocity among related businesses. Referring to the quantitative analysis method of Phelps [73], we used the total number of times enterprises and their partners participated in green patent cooperation to measure the strong tie.

Structural embeddedness (NS) refers to the impact of the relative position of enterprises in the green innovation network on enterprises. We used centrality as a measurement indicator. The higher the centrality, the more enterprises are in the core position of the network. The commonly used indicators of centrality include degree centrality, intermediate centrality, and proximity centrality. Any of the three can be chosen because the calculation results are close. This paper selected the commonly used degree of centrality to measure network centrality. The calculation formula is $C(n_i) = d(n_i)/n - 1$, where n is the total number of nodes, $d(n_i) = \Sigma j X_{ij}$ when the node n_i is not adjacent to n_j, and $X_{ij} = 0$ when node n_i is not adjacent to n_j, $X_{ij} = 1$.

3. Green reputation

We describe corporate green reputation as an overall assessment of past green environmental conduct and firms' future prospects based on stakeholders' reference to particular standards based on Fombrun's [52] definition of corporate reputation. Fortune first conducted the ranking of corporate reputation in the United States. It was scored by external directors, financial analysts, and senior managers based on eight perspectives. This survey method has been adopted in many academic studies. This paper uses the environmental management system certification score to measure green reputation. Generally speaking, the better the green reputation is, the higher the environmental management system certification score will be.

4. Ethical leadership

The concept of "ethical leadership" was created due to ethics' increasing value in both industry and academics. Ethics is seen as a crucial component of a leader's qualities. According to Brown et al. [74], ethical leadership is a leadership style that models normative and preferable behaviors through personal acts and interpersonal interactions and encourages similar behaviors in followers through two-way communication, positive reinforcement, and decision making. We measured ethical leadership from two aspects, including the humanistic care orientation and ecologically sustainable development orientation, drawing on research from Jones et al. [75], Brown et al. [74], and Wang et al. [76]. The total score of ethical leadership was obtained by summing up the scores of all indicators of the two dimensions.

5. Control variablesp

Regarding the related research [77–79], this paper controls three variables that affect corporate environmental responsibility. (1) The first category is the essential characteristics of the company, including the firm age, capital structure, cash holdings, and company value. In fulfilling their social responsibilities, older enterprises attract more attention from the public and often shoulder more social responsibilities. At the same time, CSR will waste capital and other resources and put the company at a competitive disadvantage compared to companies that undertake more CSR activities, thus reducing the value of the enterprise [80]. This paper obtained the age data of enterprises by calculating the length of time from the establishment date to the observation period (Table 1). The asset–liability ratio measures the capital structure. Cash holdings were calculated as the difference between the book value of all assets minus the short-term investments of monetary funds and the sum of cash and short-term investments. The company value was measured by the ratio of the market value of the owner's equity and liabilities to the company's total assets. (2) The second category is company performance characteristics, including the operating profit ratio and enterprise growth. The financial performance of an enterprise may affect its investment in research and development, employee compensation and welfare, environmental protection, etc., thus affecting the fulfillment of environmental responsibilities. Enterprise growth is determined by the operating revenue growth rate year over year, and the operating profit determines the operating profit ratio to the total operating revenue ratio. (3) The third type is corporate governance characteristics, including ownership, the proportion of independent directors, and board size. CSR is an extension of firms' efforts to foster effective corporate governance, ensuring sustainability via sound business practices that promote accountability and transparency [81]. These three variables can control different levels of corporate governance. Specifically, "1" refers to a state-owned enterprise, and "0" refers to a non-state-owned enterprise. The proportion of independent directors is defined as the ratio of independent directors to the number of directors. The number of directors measures the board size.

Table 1. Variable definition and measurement.

Variable Type	Variable Name	Variable Symbol	Measurement	Data Sources
Dependent variable	Corporate environmental responsibility	CER	Environmental responsibility scores in the social responsibility report evaluation system of Hexun	www.hexun.com
Independent variable	Relational embeddedness	NR	Number of patent applications jointly filed by enterprises and partners in the network	IncoPat Patent Database
	Structural embeddedness	NS	For degree centrality, the calculation formula is $C(n_i) = d(n_i)/n - 1$	IncoPat Patent Database
Mediating variable	Green reputation	GR	Environmental management system certification score	www.hexun.com
Moderating variable	Ethical leadership	Eth	Sum of scores for humanistic care orientation and environmentally sustainable development orientation	www.hexun.com
Control variable	Firm age	Age	Length of time from the establishment date to the observation period	WIND Database
	Capital structure	Lev	The ratio of total liabilities to total assets	
	Ownership	Soe	For state-owned enterprises, the assigned value is 1, and for non-state-owned enterprises, the assigned value is 0	
	Operating profit ratio	Profit	The ratio of operating profit to total operating income	
	Enterprise growth	Grow	The year-on-year growth rate of operating income	
	The proportion of independent directors	Indep	The ratio of the number of independent directors to the number of directors	
	Company value	CV	It is measured by the ratio of the market value of the owner's equity and liabilities to the company's total assets.	
	Cash holdings	Cash	The ratio between the sum of monetary capital and short-term investment and the book value of total assets minus the difference between short-term investment of monetary capital	
	Board size	Board	Number of Directors	
	Year	Year	Year dummy variable	
	Industry	Ind	Industry dummy variable	

4. Empirical Results and Analysis

4.1. Descriptive Statistics

Table 2 reports the descriptive statistics of the main variables. The primary variables' standard deviations were within the normal range, according to the statistical results of the complete sample description, which showed that the variables were less affected by extreme values. The average value of corporate environmental responsibility was 45.30, the maximum value was 90.87, and the minimum value was 8.54, which shows that each enterprise had significant differences in the performance of corporate environmental responsibility. The average value of relational embeddedness was 6.517, the maximum value was 2881, and the minimum value was 0. The average value of network structural embeddedness was 0.802, the maximum value was 68.055, and the minimum value was 0. This shows significant differences in enterprises' network embeddedness levels in the green innovation network. The green reputation of most businesses was still at a low level, as seen by the average value of 0.415, the maximum value of five, and the minimum value of 0. Other variables were within the normal range. In addition, this paper tested the variables' variance expansion coefficient (VIF). The maximum value was 5.59, and the minimum value was 1.01, both of which were less than six, indicating no multicollinearity problem.

Table 2. The statistical description of variables.

Variable	N	Mean	Sd	Min	P50	Max
CER	7639	45.30	16.71	8.54	31.56	90.87
NR	7639	6.517	58.62	0	2	2881
NS	7639	0.802	2.804	0	0.725	68.055
GR	7639	0.415	1.204	0	0	5
Eth	7639	5.752	15.28	0	1	52
Age	7639	16.83	5.679	3	17	33
Soe	7639	0.402	0.490	0	0	1
Lev	7639	0.451	0.197	0.074	0.449	0.896
Grow	7639	15.35	22.20	−20.08	12.47	71.03
Indep	7639	0.369	0.068	0	0.333	0.571
Board	7639	8.379	2.526	0	9	15
Cash	7639	0.991	1.504	0.005	0.517	12.00
CV	7639	1	0.004	0.882	1	1.072
Profit	7639	0.093	0.129	−0.474	0.083	0.484

4.2. Correlation Analysis

The correlation coefficients between the variables are reported in Table 3. The findings indicated that relational embeddedness and corporate environmental responsibility had a positive correlation coefficient ($\beta = 0.049$), which was substantially positive at the 1% level ($p < 0.01$) and so preliminarily supported Hypothesis 1a. The relationship between structural embeddedness and corporate environmental responsibility had a positive correlation coefficient ($\beta = 0.140$), which was also statistically positive at the 1% level ($p < 0.01$), which first supported Hypothesis 1b.

Secondly, the correlation coefficient between green reputation and corporate environmental responsibility was positive ($\beta = 0.804$), and it was significantly positive at the level of 1% ($p < 0.01$), indicating that the mediating effect may have existed. In addition, most control variables had significant correlation coefficients with the explained variables, showing that the choice of controls in this study was compelling. Finally, each variable's correlation coefficient in this table was lower than one, which also showed no significant multicollinearity issues with this study and was consistent with the findings of the VIF in the above analysis.

4.3. Basis Analysis Results

Table 4 shows the basis analysis results. Columns (1) and (3) are the regression results of network embeddedness and corporate environmental responsibility without adding control variables. Columns (2) and (4) add control variables. Column (2) shows that the coefficient of relational embeddedness was significantly positive ($\beta = 0.015$, $p < 0.01$), indicating that relational embeddedness was positively related to corporate environmental responsibility. Hypothesis 1a was supported. Column (4) shows that the coefficient of structural embeddedness was significantly positive ($\beta = 0.729$, $p < 0.01$), indicating that structural embeddedness was positively related to corporate environmental responsibility. Hypothesis 1b was supported.

Table 3. Pearson correlation analysis.

	CER	NR	NS	GR	Eth	Age	Soe	Lev	Grow	Indep	Board	Cash	CV	Profit
CER	1													
NR	0.049 ***	1												
NS	0.140 ***	0.608 ***	1											
GR	0.804 ***	0.056 ***	0.126 ***	1										
Eth	0.865 ***	0.038 ***	0.121 ***	0.920 ***	1									
Age	−0.055 ***	0.025 **	0.016	−0.053 ***	−0.068 ***	1								
Soe	0.127 ***	0.030 ***	0.071 ***	0.186 ***	0.200 ***	0.101 ***	1							
Lev	−0.051 ***	0.026 **	0.044 ***	0.124 ***	0.118 ***	0.068 ***	0.355 ***	1						
Grow	0.084 ***	−0.018	−0.020 *	−0.025 **	−0.012	−0.176 ***	−0.142 ***	−0.052 ***	1					
Indep	0.015	0.001	0.012	0.014	0.016	0.026 **	0.036 ***	0.020 *	−0.040 ***	1				
Board	0.172 ***	0.060 ***	0.090 ***	0.156 ***	0.165 ***	0.178 ***	0.269 ***	0.083 ***	−0.085 ***	−0.112 ***	1			
Cash	0.061 ***	−0.027 **	−0.050 ***	−0.044 ***	−0.035 ***	−0.147 ***	−0.174 ***	−0.575 ***	0.015	−0.022 *	−0.019 *	1		
CV	−0.021 *	−0.054 ***	−0.070 ***	−0.015	−0.013	0.001	−0.007	−0.037 ***	−0.005	0.015	−0.044 ***	0.017	1	
Profit	0.306 ***	−0.007	−0.011	−0.025 **	−0.012	−0.101 ***	−0.153 ***	−0.419 ***	0.268 ***	−0.063 ***	−0.071 ***	0.311 ***	0.005	1

Note: * $p < 0.1$, ** $p < 0.05$, *** $p < 0.01$.

Table 4. The results of the primary effect test.

	(1)	(2)	(3)	(4)
	CER	**CER**	**CER**	**CER**
NR	0.015 *** (4.88)	0.012 *** (4.28)		
NS			0.729 *** (11.36)	0.605 *** (10.23)
Age		0.252 *** (7.32)		0.245 *** (7.14)
Soe		3.611 *** (9.15)		3.556 *** (9.07)
Lev		5.710 *** (4.66)		5.689 *** (4.68)
Grow		−0.007 (−0.87)		−0.006 (−0.68)
Indep		15.392 *** (6.15)		15.073 *** (6.06)
Board		0.878 *** (9.19)		0.862 *** (9.08)
CV		−56.177 (−1.21)		−37.137 (−0.81)
Cash		−0.286 ** (−2.01)		−0.236 * (−1.67)
Profit		30.059 *** (12.73)		29.902 *** (12.74)
cons	36.712 *** (28.18)	61.666 (1.33)	34.208 *** (25.99)	40.949 (0.89)
Year	YES	YES	YES	YES
Ind	YES	YES	YES	YES
N	7639	7639	7639	7639
R^2	0.127	0.292	0.139	0.301

Note: *t* statistics in parentheses; * $p < 0.1$, ** $p < 0.05$, *** $p < 0.01$.

4.4. Mediating Effect Test

Table 5 shows the test results of the mediating effect of green reputation. Column (1) shows that relational embeddedness was directly related to corporate environmental responsibility. The results showed that the coefficient of relational embeddedness was positive and significant at 1%, indicating that green innovation network relational embeddedness can significantly improve corporate environmental responsibility. The relational embeddedness and green reputation coefficients were statistically positive at 1% in column (2). In column (3), relational embeddedness and green reputation were positively related to corporate environmental responsibility at 1%. Existing research indicates that the mediating effect is significant when the coefficient of relational embeddedness in column (1), the coefficient of relational embeddedness in column (2), and the coefficient of green reputation in column (3) are all significantly positive. In addition, the Sobel test and Bootstrap test were further used in this paper. The Bootstrap test had 500 samples, a Z value of 4.405 for the Sobel test, a *p* value of 0.05 or less, and a BC confidence interval after a deviation adjustment of [0.0018053, 0.0222497], excluding 0. Therefore, the research results of this search were robust. To sum up, relational embeddedness had a significant indirect effect on corporate environmental responsibility through green reputation. Hypothesis 2a was supported.

Table 5. The results of mediating effect test.

	(1)	(2)	(3)	(4)	(5)	(6)
	CER	**GR**	**CER**	**CER**	**GR**	**CER**
NR	0.012 *** (4.28)	0.001 *** (4.41)	0.005 *** (2.67)			
NS				0.605 *** (10.23)	0.037 *** (7.82)	0.221 *** (6.67)
GR			10.714 *** (127.73)			10.674 *** (127.21)
Age	0.252 *** (7.32)	0.010 *** (3.86)	0.140 *** (7.29)	0.245 *** (7.14)	0.010 *** (3.72)	0.138 *** (7.19)
Soe	3.611 *** (9.15)	0.255 *** (8.19)	0.762 *** (3.44)	3.556 *** (9.07)	0.251 *** (8.08)	0.749 *** (3.39)
Lev	5.710 *** (4.66)	0.702 *** (7.29)	−1.580 ** (−2.30)	5.689 *** (4.68)	0.704 *** (7.33)	−1.553 ** (−2.27)
Grow	−0.007 (−0.87)	−0.001 (−1.64)	0.004 (0.79)	−0.006 (−0.68)	−0.001 (−1.53)	0.004 (0.90)
Indep	15.392 *** (6.15)	0.499 ** (2.54)	9.902 *** (7.09)	15.073 *** (6.06)	0.480 ** (2.45)	9.802 *** (7.04)
Board	0.878 *** (9.19)	0.042 *** (5.63)	0.396 *** (7.41)	0.862 *** (9.08)	0.042 *** (5.60)	0.393 *** (7.38)
CV	−56.177 (−1.21)	−0.539 (−0.15)	−45.434 * (−1.76)	−37.137 (−0.81)	0.246 (0.07)	−39.104 (−1.52)
Cash	−0.286 ** (−2.01)	−0.010 (−0.88)	−0.183 ** (−2.31)	−0.236 * (−1.67)	−0.007 (−0.62)	−0.165 ** (−2.09)
Profit	30.059 *** (12.73)	0.603 *** (3.26)	23.857 *** (18.15)	29.902 *** (12.74)	0.595 *** (3.22)	23.830 *** (18.17)
Constant	61.666 (1.33)	0.366 (0.10)	53.245 ** (2.07)	40.949 (0.89)	−0.518 (−0.14)	46.271 * (1.80)
Year	YES	YES	YES	YES	YES	YES
Ind	YES	YES	YES	YES	YES	YES
N	7639	7639	7639	7639	7639	7639
R^2	0.292	0.186	0.784	0.301	0.190	0.785
Sobel-Z		4.405			7.8	
Goodman-P		0			0	
Intermediary effect		71%			63%	
Bootstrap (BC)		[0.0018053, 0.0222497]			[0.2927432, 0.6830738]	

Note: *t* statistics in parentheses; * $p < 0.1$, ** $p < 0.05$, *** $p < 0.01$.

Column (4) shows the direct effect of structural embeddedness on corporate environmental responsibility. The results showed that the coefficient of structural embeddedness was positive and significant at 1%, indicating that structural embeddedness can significantly improve corporate environmental responsibility. In column (5), structural embeddedness positively related to green reputation at 1%. In column (6), structural embeddedness and green reputation were positively related to corporate environmental responsibility at 1%. According to previous studies, the mediating effect is substantial when the coefficients of structural embeddedness in column (4), column (5), and column (6) are significantly positive, as well as when the coefficient of green reputation in column (4) and column (5) are significantly positive. The Sobel test and the Bootstrap test were also utilized in this paper. The Bootstrap test had 500 samples, a Z value of 7.8 for the Sobel test, a p value of 0 or less than 0.05, and a BC confidence interval after a deviation adjustment of [0.2927432, 0.6830738], excluding 0. Therefore, the research results of this research were robust. To sum up, structural embeddedness had a significant indirect effect on corporate environmental responsibility through green reputation. Hypothesis 2b was supported.

4.5. Moderating Effect Test

Table 6 shows the test results of the moderating effect of ethical leadership. The results in column (1) show that the interaction item of relational embeddedness and ethical leadership (NR × Eth) had a significant positive impact on green reputation. Therefore, ethical leadership positively moderated the relationship between relational embeddedness and green reputation. Thus, Hypothesis 3a was supported. The results of column (2) show that the interaction item of structural embeddedness and ethical leadership (NS × Eth) had a significant positive impact on green reputation. Therefore, ethical leadership positively moderated the relationship between structural embeddedness and green reputation. Therefore, Hypothesis 3b was supported.

Table 6. The results of moderating effect test.

	(1)	(2)
	GR	GR
NR	0.004 *** (2.61)	
NS		0.237 *** (7.49)
Eth	0.918 *** (155.11)	0.913 *** (150.18)
Interact	0.001 ** (2.38)	1.737 * (1.70)
Age	0.080 *** (5.46)	0.078 *** (5.36)
Soe	−0.080 (−0.47)	−0.084 (−0.50)
Lev	−3.951 *** (−7.75)	−3.913 *** (−7.72)
Grow	0.018 *** (5.07)	0.019 *** (5.29)
Indep	5.785 *** (5.51)	5.655 *** (5.42)
Board	0.356 *** (8.77)	0.351 *** (8.70)
CV	−39.417 ** (−2.01)	−32.634 * (−1.67)
Cash	−0.355 *** (−6.06)	−0.333 *** (−5.73)
Profit	40.880 *** (61.39)	40.699 *** (61.47)
Constant	49.549 ** (2.52)	42.101 ** (2.15)
Year	YES	YES
Ind	YES	YES
N	7639	7639
R2	0.834	0.836

Note: t statistics in parentheses; * $p < 0.1$, ** $p < 0.05$, *** $p < 0.01$.

4.6. Robustness Test

1. Endogenous test

Although the benchmark regression results showed that the higher the degree of network embeddedness was, the higher the corporate environmental responsibility was, endogenous issues still might have an impact on this outcome. In order to verify the robustness of the conclusions of this research, the following methods were adopted to control the potential endogenous problem. The benchmark regression controlled the year effect and industry effect by employing a fixed effect model. As a result, the endogenetic issues that missing factors could bring about were somewhat under control. However, there may have still been interfered with by the reverse causal problems. This means that firms with high environmental responsibility may have tended to embed in the green innovation network. Therefore, the lag-independent variable regression method used in this research could ensure causality to a certain extent. The results are shown in Table 7. The results showed that the regression results of the lagged independent variables were still robust. It can be seen that the impact of the two dimensions of network embeddedness on corporate environmental responsibility was gradually waning by comparing the coefficients of network embeddedness in various lagged years.

Table 7. The results of the endogenous test.

	(1)	(2)	(3)	(4)	(5)	(6)
	CER	CER	CER	CER	CER	CER
L.NR	0.023 *** (4.81)					
L2.NR		0.017 *** (2.96)				
L3.NR			0.010 (1.48)			
L.NS				0.672 *** (9.55)		
L2.NS					0.460 *** (6.57)	
L3.NS						0.259 *** (3.44)
Age	0.229 *** (6.42)	0.220 *** (6.04)	0.174 *** (4.71)	0.221 *** (6.23)	0.213 *** (5.88)	0.170 *** (4.60)
Soe	3.184 *** (7.82)	2.683 *** (6.51)	2.066 *** (5.01)	3.112 *** (7.68)	2.628 *** (6.40)	2.036 *** (4.94)
Lev	5.699 *** (4.51)	4.187 *** (3.27)	3.553 *** (2.76)	5.619 *** (4.47)	4.119 *** (3.23)	3.504 *** (2.72)
Grow	−0.009 (−1.00)	−0.012 (−1.31)	−0.013 (−1.45)	−0.007 (−0.83)	−0.010 (−1.16)	−0.012 (−1.30)
Indep	16.139 *** (5.74)	15.571 *** (4.82)	11.777 *** (3.28)	15.557 *** (5.56)	15.068 *** (4.68)	11.332 *** (3.16)
Board	0.791 *** (7.63)	0.707 *** (6.26)	0.526 *** (4.28)	0.791 *** (7.68)	0.709 *** (6.30)	0.526 *** (4.29)
CV	−92.941 * (−1.78)	−127.424 ** (−2.23)	−111.302 ** (−2.03)	−72.298 (−1.39)	−112.995 ** (−1.98)	−105.388 * (−1.92)
Cash	−0.228 (−1.46)	−0.246 (−1.45)	−0.224 (−1.22)	−0.170 (−1.09)	−0.202 (−1.19)	−0.199 (−1.08)
Profit	26.321 *** (10.68)	23.301 *** (9.33)	19.984 *** (7.94)	26.220 *** (10.69)	23.241 *** (9.33)	20.006 *** (7.96)
_cons	101.993 * (1.95)	140.722 ** (2.46)	127.027 ** (2.31)	79.679 (1.53)	125.208 ** (2.19)	120.611 ** (2.20)
Year	YES	YES	YES	YES	YES	YES
Ind	YES	YES	YES	YES	YES	YES
N	6586	5879	5099	6586	5879	5099
R^2	0.305	0.311	0.328	0.312	0.315	0.329

Note: t statistics in parentheses; * $p < 0.1$, ** $p < 0.05$, *** $p < 0.01$.

2. Sensitivity test of key variables

In order to further verify the robustness of the research, we used substitution independent variable and dependent variable measurement methods. First of all, Table 8 displays the outcomes of calculating the green innovation network embeddedness index using the annual time unit. In columns (1) and (4), the two dimensions of network embeddedness had significant positive effects on corporate environmental responsibility. Hypothesis 1a and 1b were supported again. In columns (2) and (5), the two dimensions of network embeddedness had a significant positive impact on green reputation. In columns (3) and (6), both the two dimensions of network embeddedness and green reputation had significant positive effects on corporate environmental responsibility, indicating that green reputation played a mediating role between network embeddedness and corporate environmental responsibility. Hypothesis 2a and 2b were supported again.

Table 8. Sensitivity test of the independent variable.

	(1)	(2)	(3)	(4)	(5)	(6)
	CER	GR	CER	CER	GR	CER
NR1	0.044 *** (5.07)	0.003 *** (4.59)	0.017 *** (3.14)			
NS1			10.711 *** (127.71)	0.480 *** (9.45)	0.0305 *** (7.61)	0.158 *** (5.52) 10.684 *** (127.24)
Age	0.253 *** (7.33)	0.010 *** (3.88)	0.140 *** (7.30)	0.247 *** (7.20)	0.010 *** (3.75)	0.139 *** (7.23)
Soe	3.619 *** (9.18)	0.255 *** (8.20)	0.765 *** (3.45)	3.590 *** (9.14)	0.252 *** (8.14)	0.759 *** (3.43)
Lev	5.564 *** (4.55)	0.692 *** (7.18)	−1.632 ** (−2.38)	5.539 *** (4.55)	0.694 *** (7.22)	−1.607 ** (−2.35)
Grow	−0.008 (−0.89)	−0.001 * (−1.68)	0.004 (0.77)	−0.007 (−0.80)	−0.001 (−1.61)	0.004 (0.82)
Indep	15.455 *** (6.18)	0.504 ** (2.56)	9.926 *** (7.11)	15.345 *** (6.16)	0.495 ** (2.52)	9.894 *** (7.10)
Board	0.877 *** (9.19)	0.043 *** (5.67)	0.397 *** (7.43)	0.855 *** (9.00)	0.041 *** (5.53)	0.392 *** (7.34)
CV	−54.545 (−1.18)	−0.589 (−0.16)	−45.165 * (−1.75)	−40.836 (−0.89)	0.108 (0.03)	−41.335 (−1.61)
Cash	−0.290 ** (−2.04)	−0.010 (−0.91)	−0.185 ** (−2.33)	−0.266 * (−1.88)	−0.009 (−0.78)	−0.177 ** (−2.24)
Roa	61.773 *** (11.93)	0.309 (0.76)	57.883 *** (20.06)	60.984 *** (11.83)	0.263 (0.65)	57.640 *** (20.00)
Profit	30.054 *** (12.74)	0.603 *** (3.26)	23.857 *** (18.15)	30.044 *** (12.79)	0.603 *** (3.27)	23.877 *** (18.19)
Constant	60.105 (1.30)	0.424 (0.12)	53.004 ** (2.06)	45.382 (0.98)	−0.339 (−0.09)	48.826 * (1.90)
Year	YES	YES	YES	YES	YES	YES
Ind	YES	YES	YES	YES	YES	YES
N	7639	7639	7639	7639	7639	7639
R^2	0.293	0.186	0.784	0.299	0.190	0.785

Note: *t* statistics in parentheses; * $p < 0.1$, ** $p < 0.05$, *** $p < 0.01$.

Secondly, referring to the existing research, we used Bloomberg ESG Disclosure Scores as the alternative variable of corporate environmental responsibility. The Bloomberg data came from transparent data of Bloomberg and the third party, which could objectively reflect the ESG situation of the enterprise. The results are shown in Table 9. In columns (1) and (4), the two dimensions of network embeddedness had significant positive effects on corporate environmental responsibility. Hypothesis 1a and 1b were supported again. In columns (2) and (5), the two dimensions of network embeddedness had a significant positive impact on the green reputation. In columns (3) and (6), both the two dimensions of network embeddedness and green reputation had significantly favorable effects on corporate environmental responsibility, indicating that green reputation played a mediating role between network embeddedness and corporate environmental responsibility. Hypothesis 2a and 2b were supported again.

Table 9. Sensitivity test of the dependent variable.

	(1)	(2)	(3)	(4)	(5)	(6)
	Social	GR	Social	Social	GR	Social
NR	0.017 *** (6.41)	0.001 *** (4.42)	0.016 *** (5.19)			
NS				0.711 *** (11.59)	0.0367 *** (7.84)	0.597 *** (10.07)
GR			3.544 *** (24.69)			3.474 *** (24.31)
Age	0.264 *** (7.02)	0.010 *** (3.85)	0.230 *** (6.37)	0.251 *** (6.73)	0.010 *** (3.71)	0.220 *** (6.14)
Soe	5.549 *** (13.71)	0.254 *** (8.17)	4.632 *** (11.86)	5.474 *** (13.62)	0.250 *** (8.06)	4.579 *** (11.80)
Lev	15.963 *** (12.80)	0.688 *** (7.27)	13.305 *** (11.07)	15.974 *** (12.91)	0.691 *** (7.32)	13.391 *** (11.21)
Grow	−0.026 *** (−2.95)	−0.001 (−1.55)	−0.022 ** (−2.55)	−0.024 *** (−2.80)	−0.001 (−1.44)	−0.021 ** (−2.44)
Indep	25.591 *** (8.91)	0.485 ** (2.48)	23.568 *** (8.57)	25.070 *** (8.80)	0.467 ** (2.39)	23.155 *** (8.47)
Board	1.438 *** (13.81)	0.042 *** (5.59)	1.280 *** (12.82)	1.429 *** (13.84)	0.042 *** (5.57)	1.279 *** (12.90)
CV	−262.666 *** (−4.51)	−0.531 (−0.15)	−238.797 *** (−4.27)	−232.988 *** (−4.03)	0.254 (0.07)	−217.196 *** (−3.91)
Cash	0.115 (0.74)	−0.011 (−1.04)	0.180 (1.21)	0.183 (1.19)	−0.008 (−0.77)	0.236 (1.59)
Profit	19.857 *** (12.46)	0.713 *** (5.81)	16.890 *** (11.03)	19.320 *** (12.21)	0.694 *** (5.67)	16.519 *** (10.85)
Constant	238.727 *** (4.10)	0.381 (0.10)	215.262 *** (3.85)	207.639 *** (3.59)	−0.505 (−0.14)	192.361 *** (3.46)
Year	YES	YES	YES	YES	YES	YES
Ind	YES	YES	YES	YES	YES	YES
N	7639	7639	7639	7639	7639	7639
R^2	0.225	0.186	0.297	0.237	0.190	0.306

Note: t statistics in parentheses; ** $p < 0.05$, *** $p < 0.01$.

4.7. Further Analysis

1. Political connection

First, as a crucial relational resource [82], resource dependence theory states that the corporation must take more precautions to minimize the danger of losing resources the more valuable the resources are to the organization [83]. Compared with companies with lower political affiliations, companies with higher political affiliations can benefit from the convenience of financing and relevant policy support. As these companies gain a lot from political connections, government departments have higher expectations for companies with political connections [84]. Therefore, high-level political-related businesses will interact with government agencies more frequently than low-level businesses, and they will be more driven to take on social responsibility as networks for green innovation become more embedded. If the company's chairman or general manager was selected by the government, is a current or former government official, serves as a deputy in the National People's Congress, or is a member of the CPPCC, the political connection was measured as 1; otherwise, it was measured as 0.

This paper distinguishes enterprise samples according to the political association level of enterprises and explored the impact of green innovation network embeddedness on corporate environmental responsibility under different political association levels. The regression results are shown in Table 10. The findings of columns (1) and (3) demonstrate that the sample of high-level political associations had a more significant coefficient of relational embeddedness. The results of columns (2) and (4) demonstrated that the sample of high-level political associations had a more significant coefficient of structural embeddedness. The regression results showed that enterprises with high-level political connections embedded in the green innovation network paid more attention to undertaking corporate environmental responsibility.

Table 10. The impact of political linkages.

	Low-Level Political Connection		High-Level Political Connection	
	(1)	(2)	(3)	(4)
	CER	CER	CER	CER
NR	0.016 *** (3.66)		0.056 *** (2.67)	
NS		0.483 *** (6.67)		0.956 *** (4.26)
Age	0.239 *** (4.59)	0.235 *** (4.52)	0.118 (0.99)	0.112 (0.95)
Soe	3.809 *** (6.83)	3.772 *** (6.79)	5.073 *** (3.57)	4.769 *** (3.37)
Lev	7.954 *** (4.62)	7.983 *** (4.66)	24.249 *** (6.14)	23.407 *** (5.94)
Grow	−0.010 (−0.86)	−0.009 (−0.78)	−0.070 *** (−2.98)	−0.066 *** (−2.82)
Indep	14.180 *** (3.02)	14.030 *** (3.00)	26.136 ** (2.31)	27.344 ** (2.43)
Board	0.791 *** (5.05)	0.781 *** (5.01)	1.243 *** (3.32)	1.344 *** (3.59)
CV	−293.28 *** (−2.93)	−257.07 ** (−2.57)	124.161 (0.96)	135.346 (1.05)
Cash	−0.478 ** (−2.45)	−0.431 ** (−2.21)	0.284 (0.75)	0.316 (0.84)
Profit	25.699 *** (7.51)	25.625 *** (7.52)	13.114 (1.47)	15.543 * (1.75)
Constant	300.531 *** (2.99)	262.894 *** (2.63)	−142.992 (−1.10)	−156.898 (−1.21)
Year	YES	YES	YES	YES
Ind	YES	YES	YES	YES
N	4370	4370	3269	3269
R^2	0.284	0.289	0.374	0.381

Note: *t* statistics in parentheses; * $p < 0.1$, ** $p < 0.05$, *** $p < 0.01$.

2. Financing constraints

The financing constraints faced by enterprises will have a significant impact on their behavior. High financial limitations force an organization to focus on its core business, lowering its investment in corporate social responsibility. When the enterprise faces low financing constraints, it can have sufficient funds so that it will pay more attention to long-term development. At this time, the enterprise has the ability to invest funds to undertake social responsibility. Therefore, this paper predicts that for enterprises with low financing constraints, green innovation network embeddedness may have a more significant effect on promoting corporate environmental responsibility.

Therefore, the sample was split into two groups based on the annual industry average of the company size: high and low financing limitations. This article used company size as an indicator to measure the level of corporate finance restraints. The regression results are shown in Table 11. Only structural embeddedness was significantly positively correlated with corporate environmental responsibility in the high financing constraint samples, and its coefficient was lower than that of the low financing constraint samples. In the low financing constraint samples, network embeddedness was significantly positively correlated with corporate environmental responsibility. This shows that in the lower sample of financing constraints, enterprises embedded in green innovation networks were more conducive to social responsibility.

Table 11. The impact of financing constraints.

	High Financing Constraints		Low Financing Constraints	
	(1)	(2)	(3)	(4)
	CER	CER	CER	CER
NR	0.001 (0.13)		0.009 *** (2.92)	
NS		0.434 ** (2.06)		0.383 *** (5.61)
Age	0.063 * (1.68)	0.064 * (1.71)	0.272 *** (5.18)	0.263 *** (5.02)
Soe	1.472 *** (3.17)	1.476 *** (3.18)	2.566 *** (4.64)	2.566 *** (4.66)
Lev	−3.496 *** (−2.61)	−3.412 ** (−2.54)	−0.569 (−0.29)	−0.177 (−0.09)
Grow	−0.005 (−0.64)	−0.005 (−0.56)	0.005 (0.43)	0.007 (0.51)
Indep	3.612 (1.48)	3.533 (1.44)	13.917 *** (3.45)	13.844 *** (3.45)
Board	0.413 *** (3.83)	0.424 *** (3.94)	0.735 *** (5.36)	0.727 *** (5.32)
CV	41.724 (0.50)	38.592 (0.46)	−26.157 (−0.47)	−15.816 (−0.29)
Cash	0.008 (0.07)	0.020 (0.16)	−0.692 ** (−2.48)	−0.623 ** (−2.24)
Profit	28.073 *** (11.81)	27.920 *** (11.74)	22.303 *** (6.13)	22.422 *** (6.18)
Constant	−24.810 (−0.30)	−22.331 (−0.27)	45.996 (0.83)	33.485 (0.61)
Year	Year	Year	Year	Year
Ind	Year	Year	Year	Year
N	3522	3522	4117	4117
R^2	0.284	0.285	0.357	0.360

Note: t statistics in parentheses; * $p < 0.1$, ** $p < 0.05$, *** $p < 0.01$.

3. Ownership

The current statistics showed that since 2012, the level of social responsibility of state-owned enterprises was almost more significant than that of non-state-owned enterprises. For state-owned enterprises, due to their solid political atmosphere, they will promote enterprises to actively assume social responsibility [85]. Non-state-owned businesses primarily engage in corporate environmental responsibility to improve their brand recognition or build political connections to lower the cost of debt financing [86]. This paper argues that different ownership determines different incentive mechanisms, which may lead to different effects on enterprises' embedded networks.

Therefore, this paper distinguished the samples according to the ownership of enterprises and explored the impact of green innovation network embeddedness on corporate environmental responsibility under different ownership samples. The regression results are shown in Table 12. The results of column (1) and column (3) show that the coefficient of relational embeddedness of non-state-owned enterprises was greater than that of state-owned enterprises. The results of column (2) and column (4) show that the coefficient of structural embeddedness of non-state-owned enterprises was more significant than that of state-owned enterprises, and the coefficient was more significant. The results showed that non-state-owned enterprises paid more attention to the long-term development of enterprises by green innovation network embeddedness to actively assume corporate environmental responsibility.

Table 12. The impact of ownership.

	Non-State-Owned Enterprises		State-Owned Enterprise	
	(1)	(2)	(3)	(4)
	CER	CER	CER	CER
NR	0.012 ** (2.28)		0.011 *** (2.99)	
NS		0.699 *** (7.31)		0.486 *** (6.11)
Age	0.353 *** (9.08)	0.324 *** (8.33)	0.140 ** (2.20)	0.155 ** (2.43)
Lev	8.669 *** (5.99)	8.231 *** (5.72)	2.653 (1.21)	3.030 (1.38)
Grow	−0.014 (−1.46)	−0.013 (−1.41)	0.009 (0.56)	0.011 (0.71)
Indep	12.663 *** (4.53)	12.464 *** (4.48)	20.219 *** (4.43)	19.967 *** (4.39)
Board	0.731 *** (6.20)	0.765 *** (6.52)	0.997 *** (6.58)	0.949 *** (6.28)
CV	−65.777 (−0.99)	−47.417 (−0.72)	−44.683 (−0.68)	−29.119 (−0.45)
Cash	0.197 (1.40)	0.205 (1.47)	−0.703 * (−1.68)	−0.553 (−1.32)
Profit	23.794 *** (8.69)	23.714 *** (8.71)	31.196 *** (7.53)	31.206 *** (7.57)
Constant	75.290 (1.13)	56.764 (0.86)	55.742 (0.85)	38.585 (0.59)
Year	YES	YES	YES	YES
Ind	YES	YES	YES	YES
N	4301	4301	3308	3308
R^2	0.240	0.248	0.363	0.369

Note: *t* statistics in parentheses; * $p < 0.1$, ** $p < 0.05$, *** $p < 0.01$.

5. Conclusions and Discussion

Chinese businesses must balance their own goals with the broader interests of society while engaging in corporate environmental responsibilities against the backdrop of institutional and economic upheaval. Therefore, how to better undertake environmental responsibility is a common concern for academia and industries. Recently, an increasing number of businesses have incorporated the concept of "green innovation" into their concept of corporate social responsibility and engage in corporate environmental responsibility by incorporating the green innovation network. Therefore, in order to further illuminate the internal workings of the advantages of green innovation, this study developed the research framework of "green innovation network embeddedness green reputation corporate environmental responsibility" and extensively examined the marginal change in the indirect effect of green reputation on the relationship between green innovation network embeddedness and corporate environmental responsibility within the constraints of ethical leadership. We took Y02 and Y04 classifications as essential indicators of green patents and built a green innovation network based on the green patents issued by Chinese A-share listed companies from 2010 to 2020. The result showed that relational and structural embeddedness could significantly promote corporate environmental responsibility. In terms of a mediating mechanism, green innovation network embeddedness aided in enhancing enterprises' green reputation, and green reputation supported firms' commitment to corporate environmental responsibility. Further analysis showed that ethical leadership positively regulated the intermediary effect of green reputation between green innovation network embeddedness and corporate environmental responsibility. In addition, embeddedness in green innovation networks in fostering corporate environmental responsibility was especially apparent in enterprises with high-level political ties, no financial constraints, and nonstate ownership.

5.1. Implications for Theory

First, the study of network embeddedness theory in green innovation was broadened by proposing the idea of a "green innovation network" and employing extensive sample data for empirical testing. The existing research mainly focuses on the impact of network embeddedness on green innovation [87,88]. Few scholars combined "green innovation" with "innovation network" in concept, focusing on the concept of "green innovation network". In addition, even though the benefits of network embeddedness in enhancing business performance was extensively covered in the literature [16,17,20], few studies used the method of extensive sample empirical tests to give empirical evidence for this aim; instead, they still relied primarily on theoretical analysis and questionnaire surveys. Based on the sample data of Chinese A-share listed companies engaged in green innovation, we built a green innovation network based on the green patents between enterprises. We empirically tested the economic consequences of the embeddedness of green innovation networks. Therefore, based on the perspective of green innovation, this study provides new research ideas for innovation networks. Through empirical research, it tested how green innovation network embeddedness improves corporate environmental responsibility, expanding the research of network embeddedness theory in green innovation.

Second, from the perspective of corporate environmental responsibility, this study expanded the research on the mechanism of green innovation network embeddedness on corporate performance. Previous studies mainly focused on the economic performance of enterprises. First, financial performance includes indicators such as net profit rate, return on equity, and return on assets [16,17]. Second, innovation performance indicators include the number of new items on the market, the productivity of new products in sales, the effectiveness of innovation, and the number of patents [18,20]. There is still a lack of research on the impact of network embeddedness on corporate social performance. This study found that green innovation network embeddedness can significantly improve corporate environmental responsibility performance. Therefore, we broadened both the theoretical application of network embeddedness and the research on the impact of network embeddedness on enterprise performance.

Third, this study put forward the research framework of "green innovation network embeddedness, green reputation, corporate environmental responsibility". It compensated for the current dearth of research on the impact of green innovation network embeddedness on corporate environmental responsibility and its mechanism by naturally merging resource-based theory, network embeddedness theory, and corporate environmental responsibility. Previous research on network embeddedness mainly explained the formation reason of network embeddedness from the perspective of "resource" acquisition. However, it did not reveal the mechanism of network embeddedness for enterprises to acquire "green resources". This study offers a fresh viewpoint on how corporate environmental responsibility and embeddedness in green innovation networks interact internally. It also thoroughly explains the transmission function of green reputation in these two domains.

Finally, this study further explored the boundary mechanism of ethical leadership in the green innovation network embeddedness to affect corporate environmental responsibility through green reputation. The existing research on ethical leadership primarily focuses on how managers' honesty, altruism, trustworthiness, and other qualities affect the creativity of specific employees and teams [68,69] and how to enhance corporate culture [89]. It pays less attention to the research on ethical leadership in network embeddedness. This study included ethical leadership in the logical framework of "green innovation network embeddedness, green reputation, corporate environmental responsibility". The study confirmed Fu's [32] view that in the sustainable development strategy, the individual differences of leaders can affect the extent to which enterprises make beneficial activities by demonstrating that the intermediary effect of green reputation between embeddedness in green innovation networks and environmental responsibility will change as a result of leader characteristics. Furthermore, this paper also expands the application of leadership behavior theory in network embeddedness research.

5.2. Practical Implications

Firstly, enterprises should attach great importance to the network embedding strategy of green innovation for corporate environmental responsibility, actively integrate the concept of green development into network relationship embedding and network structure embedding, and constantly improve the degree of network relationship embedding and network embedding. In addition, given that the process of green innovation embedding is also a process in which enterprises' green innovation practice transitions from "light green" to "dark green", enterprises should reasonably weigh resource input according to their strategic orientation, optimize the strategic combination of green innovation network embedding according to the connotation difference between relational embedding and structural embedding, and adopt scientific and reasonable strategic decisions on green innovation networks to ensure the effective development of green innovation network activities. Especially, the enterprises with low-level political ties, high financing restrictions, and state ownership should focus on environmental responsibility in order to better achieve the sustainable development of the company.

Secondly, enterprises should fully realize the scientific nature of integrating green innovation network embedding, green reputation, ethical leadership, and corporate environmental responsibility into the enterprise performance management framework; adopt scientific and reasonable strategic decisions; pay attention to the construction of external cooperation network; establish and improve communication channels with the government, customers, and other stakeholders; ensure the effective implementation of green innovation network activities; actively establish a good green image; and strive to clarify the requirements and expectations of social participants for corporate environmental responsibility. In addition, enterprises should adopt the strategy of embedding a green innovation network with highly integrated ethics and environmental regulations, create a good cultural environment for corporate environmental responsibility, consciously cultivate and promote ethical leaders with environmental and humanistic perspectives, and encourage these leaders to spread such views to their employees.

Thirdly, the development of enterprises cannot be separated from the support of the external institutional environment. Government departments should fully realize that they should provide necessary environmental incentive policies according to the development needs of enterprises. While green innovation is environmentally sound, it requires significant upfront investment and takes time to transform business models. Therefore, in addition to environmental regulations, the government should also guide entrepreneurs to actively participate in various environmental protection associations and green management training through government–enterprise cooperation, help enterprises establish and strengthen their sense of social responsibility, formulate various incentive policies to constantly encourage enterprises to invest more resources in long-term green innovation network activities, and create a favorable policy environment for the sustainable development of a green innovation network. Thus, enterprises are encouraged to take the initiative to fulfill their environmental and social responsibilities. In addition, the government needs to realize that the effective implementation of environmental mechanism policies such as green subsidies cannot be separated from the cultivation and improvement of the environmental ethics of business leaders. Therefore, the government should incorporate business leaders' ethical and moral evaluation into the allocation process of incentive policies such as green subsidies.

5.3. Limitations and Suggestions for Future Research

There are still some limitations in this study that need to be discussed in the future. Firstly, this study only explored the impact of solid ties and centrality on corporate environmental responsibility. On this basis, subsequent studies can introduce network characteristic variables such as relationship quality, network size, and network density to improve the scope of this study. Secondly, this study only discussed the social performance of network embeddedness from the perspective of corporate environmental responsibility. Subsequent studies can explore how network embeddedness affects corporate social performance, such as green governance, service quality, and employee responsibility. These factors also play an essential role in the long-term development of enterprises. Thirdly, in practice, this study can also include some other significant contingency factors that will affect the boundary conditions of green innovation network embeddedness and corporate environmental responsibility, such as green certification, media attention, and government regulation, to further explore the boundary conditions for enterprises to enhance the green reputation and corporate environmental responsibility. This study only examined the boundary conditions of network embeddedness from ethical leaders. Finally, the samples of this paper were listed companies engaged in green innovation. Different industries have different incentives to participate in the green innovation network. For example, one of the reasons for the rapid development of the new energy vehicle industry in China is government subsidies, which will lead to specific issues related to the green innovation network embeddedness. Therefore, subsequent research can focus on a specific industry.

Author Contributions: Writing—original draft, J.W.; Supervision, W.L. All authors have read and agreed to the published version of the manuscript.

Funding: This research is funded by "the Postgraduate Innovation Research Fund" of University of International Business and Economics: 202242.

Institutional Review Board Statement: No applicable.

Informed Consent Statement: No applicable.

Data Availability Statement: IncoPat Patent Database; www.hexun.com; WIND Datebase.

Conflicts of Interest: The authors declare no conflict of interest.

References

1. Sheehy, B.; Farneti, F. Corporate social responsibility, sustainability and corporate sustainability: What is the difference? And, does it matter? *Sustainability* **2021**, *13*, 5967. [CrossRef]
2. Babiak, K.; Trendafilova, S. CSR and Environmental Responsibility: Motives and Pressures to Adopt Green Management Practices. *Corp. Soc. Responsib. Environ. Manag.* **2011**, *18*, 11–24. [CrossRef]
3. Huang, J.W.; Li, Y.H. Green innovation and performance: The view of organizational capability and social reciprocity. *J. Bus. Ethics* **2017**, *145*, 309–324. [CrossRef]
4. Khan, H.Z.; Bose, S.; Ben, S.; Ali, Q. Green banking disclosure, firm value and the moderating role of a contextual factor: Evidence from a distinctive regulatory setting. *Bus. Strategy Environ.* **2021**, *30*, 3651–3670. [CrossRef]
5. Hansen, M.T. Knowledge networks: Explaining effective knowledge sharing in multiunit companies. *Organ. Sci.* **2002**, *13*, 232–248. [CrossRef]
6. Bar, E.S. A case study of obstacles and enablers for green innovation within the fish processing equipment industry. *J. Clean. Prod.* **2015**, *90*, 234–243.
7. Hojnik, J.; Ruzzier, M. What drives eco-innovation? A review of an emerging literature. *Environ. Innov. Soc. Transit.* **2016**, *19*, 31–41. [CrossRef]
8. Horbach, J. Determinants of environmental innovation: New evidence from German panel data sources. *Res. Policy* **2008**, *37*, 163–173. [CrossRef]
9. Cainelli, G.; Mazzanti, M.; Montresor, S. Environmental innovations, local networks and internationalization. *Ind. Innov.* **2012**, *19*, 697–734. [CrossRef]
10. Gilli, M.; Mazzanti, M.; Nicolli, F. Sustainability and competitiveness in evolutionary perspectives: Environmental innovations, structural change and economic dynamics in the EU. *J. Socio-Econ.* **2013**, *45*, 204–215. [CrossRef]
11. Driessen, P.H.; Hillebrand, B.; Kok, R.A.W.; Verhallen, T.M.M. Green new product development: The pivotal role of product greenness. *IEEE Trans. Eng. Manag.* **2013**, *60*, 315–326. [CrossRef]
12. Xie, X.; Huo, J.; Qi, G.; Zhu, K.X. Green process innovation and financial performance in emerging economies: Moderating effects of absorptive capacity and green subsidies. *IEEE Trans. Eng. Manag.* **2016**, *63*, 101–112. [CrossRef]
13. Yu, W.; Ramanathan, R.; Nath, P. Environmental pressures and performance: An analysis of the roles of environmental innovation strategy and marketing capability. *Technol. Forecast. Soc. Change* **2017**, *117*, 160–169. [CrossRef]
14. Chiou, T.Y.; Chan, H.K.; Lettice, F.; Chung, S.H. The influence of greening the suppliers and green innovation on environmental performance and competitive advantage in Taiwan. *Transp. Res. Part E Logist. Transp. Rev.* **2011**, *47*, 822–836. [CrossRef]
15. Wei, Z.; Shen, H.; Zhou, K.Z.; Li, J.J. How does environmental corporate social responsibility matter in a dysfunctional institutional environment? Evidence from China. *J. Bus. Ethics* **2017**, *140*, 209–223. [CrossRef]
16. Jiang, R.J.; Tao, Q.T.; Santoro, M.D. Alliance portfolio diversity and firm performance. *Soc. Sci. Electron. Publ.* **2010**, *31*, 1136–1144. [CrossRef]
17. Zhang, J.; Jiang, H.; Wu, R.; Li, J. Reconciling the Dilemma of Knowledge Sharing: A network Pluralism framework of firms' R&D alliance network and innovation performance. *J. Manag.* **2019**, *45*, 2635–2665.
18. Tsai, K.H. Collaborative networks and product innovation performance: Toward a contingency perspective. *Res. Policy* **2009**, *38*, 765–778. [CrossRef]
19. De Noni, I.; Orsi, L.; Belussi, F. The role of collaborative networks in supporting the innovation performance of lagging-behand European region. *Res. Policy* **2018**, *47*, 1–13. [CrossRef]
20. Qi, L.; Wang, L.; Li, W.A. Do mutual fund networks affect corporate social responsibility? Evidence from China. *Corp. Soc. Responsib. Environ. Manag.* **2020**, *27*, 1040–1050. [CrossRef]
21. Gulati, R. Network location and learning: The influence of network resources and firm capabilities on alliance formation. *Strateg. Manag. J.* **1999**, *20*, 397–420. [CrossRef]
22. McEvily, B.; Zaheer, A. Bridging ties: A source of firm heterogeneity in competitive capabilities. *Strateg. Manag. J.* **1999**, *20*, 1122–1156. [CrossRef]
23. Rowley, T.; Behrens, D.; Krackhardt, D. Redundant governance structures: An analysis of structural and relational embeddedness in the steel and semiconductor industries. *Strateg. Manag. J.* **2000**, *21*, 369–386. [CrossRef]
24. Barclay, P.; Barker, J.L. Greener than thou: People who protect the environment are more cooperative, compete to be environmental, and benefit from reputation. *J. Environ. Psychol.* **2020**, *72*, 101441. [CrossRef]
25. Norheim-Hanse, A. Are 'green brides' more attractive? An empirical examination of_how_prospective_partners'_environmental_reputation_affects_the_trust-based_mechanism_in_alliance_formation. *J. Bus. Ethics* **2015**, *132*, 813–830.
26. Friedman, A.L.; Miles, S. Socially responsible investment and corporate social and environmental reporting in the UK: An exploratory study. *Br. Account. Rev.* **2001**, *33*, 523–548. [CrossRef]
27. Little, P.L.; Little, B.L. Do perceptions of corporate social responsibility contribute to explaining differences in corporate price-earning ratios? A research note. *Corp. Reput. Rev.* **2000**, *3*, 137–142. [CrossRef]
28. Roberts, D. Corporate reputation and sustained superior financial performance. *Strateg. Manag. J.* **2002**, *23*, 1077–1093. [CrossRef]
29. Orlitzky, M.; Schmidt, F.L.; Rynes, S.L. Responsibility disclosure of US fortune 500 firms: An application of content analysis. *Adv. Environ. Account. Manag.* **2003**, *1*, 163–200.

30. Najafi-Tavani, S.; Najafi-Tavani, Z.; Naudé, P.; Oghazi, P.; Zeynaloo, E. How collaborative innovation networks affect new product performance: Product innovation capability, process innovation capability, and absorptive capacity. *Ind. Mark. Manag.* **2018**, *73*, 192–205. [CrossRef]
31. Chung, D.; Kim, J.; Kang, J. Influence of alliance portfolio diversity on innovation performance: The role of internal capabilities of value creation. *Rev. Manag. Sci.* **2019**, *13*, 1093–1120. [CrossRef]
32. Wu, X.; Dou, W.; Du, J.; Jiang, Y. Production network positions, innovation orientation and environmental dynamics: An empirical analysis of Chinese firms. *Int. J. Technol. Manag.* **2015**, *67*, 77–102. [CrossRef]
33. Pangarkar, N.; Lin, Y.; Hussain, S. Too much of a good thing? Alliance portfolio size and alliance expansion. *Eur. Manag. J.* **2017**, *35*, 477–485. [CrossRef]
34. Fu, R.; Tang, Y.; Chen, G. Chief sustainability officers and corporate social(Ir) responsibility. *Strateg. Manag. J.* **2020**, *41*, 656–680. [CrossRef]
35. Granovetter, M. Economic action and social structure: The problem of embeddedness. *Am. J. Sociol.* **1985**, *91*, 481–510. [CrossRef]
36. Huang, C.; Wang, Y. Evolution of network relations, enterprise learning, and cluster innovation networks: The case of the Yuyao plastics industry cluster. *Technol. Anal. Strateg. Manag.* **2018**, *30*, 158–171. [CrossRef]
37. Kim, H.S. How a firm's position in a whole network affects innovation performance. *Technol. Anal. Strateg. Manag.* **2019**, *31*, 155–168. [CrossRef]
38. Partanen, J.; Chetty, S.K.; Rajala, A. Innovation types and network relationships. *Entrep. Theory Pract.* **2014**, *38*, 1027–1055. [CrossRef]
39. Rusanen, H.; Halinen, A.; Jaakkola, E. Accessing resources for service innovation-the critical role of network relationships. *J. Serv. Manag.* **2014**, *25*, 2–29. [CrossRef]
40. Putnam, R.D. *Bowling Alone: The Collapse and Revival of American Community*; Simon & Schuster: New York, NY, USA, 2000.
41. Carney, M. The competitiveness of networked production: The role of trust and asset specificity. *J. Manag. Stud.* **1998**, *35*, 457–479. [CrossRef]
42. Deng, X.; Kang, J.K.; Low, B.S. Corporate social responsibility and stakeholder value maximization: Evidence from mergers. *J. Financ. Econ.* **2013**, *110*, 87–109. [CrossRef]
43. Wong, S.S.; Boh, W.F. The Contingent Effects of Social Network Sparseness and Centrality on Managerial Innovativeness. *J. Manag. Stud.* **2014**, *51*, 1180–1203. [CrossRef]
44. Shibin, S.; Kevin, Z.Z.; Julie, L.J. The effects of business and political ties on firm performance: Evidence from China. *J. Mark.* **2011**, *75*, 1–15.
45. Mark, C.S. Managing legitimacy: Strategic and institutional approaches. *Acad. Manag. Rev.* **1995**, *20*, 571–610.
46. Searcy, C. Corporate sustainability performance measurement systems: A review and research agenda. *J. Bus. Ethics* **2012**, *107*, 239–253. [CrossRef]
47. Wang, R.; Wijen, F.; Heugens, P.P. Government's green grip: Multifaceted state influence on corporate environmental actions in China. *Strateg. Manag. J.* **2018**, *39*, 403–428. [CrossRef]
48. Rowley, T.J. Moving beyond dyadic ties: A network theory of stakeholder influences. *Acad. Manag. Rev.* **1997**, *22*, 887–910. [CrossRef]
49. Fischer, M. Institutions and coalitions in policy processes: A cross-sectoral comparison. *J. Public Policy* **2015**, *35*, 245–268. [CrossRef]
50. Burt, R.S. Gossip and reputation. In *Globalisation Series of Seminars and Lectures*; Queen Mary University of London: London, UK, 2008.
51. Granovetter, M.S. The strength of weak ties. *Am. J. Sociol.* **1973**, *78*, 1360–1380. [CrossRef]
52. Fombrun, C.J. *Reputation: Realizing Value from the Corporate Image*; Harvard Business School Press: Boston, MA, USA, 1996.
53. Gulati, R. Does familiarity breed trust? The implications of repeated ties for contractual choice in alliances. *Acad. Manag. J.* **1995**, *38*, 85–112. [CrossRef]
54. Al-Laham, A.; Souitaris, V. Network embeddedness and new-venture internationalization: Analyzing international linkages in the German biotech industry. *J. Bus. Ventur.* **2008**, *23*, 567–586. [CrossRef]
55. Stefano, D.; Gabriele, C.; Antonella, Z. Resource-based local development and networked core-competencies for tourism excellence. *Tour. Manag.* **2010**, *31*, 260–266.
56. Leifeld, P.; Schneider, V. Information exchange in policy networks. *Am. J. Political Sci.* **2012**, *56*, 731–744. [CrossRef]
57. Choi, T.; Robertson, P.J. Deliberation and decision in collaborative governance: A simulation of approaches to mitigate power imbalance. *J. Public Adm. Res. Theory* **2013**, *24*, 495–518. [CrossRef]
58. Provan, K.G.; Milward, K. The evolution of structural embeddedness and organizational social outcomes in a centrally governed health and human services network. *J. Public Adm. Res. Theory J. Part* **2009**, *19*, 873–893. [CrossRef]
59. Gonzalez, M.A.; Triguero, A.; Saez-Martinez, F.J. Many or trusted partners for eco-innovation? The influence of breadth and depth of firms' knowledge network in the food sector. *Technol. Forecast. Soc. Chang.* **2019**, *147*, 51–62. [CrossRef]
60. Shiu, Y.M.; Yang, S.L. Does engagement in corporate social responsibility provide strategic insurance-like effects? *Strateg. Manag. J.* **2017**, *38*, 455–470. [CrossRef]
61. Martín-de Castro, G.; Amores-Salvadó, J.; Navas-López, J.E.; Balarezo-Núñez, R.M. Corporate environmental reputation: Exploring its definitional landscape. *Bus. Ethics A Eur. Rev.* **2019**, *29*, 130–142. [CrossRef]

62. Zhou, Y.; Luo, Y. Financial ecological environment, green reputation and credit financing: An empirical study based on A-share listed companies in heavily polluted industries. *South. Financ.* **2017**, *8*, 21–32.
63. Collins, J. *Good to Great and the Social Sectors*; HarperCollins: New York, NY, USA, 2005.
64. Hussainey, K.; Salama, A. The important of corporate environment reputation to investors. *J. Appl. Account. Res.* **2010**, *11*, 273–294.
65. Aintablian, S.; Mcgraw, P.A.; Roberts, G.S. Bank monitoring and environmental risk. *J. Bus. Financ. Account.* **2007**, *34*, 389–401. [CrossRef]
66. Xu, J.Z.; Guan, J.; Lin, Y. Institutional pressure, senior executives' awareness of environmental protection and enterprises' green innovation practice: Based on the perspective of new institutionalism theory and high-level theory. *Manag. Rev.* **2017**, *29*, 72–83.
67. Mayer, D.M.; Aquino, K.; Greenbaum, R.L.; Kuenzi, M. Who displays ethical leadership, and why does it matter? An examination of antecedents and consequences of ethical leadership. *Acad. Manag. J.* **2012**, *55*, 151–171. [CrossRef]
68. Mo, S.; Ling, C.D.; Xie, X.Y. The curvilinear relationship between ethical leadership and team creativity: The moderating role of team faultlines. *J. Bus. Ethics* **2019**, *154*, 229–242. [CrossRef]
69. Tu, Y.; Lu, X. How ethical leadership influence employees' innovative work behavior: A perspective of intrinsic motivation. *J. Bus. Ventur.* **2013**, *116*, 441–455. [CrossRef]
70. Maak, T.; Pless, N.M. Responsible leadership in a stakeholder society—A relational perspective. *J. Bus. Ethics* **2006**, *66*, 99–115. [CrossRef]
71. Sheehy, B. Defining CSR: Problems and solutions. *J. Bus. Ethics* **2015**, *131*, 625–648. [CrossRef]
72. Post, C.; Rahman, N.; Rubow, E. Green Governance: Boards of Directors' Composition and Environmental Corporate Social Responsibility. *Bus. Soc.* **2011**, *50*, 189–223. [CrossRef]
73. Phelps, C.C. A longitudinal study of the influence of alliance network structure and composition on firm exploratory innovation. *Acad. Manag. J.* **2010**, *53*, 890–913. [CrossRef]
74. Brown, M.E.; Trevino, L.K.; Harrison, D.A. Ethical leadership: A social learning perspective for construct development and testing. *Organ. Behav. Hum. Decis. Process.* **2005**, *97*, 117–134. [CrossRef]
75. Jones, S.A.; Michelfelder, D.; Nair, I. Engineering managers and sustainable systems: The need for and challenges of using an ethical framework for transformative leadership. *J. Clean. Prod.* **2017**, *140*, 205–212. [CrossRef]
76. Wang, D.; Feng, T.; Lawton, A. Linking ethical leadership with firm performance: A multi-dimensional perspective. *J. Bus. Ethics* **2017**, *145*, 95–109. [CrossRef]
77. Harjoto, M.A.; Jo, H. Corporate Governance and CSR Nexus. *J. Bus. Ethics* **2011**, *100*, 45–67. [CrossRef]
78. Meo, F.D.; Lara, J.M.G.; Surroca, J.A. Managerial entrenchment and earnings management. *J. Account. Public Policy* **2017**, *36*, 399–414. [CrossRef]
79. Luo, X.R.; Wang, D.; Zhang, J. Whose call to answer: Institutional complexity and firms' CSR Reporting. *Acad. Manag. J.* **2017**, *60*, 321–344. [CrossRef]
80. Aupperle, K.E.; Carroll, A.B.; Hatfield, J.D. An empirical examination of the relationship between corporate social responsibility and profitability. *Acad. Manag. J.* **1985**, *28*, 446–463. [CrossRef]
81. Jo, H.; Harjoto, M.A. Corporate governance and firm value: The impact of corporate social responsibility. *J. Bus. Ethics* **2011**, *103*, 351–383. [CrossRef]
82. Faccio, M. Politically connected firms. *Am. Econ. Rev.* **2007**, *1*, 369–386.
83. Pfeffer, J.; Salancik, G.R. *The External Control of Organizations: A Resource Dependence Perspective*; Harper and Row: New York, NY, USA, 1978.
84. Huang, M.; Li, X.; Zhu, H. An analysis of the phenomenon of enterprises being forced to donate: Whether the public is "unreasonable" or the enterprise is "unscrupulous". *Manag. World* **2008**, *10*, 115–126.
85. Ni, C. Political atmosphere and social responsibility reporting strategy of enterprises: Based on the perspective of institutional pressure. *J. Shanxi Univ. Financ. Econ.* **2019**, *41*, 100–112.
86. Shi, M.; Cai, X.; Geng, X.L. Corporate social responsibility, R&D investment and debt financing cost in a dynamic environment: An empirical study based on private listed companies in China's manufacturing industry. *J. Shanxi Univ. Financ. Econ.* **2017**, *39*, 111–124.
87. Cheng, C.C.J.; Hsu, S.H.; Sheu, C. How can green innovation from manufacturers benefit from supplier networks? *Supply Chain. Manag. Int. J.* **2022**. *ahead-of-print*. [CrossRef]
88. Xu, J.Z.; Zhai, J.Q. Research on the evaluation of green innovation capability of manufacturing enterprises in innovation network. *Sustainability* **2020**, *12*, 807. [CrossRef]
89. Pasricha, P.; Singh, B.; Verma, P. Ethical leadership, organic organizational cultures and corporate social responsibility: An empirical study in social enterprises. *J. Bus. Ethics* **2018**, *151*, 229–242.

water

MDPI

Review

Soft vs. Hard Sustainability Approach in Marine Spatial Planning: Challenges and Solutions

Fatemeh Rezaei [1,*], Pasquale Contestabile [1,2,*], Diego Vicinanza [1,2,3], Arianna Azzellino [2,4], Carlos V. C. Weiss [5] and José Juanes [5]

[1] Department of Engineering, University of Campania, Via Roma 29, 81031 Aversa, Italy; diego.vicinanza@unicampania.it
[2] Inter-University National Consortium for Marine Sciences (CoNISMa), Piazzale Flaminio, 00196 Rome, Italy; arianna.azzellino@polimi.it
[3] Department of Integrative Marine Ecology, Stazione Zoologica Anton Dohrn, 80121 Naples, Italy
[4] Civil and Environmental Engineering Department, Politecnico di Milano, Piazza Leonardo da Vinci, 32, 20133 Milan, Italy
[5] Instituto de Hidráulica Ambiental, Universidad de Cantabria, Avda. Isabel Torres, 39011 Santander, Spain; dacruzcv@unican.es (C.V.C.W.); antonio.juanes@unican.es (J.J.)
* Correspondence: fatemeh.rezaei@unicampania.it (F.R.); pasquale.contestabile@unicampania.it (P.C.)

Abstract: Hitherto, over 50% of countries with marine waters have established MSPs or launched related legal actions. However, there are still conceptual and practical challenges to be overcome in the development of MSP. In this study, we investigate two main approaches in MSP (hard vs. soft sustainability) through reports, published manuscripts and meeting proceedings in seven pioneering countries (Belgium, Netherlands, Norway, Germany, United Kingdom, Australia, and Canada). We highlight the gaps, challenges, and solutions in each of these approaches. From our findings, there are four common challenges in both soft and hard sustainability approaches as follows: (i) the political framework and inconsistent support of MSP efforts, (ii) insufficient knowledge on social dimensions, (iii) insufficient stakeholder engagement in the diversity of stakeholder's groups or in their contribution to the planning process from the initial steps, and (iv) finding a balance between environmental conservation and economic growth. We recommend that future studies should investigate how MSP can become more adaptive to long-term environmental and economic targets, how effective involving socioeconomic strata is in MSP, and how decision-making tools could help to cover the gaps in MSP. Furthermore, public forums are suggested to be developed to facilitate the systematic sharing of MSP experiences worldwide.

Keywords: marine spatial planning; soft sustainability; hard sustainability; marine zoning; ecosystem-based approach

1. Introduction

Marine ecosystems are key contributors to human food supply and welfare, energy production, global economies, tourism, biodiversity, global carbon sequestration and oxygen production [1,2]. According to reports from the United Nations (UN), marine and coastal resources and industries contribute to five percent of the global GDP and provide livelihoods for more than three billion people around the world [3]. Around 80 percent of the global transportation of goods is through marine lanes [3]. Marine environments and their resources are, therefore, critical to conserve and be sustainably managed. However, over the recent decades, marine ecosystem sustainability has been increasingly threatened by growing human activities with more demands for marine spaces, acidification, eutrophication, the overexploitation of marine resources, release of hazardous substances into the marine environments and climate change [4,5]. Therefore, the effective management of marine environments requires a comprehensive understanding of the different pressures posed on marine ecosystems. In addition, a marine management system should consider the changing nature and extent of these ecosystems to ultimately modify a marine ecosystem's structure and functions through "planning" [6,7].

To achieve marine ecosystem management goals, a sustainable planning approach must cover the gaps between societal and economic objectives alongside the state of the environment [5]. In this regard, scientists and policymakers have proposed marine spatial planning (MSP) as a holistic approach to achieving long-term ecosystem-based management goals [8,9]. According to the definition by Ehler and Douvere (2009), MSP is a "public process of analysing and allocating the spatial and temporal distribution of human activities in marine areas to achieve ecological, economic, and social objectives that are usually specified through a political process" [10,11]. The MSP process is adaptive and continuous and basically includes the following four main phases: (1) Analysis and planning: current environmental situations and human activities are assessed through research. (2) Defining and analysing future conditions: potential future alternative scenarios, as well as temporal and spatial needs for any further demands on ocean space, are identified in the planning region. Subsequently, spatial plans for the sustainable use of sea resources are generated. (3) Implementation: the proposed plan is implemented through regulations, legislation, and investments. (4) Monitoring and evaluation: the effectiveness of proposed strategies, time scales and implementation methods are assessed and modified. The results of this step act as input data for analysis and the planning phase to initiate the whole process again [10,12].

Generally, MSP originates from ocean and environmental communities with the partial involvement of other planning communities, such as land-use planning. In fact, MSP and land-use planning pursue the same goal, which is generally finding a balance between social, economic and environmental values through physical planning [10]. The first MSP was developed with the aim of conserving marine-protected areas in Australia in 1975 [13]. Afterwards, by approval of the EU maritime policy (2006) and the EU maritime roadmap (2009), MSP goals were set to resolve conflicts between the user–user or user–environment with emphasis on economic development in the sea, especially in the North Sea, as a densely used sea area. This MSP approach was implemented in the Northwest European countries of China and the United States [13–15]. The most comprehensive concept of MSP emerged in Australia in the late 1990s, followed by Canada, which considers MSP as a tool for developing human activities in the sea with ecosystem-based approaches (EBA) [13]. This approach takes into account the cumulative impacts of human activities on ecosystem services as well as the ecological integrity and biodiversity of marine ecosystems for the appropriate spatial and temporal planning of human activities in the sea [8,11]. Although EBA was subsequently included in the European Union Marine Strategy Framework Directive 2008/56/EC (MSFD), the European Parliament voted in late 2013 to emphasise economic growth in the form of the Blue Economy while downgrading the emphasis on the EBA in the objectives of the proposed Directive on MSP and integrated coastal management (ICM) [16]. The objective of MSFD is generally to achieve a good environmental status for all of the EU's marine waters by keeping marine ecosystems healthy, productive, and resilient while ensuring the more sustainable use of marine resources for current and future generations [17]. Figure 1 illustrates the objectives of ecosystem-based MSP in Australia and MSFD in European countries. Hitherto, several studies have reported that MSP could resolve issues regarding user–user conflicts (e.g., wind farms, oil and gas fields, shipping, aquaculture, conservation areas, tourism, etc.) in the southern North Sea in the Netherlands, Belgium, United Kingdom, Sweden, United States, Canada, Philippine, and Bangladesh [18–23]. It has also been reported that MSP is the key element in resolving user–environment conflicts in Australia's Great Barrier Reef Marine Park and the North Argentina Basin with a long-term perspective on the conservation of the ecosystem and biodiversity [1,2]. Such experiences demonstrate that MSP can be an effective approach for controlling/mitigating the impacts of ongoing and future human activities on marine environments.

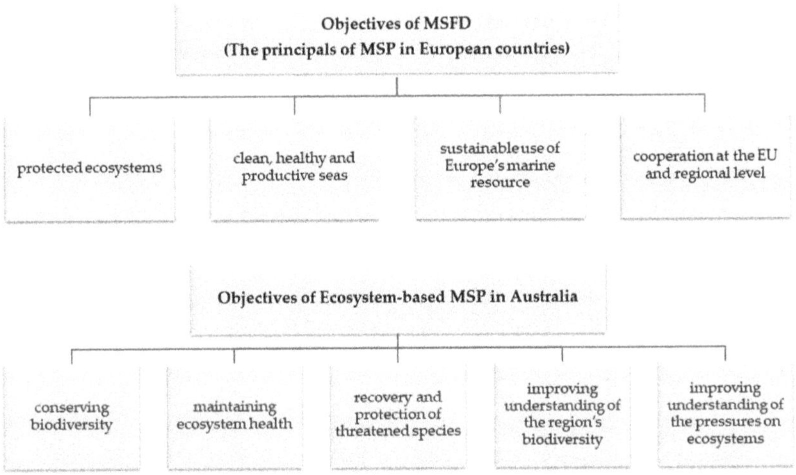

Figure 1. Comparison of MSP objectives in European countries and Australia.

So far, the Intergovernmental Oceanographic Commission of UNESCO IOC-UNESCO and the European Commission have created various guidelines on MSP, in addition to applicable national guidelines and Directives in each country. Meanwhile, more than 75 of the world's 150 countries with marine borders have adopted MSP roadmaps as a strategy for integrated marine management [24]. However, there are significant disparities in the planning process of MSP in those countries. In addition, the MSP process can be affected by the planning culture and national policy design of each country [25]. MSP in Australia and Canada, for instance, is primarily driven by ecosystem conservation objectives, with the assumption that natural resources are irreplaceable if lost (hard sustainability), whereas MSP in European countries, China and the US prioritise economic objectives, with the assumption that the loss of natural resources can be compensated through technological advancement (soft sustainability) [16]. On the other hand, there are still some countries worldwide that cannot launch MSP or are just at the beginning stages of MSP due to constraints, such as a marine data gap or a lack of linkage between authorities for the management of marine environments as a whole [13,24]. According to reports, MSP has been primarily implemented in high-income countries, while only 7 percent of coastal countries in Asia-Pacific regions have implemented MSP in their Exclusive Economic Zone (EEZ) [26]. As a result, implementing MSP at the national level and over the national borders is still challenging. Therefore, the present review study aims to address the challenges and good practices of MSP implementation with either a hard sustainability or a soft sustainability approach over the last two decades to ultimately provide insight for executives/legislators in the nations which are in the early stages of MSP implementation, as well as suggestions for future research on MSP. Accordingly, the main contribution of this study is to enhance an understanding of the obstacles and effective strategies related to implementing hard or soft sustainability approaches in MSP, as well as informing decision making regarding future MSP implementation. For this purpose, data and reports on marine spatial planning were gathered in five case study European countries (i.e., Belgium, Netherlands, Norway, Germany, and the United Kingdom) as examples of front-runners in the hard sustainability approach in MSP, as well as Australia and Canada as successful examples of the soft sustainability approach. A summary of the experiences and lessons learned from MSP implementation in each country is presented in consecutive sections, and at the end, some recommendations are made to fill the gaps for future MSP implementation.

2. MSP Challenges and Good Practices

2.1. Pioneer Countries with Soft Sustainability Approache in MSP

2.1.1. MSP in Belgium

Belgium is known as a pioneer country both in Europe and in the world for implementing MSP with a total sea area of about 0.5% of the North Sea (Royal Decree 2014). The plan cycles were drawn up for a period of every six years the first cycle (2014–2020) was completed, and the second cycle (2020–2026) is now underway [27]. In the early stages, the main driving forces of MSP in Belgium were legal and economic objectives [18]. Afterwards, a "Master Plan" was developed in response to the challenges of conflict among uses, the need for offshore energy production and natural resources conservation. The objective of the "Master Plan" was to provide a spatial vision for current and future uses, with a cross-sectoral and multi-use approach in the entire EEZ and terrestrial sea [18]. As a result, 13 uses are now considered in Belgium's MSP, including nature protection, offshore renewable energy production, shipping, ports, mineral extraction, fishing, aquaculture, underwater cultural heritage, military activities, scientific research, coastal protection, cables and pipelines, and zones for commercial and industrial activities (Royal Decree 2014) [27]. The multi-use approach was reflected in the compatibility/incompatibility uses of the sea in Belgium's marine spatial plan. For instance, normal shipping was considered incompatible in and around wind farms with a safety zone of 500 m; recreational activities could co-exist with marine-protected areas; and sand and gravel extraction were limited to fish-spawning periods (Royal Decree 2014) [18,27]. Plasman (2008), who reviewed the policy perspective of MSP in Belgium, suggested that the first step towards the effective implementation of MSP plans should be aligning scientists with politicians in decision-making processes [28]. The author mentioned that the multi-use approach of Belgium's MSP could provide a delicate balance in time and spatial scales for different activities; for example, sand extraction, fishing, and military exercises could be planned at the same place but at different time spans, whereas offshore wind farms could host clam farms simultaneously as a more creative solution for local fishermen who lost their fishing grounds [28]. Custodio et al. (2022), who studied the linkage between ecosystem services (ES) and the management of marine activities in Belgian's continental shelf, proposed that the engagement of stakeholders is crucial for an ecosystem-based MSP in complex social–ecological systems [29]. The mentioned authors suggested that different stakeholder groups should be selected based on their proportions in livelihood and the economy, and an ES priority list should be provided through stakeholder workshops. This priority list can be applied as a baseline for ES modelling and marine activity management [29]. Vanden Ede et al. (2014), who studied Marine Biological Valuation (MBV) (i.e., the value of the goods and services provided by marine ecosystems) in shallow Belgian coastal zones, suggested that MBV maps should be used along with other social, economic, political, legal and environmental maps for future decision making in MSP processes, especially where conflicts between coastal flood risks and nature-conservation uses of space exist [30]. Similar recommendations were proposed by Pascual et al. (2011) to study the MBV mapping of the Basque continental shelf [31]. Table 1 shows the adaption of MSP in Belgium to the four main stages of the MSP process.

Table 1. Adaption of MSP in Belgium to main MSP stages [10,27].

Main Stages of MSP	MSP in Belgium
Organising the process through pre-planning	The Marine Environment Act was amended in 2012. The Royal Decree of 20 March 2014 adopts MSP. The first MSP cycle was completed (2014–2020). The second MSP cycle (2020–2026) is underway.

Table 1. *Cont.*

Main Stages of MSP	MSP in Belgium
Defining and analysing existing conditions	The draft of the MSP plan includes the analysis of existing conditions. Complementary information is gained by public consultation processes, petition letters, industry, NGOs, and formal contact with neighbouring countries.
Defining and analysing future conditions	The draft of the MSP plan includes the analysis of future conditions.
Monitoring and evaluating performance	Yearly monitoring of the execution of the plan is conducted through a committee with all competent authorities. The plan is reviewed every six years.

2.1.2. MSP in The Netherlands

The Netherlands, with a total territorial water and EEZ area of about 58,000 km^2, published its first plan and policy document for the sea in 2009 [23]. This country is now in its third cycle of MSP with a priority on renewable energy development [27]. The main drivers of MSP in the Netherlands are maintaining and developing environmental status and ecological habitats on the coast and in the sea, developing wind energy, shipping and sand extraction [23]. The current marine spatial plan includes nineteen uses generally in the sectors of maritime transport, oil and gas exploitation, offshore wind farms, nature conservation, submarine pipelines and cables, aquaculture, fishing, mineral extraction, military, and underwater cultural heritage and tourism [27]. The main strength of the Netherlands' MSP policy is that this country adjusted its long-term perspectives for marine uses and sustainability in accordance with a broader international framework based on the North Sea 2050 Spatial Agenda in response to the European Commission Strategy for "Blue Growth" [27,32]. In addition, the draft of the MSP policy document was prepared by involving national stakeholders and environmental non-governmental organisations from the early stages, as well as international consultation with neighbouring countries [23,27]. However, there were some challenges in implementing MSP at the beginning, including conflict between wind farm development and the oil and gas industry, conflict between wind farms and shipping activities, and reaching a balance between the energy, ecology and food sectors (i.e., fisheries, aquaculture) [23,27]. To overcome these main challenges, a separate executive with an independent chairman was organised to reach interlinked agreements between the mentioned sectors through meetings, open discussions, preparing joint factsheets and an inter-ministerial network [27]. In addition, the multi-use of space was suggested to minimise conflict among uses (e.g., wind farms can host aquaculture, sustainable fishing, and sand extraction recovery activities) [27]. Steins et al. (2021), who studied the role of stakeholder participation in multi-use planning for marine wind farms, marine conservation and seafood in the Netherlands, reported that social, economic, technical and regulatory factors can act as the main inhibitors for multi-use in MSP [33]. The mentioned authors found that collaboration between all stakeholders would be a solution for such barriers; for instance, the Netherlands' government established an independent "Community of Practice North Sea" to achieve a balance between different stakeholders' interests. The main idea of setting up this community was to stimulate the development of multi-use pilots through sharing knowledge and experiences between different stakeholders and encouraging a cooperative culture between them in an informal setting [33]. Garcia et al. (2019) and Hees (2019) proposed that MSP has a positive role in advancing blue energy in countries such as the Netherlands [15,32]. However, since acknowledgement of the licenses and permits of large-scale electric generation projects is usually outside the responsibility of marine authorities, the mentioned authors suggested that MSP should be applied in cross-border dimensions as a multi-level governance system at an international level [15,32]. Table 2 shows the adaption of MSP in the Netherlands for the four main stages of the MSP process.

Table 2. Adaption of MSP in the Netherlands to main MSP stages [10,27].

Main Stages of MSP	MSP in The Netherlands
Organising the process through pre-planning	The National Water Act and the first plan for the sea published was in 2009. The first cycle (2009–2015) and second cycle (2016–2021) of MSP were completed, and the third cycle (2022–2027) is underway.
Defining and analysing existing conditions	The draft of the MSP plan includes the analysis of existing conditions. The National Water Plan considers all relevant land–sea interactions. For various sectoral interests, specific legislation is in place; for instance, the Electricity Law regulates offshore renewable electricity to be landed, and the Common Fisheries Policy of the EU is in place for sustainable fisheries. In addition, a Community of Practice was established in 2018 with the aim of working in sync with science and government agencies and sharing up-to-date information on the present condition. Furthermore, a consultation body, "Overleg Fysieke Leefomgeving", was established for stakeholders' engagement and understanding of existing conditions.
Defining and analysing future conditions	The draft of the MSP plan includes the analysis of future conditions. The Policy Document on the North Sea 2016–2021 includes a framework vision map regarding the Netherlands' MSP.
Monitoring and evaluating performance	A review of the Policy Document was carried out in 2018 under the National Environmental Vision to conduct further analysis into the impacts on the environment, as well as separate monitoring and the general evaluation of the good environmental status of the sea. A review of the plan started in 2022 to meet 2030 and post-2030 renewable energy targets.

2.1.3. MSP in Norway

Marine management plans in Norway cover the whole area from the shoreline to offshore. The Barents Sea sector of the Norwegian EEZ is covered by an ecosystem-based MSP plan that was authorised in 2006 and subsequently amended in 2011 and 2015. In the meantime, the Norwegian EEZ's part of the North Sea and the Norwegian Sea were covered by an MSP plan authorised in 2013 and 2009, respectively [27]. The policy design for MSP in Norway was first finalised in 2006, which serves as both a marine spatial plan and a marine strategy [25]. The core element in policy formulation in Norwegian MSP is performing an environmental impact assessment for each maritime sector before its combination into a cumulative impact assessment [25]. Surís-Regueiro et al. (2021) studied the direct economic impacts resulting from the implementation of MSP policies in Norway. The authors reported that there is a significant positive effect of MSP implementation in the Norwegian Sea in contrast to stakeholders' beliefs. From their results, MSP implementation increased the production value of marine industries in the Norwegian Sea to about EUR 2262 million in 2013-16. The authors suggest that it is critical to develop protocols and procedures for gathering and processing the information provided by stakeholders in similar studies and estimations [34,35].

Kirkfeldt et al. (2020) conducted an interview with Norwegian planners to find a balance between the interests of different sectors in Norway [25]. Their study showed that one of the main challenges of Norwegian MSP is to resolve competing interests among sectoral stakeholders and homogenising legal frameworks across sectors and geographical locations, as well as challenges regarding the lack of a legal framework for marine-protected areas (MPAs) in the EEZ. The authors suggested that MSP should be implemented in the first place alongside making some changes to policy design wherever it is needed. These changes can come through good practices and experiences that already exist in other countries with a longer MSP experience. This approach helps to make more capacity and resources available to address other challenges that may arise during MSP implementation [25]. Olsen et al. (2014) studied the challenge of integrating multiple stakeholders and governmental levels in MSP in Norway, Belgium, and the US. The authors found that both Norway and Belgium have successfully fostered horizontal integration across sectors by establishing neutral meeting spaces (i.e., round-table meetings) where all stakeholders can participate. Similarly, vertical integration between government tiers has been accomplished in both nations by aligning parliamentary processes with executive government levels while also incorporating input from stakeholders [34]. Table 3 shows the adaption of MSP in Norway to the four main stages of the MSP process.

Table 3. Adaption of MSP in Norway to main MSP stages [10,27].

Main Stages of MSP	MSP in Norway
Organising the process through pre-planning	Pre-planning started in 2002. The first generation of plans was put into place for the Barents Sea-Lofoten area in 2006. The Nature Management Act was approved in 2008–2009. A new Marine Resource Act entered into force in 2009. MSP for the Norwegian Sea and North Sea–Skagerrak was launched in 2009 and 2013, respectively.
Defining and analysing existing conditions	The draft of the MSP plan included the analysis of existing conditions. The monitoring group, the forum for integrated ocean management, a steering committee of 10 ministries led by the Ministry of Climate and Environment, and an interdisciplinary MAREANO programme for mapping the seabed in Norway's marine and coastal waters provided complementary information on analysing existing and future conditions.
Defining and analysing future conditions	The draft of the MSP plan included the analysis of future conditions.
Monitoring and evaluating performance	The monitoring group, established in 2006, is responsible for the environmental monitoring of the marine ecosystems in Norwegian sea areas. The monitoring group annually produces short-status reports of all Norwegian sea areas, where every four years, a more detailed report is produced on the environmental conditions and development of all three sea areas. In addition, a supplementary report on the environmental status of pollution in the Norwegian Sea areas is produced every four years. The MSP plan is reviewed every four years, based on an updated cross-sectoral factual basis. The last update for all areas was endorsed by Parliament in June 2020. New updates are scheduled for 2024.

2.1.4. MSP in Germany

Germany, with a territorial sea and EEZ area of about 21,400 km^2 and 33,000 km^2, respectively, first started MSP in 2009. The objectives and principles of spatial planning in the German EEZ are based on the Spatial Planning Act of 1998, which considers economic and scientific use along with ensuring the safety and ease of maritime shipping and also the protection of the marine environment. Current maritime uses include mining, fisheries, aquaculture, coastal protection, ammunition storage sites, underwater cultural heritage and radars [27]. One of the strengths of MSP in Germany is considering the possibilities of transnational cooperation programs in the North and Baltic Sea regions in the planning process. Accordingly, transnational cooperation is provided with the participation of states and federal institutions with the aim of the sustainable development of these marine areas. Another strength is the involvement of coastal stakeholders in the planning process through consultation [27]. However, some challenges in implementing MSP in the German EEZ have been reported in different studies. For instance, the accomplishment of climate targets and the creation of jobs are the main justifications for developing offshore wind farms in Germany. However, local communities disagree, alleging possible impacts (e.g., environmental and landscape impacts, conflicts with fishing and shipping and freedom and wildness of the sea). Therefore, local support for particular sea uses also depends on cultural and regional distinctions as well as local demands [36]. Kannen (2014) suggested that it is indispensable to consider interactions between the social and ecological components of MSP through the involvement of local people in the planning and decision-making process [36]. Gimpel et al. (2015), in the study of the co-location of offshore wind farms and aquaculture in the German EEZ, proposed that a geographic information system (GIS)-based framework can be an effective tool for the site selection of an activity where conflict of uses exist [37]. Stelzenmüller et al. (2016) demonstrated that the socioeconomic importance of spatial overlap is affected by planning boundaries in the German EEZ [38]. The authors suggested that an interdisciplinary bottom-up strategy that takes into account the ecological consequences of human activities on target species helps to identify potential multi-use sites. As well as this, the following major issues must be resolved in order to develop the idea of multi-uses into MSP practice: defining a legal basis; enforcing safety regulations; defining the minimal specifications for each activity to be conducted in areas of other compatible activities (capacity, quotas, technical equipment); implementing a licensing process; and identifying financial subsidies to help businesses develop compatible activities [38]. Jay et al. (2016) proposed that communication between different stakeholders and their perspectives on the ecosystem approach within the MSP process can facilitate the development and implementation of the concept [39]. Berkenhagen et al. (2010), in their study of decision bias in the marine spatial planning of offshore wind farms in the German EEZ, suggested that the singular assessment of the economic impact of different activities is a drawback in the MSP process. For instance, assessments do not consider displacement costs when fishermen are forced to concentrate their efforts on small fishing grounds left open to fishing after the installation of all wind farms, which likely result of increased competition among fishermen and a rapid decrease in catch rates and subsequently no yield benefits [40]. The authors suggested that cumulative economic impact studies are substantial in MSP, and the following points should be taken into account: indirect costs associated with the displacement of activity to other areas (e.g., higher fuel costs, etc.), marine habitats and the species affected by the wind farms, an assessment of the cumulative effects of the closure of fishing areas due to wind farms, shipping, military activities, marine-protected areas, and other uses [40]. Table 4 shows the adaption of MSP in Germany to the four main stages of the MSP process.

Table 4. Adaption of MSP in Germany to main MSP stages [10,27].

Main Stages of MSP	MSP in Germany
Organising the process through pre-planning	The national legal basis for MSP in the German EEZ is the Spatial Planning Act (i.e., ROG Act), which was last revised in 2008 and amended in 2017. In 2004, MSP was included in the law for the first time. The legal regulation on spatial planning came into effect in 2009. According to the Spatial Planning Act, the federal government is responsible for MSP in the German EEZ.
Defining and analysing existing conditions	The draft of the MSP plan includes the analysis of existing conditions. The federal government carries out the preparatory procedural steps for drawing up the spatial planning plan with the consent of the Ministry. These include the creation of preliminary drafts and plan alternatives, the implementation of environmental assessments, the preparation of environmental reports and the participation of the public, those responsible for public affairs and other stakeholders.
Defining and analysing future conditions	The draft of the MSP plan includes the analysis of future conditions.
Monitoring and evaluating performance	According to the Federal Spatial Planning Act, the MSP has to be reviewed at least every ten years. According to the MSP Ordinance 2021, the plan is to be evaluated every 5 years.

2.1.5. MSP in the United Kingdom

The legislative framework for marine spatial plans in the United Kingdom (the UK) was first submitted in 2009 with the long-term consideration of actions (i.e., 20 years) and three-year evaluation periods. Subsequently, the marine policy statement was adopted in 2011, which established a policy framework for holistic marine planning and management in the UK waters [15,27]. MSP in the UK runs independently in England (11 marine plan areas in the east and south inshore and offshore waters), Scotland (one strategic national plan and 11 regional inshore plans), Wales and Northern Ireland (one plan in each nation both for offshore and inshore waters) [24,27]. The territorial sea and EEZ area of England covers around 51,700 km^2 and 178,600 km^2, respectively [27]. The main drivers of MSP in the UK are the optimum use of space and coexistence the activities, protecting marine resources, alongside sustainable development. The MSP includes fifteen main uses of the sea, including offshore renewable energy, fisheries, aquaculture, ports, shipping, military activities, conservation, coastal protection, scientific research, marine aggregate extraction, oil and gas extraction, cables and pipelines, underwater cultural heritage, tourism and leisure, and the dredging and disposal of dredged materials [27]. The main strength of MSP in the UK is the identification and participation of different stakeholders at the early stages of planning processes through workshops and research projects by the Marine Management Organisation (MMO) [27,41,42]. As well as this, in order to avoid common issues and ensure the better development of successful approaches, MMO gathered and analysed good practices on MSP from Australia, Belgium, Canada, Germany, Netherlands and the USA alongside referencing UNESCO's "Marine Spatial Planning—A Step-by-Step Guide". Ansong et al. (2021) studied a practical approach to building capacity in MSP in European countries and reported that there is a gap between the teaching and implementation of MSP in the UK [41]. In fact, the MSP process in the UK launched before establishing structured courses and training on marine planning. Therefore, there was a lack of comparison between theory and practice by trained marine planners. To overcome this drawback, MMO and the Welsh government, as the public administrators, provided internal training for their MSP team [41]. Furthermore, the government established three academic degrees specifically on MSP, making the UK the country with the most degrees on MSP [43]. Universities and administration organisations in the UK have also participated in different transboundary projects for an effective exchange of experiences along with the enhancement of MSP implementation both at the national and international level (e.g., the collaborative MASPNOSE project [44] with Germany and the Netherlands to enhance interactions between stakeholders and fisheries, the ADRIPLAN project [45] for boosting the methodology for MSP implementation in the Mediterranean region, and the TPEA project [46] for testing a cross-border MSP in the Eastern Atlantic).

Another positive point of the UK's MSP to address uncertainties of loss and damages to the marine environments is its multi-phased approach to the co-siting of emerging activities. For instance, since the long-term environmental effects of marine renewable energies are still not clearly known, a "survey-deploy-monitor" method has been applied to licensing marine renewable energy activities in marine-protected areas. This method proposes the single-unit deployment of a device at the first step and suggests that the environmental impacts of one device may be assessed before a decision is taken on a more extensive deployment of several devices [47,48].

Brennan et al. (2014) reported some challenges and solutions of MSP in the UK. For instance, to fill the gap in data for planning processes, the authors proposed improving data collection and exchange between authorities [49]. Similarly, Rodwell et al. (2014) reported that there are data quality/availability gaps, incomplete or missing metadata, non-uniform data formats and restrictions on access and licensing in the UK's MSP. The authors suggested that these gaps necessitate high-level assistance from committees and authorities to provide proper data access and discovery for the public as well as maintain data credibility. To overcome this challenge, the UK government has put more emphasis on data standards and data gathering and management, even for short-term projects [50]. The challenge of the insufficient participation of stakeholders in the planning process was suggested to be resolved by making MSP more flexible and adaptive rather than overly prospective. A flexible MSP can provide the effective handling of changing circumstances and attract new information as well as encourage stakeholders' engagement [43,49,50]. Regarding the financial challenges and insufficient human resources, the authors suggested the reallocation of resources from other areas or seeking additional funding from external resources [49]. To solve the challenge between fisheries and marine renewable energy development, a workshop was held to discuss the concerns and solutions of the interested stakeholders. In this workshop, monitoring programs were suggested as a solution to observe and assess the impacts of marine renewable energies on fisheries with the ultimate goal of developing collaboration between these two sectors. In addition, financial impacts on the fishery were suggested to be alleviated by mitigation strategies rather than financial compensation. Accordingly, a strategic mitigation fund was established to support communities and projects on local fuel supply while local boats were engaged in survey works [15,50]. Moreover, a fisheries mitigation working group was established and is being funded by the Natural Environment Research Council Knowledge Exchange Program [50].

Gissi and Vivero (2016), in their study on the challenges of the structuring transdisciplinary nature of MSP in educational courses and training in the UK, proposed that teaching skills and content on environmental and economic perspectives on marine resources alongside knowledge on maritime affairs, legislation and laws are necessary for development of an educational course specifically focused on MSP while planning theory and practical experiences in MSP should not be neglected [43]. Table 5 shows the adaption of MSP in the UK to the four main stages of the MSP process.

Table 5. Adaption of MSP in the UK to main MSP stages [10,27].

Main Stages of MSP	MSP in the UK
Organising the process through pre-planning	Under the Marine and Coastal Access Act 2009; the Marine (Scotland) Act 2010; and the Marine Act (Northern Ireland) 2013, marine planning was introduced for the "UK marine area", which included the territorial seas and offshore area adjacent to the UK. A marine policy statement was adopted in the UK in 2011. Marine planning functions for the Scottish and Welsh marine areas were devolved by the Scottish and Welsh governments, respectively. The MMO is responsible for preparing MSP in England.
Defining and analysing existing conditions	The draft of the MSP plan includes the analysis of existing conditions. The MMO is responsible for the delivery of planning, in addition to the licensing of marine activities, fisheries management and enforcement functions. MMO maintains a marine information system, as well as a strategic scoping exercise, which allows stakeholders to view and comment on the data layers and evidence base with the aim of analysing current and future conditions and uses of the sea.
Defining and analysing future conditions	The draft of the MSP plan includes the analysis of future conditions.
Monitoring and evaluating performance	According to the Marine and Coastal Access Act 2009, the effect of the policies in the marine plan; the effectiveness of those policies in securing the objectives for the marine plan; and the progress being made to secure the objectives are revised using a logic model with indicators for policies at different stages in the logic model every three years.

2.2. *Pioneering Countries with a Hard Sustainability Approach in MSP*

2.2.1. MSP in Australia

Australia, with an EEZ area of 9 million km^2, the third largest in the world, introduced zoning for the Great Barrier Reef Marine Park (GBRMP) in 1975, which is known as the first example of MSP in the world [24,27]. In the initial plan, around 4.5% of GBRMP was designated as a protected area in 1981 [51]. However, monitoring and periodic evaluation revealed that ecosystem preservation targets had not been attained; thus, the protected zone was expanded to approximately one-third of the overall area of the GBRMP. In the late 1990s, an ecosystem-based approach to MSP in Australia was launched by establishing the platform of "marine bio-regionalization" to map the location, composition, and structure of benthic organisms for the finer-scale planning and management of the sea [24,51]. Accordingly, Australia's Commonwealth waters (from 3 to 200 nautical miles) were classified into different bioregions based on biological similarities, species distribution, and the geomorphological characteristics of the seabed [13,52]. Bioregional plans (i.e., plans for the North-West, North, South-West, and Temperate East regions) were subsequently developed considering human activities, economic benefits, legal obligations, environmental protection and potential threats to ecological sustainability, protected areas, and species [13]. Each of the marine bioregional plans contributed to the enhancement of decision-making processes under the Environment Protection and Biodiversity Conservation Act 1999 (EPBC Act), specifically with regard to the preservation of marine biodiversity and the sustainable use of oceans [53].

Generally, the ecosystem-based approach for MSP in Australia includes the following three stages: (1) a bioregional profile of the area was created to understand the ecology and biophysical characteristics of the area; (2) a draft plan was created to assess the impacts and threats of current and future human activities on marine ecosystems; (3) a bioregional plan was developed with the identification of priorities for action in each bioregion. The legal basis for the ecosystem-based management of marine areas in Australia was established in 2005 [51,52]. The main strength of Australian MSP is its richness in spatial data of important biological features, species habitats, and species group levels alongside the pressure analysis of planned activities as the underlying materials of planning. As well as this, the Minister and the Department of Environment in Australia are responsible for evaluating whether a proposed project of human activities in the sea interferes with critical ecological processes or protected marine biota. The data provided in the Australian MSP are accessible for use by marine industrial planners to understand whether a referral is necessary or not. Therefore, the whole process is closely connected to the EPBC Act (1999) [52]. Pressure analysis studies of human activities on the sea before the planning process is another strength of ecosystem-based management in Australian MSP.

According to the survey results of Zimmerhackel et al. (2023), most stakeholders are concerned about the responsibility for maintaining and decommissioning artificial structures in the Australian sea, as well as clean-up costs in the event of an accident. Their findings highlighted the significance of clear rules for both phases in order to minimise the conflicts associated with man-made marine structures. The authors proposed that for a successful MSP, it is essential to understand the relationship between people and marine areas, as well as public perceptions of artificial structures. In this regard, if the policymakers want the public's support, they must explain the environmental benefits of artificial structures. Furthermore, the authors proposed that access to man-made marine structures, job creation, business profits, and user well-being are key societal values that should be included in the MSP process [54]. Barriers to the integrated management of marine ecosystems in MSP in Australia were pointed out by Stephenson et al. (2019, 2023) which included the following: (i) the lack of efficiency of multisectoral management; (ii) disregard for the relationship between cultural, social and economic aspects; (iii) lack of knowledge of the cumulative impact of human activities; (iv) insufficient attention to other experiences and best practices around the world; (v) the lack of supporting legal frameworks; and (vi) complexity or fragmentation in policies and legal obligations [55,56]. Domínguez-Tejo and Metternicht (2019) studied coastal planning in New South Wales, Australia. The authors identified a challenge in the planning process of the study area that arose from insufficient ecological and social data. To address this issue, the authors proposed that combining Bayesian Belief Networks and computational models with GIS could comprehensively handle data limitations/uncertainties in the MSP process while also facilitating marine management through the simulation of various management scenarios [57]. In this regard, Kobryn et al. (2018) implemented an internet-based public participation GIS system for collecting spatial information on cultural ecosystem values in the Kimberly coastline, Australia. The authors suggested that the online-public participation GIS system could be useful for including socio-ecological data as well as cultural ecosystem values in MSP, especially in long, remote coastlines with different stakeholder interests, where traditional methods such as interviews and workshops are not practical [58]. Moore et al. (2017) found that at least 30% of near-shore marine-protected areas in the Kimberly coastline, Australia, had conflict potential with other marine uses. They proposed participatory mapping as a technique for incorporating social data in planning processes alongside predicting conflict potentials. Accordingly, social data were collected by extensive face-to-face interviews for the subsequent mapping of place values. The authors concluded that biodiversity, physical landscapes and aboriginal culture are the most important values in decision making, which should be classified into both compatible and incompatible categories to understand the potential of conflict in MSP. They suggested that social data should be considered along with stakeholders' participation in MSP, especially in the case of marine-protected areas [59]. Another survey in the same study area showed that local and non-local people had generally similar perspectives in mapping values in MSP, with slightly different priorities in management preferences in MSP on the wildness value and biological/conservation value for non-local people and recreational fishing values for local people. The authors proposed that future MSP should widen representative stakeholder groups to include opinions from a larger community and, thereby, public values [60].

According to Bach et al. (2019), there are still insufficient benthic habitat maps in Australia, making it challenging to establish reserved areas for biodiversity in MSP with a habitat-based management approach. The authors investigated fish assemblage compositions along an inshore-to-offshore gradient (3–23 m depth) and in two marine reserves in Western Australia, 70 km apart. Their findings revealed that the depth and distance from the shore had a substantial effect on the richness and quantity of fish species in the two studied locations [61]. Similar results were reported by McLean et al. (2016) in the Pilbara bioregion in north-western Australia [62]. It was suggested that while developing regional marine reserves, reserved area boundaries should include the entire depth gradient as well as the entire distance from the shore to offshore in order to fully account for fish migration patterns. Stratified sample methods are also beneficial for understanding the effect of the depth and distance from the shore within a habitat type in monitoring the outcomes of MSP when selecting reserved areas for biodiversity [61,62]. Table 6 shows the adaption of MSP in Australia to the four main stages of the MSP process.

Table 6. Adaption of MSP in Australia to the main MSP stages [10,53].

Main Stages of MSP	MSP in Australia
Organising the process through pre-planning	In 1998, Australia's Oceans Policy was released and established an MSP process for the entire commonwealth marine jurisdiction. Under the EPBC Act 1999, marine bioregional plans were developed by the Department of the Environment and cover the commonwealth marine area (i.e., beyond the outer edge of state/territory waters to the seaward boundary of Australia's EEZ) in each marine region.
Defining and analysing existing conditions	The draft of the MSP plan included the analysis of existing conditions. The regional pressure analysis was informed by peer-reviewed scientific literature, and its findings were subject to external review by experts in the relevant fields. The proposed marine bioregional plan was made public for a 90-day period of public comment. The views collected from stakeholders and the general public were considered in finalising the plan. Through marine bioregional plans, the Environment Minister and the Australian government had access to comprehensive information about each marine region. The information is also available to the general public and those planning activities in the Commonwealth seas or actions that have a substantial impact on the Commonwealth marine environment.
Defining and analysing future conditions	The draft of the MSP plan includes the analysis of future conditions.
Monitoring and evaluating performance	After the first five-year review of Australia's Oceans Policy, its focus changed from a broad multiple-use perspective to an environmental one. The EPBC Act was independently reviewed in October 2009 and the final report suggested a number of changes to bioregional marine planning.

2.2.2. MSP in Canada

Canada, with a territorial sea (12 nm zone) and EEZ area of 200,000 and 2,900,000 km^2, respectively, is the most pioneering country in the world to implement the Ocean Act of 1996, the world's first comprehensive law on integrated ocean management. The aim of this law is to sustainably protect and develop the oceans [24]. Afterwards, the marine ecosystem-based management policy in Canada was published in 2002 [52]. Following that, the Canadian MSP advanced from 2011 to 2016 by approving four sub-regional MSP plans, as well as a general framework plan and implementation strategy, through the development of the Marine Plan Partnership for the Canadian Pacific North Coast (MaPP) [24]. The main drivers of MSP in Canada included ecological and biological diversity conservation, economic growth, and social objectives. As a result, 16 uses are now considered in Canada's MSP, including offshore renewable energy, oil and gas, shipping, ports, military, ammunition storage sites, fisheries, aquaculture, scientific research, coastal protection, nature conservation, disposal at sea, traditional uses, tourism and leisure, underwater cultural heritage, and cables and pipelines [27].

One of the challenges in MSP in Canada has been the lack of sufficient fine-resolution special data for assessing the suitability of a habitat for conservation [63,64]. In this regard, Ban et al. (2009) studied scientific and community-based approaches in marine-protected area sites for selection in two regions in British Colombia, Canada. The authors conducted individual and group interviews to better understand the community's priorities. As well as this, Marxan software was employed as a decision support tool to identify the scientific priorities for selecting marine-protected areas based on biotic and abiotic criteria. Their findings revealed that the final maps of community-based and scientific-based priorities were strikingly comparable. Furthermore, when spatial diversity in human impacts on marine regions and commercial fishing were considered, Marxan analyses revealed different scenarios from each constituent map [65]. According to the authors, these findings emphasise the necessity of merging community-based and scientific planning techniques into conservation initiatives in order to achieve community support and maximise conservation benefits. Furthermore, people's assessments based on traditional ecological knowledge may be a reasonable substitute for scientific methods for selecting ecologically important areas [65]. Kinlan et al. (2020) implemented the maximum entropy predictive model to map deep-sea corals in North Carolina and the Gulf of Maine on Canada's border to estimate the distribution of corals. They found that this model can be useful in mapping deep-sea biodiversity in order to assign conservation zones in MSP, particularly in areas with logistical barriers and high costs for exploration [66]. Ban et al. (2010) investigated the cumulative impacts of thirty-eight human activities on benthic, shallow pelagic and deep pelagic communities in Canada's Pacific waters. Their study showed that the entire continental shelf of the Pacific marine waters of Canada was impacted by human activities to some extent. Among all the studied activities, commercial fishing, land-based activities, and maritime transportation had the highest share in the total cumulative impacts on marine waters, with almost 57%, 19%, and 18%, respectively. The authors proposed that cumulative impact maps could be useful for planning decisions, especially in allocating conservation areas in MSP and reducing strategies in human-induced stressors [63].

Bennett et al. (2018) believed that the challenge of conflict and competition between marine uses in Canada could be exacerbated in the future, particularly given the country's commitment to global marine protection targets (i.e., 10% protection of its marine areas under the Convention on Biological Diversity). In addition, MSP should provide indigenous communities with equal access to ocean and coastal resources and spaces, particularly in the fisheries area, in order to secure their well-being under integrated coastal management. However, there is currently a lack of data and information concerning the effects of access to ocean resources on indigenous community's well-being, necessitating additional research on the subject [67].

Eger and Courtenay (2021) identified some challenges in integrated coastal and marine management in the Bay of Fundy, Atlantic Canada, including insufficient diversity in stakeholder groups, policy fragmentation and a lack of a solid structure to sustain practices. Furthermore, because Canada's current marine socio-ecological management system is sector-based, with ultimate decision-making delegated to federal and provincial departments, the challenge of inconsistent commitment by legal authorities persists, with new mandates, priorities and commitments established for each new electoral cycle and political landscape [68,69]. It has been proposed that official and informal mechanisms that facilitate the horizontal and vertical integration of policies across departments can be helpful to overcome the mentioned gaps in the management of the Bay of Fundy, Canada [68,70]. McGee et al. (2022) proposed that advisory committees with representations from regional and sub-regional interests are critical to successfully engage a diverse set of stakeholders in the planning process [71]. In this regard, an MaPP organisation was recently established in British Columbia as the first national government in Canada to achieve a collaborative MSP with different provinces. This experience is considered a successful example of MSP practice in Canada, with the goal of protecting the cultural and economic values of indigenous communities while also promoting marine conservation. In this approach, sub-regional plans were initially prepared by the First Nations and then reconciled with the provincial government and stakeholders in MaPP with the supervision of the Ministry of Forest, Lands and Natural Resources. This strategy enabled comprehensive engagement from the level of the First Nations to the Ministry on all aspects of planning and implementation while safeguarding the First Nations' governance and economy, their cultural values and activities, and resource management goals [72,73]. Table 7 shows the adaption of MSP in Canada for the four main stages of the MSP process.

Table 7. Adaption of MSP in Canada to main MSP stages [10,27].

Main Stages of MSP	MSP in the Canada
Organising the process through pre-planning	The primary legal basis supporting MSP in Canada is the Oceans Act (1996). The marine ecosystem-based management policy in Canada was published in 2002. Canadian MSP advanced from 2011 to 2016 by approving four sub-regional MSP plans, as well as developing the general framework of MaPP. The national authority in charge of MSP is Fisheries and Oceans Canada.
Defining and analysing existing conditions	Pathways of the Effect National Working Group (PoE-NWG) was developed for ecosystem-based marine planning for planners and decision-makers in Canada. In addition, a PoE model was developed to analyse a fact-based relationship between human activities and their associated sub-activities and the pressures and environmental effects or impacts they may have on a specific ecological or biological function that needs protection. Also, relevant sectors are involved throughout the MSP process with the aim of the comprehensive analysis of existing conditions.
Defining and analysing future conditions	The draft of the MSP plan includes the analysis of future conditions.
Monitoring and evaluating performance	Processes for planning revision, performance monitoring and evaluation are not defined under policy/legislation.

3. Emerging Issues for Future MSP Strategies

As MSP strategies continue to evolve, several future challenges are poised to shape their effectiveness and implementation. MSP frameworks will need to adapt to these dynamic and interconnected challenges by incorporating resilience-based approaches that integrate scientific knowledge, stakeholder perspectives and governance mechanisms to ultimately promote sustainable and resilient marine ecosystems for future generations. The future challenges for MSP strategies can be summarised as follows:

- **Prioritisation**

In theory, the ultimate goal of MSP is to achieve both "development" and "conservation" in the marine area. However, not all MSP practices have been successful in achieving this objective. In fact, in many countries, MSP has been a tool for "economic development" with little sustainability or environmental conservation considerations (soft sustainability) [24]. In some other countries, by contrast, ecosystem conservation is prioritised over economic development (hard sustainability). Along with the socio-political priorities of each country, the reason could be attributed to the total surface area of the countries' EEZ zones; that is, there is less competition for space between users in countries with the largest EEZs in the world, such as Australia (6%) and Canada (3.9%), compared to countries with a dense EEZ area, such as the Netherlands (0.1%) or Belgium (0.002%). As a result, choosing the MSP approaches (i.e., hard sustainability or soft sustainability) and challenges must be explored in the context of national borders within a broad cross-border perspective. In this regard, a continuous learning process and the incorporation of lessons learned are critical to guarantee both the sustainability and integrity of MSP [24].

- **The approach**

On the other hand, many countries mistakenly employ zoning as a substitute for MSP; nevertheless, zoning is only a management tool for implementing MSP. In fact, national MSPs should be incorporated into a broader international scale to appropriately understand the cumulative impacts of their activities on marine ecosystems at a larger scale and then employ zoning in ecologically meaningful units [14,74].

- **Data and data Standardisation**

Another challenge in implementing MSP with either soft sustainability or hard sustainability approaches is the access to high-quality data and a lack of implementation tools that can effectively encourage public discussion [36,39,74]. Moreover, conceptual ambiguities in the used terms (e.g., strategy planning versus zoning-based planning, potential areas versus exclusion areas, etc.) may need to be clarified with a glossary of terms to provide effective communication among stakeholders and for cross-border cooperation.

- **Permanent education**

It is important to note that, currently, the MSP team in many countries include policymakers, GIS specialists, fisheries experts or biologists, while MSP has a transdisciplinary nature on its own and, therefore, permanent education is critical. In this regard, developing academic degrees and training experts and specialists in this field with the knowledge of environmental conservation, environmental planning, and economic and maritime law could be an effective measure. Another strategy could be developing educational courses under the UNESCO program in order to teach how countries can apply soft or hard sustainability approaches and how to manage and use data.

- **Public engagement**

In addition to the challenges above, there is a large gap in the present MSP practices, which is the lack of integration of social components of sustainability in planning processes. Therefore, as the MSP process evolves, more social science research is required to better understand the relationship between people's perceptions, attitudes, and values, as well as the roles of various players in decision-making processes [36,42].

- **Climate change and other forms of evolution**

A key cog in re-thinking MSP for the future is that current protocols implement historical as well as predicted data on several techno-economic items, but less importance is provided on the long-term forecast of social and environmental data. Temperature, rainfall, and the probability of extreme weather events as a result of climate change, as well as the migration routes of birds/aquatic animals and other large-scale social phenomena, should be considered in MSP planning. Furthermore, since these plans have a general perspective of at least 20 years, some additional environmental relationships between sea and land should be analysed and included. For instance, the direct discharge of wastewater or runoff into the sea has the potential to affect aquaculture, fisheries, and tourism. But the need for flexibility in the relocation of activities is not just an environmental matter. On the other hand, by improving technology, new activities might arise. In this vein, better integration between coastal zone management and MSP is recommended in the future, as well as the implementation of procedures to make the planning process more flexible and able to anticipate emerging activities.

- **Scale**

It is worthwhile to note that since the major MSP protocols refer to issues and uses focused on the surface of the sea or on near-shore areas, the items devoted to ecosystem preservation are based on rules and approaches with more focus on surface waters, neglecting the impacts on deep-sea ecosystems (>200 m water depth). This region is less studied despite its significant bio–geo–chemical effects on marine ecosystems [75,76]. In addition, new threats may emerge due to the development of submarine power cables, the active exploitation of resources (for instance, floating wind farms [77]) and industrial-scale deep fishing. In conclusion, more research on techniques to make MSP flexible and to incorporate coastal, marine, and ocean management as a whole into MSP is still needed. Figure 2 summarises the main challenges and solutions in the current MSP practices based on the experiences that have been achieved so far.

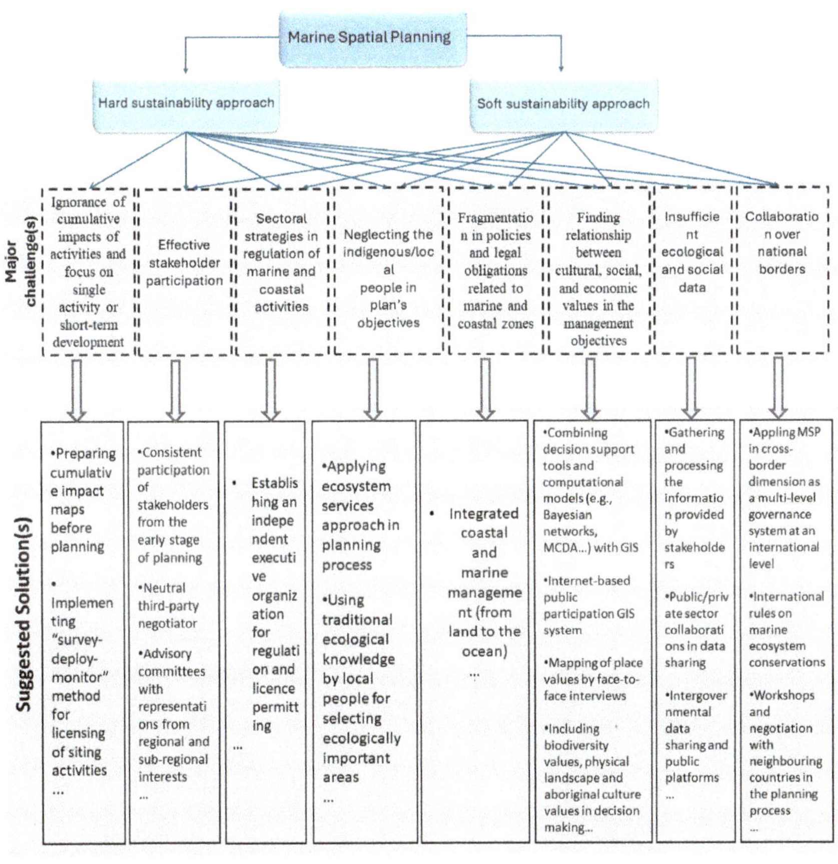

Figure 2. Major challenges and solutions in the hard and soft MSP approaches based on achieved experiences.

4. Conclusions and Recommendations

MSP is widely accepted as an essential tool for improving sustainable management and the development of marine spaces. In the meantime, different international and national Directives on MSP have been in place for more than three decades. However, MSP remains a relatively new and complex process that involves a variety of disciplines and legal contexts, as well as the engagement of various stakeholders. In this work, the study on challenges and good- practices of MSP in some pioneering countries with either soft sustainability or hard sustainability approaches (i.e., Belgium, Netherlands, Norway, Germany, United Kingdom, Australia, and Canada) highlights the following lessons in general:

1. It is essential that the academic and teaching communities provide proactive input into MSP development and implementation; this should be complemented by experiences from practical MSP implementation to inform and refine courses and training. This symbiotic relationship between academia and practical implementation is vital for several reasons. Above all, academic institutions possess valuable theoretical knowledge and research findings that can inform the design and execution of MSP. By leveraging this knowledge, practitioners can develop more robust strategies and approaches for effective MSP implementation. On the other hand, the practical experience gained through real-world MSP initiatives provides invaluable insights and lessons learned that may not be evident in theoretical frameworks alone. This experiential knowledge can help refine and validate academic theories, making them more applicable and relevant to real-world settings. Moreover, integrating practical experience into academic courses and training programs enriches the learning experience for students and professionals alike. It bridges the gap between theory and practice, equipping individuals with the skills, knowledge, and insights needed to navigate the complexities of MSP implementation effectively.
2. Providing MSP expertise and marine spatial planners who can deliver MSP in practice would increase the possibility of dealing successfully with the challenges of MSP. Trained marine spatial planners possess the necessary knowledge and skills to navigate the intricacies of MSP effectively. MSP planners are equipped to analyse marine environments, identify potential conflicts or synergies among different uses, and develop comprehensive spatial plans that balance ecological, social, and economic objectives. In addition, MSP experts and planners could enhance the capacity of decision-makers and stakeholders to make informed choices regarding marine resource management and conservation.
3. Public–private sector collaborations would be useful for sharing expensive geospatial technologies in resource-limited situations. Generally, acquiring and maintaining geospatial data can be financially burdensome, especially for organisations operating in resource-limited settings, such as developing countries or small-scale enterprises. Private sector firms often possess specialised knowledge, skills, and resources in geospatial technology development and application. Collaborating with these entities enables public organisations to leverage the expertise of industry professionals, thereby enhancing their capacity to effectively utilise geospatial tools for decision making and problem solving. In addition, leveraging public–private collaborations allows for the scalability and sustainability of geospatial initiatives. Private sector entities could invest in research and development efforts to enhance the accessibility and affordability of geospatial technologies, along with ensuring the long-term viability of geospatial applications in resource-limited contexts.
4. Permanent education by developing academic degrees on MSP or training courses under the UNESCO program could be an important step for future MSP. Students would gain a deep understanding of the principles, methodologies, and best practices of MSP through structured courses, research opportunities, and practical training. In the meantime, training courses under the auspices of UNESCO or similar international organisations could extend the capacity-building efforts of MSP globally. These courses could be designed to cater to a diverse audience, including government

officials, policymakers, researchers, and practitioners from both developed and developing countries, thereby fostering a more inclusive and collaborative approach to MSP.

More specifically, the key contributions of this study and the lessons from challenges and good practices in the studied countries highlight the following points:

5. Establishing an independent executive organisation can be useful to achieve interconnected agreements between stakeholders and the government. Furthermore, a neutral third-party negotiator could assist in reaching a compromise between the stakeholders. In fact, by operating independently of specific government departments or agencies, an independent executive organisation could prioritise MSP objectives and facilitate collaboration among stakeholders without being influenced by bureaucratic constraints or political agendas. Additionally, such an organisation could enhance transparency, accountability, and legitimacy in MSP processes by ensuring that decisions are based on scientific evidence, stakeholder input, and public participation. On the other hand, since MSP is inherently a complex process involving multiple stakeholders with competing interests, a neutral third-party negotiator could assist stakeholders in overcoming barriers to agreement, such as power imbalances, cultural differences, or different priorities. Moreover, the presence of a neutral negotiator could reduce the risk of deadlock or conflict, thereby expediting decision making and the implementation of MSP.
6. The consistent participation of stakeholders from the early stage of planning is critical for success in MSP. Early engagement allows stakeholders to contribute their insights, concerns, and priorities, which can help identify potential conflicts, opportunities, and synergies among various land and ocean uses. Moreover, by fostering ownership and buy-ins from stakeholders early on, MSP initiatives are more likely to garner support, legitimacy, and commitment throughout their planning and implementation phases. In addition, MSPs should be open to discussing new concepts, goals, and risks to remain impartial, especially in countries where there are different cultural and socio-economic backgrounds among the stakeholders. In this regard, MSP should create opportunities for participation, dialogue, and collaboration among diverse stakeholders, including indigenous communities, fishers, coastal residents, industry representatives, environmental NGOs, and government agencies. By fostering mutual understanding and cooperation among these groups, MSP initiatives could build social capital, enhance social cohesion, and promote collective action for sustainable marine governance.
7. The ecosystem services approach can provide a useful framework for connecting social and natural systems while integrating a wide range of criteria into the valuation process. Generally, ecosystem services refer to the numerous benefits that humans obtain from ecosystems, which range from providing services (e.g., food and water) to regulating services (e.g., climate regulation and flood control), as well as cultural services (e.g., recreation and spiritual enrichment). By adopting the ecosystem services approach, MSP recognises the intricate connections between ecological processes and human well-being, thereby fostering a holistic understanding of marine ecosystems and their importance to society. In addition, the ecosystem services approach expands the scope of valuation to social, cultural, and ecological dimensions. This broader perspective allows MSP practitioners to consider a wide range of criteria, including biodiversity conservation, habitat protection, cultural heritage preservation, and social equity in decision-making processes.
8. The experience in Australia demonstrates that adaptive MSP processes can achieve their goals in ecosystem conservation through the continued monitoring and evaluation of the ecosystem and the subsequent modification of plans according to changing circumstances. Ongoing monitoring and evaluation enable MSP practitioners to track changes in ecosystem health and functioning, identify emerging threats or vulnerabilities, and evaluate the effectiveness of management interventions. Since marine

ecosystems are dynamic and subject to various natural and anthropogenic pressures, MSP plans should be adaptive to accommodate uncertainties. This may involve incorporating buffer zones, setting aside areas for ecological reserves or MPAs, establishing adaptive management strategies, or implementing zoning schemes that can be adjusted over time in response to new information or changing priorities.

9. Internet-based public participation GIS systems could be a useful technique for incorporating social data in planning processes as well as predicting possible conflicts of use. Web-based mapping tools with user-friendly interfaces enable diverse stakeholders, including local communities, fishermen, recreational users, environmental groups, and government agencies, to actively participate in data collection, mapping, and decision-making processes. Internet-based public participation GIS systems can facilitate the exchange of knowledge, perspectives, and preferences among stakeholders through online surveys, mapping exercises, and interactive forums. Additionally, by overlaying social datasets onto spatial maps, stakeholders can visualise and analyse the distribution of human activities, interests, and concerns within marine areas. On the other hand, MSP practitioners can identify potential conflict hotspots and prioritise areas for further analysis or intervention by crowdsourcing information from stakeholders and mapping areas of overlapping or incompatible activities.

Aside from the lessons learnt thus far in MSP, other forthcoming challenges may include integrating the effects of climate change into the planning process and understanding how it might affect distribution and space allocation in different activities in the sea in the long-term, providing sustainable financial support for the establishment of research on MSP especially at an international level, extending MSP beyond the EEZ boundary, and ensuring effective cooperation between countries; more discussion on the future challenges of MSP can be found in [12,24,78,79].

It is recommended that international cooperation should be further developed to secure (1) an effective data-sharing experience, (2) a platform for joint legislation frameworks, and (3) properly considering the connectivity of habitats and integrity of marine ecosystems beyond national waters in the planning process. As well as this, as new sectors and industries emerge (e.g., marine renewable energies [80,81]), it is suggested to explore how they might shape MSP in the future. Finally, it is important to determine how successfully MSP can be monitored or measured since priority and planning approaches differ from country to country. It is also noteworthy to mention that the present study is based on the available reports/publications from scientific databases, and there might be scattered reports or statements about MSP challenges in the studied countries that were not included in this study due to the limited access of the authors to many of the relevant documents in the private sectors or in governmental sections. Although there have been considerable efforts for intergovernmental data sharing so far (e.g., the ocean decade website by UNESCO or MSPglobal and MSPforum by IOC-UNESCO and the European Commission), it is recommended that data sharing and public platforms be improved to share not only experiences and knowledge on MSP but also challenges and solutions in a more readily accessible and systematic manner.

Author Contributions: F.R.—Conceptualisation, methodology, investigation, original draft preparation, writing, reviewing, and editing. P.C.—Investigation, draft reviewing, and editing. D.V.—Supervision, reviewing and editing. A.A.—Supervision, draft reviewing, and editing. C.V.C.W.—writing, reviewing, and editing. J.J.—Supervision, reviewing and editing. All authors have read and agreed to the published version of the manuscript.

Funding: This research received no external funding.

Acknowledgments: P.C. and D.V. acknowledge the "Ricerca di Sistema" project (RSE,1.8 "Energia elettrica dal mare") via the Ministry of the Environment and Energy Safety. C.V.C. Weiss is grateful to the Universidad de Cantabria for the Margarita Salas Grants (RMS-04).

Conflicts of Interest: The authors declare no conflicts of interest.

References

1. Kenchington, R.A.; Day, J.C. Zoning, a fundamental cornerstone of effective Marine Spatial Planning: Lessons learnt from the Great Barrier Reef, Australia. *J. Coast. Conserv.* **2011**, *15*, 271–278. [CrossRef]
2. Socrate, J.; Verón, E. Analysis of uses and activities in the Argentine Sea. Bases for a Marine Spatial Planning in the North Argentina Basin Spatial Planning in the North Argentina Basin. *Mar. Policy* **2022**, *139*, 105014. [CrossRef]
3. United Nations Homepage. 2023. Available online: https://www.un.org/sustainabledevelopment/oceans/ (accessed on 23 December 2023).
4. Gacutan, J.; Pinarbasi, K.; Agbaglah, M.; Bradley, C.; Galparsoro, I.; Murillas, A.; Adewumi, I.; Praphotjanaporn, T.; Bordt, M.; Findlay, K.; et al. The emerging intersection between marine spatial planning and ocean accounting: A global review and case studies. *Mar. Policy* **2022**, *140*, 105055. [CrossRef]
5. Karlsson, M. Closing marine governance gaps? Sweden's marine spatial planning, the ecosystem approach to management and stakeholders' views. *Ocean. Coast. Manag.* **2019**, *179*, 104833. [CrossRef]
6. Pataki, Z.; Kitsiou, D. Marine Spatial Planning: Assessment of the intensity of conflicting activities in the marine environment of the Aegean Sea. *Ocean. Coast. Manag.* **2022**, *220*, 106079. [CrossRef]
7. Yang, S.; Fang, Q.; Ikhumhen, H.O.; Meilana, L.; Zhu, S. Marine spatial planning for transboundary issues in bays of Fujian, China: A hierarchical system. *Ecol. Indic.* **2022**, *136*, 108622. [CrossRef]
8. Galparsoro, I.; Pinarbasi, K.; Gissi, E.; Culhane, F.; Gacutan, J.; Kotta, J.; Cabana, D.; Wanke, S.; Aps, R.; Bazzucchi, D.; et al. Operationalisation of ecosystem services in support of ecosystem-based marine spatial planning: Insights into needs and recommendations. *Mar. Policy* **2021**, *131*, 104609. [CrossRef]
9. Gates, A.R.; Durden, J.M.; Richmond, M.D.; Muhando, C.A.; Khamis, Z.A.; Jones, D.O.B. Ecological considerations for marine spatial management in deep-water Tanzania. *Ocean. Coast. Manag.* **2021**, *210*, 105703. [CrossRef]
10. Ehler, C.; Douvere, F. *Marine Spatial Planning: A Step-by-Step Approach toward Ecosystem-Based Management*; Intergovernmental Oceanographic Commission and Man and the Biosphere Program. IOC Manual and Guids No. 53, ICAM Dossier No. 6; UNESCO: Paris, France, 2009.
11. Douvere, F. The importance of marine spatial planning in advancing ecosystem-based sea use management. *Mar. Policy* **2008**, *32*, 762–771. [CrossRef]
12. Douvere, F.; Ehler, C.N. The importance of monitoring and evaluation in adaptive maritime spatial planning. *J. Coast. Conserv.* **2011**, *15*, 305–311. [CrossRef]
13. Douvere, F. *Marine Spatial Planning: Concepts, Current Practice and Linkages to Other Management Approaches*; Ghent University: Ghent, Belgium, 2010.
14. Douvere, F.; Ehler, C.N. New perspectives on sea use management: Initial findings from European experience with marine spatial planning. *J. Environ. Manag.* **2009**, *90*, 77–88. [CrossRef] [PubMed]
15. Quero García, P.; García Sanabria, J.; Chica Ruiz, J.A. The role of maritime spatial planning on the advance of blue energy in the European Union. *Mar. Policy* **2019**, *99*, 123–131. [CrossRef]
16. Scarff, G.; Fitzsimmons, C.; Gray, T. The new mode of marine planning in the UK: Aspirations and challenges. *Mar. Policy* **2015**, *51*, 96–102. [CrossRef]
17. European Commission Homepage. Available online: https://environment.ec.europa.eu (accessed on 1 November 2023).
18. Douvere, F.; Maes, F.; Vanhulle, A.; Schrijvers, J. The role of marine spatial planning in sea use management: The Belgian case. *Mar. Policy* **2007**, *31*, 182–191. [CrossRef]
19. Heitmann, T.O. The Challenges for Maritime Spatial Planning in Sweden: Results from the Compliance Process with Directive 2014/89/EU on Developing and Implementing MSP. Bachelor's Thesis, University of Twente, Enschede, The Netherlands, 2018.
20. Laynesa, G. Marine Spatial Planning (MSP) in the Philippines: An Approach towards Long Term Sustainable Ocean Governance and Resolving Ruture Conflict: The Case of Balayan Bay, Batangas. Master's Thesis, World Maritime University, Malmö, Sweden, 2021.
21. McWhinnie, L.H. Aquaculture Site Selection: A GIS-based Approach to Marine Spatial Planning in Scotland. Ph.D. Thesis, Heriot-Watt University, Edinburgh, UK, 2015.
22. Röckmann, C.; Lagerveld, S.; Stavenuiter, J. Operation and maintenance costs of offshore wind farms and potential multi-use platforms in the Dutch North Sea. In *Aquaculture Perspective of Multi-Use Sites in the Open Ocean*; Springer: Cham, Switzerland, 2017. [CrossRef]
23. de Vrees, L. Adaptive marine spatial planning in the Netherlands sector of the North Sea. *Mar. Policy* **2021**, *132*, 103418. [CrossRef]
24. Ehler, C.N. Two decades of progress in Marine Spatial Planning. *Mar. Policy* **2021**, *132*, 104134. [CrossRef]
25. Kirkfeldt, T.S.; van Tatenhove, J.P.M.; Nielsen, H.N.; Vammen Larsen, S. An ocean of ambiguity in Northern European marine spatial planning policy designs. *Mar. Policy* **2020**, *119*, 104063. [CrossRef]
26. Nakornchai, P.; Bordt, M.; Pitaksereekul, N.; Praphotjanaporn, T. Asia-Pacific Marine Spatial Planning Snapshot. **2019**, *2*, 1–20. Available online: https://documents1.worldbank.org/curated/en/099813206062230702/pdf/IDU0afe34d600494f04ee009e8c0edf0292c1a96.pdf (accessed on 1 January 2024).
27. European MSP Platform. European Marine Spatial Platform. Available online: https://www.mspglobal2030.org/ (accessed on 1 January 2024).

28. Plasman, I.C. Implementing marine spatial planning: A policy perspective. *Mar. Policy* **2008**, *32*, 811–815. [CrossRef]
29. Custodio, M.; Moulaert, I.; Asselman, J.; van der Biest, K.; van de Pol, L.; Drouillon, M.; Lucas, S.H.; Taelman, S.E.; Everaert, G. Prioritizing ecosystem services for marine management through stakeholder engagement. *Ocean Coast. Manag.* **2022**, *225*, 106228. [CrossRef]
30. Eede, S.V.; Laporta, L.; Deneudt, K.; Stienen, E.; Derous, S.; Degraer, S.; Vincx, M. Marine biological valuation of the shallow Belgian coastal zone: A space-use conflict example within the context of marine spatial planning. *Ocean Coast. Manag.* **2014**, *96*, 61–72. [CrossRef]
31. Pascual, M.; Borja, A.; Eede, S.V.; Deneudt, K.; Vincx, M.; Galparsoro, I.; Legorburu, I. Marine biological valuation mapping of the Basque continental shelf (Bay of Biscay), within the context of marine spatial planning. *Estuar. Coast. Shelf Sci.* **2011**, *95*, 186–198. [CrossRef]
32. van Hees, S. Increased integration between innovative ocean energy and the EU habitats, species and water protection rules through Maritime Spatial Planning. *Mar. Policy* **2019**, *100*, 27–42. [CrossRef]
33. Steins, N.A.; Veraart, J.A.; Klostermann, J.E.M.; Poelman, M. Combining offshore wind farms, nature conservation and seafood: Lessons from a Dutch community of practice. *Mar. Policy* **2021**, *126*, 104371. [CrossRef]
34. Olsen, E.; Fluharty, D.; Hoel, A.H.; Hostens, K.; Maes, F.; Pecceu, E. Integration at the round table: Marine spatial planning in multi-stakeholder settings. *PLoS ONE* **2014**, *9*, e109964. [CrossRef] [PubMed]
35. Surís-Regueiro, J.C.; Santiago, J.L.; González-Martínez, X.M.; Garza-Gil, M.D. Estimating economic impacts linked to Marine Spatial Planning with input-output techniques. Application to three case studies. *Mar. Policy* **2021**, *129*, 104541. [CrossRef]
36. Kannen, A. Challenges for marine spatial planning in the context of multiple sea uses, policy arenas and actors based on experiences from the German North Sea. *Reg. Environ. Chang.* **2014**, *14*, 2139–2150. [CrossRef]
37. Gimpel, A.; Stelzenmüller, V.; Grote, B.; Buck, B.H.; Floeter, J.; Núñez-Riboni, I.; Pogoda, B.; Temming, A. A GIS modelling framework to evaluate marine spatial planning scenarios: Co-location of offshore wind farms and aquaculture in the German EEZ. *Mar. Policy* **2015**, *55*, 102–115. [CrossRef]
38. Stelzenmüller, V.; Diekmann, R.; Bastardie, F.; Schulze, T.; Berkenhagen, J.; Kloppmann, M.; Krause, G.; Pogoda, B.; Buck, B. Co-location of passive gear fisheries in offshore wind farms in the German EEZ of the North Sea: A first socio-economic scoping. *J. Environ. Manag.* **2016**, *183*, 794–805. [CrossRef]
39. Jay, S.; Klenke, T.; Janßen, H. Consensus and variance in the ecosystem approach to marine spatial planning: German perspectives and multi-actor implications. *Land Use Policy* **2016**, *54*, 129–138. [CrossRef]
40. Berkenhagen, J.; Döring, R.; Fock, H.O.; Kloppmann, M.H.F.; Pedersen, S.A.; Schulze, T. Decision bias in marine spatial planning of offshore wind farms: Problems of singular versus cumulative assessments of economic impacts on fisheries. *Mar. Policy* **2010**, *34*, 733–736. [CrossRef]
41. Ansong, J.; Calado, H.; Gilliland, P.M. A multifaceted approach to building capacity for marine/maritime spatial planning based on European experience. *Mar. Policy* **2021**, *132*, 103422. [CrossRef]
42. Rivers, N.; Truter, H.; Strand, M.; Jay, S.; Portman, M.; Lombard, A.; Amir, D.; Boyd, A.; Brown, R.; Cawthra, H.; et al. Shared visions for marine spatial planning: Insights from Israel, South Africa, and the United Kingdom. *Ocean Coast. Manag.* **2022**, *220*, 106069. [CrossRef]
43. Gissi, E.; de Vivero, J.L.S. Exploring marine spatial planning education: Challenges in structuring transdisciplinarity. *Mar. Policy* **2016**, *74*, 43–57. [CrossRef]
44. MASPNOSE Project Homepage. Available online: https://maritime-spatial-planning.ec.europa.eu/projects/preparatory-action-marine-spatial-planning-north-sea (accessed on 21 December 2023).
45. Adriplan Project Homepage. Available online: http://adriplan.eu/ (accessed on 11 October 2023).
46. TPEA Project Homepage. Available online: https://maritime-spatial-planning.ec.europa.eu/practices/tpea-evaluation-report (accessed on 15 October 2023).
47. Kyriazi, Z.; Maes, F.; Degraer, S. Coexistence dilemmas in European marine spatial planning practices. The case of marine renewables and marine protected areas. *Energy Policy* **2016**, *97*, 391–399. [CrossRef]
48. Wright, G. Marine governance in an industrialised ocean: A case study of the emerging marine renewable energy industry. *Mar. Policy* **2015**, *52*, 77–84. [CrossRef]
49. Brennan, J.; Fitzsimmons, C.; Gray, T.; Raggatt, L. EU Marine Strategy Framework Directive (MSFD) and Marine Spatial Planning (MSP): Which is the more dominant and practicable contributor to maritime policy in the UK? *Mar. Policy* **2014**, *43*, 359–366. [CrossRef]
50. Rodwell, L.; Fletcher, S.; Glegg, G.; Campbell, M.; Rees, S.; Ashley, M.; Linley, E.; Frost, M.; Earll, B.; Wynn, R.; et al. Marine and coastal policy in the UK: Challenges and opportunities in a new era. *Mar. Policy* **2014**, *45*, 251–258. [CrossRef]
51. Vince, J. Marine bioregional plans and implementation issues: Australia's oceans policy process. *Mar. Policy* **2013**, *38*, 325–329. [CrossRef]
52. Rodriguez, N.J.I. A comparative analysis of holistic marine management regimes and ecosystem approach in marine spatial planning in developed countries. *Ocean Coast. Manag.* **2017**, *137*, 185–197. [CrossRef]

53. Australian Government, Department of Climate Change, Energy, the Environment and Water. Available online: https://www.dcceew.gov.au/environment/marine/marine-bioregional-plans (accessed on 1 January 2024).
54. Zimmerhackel, J.; Clifton, J.; Ackermann, F.; Burton, M.; Elrick-Barr, C.; Hill, G.; Harvey, E. A framework for the integrated assessment of social and economic values associated with man-made marine structures. *Mar. Policy* **2023**, *152*, 105612. [CrossRef]
55. Stephenson, R.L.; Hobday, A.J.; Butler, I.; Cannard, T.; Cowlishaw, M.; Cresswell, I.; Cvitanovic, C.; Day, J.C.; Dobbs, K.; Dutra, L.X.; et al. Integrating management of marine activities in Australia. *Ocean Coast. Manag.* **2023**, *234*, 106465. [CrossRef]
56. Stephenson, R.L.; Hobday, A.J.; Cvitanovic, C.; Alexander, K.A.; Begg, G.A.; Bustamante, R.H.; Dunstan, P.K.; Frusher, S.; Fudge, M.; Fulton, E.A.; et al. A practical framework for implementing and evaluating integrated management of marine activities. *Ocean Coast. Manag.* **2019**, *177*, 127–138. [CrossRef]
57. Domínguez-Tejo, E.; Metternicht, G. An ecosystem-based approach and Bayesian modelling to inform coastal planning: A case study of Manly, Australia. *Environ. Sci. Policy* **2019**, *101*, 72–86. [CrossRef]
58. Kobryn, H.T.; Brown, G.; Munro, J.; Moore, S.A. Cultural ecosystem values of the Kimberley coastline: An empirical analysis with implications for coastal and marine policy. *Ocean Coast. Manag.* **2018**, *162*, 71–84. [CrossRef]
59. Moore, S.A.; Brown, G.; Kobryn, H.; Strickland-Munro, J. Identifying conflict potential in a coastal and marine environment using participatory mapping. *J. Environ. Manag.* **2017**, *197*, 706–718. [CrossRef] [PubMed]
60. Munro, J.; Pearce, J.; Brown, G.; Kobryn, H.; Moore, S.A. Identifying 'public values' for marine and coastal planning: Are residents and non-residents really so different? *Ocean Coast. Manag.* **2017**, *148*, 9–21. [CrossRef]
61. Bach, L.L.; Saunders, B.J.; Newman, S.J.; Holmes, T.H.; Harvey, E.S. Cross and long-shore variations in reef fish assemblage structure and implications for biodiversity management. *Estuar Coast Shelf Sci* **2019**, *218*, 246–257. [CrossRef]
62. McLean, D.L.; Langlois, T.J.; Newman, S.J.; Holmes, T.H.; Birt, M.J.; Bornt, K.R.; Bond, T.; Collins, D.L.; Evans, S.N.; Travers, M.J.; et al. Distribution, abundance, diversity and habitat associations of fishes across a bioregion experiencing rapid coastal development. *Estuar. Coast. Shelf Sci.* **2016**, *178*, 36–47. [CrossRef]
63. Ban, N.C.; Alidina, H.M.; Ardron, J.A. Cumulative impact mapping: Advances, relevance and limitations to marine management and conservation, using Canada's Pacific waters as a case study. *Mar. Policy* **2010**, *34*, 876–886. [CrossRef]
64. Haggarty, D.; Yamanaka, L. Evaluating Rockfish Conservation Areas in southern British Columbia, Canada using a Random Forest model of rocky reef habitat. *Estuar. Coast. Shelf Sci.* **2018**, *208*, 191–204. [CrossRef]
65. Ban, N.C.; Picard, C.R.; Vincent, A.C.J. Comparing and Integrating Community-Based and Science-Based Approaches to Prioritizing Marine Areas for Protection. *Conserv. Biol.* **2009**, *23*, 899–910. [CrossRef] [PubMed]
66. Kinlan, B.P.; Poti, M.; Drohan, A.F.; Packer, D.B.; Dorfman, D.S.; Nizinski, M.S. Predictive modeling of suitable habitat for deep-sea corals offshore the Northeast United States. *Deep. Sea Res. Part I Oceanogr. Res. Pap.* **2020**, *158*, 103229. [CrossRef]
67. Bennett, N.J.; Kaplan-Hallam, M.; Augustine, G.; Ban, N.; Belhabib, D.; Brueckner-Irwin, I.; Charles, A.; Couture, J.; Eger, S.; Fanning, L.; et al. Coastal and Indigenous community access to marine resources and the ocean: A policy imperative for Canada. *Mar. Policy* **2018**, *87*, 186–193. [CrossRef]
68. Eger, S.L.; Courtenay, S.C. Integrated coastal and marine management: Insights from lived experiences in the Bay of Fundy, Atlantic Canada. *Ocean Coast. Manag.* **2021**, *204*, 105457. [CrossRef]
69. Wiber, M.G.; Mather, C.; Knott, C.; Gómez, M.A.L. Regulating the Blue Economy? Challenges to an effective Canadian aquaculture act. *Mar. Policy* **2021**, *131*, 104700. [CrossRef]
70. Kapoor, A.; Fraser, G.S.; Carter, A. Marine conservation versus offshore oil and gas extraction: Reconciling an intensifying dilemma in Atlantic Canada. *Extr. Ind. Soc.* **2021**, *8*, 100978. [CrossRef]
71. McGee, G.; Byington, J.; Bones, J.; Cargill, S.; Dickinson, M.; Wozniak, K.; Pawluk, K.A. Marine Plan Partnership for the North Pacific Coast: Engagement and communication with stakeholders and the public. *Mar. Policy* **2022**, *142*, 104613. [CrossRef]
72. Diggon, S.; Butler, C.; Heidt, A.; Bones, J.; Jones, R.; Outhet, C. The Marine Plan Partnership: Indigenous community-based marine spatial planning. *Mar. Policy* **2021**, *132*, 103510. [CrossRef]
73. Diggon, S.; Bones, J.; Short, C.J.; Smith, J.L.; Dickinson, M.; Wozniak, K.; Topelko, K.; Pawluk, K.A. The Marine Plan Partnership for the North Pacific Coast—MaPP: A collaborative and co-led marine planning process in British Columbia. *Mar. Policy* **2022**, *142*, 104065. [CrossRef]
74. Calado, H.; Ng, K.; Johnson, D.; Sousa, L.; Phillips, M.; Alves, F. Marine spatial planning: Lessons learned from the Portuguese debate. *Mar. Policy* **2010**, *34*, 1341–1349. [CrossRef]
75. Levin, L.A.; Alfaro-Lucas, J.M.; Colaço, A.; Cordes, E.E.; Craik, N.; Danovaro, R.; Hoving, H.-J.; Ingels, J.; Mestre, N.C.; Seabrook, S.; et al. Deep-sea impacts of climate interventions. *Science* **2023**, *379*, 978–981. [CrossRef] [PubMed]
76. Carlesi, L.; Dudinskaya, E.C.; Danovaro, R.; D'Onghia, G.; Mandolesi, S.; Naspetti, S.; Zanoli, R. Estimating preferences for Mediterranean deep-sea ecosystem services: A discrete choice experiment. *Mar. Policy* **2023**, *151*, 105593. [CrossRef]
77. Rezaei, F.; Contestabile, P.; Vicinanza, D.; Azzellino, A. Towards understanding environmental and cumulative impacts of floating wind farms: Lessons learned from the fixed-bottom offshore wind farms. *Ocean Coast. Manag.* **2023**, *243*, 106772. [CrossRef]
78. Ehler, C.N. Pan-Arctic Marine Spatial Planning: An Idea Whose Time Has Come. In *Arctic Marine Governance*; Springer: Berlin/Heidelberg, Germany, 2014; pp. 199–213. [CrossRef]
79. Frazão Santos, C.; Agardy, T.; Andrade, F.; Crowder, L.B.; Ehler, C.N.; Orbach, M.K. Major challenges in developing marine spatial planning. *Mar. Policy* **2021**, *132*, 103248. [CrossRef]

80. Azzellino, A.; Lanfredi, C.; Riefolo, L.; De Santis, V.; Contestabile, P.; Vicinanza, D. Combined Exploitation of Offshore Wind and Wave Energy in the Italian Seas: A Spatial Planning Approach. *Front. Energy Res.* **2019**, *7*, 42. [CrossRef]
81. Weiss, C.V.; Ondiviela, B.; Guinda, X.; del Jesus, F.; González, J.; Guanche, R.; Juanes, J.A. Co-location opportunities for renewable energies and aquaculture facilities in the Canary Archipelago. *Ocean Coast. Manag.* **2018**, *166*, 62–71. [CrossRef]

energies

MDPI

Review

Energy Literacy: A Systematic Review of the Scientific Literature

Oscar S. Santillán [1,2] and Karla G. Cedano [2,*]

1 Subdirección de Negocios de Redes, Dirección Corporativa de Operaciones, Comisión Federal de Electricidad, Río Ródano #14, 7mo. piso, Alcaldía Cuauhtémoc, Ciudad de México 06598, Mexico; oss@ier.unam.mx
2 Instituto de Energías Renovables, Universidad Nacional Autónoma de México, Priv. Xochicalco s/n, Col. Centro, Temixco 62580, Mexico
* Correspondence: kcedano@ier.unam.mx; Tel.: +52-777-3620090

Abstract: Amidst the global energy crisis, governments are pursuing transitions towards low-carbon energy systems. In addition to physical infrastructure, political and regulatory enablers, and knowledge and capacities, changes in the energy systems require an energy-literate citizenship. Energy literacy is the understanding of how energy is generated, transported, stored, distributed, and used; awareness about its environmental and social impacts; and the knowledge to use it efficiently. The objective of the study is to provide a systematic review of the literature concerning energy literacy. In the methodology followed, the 138 papers found were categorized and subcategorized according to the research field and the main research objective, respectively. The papers are later described together with similar studies. Results show that most of the work performed around energy literacy addresses its evaluation among different groups, particularly students at different levels, and the construction, application, and evaluation of tools for improving energy literacy. Also studied are the influence of energy literacy in decision-making, its drivers, and conceptual research about the topic. The discussion highlights the debate on the link between energy-literate persons and efficient energy use, the under-researched areas of energy literacy, and the key role of energy literacy in addressing the energy crisis.

Keywords: energy literacy; energy reviews; energy-related knowledge; energy transitions; energy education

1. Introduction

Energy drives economies and sustains societies (UNEP, 2023) [1]. Energy is fundamental for human development, key to addressing several of the challenges that humanity faces and necessary to achieve the United Nations agenda for 2030 (UN & UN-Energy, 2022) [2]. On the other hand, around 75% of greenhouse gas emissions come from the energy sector, highlighting that the production and use of energy are the main drivers of current climate change (UNDP, 2023) [3].

Climate change has serious negative effects globally, making it one of the most urgent problems that people and countries face today (UN, 2023) [4]. All scientific evidence indicates that human activity is the main cause of current climate change, particularly, the burning of fossil fuels, such as coal, oil, and natural gas (UN, 2023) [4]. Despite its importance, governments and civil society need to take action not only to combat climate change, an environmental issue related to energy, but also to take actions that also address political and economic issues related to energy, such as energy security and access to domestic energy services (Heffron et al., 2015) [5]. Frequently, the instruments used to achieve goals related to these three issues compete in a phenomenon called the energy trilemma (Gunningham, 2013) [6], which can be defined as "the conflicting goals that governments face to ensure energy supply, provide universal access to energy services and promote environmental protection" (Figure 1) (Gunningham, 2013) [6].

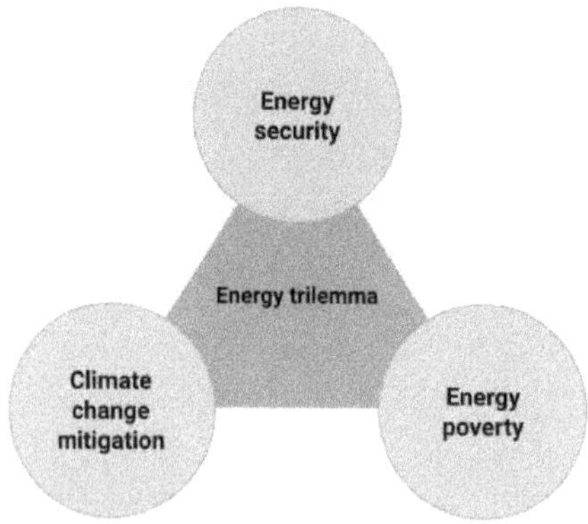

Figure 1. The energy trilemma.

The lack of access to energy services is known as energy poverty. It occurs when a household cannot ensure a level of domestic energy services that would allow the household to fully participate in the customs and activities that define membership in a specific society (Thomson et al., 2019) [7]. In the trilemma, energy poverty (sometimes included in the broader concept of energy justice) is the aspect that generally receives the least attention (Tomei & Gent, 2015) [8]. Although the incidence and intensity of the phenomenon, as well as the degree of study, vary considerably from one country to another, Latin America seriously suffers from energy poverty (Thompson et al., 2022) [9].

In general terms, energy security is the supply of adequate and stable energy to meet the demand of all economic sectors within a country (Kanchana & Unesaki, 2015) [10]. Furthermore, the energy supply must be reliable, uninterrupted, sufficient, and affordable, and for countries that export energy products, energy security also considers the certainty of demand from abroad (Kanchana & UnesakiYou have a sack, 2015) [10]. Additionally, the flexibility of systems and the diversification of energy products play an important role in energy security.

Currently, energy systems around the world are under pressure, and governments are seeking to achieve universal access to energy services, ensure energy security, and mitigate climate change, all while immersed in a global economic recession, suffering the consequences of the COVID-19 pandemic and during Russia's war in Ukraine, which may further escalate and exacerbate the global energy crisis. On the other hand, the effects of the crisis on energy markets have led to a decrease in collaboration between countries, which is a key element to achieving a net zero emissions energy system (IEA, 2022) [11].

In this regard, countries in The Organization for Economic Cooperation and Development (OECD), including Mexico, are facing security risks and using mitigation measures, such as diversification of the energy matrix as well as transitions to low-carbon energy systems (Cergibozan, 2021) [12]. Another case is found in the countries near the conflict area in Ukraine, such as Poland and Lithuania, which are also facing security risks and using energy transitions as mitigation instruments (Chomać-Pierzecka et al., 2022) [13]. In the latter case, the work of Chomać-Pierzecka et al. (2022) [13] concludes that increased social awareness determines the popularization of renewable energy solutions.

The energy crisis has also highlighted the importance of energy efficiency and behavioral measures in keeping energy supply and demand in balance. In this sense, the International Energy Agency (IEA) recognizes the crucial participation of governments, companies, and citizens to keep the increase in global temperature below 1.7 °C in the Scenario of Announced Commitments and below 1.5 °C in the Net Zero Emissions Scenario in the year 2100 (IEA, 2022) [11]. Effectively addressing the current energy crisis will require, among other things, a well-informed and participatory citizenship.

In the context of citizenship, information plays a key role for people to acquire knowledge and to find out about events that are taking place, both in their immediate environment and around the world. Access to information allows citizens to know their rights; educate themselves; and find out about health services, housing, employment alternatives and public programs and policies (Chávez, 2015) [14]. The information also enables people to understand domestic energy use, its importance for economic development, and its environmental impacts.

To refer to citizens informed in terms of energy, the concept of energy literacy is frequently used in the academic literature. DeWaters and Powers, authors with great influence in the research field of energy literacy, define it as "the citizen understanding of energy that encompasses broad knowledge of the subject, as well as affective and behavioral aspects" (DeWaters & Powers, 2013) [15]. According to Van den Broek (2019) [16], an energy literate person can be someone who knows the energy consumption of his appliances, what actions can save energy in his homes, how to make energy efficient decisions or knows the relationship between energy consumption and climate change. Thus, a person can be energy literate in one or more aspects and not so in others. In another definition, Wang and collaborators define energy literacy as the ability of people to understand the roles of energy and energy knowledge to ensure environmental sustainability (Wang et al., 2021) [17].

In the energy literacy research field, there is a gap in a review that describes how energy literacy is addressed in the scientific literature. Moreover, there are only two reviews of energy literacy, both constructed from a conceptual focus, one regarding the approaches and the other concerning the dimensions that energy literacy encompasses. This work seeks to tackle this gap by answering the research question: What are the aspects and methodologies of energy literacy that are addressed in the scientific literature? This work aims to generate knowledge regarding energy literacy and contribute to the training of energy-literate citizens, who engage in addressing the current energy crisis. Furthermore, this article can be used as a starting point to design and implement specific tools that increase energy literacy, especially in the Mexican and Latin American population. Its contribution to the field of study is related to three main points: (1) There are few systematic reviews on energy literacy in the literature. This article seeks to contribute to the topic by approaching it differently, paying special attention to the main objective of the studies. (2) The information shown has the potential to contribute to closing existing gaps in energy literacy, such as the lack of information on the subject. (3) Finally, Latin America has few publications on energy literacy. In particular, the review found only one article that explicitly mentions that it was carried out in the Mexican context.

For the review, 138 documents were analyzed, which were published in journals (110), conferences (27) and books (1). From these, seven had an empty abstract. The findings show that most of the scientific literature addresses the evaluation of energy literacy, in particular the assessment per se and the construction of measures; the construction, implementation, and evaluation of improvement tools; the influence of energy literacy in decision-making and other variables; and its research in other fields, such as education, factors and theoretical work. Most of the authors agree on the importance of energy literacy to improve energy consumption; nevertheless, there is no general consensus on the efficient use of energy among energy-literate persons.

Regarding its limitations, the bibliographic review was carried out on the platform Web of Science; therefore, the documents that are not included in this database were outside the scope of the work. This is particularly true for gray literature, that is, information that is not published in the regular media. Also, classifying documents according to their main objective is not a simple task, especially because they often have different objectives with similar importance. The classification is performed in an illustrative manner, with the aim of making information consultation more accessible. Energy literacy is dependent on economic, cultural, and sociodemographic aspects, and its evaluation is complicated. The information presented in this article has the potential to positively influence research related to energy literacy.

The article is structured as follows: in Section 2, the methodology followed to carry out the bibliographic review is described. In Section 3, the results of the review are presented and categorized according to the main objective of each reviewed article. The categories included in this section are Reviews, Measurement, Improvement Tools, Influence, and Other Lines of Research. Finally, Section 4 discusses the results from the review, the main gaps found in the literature, and what is next for this research field. The most relevant conclusions are also presented in this section.

2. Materials and Methods

The citation mining methodology is based on the application of a combination of bibliometric techniques and text mining for the analysis of bibliographic data (Kostoff et al., 2001; del Río et al., 2002) [18,19]. In this case of study, the objective has been defined as research articles on energy literacy with the following search criteria: TS = ("energy literacy" OR "energy alphabet*") written until February 2023 that are part of the Web of Science's Core Collection. These include Science Citation Index Expanded (SCI-Expanded), Social Sciences Citation Index (SSCI), Arts and Humanities Citation Index (A&HCI), Conference Proceedings Citation Index-Sciences (CPCI-S), Conference Proceedings Citation Index-Social Sciences and Humanities (CPCI-SSH), Book Citation Index-Sciences (BKCI-S), Book Citation Index-Social Sciences and Humanities (BKCI-SSH), and Citation Index from emerging sources (ESCI).

This set was analyzed using the computational tool that our research group has developed for this purpose, a text mining algorithm (Cortés et al., 2008) [20]. While the bibliometric stage is performed exclusively by counting similar data from different fields in said bibliographic records, the text mining stage uses an entropy-based algorithm to find the most relevant words in the record summaries. This algorithm is based on research conducted by Ortuño et al. (2002) [21]. The distance between two occurrences of a particular word appearing in the text of an abstract was compared to the standard deviation of all words in all abstracts. A normalized standard deviation greater than 1 indicates that the word distribution within a particular abstract is not random, allowing us to determine which words or strings of words can be considered relevant to that text.

The reasoning behind this assumption is that standard deviation is an indicator analogous to entropy (Reiss et al., 1986) [22] and can sometimes play a role as a measure of order (or disorder). The advantage of this technique is that it does not require a laborious review of individual words to extract keywords from a text but rather provides a prepared list of the most frequently occurring words and word strings, the distribution of which within a text is not random and therefore is likely to be significant. This technique has been used to analyze highly visible science topics (Russell et al., 2007) [23].

After this, a csv file containing information on all the 138 articles that appeared on the search was retrieved from the platform Web of Science. The file included the title of the paper, authors, year of publication, institution, and abstract (except for seven papers), among other data. The next step was to read all the information, with special attention to the abstracts, to ensure that all the papers in the group in fact addressed energy literacy, as well as to get acquainted with the field. This first run showed that the search captured an interesting and relevant set of articles, except for 8 that do not address energy literacy directly and 7 that had a lot of missing information, to carry out the review.

Later, the abstracts were re-read with the aim to summarize with one or few phrases the core of each of the articles. With these phrases, the principal objective of each paper was identified. After this, the papers with similar objectives were grouped. It highlighted that a significant set of papers (48 from a total of 138) dealt with the measurement of energy literacy and the evaluation, construction, comparison, and adaptation of measures. Thereby, the authors considered that for a well-structured and -described review, wider categories were needed. These categories include the research field, and, thereby, the subcategories include the main research objectives. A third run was carried out to ensure that the main objective of the article was correct; otherwise, it was replaced. Moreover, this re-read allowed us to identify the authors with more influence in the research of energy literacy and the more relevant papers within the field.

3. Results

As of February 2023, 138 articles that talk about energy literacy (defined according to the criteria mentioned in the Materials and Methods section) were found on the platform Web of Science. In the review, 6 categories were identified, including Reviews, Measurement, Improvement Tools, Influence, Other Lines of Research, and No Direct Relationship, as can be seen in Table 1. Additionally, 7 documents were found with abstracts that could not be read from the platform. As the table shows, each category is subdivided according to the main research objective pursued in the work, identified through the information provided in the abstract.

Table 1. Categories and main research objective of the analyzed articles. Information until February 2023.

Category	Main Research Objective	Number of Documents
Reviews	Review	2
Measurement	Adaptation of measures	1
	Comparison	2
	Construction of measures	6
	Assessment	35
	Measures evaluation	2
Improvement tools	Evaluation of improvement tools	17
	Improvement tools	13
Influence	Model construction	6
	Direct influence	15
Other lines of research	Education	8
	Factors	6
	Conceptual research	10
No direct relationship	No energy literacy	8
Empty	Empty	7

It is important to mention that several articles can be classified into more than one subcategory; however, they were grouped according to the main objective defined in the work abstract, sometimes implicitly. A clear example of this is given in the Measurement category, where several documents evaluate the energy literacy of specific groups of people, thus placing themselves in the Evaluation objective, while the authors also construct the evaluation measures, generating the alternative to place them in the Construction of measures objective. This is the case for most of the works that evaluate energy literacy since the way to evaluate it is also designed in the research methodology. As mentioned before, for subdivision, we take just one of the objectives.

3.1. *Reviews*

Of the articles analyzed, only two have as their main objective to carry out reviews of the scientific literature. In the first of them, published in 2019 (Van den Broek, 2019) [16], a bibliographic review of the different related approaches and methodologies is completed. The author identified four general approaches: device energy literacy, action energy literacy, financial energy literacy, and multifaceted energy literacy. This is one of the few articles that discusses approaches and not aspects of energy literacy as most of the literature does.

In the second review article, published three months later in February 2020 (Martins et al., 2019) [24], the authors also carry out a literature review on energy literacy. Unlike the first, this review is built around the work of DeWaters and Powers (2013) [15] on the dimensions that energy literacy encompasses: knowledge, emotional or affective, and behavioral. Additionally, the authors identified financial knowledge as another important aspect of energy literacy.

3.2. Measurement

Until the date of the review, 47 documents were found with a main objective related to the measurement of energy literacy. These were subdivided into five classes of objectives: Adaptation of measures, Comparison, Construction of measures, Assessment, and Measures Evaluation. Of the group of documents analyzed, only one had the main objective of adapting a methodology (Güven et al., 2019) [25], in which an energy literacy evaluation scale originally designed in English is adjusted to the Turkish context. The reliability of the tool is also validated.

On the other hand, two documents made a comparison of energy literacy, one between different universities in the same country (UK) and another between different countries (UK and China). Both works have several authors in common, including the main author. In the first (Cotton et al., 2017) [26], a comparison of energy literacy is carried out between university students in different positions of the Green league in the United Kingdom; in the second, energy literacy is evaluated among students from universities in China and the United Kingdom (Cotton et al., 2020) [27].

Regarding the Construction of measures, DeWaters and Powers (2013) [15] propose explicit criteria to develop measurable objectives related to energy literacy in three dimensions: cognitive, affective or emotional, and behavioral. At the same time, these authors and other colleagues (DeWaters et al., 2013) [28] published their work on the development of a measurement scale for evaluating energy literacy and its application to middle and high school students in the United States. The instrument developed in these two works used psychometric principles of educational and social psychology, making the work have, since its publication and to date, a significant influence on the study of energy literacy; several subsequent publications derive from the works of DeWaters, Powers, Qaqish and Graham.

Continuing with the objective Construction of measures, Turner et al. (2014) [29] designed a survey to evaluate energy literacy related to electricity; Yusup et al. (2017) [30] built a structure for the energy literacy assessment of future physics teachers; Martins et al. (2020) [31] developed an energy literacy index as well as an index for each of the dimensions knowledge, financial calculations, attitudes, and behavior; and Das and Richman (2022) [32] built and applied a public instrument to measure energy literacy in three dimensions, including cognitive, attitudinal, and behavioral. It should be noted that in some works, such as that of Das and Richmam and that of Martins et al., the affective or emotional dimension is replaced by that of attitudes, although the criteria that characterize them are similar.

Most articles on energy literacy evaluate it among different groups of people, notably focusing on students of different levels. As mentioned above, in several of the works, the tools to carry out the evaluation are also designed. Table 2 shows the 35 documents with the main objective of evaluating energy literacy identified in the literature review. The table shows the region, country, state, or city; the population evaluated; and the reference to the document. Of these, several analyze the factors that influence energy literacy (in italics the reference in the table), including the gender or sex of the people evaluated (in italics and bold the reference in the table), of which the work of Lee et al. [33–36] stands out.

Table 2. Documents with the main objective of evaluating energy literacy. Data until February 2023.

Region	Population	Reference
New York State	Middle and high school students	DeWaters and Powers, 2008 [37]
New York State	Middle and high school students	DeWaters and Powers, 2011 [38]
New York State	Secondary students	DeWaters and Powers, 2011 [39]
State of Pennsylvania	Urban 8th grade students	Bodzin, 2012 [40]
Netherlands	Private homes	Brounen et al., 2013 [41]

Table 2. *Cont.*

Region	Population	Reference
New Zealand	Children (9–10 years)	Aguirre-Bielschowsky et al., 2015 [42]
Taiwan	Vocational high school students	***Lee et al., 2015*** [33]
Taiwan	Secondary students	***Lee et al., 2015*** [34]
United Kingdom	University students	Cotton et al., 2015 [43]
Denmark	General population	Sovacool and Blyth, 2015 [44]
Denmark	General population	Sovacool, 2016 [45]
Greece	High school students	Keramitsoglou, 2016 [46]
Taiwan	First year high school students	Yeh et al., 2017 [47]
Indonesia	Future physics teachers	Yusup et al., 2017 [48]
Indonesia	Future physics teachers	Yusup, et al., 2018 [49]
Portugal	University community	Martins et al., 2019 [50]
Taiwan	Nursing students	*Lee et al., 2019* [35]
State of Virginia	Net-Zero building residents	Paige et al., 2019 [51]
Poland	General population	***Gołębiowska, 2020*** [52]
Portugal	University community	***Martins et al., 2020*** [53]
China	Peasant tourist houses	Zhang and Zhang, 2020 [54]
Bilbao, Spain	University community	Lasuen et al., 2020 [55]
Taiwan	Adult population	Hsu, 2020 [56]
Nepal	Urban homes	Filippini et al., 2020 [57]
Indonesia	Students and teachers of different levels	Laliyo, 2020 [58]
Poland	Rural community	Chodkowska-Miszczuk et al., 2021 [59]
Mashhad, Iran	General population	Sayarkhalaj and Khesal, 2022 [60]
Poland	University students	***Białynicki-Birula et al., 2022*** [61]
Vietnam	High school students (12th grade)	*Lee et al., 2022* [36]
South Africa	General population	Force and Longe, 2022 [62]
Brazil and Belgium	University students	*Franco et al., 2022* [63]
State of California	Energy users	Zanocco et al., 2022 [64]
New York State	General population	Gervich, 2022 [65]
China	Ethnic residents	Wu et al., 2022 [66]
Arizona State	Future primary school teachers	Merritt et al., 2023 [67]

From the documents, it stands out that Cotton et al. (2015) [43] suggest measures to improve energy literacy among students, while Yeh et al. (2017) [47] identify some misconceptions that they have about energy. Paige et al. (2019) [51] perform an analysis on how energy literacy affects energy consumption in buildings that do not behave like Net-Zero, even though they were designed that way. Zhang and Zhang (2020) [54] mention the importance of analyzing in more detail the energy literacy results obtained from questionnaires by conducting interviews. Zanocco et al. (2022) [64] introduce the concept of "load shape" when analyzing energy literacy, considering the daily timing of electricity demand. Wu et al. (2022) [66] explore the relationship between residents' energy literacy and sustainable tourism in ethnic areas of China.

There are two documents for which the main objective is related to the evaluation of energy literacy measurements. That is, they analyze their effectiveness and applicability. Langfitt et al. (2015) [68] analyzed the applicability of an energy literacy measure based on competency, course deliverables, or artifacts. Van der Horst et al. (2015) [69] evaluated the pedagogical aspects of field work carried out by university students in their homes when evaluating energy technologies and habits.

3.3. Improvement Tools

Of the documents analyzed, 30 have as their main objective the analysis of tools to improve energy literacy. Thirteen are related to their design and application, and seventeen evaluate the effectiveness of these tools. As in the case of evaluations, in this category, there are also several works that design, apply, and evaluate improvement tools. It highlights that several of the tools are intended to improve the energy literacy of children or young students. Some others are intended for energy users or the general population.

For example, Huang et al. (2012) [70] designed and implemented an energy literacy program aimed at elementary school students, and similarly, Merritt et al. (2019) [71] developed a curriculum on energy resources aimed at fourth grade students. Jeng et al. (2013) [72] set out to teach children concepts related to energy literacy through computer games, and Fraternali and Gonzalez (2019) [73] describe an augmented reality tool to improve energy-saving behavior in children.

Chen et al. (2013) [74] propose a framework for energy education that captures the concept of carbon saving and reduction in Taiwan. Tarabieh et al. (2015) [75] designed a user-friendly data interface to simplify and support the implementation of energy literacy programs on a university campus. Wahyudi et al. (2019) [76] sought to improve energy literacy related to geothermal energy in Indonesia through information contained in vocational high school textbooks. Ilmi et al. (2021) [77] designed and implemented a program to improve energy literacy in high school students and evaluated energy literacy before and after the implementation of the program.

Regarding the general population, Moret et al. (2014) [78] created an energy calculator within an online learning platform with the purpose of supporting citizens and decision-makers to understand the energy system; Moreno et al. (2015) [79] sought to increase energy literacy through a user-centered building management system, which provides personalized actions to save energy; Zapico and Hedin (2017) [80] created an interactive tool to increase energy literacy; and Mogles et al. (2017) [81] analyzed the effect of providing more detailed information in smart energy meters on energy and monetary savings. Spence et al. (2017) [82] developed a tool with the purpose of motivating the occupants of a workspace with energy data and supporting them to take actions that reduce energy consumption.

Table 3 shows a synthesis of the documents for which the main objective is to evaluate tools to improve energy literacy. This contains the type of tool, the population to which it was applied, a summary of the evaluation carried out, and the reference. As can be seen in the table, the tools identified in the review include education programs (formal, aimed at students at different levels, and informal, for the general population), serious games, presentation of information, interactive activities, web pages, and technologies, such as smart meters and augmented reality. The populations that use the tools are mostly students, the general population, and people with some type of specific technology, such as smart meters.

Table 3. Energy literacy assessment tools, target population, type of assessment, and reference. Data until February 2023.

Tool	Population	Evaluation Summary	Reference
Formal education	High school students	Comprehensive evaluation of the impacts that an energy module had on knowledge, attitudes, behaviors, and self-efficacy.	DeWaters and Powers (2006) [83]
Formal education	8th grade high school students	Evaluation of the effects of two programs, one with a focus on geospatial technologies and the other Business as Usual.	Bodzin et al. (2013) [84]
Serious games	Serious game users	Definition of evaluation criteria of the impacts of serious games.	Wood et al. (2014) [85]
Formal education	Secondary students	Evaluation of the hypothesis that project-based energy learning does not improve energy-related knowledge, attitudes, behavior, and beliefs.	Karpudewan et al. (2015) [86]
Presentation of information	General population	Analysis of the interpretation of the electricity bill according to different ways of displaying the information.	Canfield et al. (2016) [87]

Table 3. *Cont.*

Tool	Population	Evaluation Summary	Reference
Digital serious games	University students	Analysis of locus of control effects on behavioral intention and performance in game-based energy learning.	Yang et al. (2017) [88]
Presentation of information	General population	Analysis of how people understand energy information and interpret feedback through different ways of viewing data on smart meters.	Herrmann et al. (2017) [89]
Presentation of information	University database	Evaluating the impact of different forms of data visualization on learning about household energy consumption.	Herrmann et al. (2017) [90]
Interactive activities	Engineering students	Evaluation of the effects of an interactive learning tool immediately, one week and six months after using it.	Hedin and Zapico (2018) [91]
Smart meters	Population with energy consumption monitors	Evaluation of the effects of home energy monitors after 10 years.	Snow et al. (2019) [92]
Augmented reality	General population	Evaluation of four augmented reality methods for representing energy consumption in air conditioners.	García-Manzano et al. (2019) [93]
Informal education	General population	Evaluating the effects of two versions of a one-hour museum visit: collaborative or competitive.	Applebaum et al. (2021) [94]
Formal education	Fourth grade primary school students	Analysis of the effectiveness of a service-based learning program.	Rimm-Kaufman et al. (2021) [95]
Informal education	University students	Evaluation of the impact of an energy awareness campaign.	Ntouros et al. (2021) [96]
Web pages	General population	Self-assessment of improvement in knowledge after interacting with two web pages: one animated and one static.	Henni et al. (2022) [97]
Formal education	Middle and high school students	Evaluation of the effects of a workshop in the short (few days) and long term (one year) after having participated.	Keller et al. (2022) [98]
Informal education	Building occupants	Analysis of the impact of different educational interventions on energy.	Ramallo-González et al. (2022) [99]

Some evaluations explicitly analyze the impacts that tools have on energy-related knowledge, attitudes, and behavior, such as DeWaters and Powers (2006) [83] and Karpudewan et al. (2015). [86] Other documents describe the impacts of the tools in different periods of time (Hedin and Zapico (2018) [91], Snow et al. (2019) [92] and Keller et al. (2022) [98]). Several of the works investigate the improvement in energy literacy using different educational programs, of which the research carried out by Bodzin et al. stands out (2013) [84], in which a program with a focus on geospatial technologies is compared to the Business as Usual, and that of Rimm-Kaufman et al. (2021) [95], with the evaluation of the impacts of a service-based energy learning program, also against a Business as Usual scenario.

3.4. *Influence*

Within the group of documents reviewed, a category is proposed that includes works that investigate the influence of energy literacy on people's decision-making or behavior. This categorization was divided into two subsets: Direct Influence and Model Construction. The first subset includes articles for which the main objective is related to analyzing how energy literacy affects other aspects of the lives of people and societies, while the second contains documents in which energy literacy is used as a variable in the construction of models designed to explain and predict phenomena related to knowledge, environment, and energy consumption.

One of the aspects in which the influence of energy literacy is most addressed in literature, of course, is energy use. For example, this topic is addressed in the context of energy consumption of employees working in educational, health, and government buildings (Medojevic et al., 2016) [100]; electricity consumption (Blasch et al., 2017) [101]; and excessive energy consumption (Broberg and Kažukauskas, 2020) [102], both at the household level. It is also analyzed in the context of the barriers to the provision of solar energy systems in homes (Thomas et al., 2021) [103] and the implementation of renewable energies (Mehmood et al., 2022) [104]; activities that can change schedule due to the cost of energy and impacts on the environment (Walker and Hope, 2020) [105]; the willingness to adopt temporary tariffs on energy consumption (Reis et al., 2021) [106]; and the motivations and barriers to increase flexibility in individual electricity demand (Bohdanowicz et al., 2021) [107].

Similarly, the acceptance of energy generation technologies is analyzed (Sherren et al., 2019) [108], and preferences for efficient devices when knowing their emissions (He et al., 2022) [109] and refrigerators according to their energy characteristics (Olsthoorn et al., 2023) [110] are also assessed. Other contexts within which the influence of energy literacy is analyzed include community renewable energy projects (Cloke et al., 2017) [111], consumer awareness viewing efficiency labels when purchasing household appliances (He et al., 2022a) [112], and referring to the heuristics in household energy use (Van den Broek and Walker, 2019) [113] and the heuristics of energy experts (Kantenbacher and Attari, 2021) [114].

Regarding model construction, Mogles et al. (2017) [115] analyze the way in which different variables relate to each other and affect energy consumption. Within the same context, Satre-Meloy (2019) [116] applies different regularization series to regression models to analyze electricity consumption using information contained in a survey; Motz (2021) [117] analyzes the impact that demographic, behavioral, and attitudinal factors have on preferences regarding the price, origin, and reliability of electricity supply; and Reis et al. (2022) [118] analyze the behavior of a community with prosumers (a word that indicates those consumers involved in energy generation and storage activities) and vulnerable consumers. Chen et al. (2015) [119] evaluate the interaction that exists between knowledge, attitudes, self-efficacy, and behaviors related to energy, and Wang et al. (2021) [17] analyze the dynamics of environmental regulation, including regulatory authorities, companies, and civil society.

3.5. Other Lines of Research

In addition to the construction and application of measurement tools as well as the evaluation and impact of energy literacy in different sectors, there are other lines of research that are also frequently addressed in the literature. Through the review carried out, three main lines of research were found: Factors, referring to those variable elements that influence energy literacy; Education; exploring how energy-related teaching and learning are addressed in different educational programs; and Conceptual research, analyzing some concepts related to the study of energy literacy.

Regarding research related to factors, works were identified that explore the impact of practical activities among students in their final year of high school (Lin and Lu, 2018) [120] and the field of education (Martins et al., 2019) [121] on the level of energy literacy. Also, the effects of technological factors, such as smart energy sensors, are analyzed (de Leon Barido et al., 2018) [122]; the effects of geographical and contextual factors (such as socioeconomic situation, social practices and access to fuel and appliances) on children (Lusinga and de Groot, 2019) [123] and the effects of prices and transparency in the balance of markets (Numminen et al., 2022) [124] on energy literacy are also assessed. On the other hand, the common determinants among education, financial knowledge, and energy literacy among members of a university in Portugal are studied (Martins et al., 2022) [125].

The only work that explicitly mentions that the research was developed within the Mexican context (relevant due to the nationality and institutions of the authors) has to do with education (Castañeda-Garza and Valerio-Ureña, 2022) [126]. The document describes the review of related content with energy literacy in textbooks for primary education in Mexico. Also related to national education, Bogovic et al. (2013) [127] present the activities developed within the framework of two competitions on energy literacy in primary and secondary schools in Slovenia; Balouktsis and Kekkeris (2013) [128] describe the aspects of energy-related education at different levels of education in Greece; and Mažeikienė and Norkutė (2021) [129] analyze how energy topics are presented in the geography curriculum in Lithuania.

In other works related to energy literacy in education, Cotton et al. (2017) [130] analyze the potential risks of directing energy saving efforts solely to behavioral change and individual actions without increasing knowledge on energy issues; Van der Horst and Staddon (2017) [131] analyze the relationship that exists between research on energy behavior and management of energy demand in the home and educational research on learning processes;Salvia et al. (2020) [132] investigate the extent to which energy sustainability is considered in educational programs and dissemination activities in 36 universities around the world; and Pestana et al. (2021) [133] investigate the impact of some citizen energy education initiatives.

The last of the three main lines of research identified in the literature is on important concepts related to the analysis of energy literacy. Some of these works propose methodological frameworks that include different or more specific aspects than those normally included in the study of energy literacy. For example, Kavčič and Drevenšek (2014) [134] describe a methodological framework for energy literacy that includes nuclear energy; Lowan-Trudeau and Fowler (2021) [135] propose the concept of critical energy literacy, which considers social, environmental, political, economic, and technological aspects; and Gladwin and Ellis (2023) [136] propose a conceptual framework on energy literacy taking into account theoretical ideas and concepts to understand energy holistically.

Also, documents that analyze energy literacy among specific groups or sectors were identified. Such is the case of the works of Aguirre-Bielschowsky et al. (2018) [137], who investigate how nine- and ten-year-old children learn about electricity and how they consume it; Adams et al. (2022) [138] investigate the concept of energy literacy among vulnerable users, paying special attention to the specific dynamics of each place; and Plets and Kuijt (2022) [139], who analyze the meaning and impact of private financing in the Dutch heritage and museum sectors, and assess how it affects energy literacy.

Other works related to the conceptual research of energy literacy address the issues of analyzing the depth of energy knowledge gaps (Holasova, 2018) [140]; analysis of how to display information on the energy consumption of electrical devices in monetary terms, rather than in physical units, increases the likelihood that an individual will make a calculation and identify the device with the lowest cost throughout its life cycle (Blasch et al., 2019) [141]; promoting the sociocultural aspects of energy literacy as a basis for energy and climate justice (Gladwin et al., 2022) [142]; and the analysis of the contribution that energy cooperatives have when implementing and disseminating programs that promote energy literacy (Meira et al., 2022) [143].

4. Discussion

Amidst several global challenges, the world is facing an energy crisis. To face it, governments aim to guarantee universal energy access and maintain energy security. At the same time, there is a pressing need to mitigate climate change, and this requires reducing the burning of fossil fuels—oil, natural gas, and coal—which can make it difficult to achieve energy security and universal access to energy services. Furthermore, the negative effects of climate change, such as natural disasters and extreme temperatures, are adding extra pressure to global energy systems, making the current situation even more complicated.

One of the possible solutions is the energy transition (or transitions, since each country or region will have one), in which the aim is to move from an energy matrix based on fossil fuels to a low-carbon matrix with energy efficiency, renewable sources, and distributed generation as pillars. In this sense, variable renewable energies (such as solar photovoltaic and wind) require large investments for the digitalization of electricity transmission and distribution networks; however, decentralized electricity generation models can eliminate this obstacle and leverage local capabilities to create socio-technical energy systems where communities produce, manage, and generate their energy.

According to the International Renewable Energy Agency (IRENA), energy transitions around the world require (1) physical infrastructure; (2) public policy and regulatory enablers; and (3) knowledge and capabilities (IRENA, 2023) [144]. Renewable energies and clean technologies are not only available but also affordable. However, the speed at which energy decentralization is progressing is still very low. One of the reasons for this is the lack of knowledge on the technology and its economic and environmental advantages. Getting people to adopt technology, no matter how attractive it may seem, requires overcoming the threshold of mistrust and resistance to change. This is only possible through strategies that bring knowledge closer to people. This knowledge, which should not be theoretical, must be internalized by people so that it moves them to action. Without this current knowledge, technological adoption will continue to advance slowly, and we will fail to address the climate emergency due to the generation of electricity with fossil and polluting sources (O'Neill-Carrillo et al., 2018) [145]. Therefore, the change in energy systems will require a well-informed and participatory citizenship, who understands the importance of transitions and can be part of the decisions when investing in infrastructure, encouraging a low-carbon matrix through public policies, and of course, being able to participate first-hand as trained human resources. It is within this context that the research field of energy literacy develops.

In the scientific literature, there are several definitions of energy literacy. Broadly speaking, it can be said that energy literacy is the understanding of how energy is generated, transported, stored, distributed, and used; awareness about the environmental and social impacts of its generation and use; and the knowledge to use it efficiently in the different sectors of the economy. The present work seeks to contribute to the study of energy literacy by answering the research question, what are the aspects and methodologies on energy literacy that are addressed in the scientific literature? To answer it, a systematic review of the literature was carried out, and the documents were grouped according to the main objective they pursue.

According to the review carried out, the research question can be answered by saying that most of the scientific knowledge on energy literacy addresses the evaluation of energy literacy in different populations, in particular students of different levels, and the construction, application, and evaluation of tools to improve energy literacy. Other frequent topics studied are the influence of energy literacy on decision-making, the factors that impact the level of energy literacy, and the conceptual study of energy literacy. It is important to mention that the objective of this work is to propose a qualitative classification of the articles that address the topic of energy literacy and then broadly describe the contribution they have to this field of research. However, this classification can be improved and adapted to incorporate other types of documents, for example those that address more than one main objective.

As for the correlation with the other two reviews on the subject, this work is similar to the one carried out by Martins et al. (2020) [24], although with different approaches. They also organize the literature but focus on the dimensions of knowledge, affectivity, and behavior. The researchers conclude that most of the documents seek to assess the level of energy literacy, as it is noted in the present paper. Our work aimed to go further, defining categories and subcategories and describing the main features of each work together with similar studies. The review performed by Van Den Broek (2019) [16] varies from this one as it organizes the literature in a conceptual and methodological manner. For this reason, they discuss and conclude their findings from another perspective.

The bibliographic review carried out also allows us to grasp the importance of energy literacy in populations, so that final consumers can make efficient use of energy and they can participate in decision-making regarding energy technologies generation, transmission, and distribution of energy vectors. Above all, from the authors' point of view, it is important that people can participate in energy transitions worldwide, especially the energy transition specific to their sociocultural context. Many things are required to have just energy systems, and energy literacy is a fundamental factor in achieving them.

From the review, it can be said that energy literacy is a research topic that has gained strength in recent years. Most researchers agree on its potential to promote efficient energy consumption and ensure sustainable development. For example, Lee et al. (2015) [34] state that energy literacy can empower people to make wise decisions and take responsible actions. In the same direction, Gołębiowska (2020) [52] mentions that increasing energy literacy can be key to ensuring sustainable development in the coming years. Martins and collaborators (2020) [53] expect that basic knowledge about energy and financial knowledge will serve to provide citizens with the tools and motivation to make the necessary changes and achieve efficient consumption habits.

However, the same authors mention that most studies show that even in populations that have adequate knowledge about energy, this knowledge does not transform into a strong motivation leading to behavioral changes (Martins et al., 2020) [53]. Although in most of the documents reviewed there is consensus on the importance of energy literacy, the disagreements are whether energy literacy translates into efficient energy consumption. In this regard, Adams et al. (2022) [138] mention that several researchers have found that the data suggest that there is no correlation between an increase in energy literacy and an increase in sustainability. This finding coincides with the statement of Van den Broek (2019) [16], who says that the literature shows little evidence of the impact of energy literacy on energy consumption habits.

Although the way energy is used may not be strongly related to the level of knowledge, an informed and energy-literate citizen is more likely to participate in decision-making processes and will be better prepared to choose and act responsibly on energy issues (DeWaters and Powers, 2013) [15]. The authors of the present paper consider that regardless of whether the increase in knowledge about energy makes an impact on actions aimed at having efficient energy consumption, energy literacy is essential for people who seek to improve their energy use habits and contribute to sustainable development can do it effectively. Most likely, knowing the environmental and social impacts of the generation, transportation, and use of energy can encourage people to change their consumption habits.

One of the main gaps found in the literature is how to encourage energy-literate persons to use energy efficiently. This is a significant issue since it is necessary not only to improve energy literacy among the population but also for energy-literate citizens to actively participate in tackling the energy crisis. Another important gap is the heuristics on energy consumption, with few papers researching this field (Van Den Broek & Walker, 2019 [113] and Kantenbacher & Attari, 2021 [114]). There are also gaps when analyzing the energy poverty—energy literacy link. In this regard, Białynicki-Birula et al. (2022) [61] acknowledges energy poverty as one of the main factors determining energy literacy. However, energy literacy has the potential to aid within the homes facing energy poverty, and there is a clear lack of research on this link. Finally, there is a gap in the literature concerning the research carried out in Latin American countries. We hope that this paper can bring the discussion to these countries, and especially, that this work can contribute to the implementation of improvement tools.

The present work seeks to fill a gap in the literature concerning a review on how energy literacy is addressed in the scientific literature. It contributes to the research field by providing a systematic arrangement of the documents researching energy literacy and describing its main features and contributions. The work also highlights a non-consensus topic: the link between energy-literate persons and efficient energy use. Finally, this paper presents important aspects of energy literacy that are under-researched and can be used as a starting point to undertake these fields. As for future work, it is important that spaces be built to improve energy literacy, especially in Latin America, both with formal education programs at different educational levels and with non-formal education projects to be able to access the majority of the population. Also, it is necessary to build tools for the evaluation of energy literacy in people and know what the main issues are to improve. Another important research opportunity is found in the correlation between energy literacy and the efficient use of energy, particularly on the mechanisms that can improve this relationship and ensure that people with advanced knowledge about energy are precisely those who consume it efficiently.

The energy crisis poses a risk for homes and societies, and energy literacy plays a key role in addressing this problem. Energy-literate people can make efficient use of energy, participate and promote energy-related decision-making, and contribute to the construction of more just energy futures. It is essential that governments and the private sector, particularly energy companies, become involved in improving the level of energy literacy of civil society; however, academia must be the protagonist in this sense. Scientists working on energy issues must actively participate in developing citizenship with a high level of energy literacy who are capable of participating in an informed manner in decisions related to energy and exhibit sustainable energy consumption, both individually as well as collectively.

Author Contributions: Conceptualization, O.S.S.; methodology, O.S.S. and K.G.C.; software, K.G.C.; validation, O.S.S. and K.G.C.; formal analysis, O.S.S.; investigation, O.S.S. and K.G.C.; resources, O.S.S. and K.G.C.; data curation, O.S.S.; writing—original draft preparation, O.S.S.; writing—review and editing, O.S.S. and K.G.C.; visualization, O.S.S.; supervision, O.S.S. and K.G.C.; project administration, O.S.S. All authors have read and agreed to the published version of the manuscript.

Funding: The author O.S.S. thanks the Mexican National Council of Humanities, Sciences and Technologies (CONAHCyT) for the postdoctoral grant (application number 4198764) provided for the completion of this work.

Data Availability Statement: No new data were created or analyzed in this study. Data sharing is not applicable to this article.

Conflicts of Interest: The authors declare no conflict of interest.

References

1. UNEP (United Nations Environment Programme). Why Does Energy Matter? 2023. Available online: https://www.unep.org/explore-topics/energy/why-does-energy-matter (accessed on 15 February 2023).
2. UN (United Nations); UN-Energy. UN-Energy Highlights Implementing the UN-Energy Plan of Action towards 2025. 2022. Available online: https://www.un.org/sites/un2.un.org/files/un-energy_highlights-092222.pdf (accessed on 1 July 2023).
3. UNDP (United Nations Development Programme). Energy and Climate. 2023. Available online: https://www.undp.org/energy/our-work-areas/energy-and-climate (accessed on 15 February 2023).
4. UN (United Nations). What Is Climate Change? 2023. Available online: https://www.un.org/en/climatechange/what-is-climate-change (accessed on 23 January 2023).
5. Heffron, R.J.; McCauley, D.; Sovacool, B.K. Resolving society's energy trilemma through the Energy Justice Metric. *Energy Policy* **2015**, *87*, 168–176. [CrossRef]
6. Gunningham, N. Managing the energy trilemma: The case of Indonesia. *Energy Policy* **2013**, *54*, 184–193. [CrossRef]
7. Thomson, H.; Simcock, N.; Bouzarovski, S.; Petrova, S. Energy poverty and indoor cooling: An overlooked issue in Europe. *Energy Build.* **2019**, *196*, 21–29. [CrossRef]
8. Tomei, J.; Gent, D. Equity and the Energy Trilemma Delivering Sustainable Energy Access in Low-Income Communities. International Institute for Environment and Development. 2015. Available online: https://www.iied.org/sites/default/files/pdfs/migrate/16046IIED.pdf (accessed on 1 July 2023).
9. Thomson, H.; Day, R.; Ricalde, K.; Brand-Correa, L.I.; Cedano, K.; Martinez, M.; Santillán, O.; Triana, Y.D.; Cordova, J.G.L.; Gómez, J.F.M.; et al. Understanding, recognizing, and sharing energy poverty knowledge and gaps in Latin America and the Caribbean—because conocer es resolver. *Energy Res. Soc. Sci.* **2022**, *87*, 102475. [CrossRef]
10. Kanchana, K.; Unesaki, H. Assessing Energy Security Using Indicator-Based Analysis: The Case of ASEAN Member Countries. *Soc. Sci.* **2015**, *4*, 1269–1315. [CrossRef]
11. IEA (International Energy Agency). World Energy Outlook 2022. 2022. Available online: https://www.iea.org/reports/world-energy-outlook-2022 (accessed on 1 June 2023).

12. Cergibozan, R. Renewable energy sources as a solution for energy security risk: Empirical evidence from OECD countries. *Renew. Energy* **2021**, *183*, 617–626. [CrossRef]
13. Chomać-Pierzecka, E.; Sobczak, A.; Urbańczyk, E. RES Market Development and Public Awareness of the Economic and Environmental Dimension of the Energy Transformation in Poland and Lithuania. *Energies* **2022**, *15*, 5461. [CrossRef]
14. Chávez, H.A.R. Información y ciudadanía, una propuesta desde la gobernanza. *Investig. Bibl. Arch. Bibl. E Inf.* **2015**, *29*, 113–140. [CrossRef]
15. DeWaters, J.; Powers, S. Establishing Measurement Criteria for an Energy Literacy Questionnaire. *J. Environ. Educ.* **2013**, *44*, 38–55. [CrossRef]
16. Van den Broek, K.L. Household energy literacy: A critical review and a conceptual typology. *Energy Res. Soc. Sci.* **2019**, *57*, 57. [CrossRef]
17. Wang, M.; Hou, G.; Wang, P.; You, Z. Research of energy literacy and environmental regulation research based on tripartite deterrence game model. International Conference on Energy Engineering and Power Systems (EEPS). *Energy Rep.* **2021**, *7*, 1084–1091. [CrossRef]
18. del Río, J.A.; Garcia, E.O.; Ramirez, A.M.; Humenik, J.A. Phenomenological approach to profile impact of scientific research: Citation mining. *Adv. Complex Syst.* **2002**, *5*, 19–42. [CrossRef]
19. Kostoff, R.N.; del Río, J.A.; Humenik, J.A.; García, E.O.; Ramírez, A.M. Citation mining: Integrating text mining and bibliometrics for research user profiling. *J. Am. Soc. Inf. Sci. Technol.* **2001**, *52*, 1148–1156. [CrossRef]
20. Cortés, H.D.; del Rio, J.A.; Garcia, E.O.; Robles, M. Web application to profiling scientific institutions through citation mining. *Int. J. Comput. Informat. Eng.* **2008**, *2*, 3213–3217.
21. Ortuño, M.; Carpena, P.; Bernaola-Galván, P.; Muñoz, E.; Somoza, A.M. Keyword detection in natural languages and DNA. *EPL Europhys. Lett.* **2002**, *57*, 759–764. [CrossRef]
22. Reiss, H.; Hammerich, A.D.; Montroll, E.W. Thermodynamic treatment of nonphysical systems: Formalism and an example (Single-lane traffic). *J. Stat. Phys.* **1986**, *42*, 647–687. [CrossRef]
23. Russell, J.M.; Del Río, J.A.; Cortés, H.D. Highly visible science: A look at three decades of research from Argentina, Brazil, Mexico and Spain. *Interciencia* **2007**, *32*, 629–634.
24. Martins, A.; Madaleno, M.; Dias, M.F. Energy literacy: What is out there to know? *Energy Rep.* **2019**, *6*, 454–459. [CrossRef]
25. Güven, G.; Yakar, A.; Sülün, Y. Adaptation of the energy literacy scale into turkish: A validity and reliability study. *Cukurova Univ. Facult. Educ. J.* **2019**, *48*, 821–857.
26. Cotton, D.R.E.; Winter, J.; Miller, W.; Valle, L.D. Is students' energy literacy related to their university's position in a sustainability ranking? *Environ. Educ. Res.* **2017**, *24*, 1611–1626. [CrossRef]
27. Cotton, D.; Zhai, J.; Miller, W.; Valle, L.D.; Winter, J. Reducing energy demand in China and the United Kingdom: The importance of energy literacy. *J. Clean. Prod.* **2020**, *278*, 123876. [CrossRef]
28. De Waters, J.; Qaqish, B.; Graham, M.; Powers, S. Designing an Energy Literacy Questionnaire for Middle and High School Youth. *J. Environ. Educ.* **2013**, *44*, 56–78. [CrossRef]
29. Turner, M.; Foreman, C.; Perusich, K. Development of an electric energy literacy survey. In Proceedings of the 2014 International Energy and Sustainability Conference (IESC), New York, NY, USA, 5–6 October 2015; pp. 1–5.
30. Yusup, M.; Setiawan, A.; Rustaman, N.Y.; Kaniawati, I. Developing a Framework for The Assessment of Pre-Service Physics Teachers' Energy Literacy. *J. Phys.* **2017**, *877*, 012014. [CrossRef]
31. Martins, A.; Madaleno, M.; Dias, M.F. Energy literacy assessment among Portuguese university members: Knowledge, attitude, and behavior. *Energy Rep.* **2020**, *6*, 243–249. [CrossRef]
32. Das, R.R.; Richman, R. The Development and Application of a Public Energy Literacy Instrument. *Can. J. Sci. Math. Technol. Educ.* **2022**, *22*, 42–67. [CrossRef]
33. Lee, L.-S.; Chang, L.-T.; Lai, C.-C.; Guu, Y.-H.; Lin, K.-Y. Energy literacy of vocational students in Taiwan. *Environ. Educ. Res.* **2015**, *23*, 855–873. [CrossRef]
34. Lee, L.-S.; Lee, Y.-F.; Altschuld, J.W.; Pan, Y.-J. Energy literacy: Evaluating knowledge, affect, and behavior of students in Taiwan. *Energy Policy* **2015**, *76*, 98–106. [CrossRef]
35. Lee, L.-S.; Lee, Y.-F.; Wu, M.-J.; Pan, Y.-J. A study of energy literacy among nursing students to examine implications on energy conservation efforts in Taiwan. *Energy Policy* **2019**, *135*, 111005. [CrossRef]
36. Lee, Y.-F.; Nguyen, H.B.N.; Sung, H.-T. Energy literacy of high school students in Vietnam and determinants of their energy-saving behavior. *Environ. Educ. Res.* **2022**, *28*, 907–924. [CrossRef]
37. De Waters, J.; Powers, S. Energy literacy among middle and high school youth. In Proceedings of the 38th Annual 2008 IEEE Frontiers in Education Conference (FIE), Saratoga Springs, NY, USA, 22–25 October 2008; pp. T2F-6–T2F-11.
38. DeWaters, J.E.; Powers, S.E. Energy literacy of secondary students in New York State (USA): A measure of knowledge, affect, and behavior. *Energy Policy* **2011**, *39*, 1699–1710. [CrossRef]
39. DeWaters, J.E.; Powers, S.E. Improving energy literacy among middle school youth with project-based learning pedagogies. In Proceedings of the 2011 Frontiers in Education Conference (FIE), Rapid City, SD, USA, 13–15 October 2011; pp. T1D-1–T1D-7.
40. Bodzin, A. Investigating Urban Eighth-Grade Students' Knowledge of Energy Resources. *Int. J. Sci. Educ.* **2012**, *34*, 1255–1275. [CrossRef]

41. Brounen, D.; Kok, N.; Quigley, J.M. Energy literacy, awareness, and conservation behavior of residential households. *Energy Econ.* **2013**, *38*, 42–50. [CrossRef]
42. Aguirre-Bielschowsky, I.; Lawson, R.; Stephenson, J.; Todd, S. Energy literacy and agency of New Zealand children. *Environ. Educ. Res.* **2015**, *23*, 832–854. [CrossRef]
43. E Cotton, D.R.; Miller, W.; Winter, J.; Bailey, I.; Sterling, S. Developing students' energy literacy in higher education. *Int. J. Sustain. High. Educ.* **2015**, *16*, 456–473. [CrossRef]
44. Sovacool, B.K.; Blyth, P.L. Energy and environmental attitudes in the green state of Denmark: Implications for energy democracy, low carbon transitions, and energy literacy. *Environ. Sci. Policy* **2015**, *54*, 304–315. [CrossRef]
45. Sovacool, B.K. Differing cultures of energy security: An international comparison of public perceptions. *Renew. Sustain. Energy Rev.* **2016**, *55*, 811–822. [CrossRef]
46. Keramitsoglou, K.M. Exploring adolescents' knowledge, perceptions and attitudes towards Renewable Energy Sources: A colour choice approach. *Renew. Sustain. Energy Rev.* **2016**, *59*, 1159–1169. [CrossRef]
47. Yeh, S.-C.; Huang, J.-Y.; Yu, H.-C. Analysis of Energy Literacy and Misconceptions of Junior High Students in Taiwan. *Sustainability* **2017**, *9*, 423. [CrossRef]
48. Yusup, M.; Setiawan, A.; Rustaman, N.Y.; Kaniawati, I. Assessing Pre-Service Physics Teachers' Energy Literacy: An Application of Rasch measurement. *J. Physics: Conf. Ser.* **2017**, *895*, 012161. [CrossRef]
49. Yusup, M.; Setiawan, A.; Rustaman, N.; Kaniawati, I. Pre-service Physics Teachers' Knowledge, Decision Making, and Self-system Toward Energy Conservation. *J. Pendidik. Fis. Indones.* **2018**, *14*, 60–64. [CrossRef]
50. Martins, A.; Madaleno, M.; Dias, M.F. Energy Literacy: Knowledge, affect, and behavior of university members in Portugal. In Proceedings of the 2019 16th International Conference on the European Energy Market (EEM), Ljubljana, Slovenia, 18–20 September 2019.
51. Paige, F.; Agee, P.; Jazizadeh, F. flEECe, an energy use and occupant behavior dataset for net-zero energy affordable senior residential buildings. *Sci. Data* **2019**, *6*, 1–9. [CrossRef] [PubMed]
52. Gołębiowska, B. Energy literacy in Poland. *Econ. Environ.* **2020**, *73*, 23. [CrossRef]
53. Martins, A.; Madaleno, M.; Dias, M.F. Financial Knowledge's Role in Portuguese Energy Literacy. *Energies* **2020**, *13*, 3412. [CrossRef]
54. Zhang, J.; Zhang, Y. Examining the energy literacy of tourism peasant households in rural tourism destinations. *Asia Pac. J. Tour. Res.* **2020**, *25*, 441–456. [CrossRef]
55. Lasuen, U.O.; Iragorri, M.A.O.; Diez, J.R. Towards energy transition at the Faculty of Education of Bilbao (UPV/EHU): Diagnosing community and building. *Int. J. Sustain. High. Educ.* **2020**, *21*, 1277–1296. [CrossRef]
56. Hsu, Y.-C. A Pilot Study to Incorporate Collaboration and Energy Competency into an Engineering Ethics Course. *Educ. Sci.* **2020**, *10*, 72. [CrossRef]
57. Filippini, M.; Kumar, N.; Srinivasan, S. Energy-related financial literacy and bounded rationality in appliance replacement attitudes: Evidence from Nepal. *Environ. Dev. Econ.* **2020**, *25*, 399–422. [CrossRef]
58. Laliyo, L.A.R.; Puluhulawa, F.U.; Eraku, S.; Salimi, Y.K. The Prevalence of Students and Teachers' Ideas about Global Warming and the Use of Renewable Energy Technology. *J. Environ. Account. Manag.* **2020**, *8*, 243–256. [CrossRef]
59. Chodkowska-Miszczuk, J.; Kola-Bezka, M.; Lewandowska, A.; Martinát, S. Local Communities' Energy Literacy as a Way to Rural Resilience—An Insight from Inner Peripheries. *Energies* **2021**, *14*, 2575. [CrossRef]
60. Sayarkhalaj, H.; Khesal, M.F. Investigating energy literacy and its structural model for citizens of Mashhad. *Heliyon* **2022**, *8*, e11449. [CrossRef] [PubMed]
61. Białynicki-Birula, P.; Makieła, K.; Mamica, Ł. Energy Literacy and Its Determinants among Students within the Context of Public Intervention in Poland. *Energies* **2022**, *15*, 5368. [CrossRef]
62. Force, T.S.; Longe, O.M. Impact of Energy Literacy on Energy Consumption, Expenditure and Management. In Proceedings of the 2022 IEEE Nigeria 4th International Conference on Disruptive Technologies for Sustainable Development (NIGERCON), Lagos, Nigeria, 5–7 April 2022; pp. 1–5.
63. Franco, D.; Macke, J.; Cotton, D.; Paço, A.; Segers, J.-P.; Franco, L. Student energy-saving in higher education tackling the challenge of decarbonisation. *Int. J. Sustain. High. Educ.* **2022**, *23*, 1648–1666. [CrossRef]
64. Zanocco, C.; Sun, T.; Stelmach, G.; Flora, J.; Rajagopal, R.; Boudet, H. Assessing Californians' awareness of their daily electricity use patterns. *Nat. Energy* **2022**, *7*, 1191–1199. [CrossRef]
65. Gervich, C.D. The Impact of Extremely Low Electric Rates on Energy Conservation, Planning and Resilience: A Case Study of Plattsburgh, New York. *Public Work. Manag. Policy* **2022**, *28*, 476–517. [CrossRef]
66. Wu, S.; Li, Y.; Fang, C.; Ju, P. Energy Literacy of Residents and Sustainable Tourism Interaction in Ethnic Tourism: A Study of the Longji Terraces in Guilin, China. *Energies* **2022**, *16*, 259. [CrossRef]
67. Merritt, E.G.; Weinberg, A.E.; Archambault, L. Exploring Energy Through the Lens of Equity: Funds of Knowledge Conveyed Through Video-Based Discussion. *Int. J. Sci. Math. Educ.* **2023**, 1–24. [CrossRef]
68. Langfitt, Q.; Haselbach, L.; Hougham, R.J. Artifact-Based Energy Literacy Assessment Utilizing Rubric Scoring. *J. Prof. Issues Eng. Educ. Pr.* **2015**, *141*, C5014002. [CrossRef]
69. van der Horst, D.; Harrison, C.; Staddon, S.; Wood, G. Improving energy literacy through student-led fieldwork—At home. *J. Geogr. High. Educ.* **2015**, *40*, 67–76. [CrossRef]

70. Huang, Y.; Chou, Y.-C.; Yen, H.-W.; Bai, H.-C. Developing an Innovative Educational Program for Energy Saving and Carbon Reduction: An Elementary School Example. *Procedia Soc. Behav. Sci.* **2012**, *51*, 840–848. [CrossRef]
71. Merritt, E.G.; Bowers, N.; Rimm-Kaufman, S.E. Making connections: Elementary students' ideas about electricity and energy resources. *Renew. Energy* **2019**, *138*, 1078–1086. [CrossRef]
72. Jeng, C.C.; Shin, C.Y.; Ho, W.C.; Han, C. A Computer Game Designed for Helping Kids to Make Informed Energy Decision. In Proceedings of the 2013 International Conference on Advanced Computer Science and Electronics Information (ICACSEI 2013), Beijing, China, 25–26 August 2013; Atlantis Press: Dordrecht, The Netherlands, 2013; pp. 403–407.
73. Fraternali, P.; Gonzalez, S.L.H. An Augmented Reality Game for Energy Awareness. In *Computer Vision Systems: 12th International Conference, ICVS 2019, Thessaloniki, Greece, 23–25 September 2019*; Springer International Publishing: Berlin/Heidelberg, Germany, 2019; pp. 629–638.
74. Chen, K.-L.; Huang, S.-H.; Liu, S.-Y. Devising a framework for energy education in Taiwan using the analytic hierarchy process. *Energy Policy* **2013**, *55*, 396–403. [CrossRef]
75. Tarabieh, K.A.; Elnabarawy, I.O.; Mashaly, I.A.; Rashed, Y.M. The power of data visualization: A prototype energy performance map for a university campus. In *Sustainable Human–Building Ecosystems*; American Society of Civil Engineers: Reston, VA, USA, 2015; pp. 194–203. [CrossRef]
76. Wahyudi, W.; A Pambudi, N.; Biddinika, M.K.; Ranto, R.; Rudiyanto, B. Readability of geothermal energy information in vocational textbooks. *J. Phys. Conf. Ser.* **2019**, *1402*, 044060. [CrossRef]
77. Ilmi, N.; Sanjaya, L.A.; Budi, A.S.; Astra, I.M.; Puspa, D.R.W.; Dinata, F.A.; Putri, R.A.; Winarko, H.B.; Pertiwi, W.A.; Rasmi, D.P. Project based learning: Model electric power plants MaS WaWi (biomass, sun, water, and wind) to improve student energy literacy. *AIP Conf. Proc.* **2021**, *2912*, 010001.
78. Moret, S.; Codina Gironès, V.; Maréchal, F.; Favrat, D. Swiss-energyscope. ch: A platform to widely spread energy literacy and aid decision-making. In Proceedings of the 17th Conference on Process Integration, Modelling and Optimisation for Energy Saving and Pollution Reduction (PRES 2014), Prague, Czech Republic, 23–27 August 2014; Volume 39, pp. 877–882. [CrossRef]
79. Moreno, M.V.; Zamora, M.A.; Skarmeta, A.F. An IoT based framework for user-centric smart building services. *Int. J. Web Grid Serv.* **2015**, *11*, 78. [CrossRef]
80. Zapico, J.L.; Hedin, B. Energy weight: Tangible interface for increasing energy literacy. In Proceedings of the 2017 Sustainable Internet and ICT for Sustainability (SustainIT), Funchal, Portugal, 6–7 December 2017.
81. Mogles, N.; Walker, I.; Ramallo-González, A.P.; Lee, J.; Natarajan, S.; Padget, J.; Gabe-Thomas, E.; Lovett, T.; Ren, G.; Hyniewska, S.; et al. How smart do smart meters need to be? *Build. Environ.* **2017**, *125*, 439–450. [CrossRef]
82. Spence, A.; Goulden, M.; Leygue, C.; Banks, N.; Bedwell, B.; Jewell, M.; Yang, R.; Ferguson, E. Digital energy visualizations in the workplace: The e-Genie tool. *Build. Res. Inf.* **2017**, *46*, 272–283. [CrossRef]
83. DeWaters, J.; Powers, S. Work in Progress: A Pilot Study to Assess the Impact of a Special Topics Energy Module on Improving Energy Literacy of High School Youth. Proceedings. Frontiers in Education. In Proceedings of the 36th Annual Conference Frontiers in Education, San Diego, CA, USA, 27–31 October 2006; pp. 11–12.
84. Bodzin, A.M.; Fu, Q.; Peffer, T.E.; Kulo, V. Developing Energy Literacy in US Middle-Level Students Using the Geospatial Curriculum Approach. *Int. J. Sci. Educ.* **2013**, *35*, 1561–1589. [CrossRef]
85. Wood, G.; van der Horst, D.; Day, R.; Bakaoukas, A.G.; Petridis, P.; Liu, S.; Jalil, L.; Gaterell, M.; Smithson, E.; Barnham, J.; et al. Serious games for energy social science research. *Technol. Anal. Strat. Manag.* **2014**, *26*, 1212–1227. [CrossRef]
86. Karpudewan, M.; Ponniah, J.; Zain, A.N.M. Project-Based Learning: An Approach to Promote Energy Literacy Among Secondary School Students. *Asia-Pacific Educ. Res.* **2015**, *25*, 229–237. [CrossRef]
87. Canfield, C.; de Bruin, W.B.; Wong-Parodi, G. Perceptions of electricity-use communications: Effects of information, format, and individual differences. *J. Risk Res.* **2016**, *20*, 1132–1153. [CrossRef]
88. Yang, J.C.; Lin, Y.L.; Liu, Y.-C. Effects of locus of control on behavioral intention and learning performance of energy knowledge in game-based learning. *Environ. Educ. Res.* **2017**, *23*, 886–899. [CrossRef]
89. Herrmann, M.R.; Brumby, D.P.; Oreszczyn, T. Watts your usage? A field study of householders' literacy for residential electricity data. *Energy Effic.* **2017**, *11*, 1703–1719. [CrossRef]
90. Herrmann, M.R.; Brumby, D.P.; Oreszczyn, T.; Gilbert, X.M.P. Does data visualization affect users' understanding of electricity consumption? *Build. Res. Inf.* **2017**, *46*, 238–250. [CrossRef]
91. Hedin, B.; Zapico, J.L. What Can You Do with 100 kWh? A Longitudinal Study of Using an Interactive Energy Comparison Tool to Increase Energy Awareness. *Sustainability* **2018**, *10*, 2269. [CrossRef]
92. Snow, S.; Viller, S.; Glencross, M.; Horrocks, N. Where Are They Now? Revisiting energy use feedback a decade after deployment. In Proceedings of the Ozchi'19: 31st Australian Conference on Human-Computer-Interaction, New York, NY, USA, 2–5 December 2019; pp. 397–401.
93. García-Manzano, R.; Ramallo-González, A.P.; Sanchez-Iborra, R. Energy Representation Tool for Air Conditioning that Enhance Energy Savings and Improve Energy Literacy on Users. In Proceedings of the 16th IBPSA Conference, Rome, Italy, 2–4 September 2019. [CrossRef]
94. Applebaum, L.R.; Price, C.A.; Foster, A.Y. Collaboration and Competition in Exhibit Facilitation Around Energy Literacy. *J. Mus. Educ.* **2021**, *46*, 113–126. [CrossRef]

95. Rimm-Kaufman, S.E.; Merritt, E.G.; Lapan, C.; De Coster, J.; Hunt, A.; Bowers, N. Can service-learning boost science achievement, civic engagement, and social skills? A randomized controlled trial of Connect Science. *J. Appl. Dev. Psychol.* **2021**, *74*, 101236. [CrossRef]
96. Ntouros, V.; Romanowicz, J.; Charalambous, C.; Kousis, I.; Laskari, M.; Assimakopoulos, M.N. Empowering Students to Save Energy through a Behavioural Change Campaign in University Accommodation. In Proceedings of the Resilience and economic intelligence through digitalizationand big data analytics, Bucharest, Romania, 10–11 June 2021. [CrossRef]
97. Henni, S.; Franz, P.; Staudt, P.; Peukert, C.; Weinhardt, C. Evaluation of an interactive visualization tool to increase energy literacy in the building sector. *Energy Build.* **2022**, *266*, 112116. [CrossRef]
98. Keller, L.; Riede, M.; Link, S.; Hüfner, K.; Stötter, J. Can Education Save Money, Energy, and the Climate?—Assessing the Potential Impacts of Climate Change Education on Energy Literacy and Energy Consumption in the Light of the EU Energy Efficiency Directive and the Austrian Energy Efficiency Act. *Energies* **2022**, *15*, 1118. [CrossRef]
99. Ramallo-González, A.P.; Bardaki, C.; Kotsopoulos, D.; Tomat, V.; Vidal, A.G.; Ruiz, P.J.F.; Gómez, A.S. Reducing Energy Consumption in the Workplace via IoT-Allowed Behavioural Change Interventions. *Buildings* **2022**, *12*, 708. [CrossRef]
100. Medojevic, M.; Medojevic, M.; Petrovic, J.; Delic, M. Energy Consumption Indicators of Public Facilities in AP Vojvodina with Focus on Human Factor and Factors Influencing Energy-Saving Behavior of Employees. In Proceedings of the 6th International Conference on Energy Research and Development, Kuwait City, Kuwait, 14–16 March 2016; pp. 17–24.
101. Blasch, J.; Boogen, N.; Filippini, M.; Kumar, N. Explaining electricity demand and the role of energy and investment literacy on end-use efficiency of Swiss households. *Energy Econ.* **2017**, *68*, 89–102. [CrossRef]
102. Broberg, T.; Kažukauskas, A. Information policies and biased cost perceptions—The case of Swedish residential energy consumption. *Energy Policy* **2020**, *149*, 112095. [CrossRef]
103. Thomas, P.; Sandwell, P.; Williamson, S.; Harper, P. A PESTLE analysis of solar home systems in refugee camps in Rwanda. *Renew. Sustain. Energy Rev.* **2021**, *143*, 110872. [CrossRef]
104. Mehmood, A.; Lin, R.; Zhang, L.; Lee, C.K.; Ren, J. Qualitative mapping of barriers to the renewables' development against energy literacy dimensions: A case study of Pakistan. *Energy Rep.* **2022**, *8*, 332–337. [CrossRef]
105. Walker, I.; Hope, A. Householders' readiness for demand-side response: A qualitative study of how domestic tasks might be shifted in time. *Energy Build.* **2020**, *215*, 109888. [CrossRef]
106. Reis, I.F.G.; Lopes, M.A.R.; Antunes, C.H. Energy literacy: An overlooked concept to end users' adoption of time-differentiated tariffs. *Energy Effic.* **2021**, *14*, 1–28. [CrossRef]
107. Bohdanowicz, Z.; Kowalski, J.; Kobyliński, P. Engaging Electricity Users in Italy, Denmark, Spain, and France in Demand-Side Management Solutions. In Proceedings of the Conference on Multimedia, Interaction, Design and Innovation, Warsaw, Poland, 9–10 December 2021; Cham: Geneva, Switzerland, 2021; pp. 171–178.
108. Sherren, K.; Parkins, J.R.; Owen, T.; Terashima, M. Does noticing energy infrastructure influence public support for energy development? Evidence from a national survey in Canada. *Energy Res. Soc. Sci.* **2019**, *51*, 176–186. [CrossRef]
109. He, S.; Blasch, J.; van Beukering, P. How does information on environmental emissions influence appliance choice? The role of values and perceived environmental impacts. *Energy Policy* **2022**, *168*, 113142. [CrossRef]
110. Olsthoorn, M.; Schleich, J.; Guetlein, M.-C.; Durand, A.; Faure, C. Beyond energy efficiency: Do consumers care about life-cycle properties of household appliances? *Energy Policy* **2023**, *174*, 113430. [CrossRef]
111. Cloke, J.; Mohr, A.; Brown, E. Imagining renewable energy: Towards a Social Energy Systems approach to community renewable energy projects in the Global South. *Energy Res. Soc. Sci.* **2017**, *31*, 263–272. [CrossRef]
112. He, S.; Blasch, J.; van Beukering, P.; Wang, J. Energy labels and heuristic decision-making: The role of cognition and energy literacy. *Energy Econ.* **2022**, *114*, 106279. [CrossRef]
113. Broek, K.L.v.D.; Walker, I. Heuristics in energy judgement tasks. *J. Environ. Psychol.* **2019**, *62*, 95–104. [CrossRef]
114. Kantenbacher, J.; Attari, S.Z. Better rules for judging joules: Exploring how experts make decisions about household energy use. *Energy Res. Soc. Sci.* **2021**, *73*, 101911. [CrossRef]
115. Mogles, N.; Padget, J.; Gabe-Thomas, E.; Walker, I.; Lee, J. A computational model for designing energy behaviour change interventions. *User Model. User-Adapted Interact.* **2017**, *28*, 1–34. [CrossRef] [PubMed]
116. Satre-Meloy, A. Investigating structural and occupant drivers of annual residential electricity consumption using regularization in regression models. *Energy* **2019**, *174*, 148–168. [CrossRef]
117. Motz, A. Consumer acceptance of the energy transition in Switzerland: The role of attitudes explained through a hybrid discrete choice model. *Energy Policy* **2021**, *151*, 112152. [CrossRef]
118. Reis, I.F.; Gonçalves, I.; Lopes, M.A.; Antunes, C.H. Towards inclusive community-based energy markets: A multiagent framework. *Appl. Energy* **2022**, *307*, 118115. [CrossRef]
119. Chen, S.-J.; Chou, Y.-C.; Yen, H.-Y.; Chao, Y.-L. Investigating and structural modeling energy literacy of high school students in Taiwan. *Energy Effic.* **2015**, *8*, 791–808. [CrossRef]
120. Lin, K.-Y.; Lu, S.-C. Effects of project-based activities in developing high school students' energy literacy. *J. Balt. Sci. Educ.* **2018**, *17*, 867–877. [CrossRef]
121. Martins, A.; Madaleno, M.; Dias, M.F. Energy Literacy: Does education field matter? In Proceedings of the TEEM'19: Technological Ecosystems for Enhancing Multiculturality, Salamanca, Spain, 19–21 October 2019; pp. 494–499. [CrossRef]

122. Barido, D.P.d.L.; Suffian, S.; Kammen, D.M.; Callaway, D. Opportunities for behavioral energy efficiency and flexible demand in data-limited low-carbon resource constrained environments. *Appl. Energy* **2018**, *228*, 512–523. [CrossRef]
123. Lusinga, S.; de Groot, J. Energy consumption behaviours of children in low-income communities: A case study of Khayelitsha, South Africa. *Energy Res. Soc. Sci.* **2019**, *54*, 199–210. [CrossRef]
124. Numminen, S.; Ruggiero, S.; Jalas, M. Locked in flat tariffs? An analysis of electricity retailers' dynamic price offerings and attitudes to consumer engagement in demand response. *Appl. Energy* **2022**, *326*, 120002. [CrossRef]
125. Martins, A.; Madaleno, M.; Dias, M.F. Are the energy literacy, financial knowledge, and education level faces of the same coin? *Energy Rep.* **2022**, *8*, 172–178. [CrossRef]
126. Castañeda-Garza, G.; Valerio-Ureña, G. Energy literacy in elementary school textbooks in Mexico. *Environ. Educ. Res.* **2022**, *29*, 410–422. [CrossRef]
127. Bogovic, K.; Jarkovic, T.; Kavcic, M.L.; Kosinac, G. The Young in the World of Energy—A Communication Project to Promote and Educate on Energy-related Topics Among the Younger Generations. In Proceedings of the 22nd International Conference Nuclear Energy for New Europe, (NENE 2013), Bled, Slovenia, 9–12 September 2013. Available online: https://arhiv.djs.si/proc/nene2013/pdf/NENE2013_1311.pdf (accessed on 1 July 2023).
128. Balouktsis, I.; Kekkeris, G. Energy education in Greece: Learning about renewable electrical energy perspectives. In Proceedings of the 2013 24th EAEEIE Annual Conference (EAEEIE 2013), Chania, Greece, 30–31 May 2013; pp. 128–132. [CrossRef]
129. Mažeikienė, N.; Norkutė, O. Toward a new energy paradigm in geography: Revisiting the curriculum and teaching practices. *J. Educ. Cult. Soc.* **2021**, *12*, 131–150. [CrossRef]
130. Cotton, D.; Miller, W.; Winter, J.; Bailey, I.; Sterling, S. Knowledge, agency and collective action as barriers to energy-saving behaviour. *Local Environ.* **2015**, *21*, 883–897. [CrossRef]
131. van der Horst, D.; Staddon, S. Types of learning identified in reflective energy diaries of post-graduate students. *Energy Effic.* **2017**, *11*, 1783–1795. [CrossRef]
132. Salvia, A.L.; Brandli, L.L.; Filho, W.L.; Rebelatto, B.G.; Reginatto, G. Energy sustainability in teaching and outreach initiatives and the contribution to the 2030 Agenda. *Int. J. Sustain. High. Educ.* **2020**, *21*, 1607–1624. [CrossRef]
133. Pestana, C.; Barros, L.; Scuri, S.; Barreto, M. Can HCI Help Increase People's Engagement in Sustainable Development? A Case Study on Energy Literacy. *Sustainability* **2021**, *13*, 7543. [CrossRef]
134. Kavčič, M.L.; Drevenšek, M. Improving Nuclear Energy Literacy through e-Communication: The eWorld Web Project. In Proceedings of the 23RD International Conference Nuclear Energy for New Europe, (NENE 2014), Portorož, Slovenia, 8–11 September 2014. Available online: https://arhiv.djs.si/proc/nene2014/pdf/NENE2014_204.pdf (accessed on 1 June 2023).
135. Lowan-Trudeau, G.; Fowler, T.A. Towards a theory of critical energy literacy: The Youth Strike for Climate, renewable energy and beyond–CORRIGENDUM. *Aust. J. Environ. Educ.* **2021**, *38*, 58–68. [CrossRef]
136. Gladwin, D.; Ellis, N. Energy literacy: Towards a conceptual framework for energy transition. *Environ. Educ. Res.* **2023**, 1–15. [CrossRef]
137. Aguirre-Bielschowsky, I.; Lawson, R.; Stephenson, J.; Todd, S. Kids and Kilowatts: Socialisation, energy efficiency, and electricity consumption in New Zealand. *Energy Res. Soc. Sci.* **2018**, *44*, 178–186. [CrossRef]
138. Adams, J.; Kenner, A.; Leone, B.; Rosenthal, A.; Sarao, M.; Boi-Doku, T. What is energy literacy? Responding to vulnerability in Philadelphia's energy ecologies. *Energy Res. Soc. Sci.* **2022**, *91*, 102718. [CrossRef]
139. Plets, G.; Kuijt, M. Gas, Oil and Heritage: Well-oiled Histories and Corporate Sponsorship in Dutch Museums (1990–2021). *BMGN—Low Ctries. Hist. Rev.* **2022**, *137*, 50–77. [CrossRef]
140. Holasova, A. Energy Literacy in the Czech Republic. *Energy Ecol. Econ.* 2018. Available online: https://www.webofscience.com/wos/woscc/full-record/WOS:000485101200016 (accessed on 1 September 2023).
141. Blasch, J.; Filippini, M.; Kumar, N. Boundedly rational consumers, energy and investment literacy, and the display of information on household appliances. *Resour. Energy Econ.* **2019**, *56*, 39–58. [CrossRef]
142. Gladwin, D.; Karsgaard, C.; Shultz, L. Collaborative learning on energy justice: International youth perspectives on energy literacy and climate justice. *J. Environ. Educ.* **2022**, *53*, 251–260. [CrossRef]
143. Meira, D.; Guerra-Mota, M.; Cunha, R. Energy communities in cooperative form as an ideal scenario in the fight against energy poverty. In Proceedings of the ICEE International Conference on Energy & Environment, Porto, Portugal, 2–3 June 2022.
144. IRENA, World Energy Transitions Outlook 2023: 1.5 °C Pathway, Volume 1, International Renewable Energy Agency, Abu Dhabi. 2023. Available online: http://www.irena.org/Publications (accessed on 1 September 2023).
145. O'Neill-Carrillo, E.; Mercado, E.; Luhring, O.; Jordan, I. Local socio-economic development through community-based distributed energy resources. In Proceedings of the 2018 IEEE International Symposium on Technology and Society (ISTAS), Washington, DC, USA, 13–14 November 2018; pp. 8–13.

resources

MDPI

Review

Understanding Health Outcomes from Exposure to Blue Space Resources: Towards a Mixed Methods Framework for Analysis

Megan J. Grace [1], Jen Dickie [1], Phil Bartie [2], Caroline Brown [3] and David M. Oliver [1,*]

[1] Biological and Environmental Sciences, Faculty of Natural Sciences, University of Stirling, Stirling FK9 4LA, UK; m.j.grace@stir.ac.uk (M.J.G.); j.a.dickie@stir.ac.uk (J.D.)

[2] School of Mathematical and Computer Sciences, Heriot Watt University, Edinburgh EH14 4AS, UK; phil.bartie@hw.ac.uk

[3] The Urban Institute, School of Energy, Geoscience, Infrastructure and Society, Heriot Watt University, Edinburgh EH14 4AS, UK; c.j.brown@hw.ac.uk

* Correspondence: david.oliver@stir.ac.uk

Abstract: With healthcare systems facing growing pressure from ageing populations and associated complex care needs, attention is increasingly being focused on sustainable strategies to improve health outcomes across populations. Encouraging access to natural environments is one form of preventative public health strategy that has been shown to lead to improved physical and mental health outcomes at the population level. A significant body of research has documented the health benefits of accessing a wide range of natural environments, including green space and coastal areas. However, freshwater resources, or inland blue spaces, have received less attention in the field of human–environment interactions. This critical review highlights current research opportunities for developing rich and nuanced insight into inland blue space experiences. Future research must take steps to account for the dynamic and unique nature of inland blue spaces through the application of a wide range of flexible and sensitive research methodologies alongside the application of broader mixed methods research approaches. To effectively utilise inland blue spaces as public health resources, it is vital that research captures the influence of temporal changes on blue space interactions and considers the overarching impact of context-specific factors. Addressing current research gaps in combination with advancing research methodologies offers the potential to consolidate inland blue space findings and create a robust evidence base for the implementation of effective public health policies.

Keywords: blue health; blue–green space; public health; wellbeing; mixed methods research

1. Introduction

Water environments constitute an important natural resource that supplies drinking water, supports biodiversity, enables food and energy production, and provides recreational opportunities [1–5]. Visiting blue spaces, such as lakes, rivers, and oceans, also has the potential to make a positive impact on population health and wellbeing. All visible surface waters, including both marine and freshwater environments, can be described as blue spaces [6]. Exposure to blue space, in the form of viewing waterbodies from indoor settings, submersion in the water, and close proximity to water-based environments, is associated with numerous health-related benefits. These positive health outcomes include higher levels of wellbeing [7–9], higher levels of self-reported health [10], and a better quality of life [11]. In recognition of the benefits of nature, interest in nature-based interventions, such as the promotion of blue space exposure, is increasingly growing, with governments and organisations beginning to introduce these interventions into policy and practice [12,13]. A significant advantage of nature-based interventions is that they act in a preventative manner, helping to mitigate the development of non-communicable diseases and consequently reducing pressure on healthcare systems [14,15]. However, despite the proven potential of exposure to nature for improving both mental and physical health outcomes, research findings from blue space studies are only just beginning to be translated into policy. Blue spaces along with other natural environments are currently undervalued resources that remain significantly underutilised in public health policy [14,16,17].

Whilst the term blue space applies to a diverse range of environments, currently, not all of these environments have received equal attention in the field of blue space research. Most research on the benefits of blue space exposure has focused on coastal areas and the benefits associated with accessing marine environments; this can help inform the effective management of coastal areas but is not applicable for managing the health and wellbeing benefits of freshwater areas that differ substantially from coastal environments in terms of biodiversity, ecosystem services, and aesthetic values. An extensive range of freshwater types are categorised as part of inland blue space, including canals, waterfalls, rivers, lakes, and reservoirs. Compared to marine blue space, often, inland blue space incorporates a greater diversity of environments with vastly different aesthetic characteristics, contrasting spatial scales, and distinct differences in the flow of water and presence of nearby vegetation.

To further consolidate the blue space evidence base and help promote a range of blue spaces as public health assets, inland blue space exposure should be considered in more detail. Proximity to natural environments is a key factor that influences environmental usage, with individuals more likely to frequently visit nearby blue space environments [18]. Given that inland blue spaces typically have broad geographic coverage and national distribution, with correct management in place, freshwater areas therefore have the potential to regularly benefit a significant proportion of the population. A greater focus on inland blue space research is warranted to better understand the complex relationship between inland blue space exposure and health and wellbeing outcomes. Initial research has identified differences in the wellbeing outcomes associated with recreational activities at different blue space types, with coastal recreation often leading to higher wellbeing outcomes for visitors than activities at inland blue spaces [19–21]. Differences have also been identified in user groups between blue space types, with inland blue spaces shown to attract visits from individuals with higher socio-economic statuses compared to coastal areas [22]. To encourage equitable access to inland blue space and help promote positive health and wellbeing outcomes for all user groups, greater insight is required to gather definitive conclusions and effectively translate blue space research into policy.

A particular challenge when considering inland blue space exposure is to adequately account for the wide range of variables that impact on blue space experiences and consequently affect health-related exposure outcomes. This insight is vital for informing land management strategies, but due to the dynamic and subjective nature of personal experiences with nature, key information on the impact of environmental and socio-economic variables can be difficult to obtain. A variety of different but complementary research approaches are required to develop an intrinsic understanding of user experiences at blue spaces and determine definitive patterns in exposure outcomes [23–25]. The available evidence and future research directions for two broad categories of blue space exposure variables will be considered in this critical review: temporal and spatial factors. Within these broad categories, the effects of both environmental and personal determinants on blue space experiences and health outcomes will be discussed.

This critical review will focus on inland blue spaces and highlight the research limitations that need to be addressed to help facilitate efficient policy decision making. The three key objectives of this critical review are to: (i) evaluate, critically, the current gaps in our understanding of the impact of spatial and temporal factors on inland blue space usage and exposure outcomes; (ii) determine how our understanding of inland blue space exposure outcomes could benefit from the evidence base associated with human interactions with other environments, for example, green space; and (iii) identify the potential for an enhanced analytical framework focusing on qualitative and mixed methods approaches to provide novel insight into the importance of blue space environments through more nuanced and richer accounts of human experiences.

2. Sense of Place and Temporality

Globally, freshwater environments are utilised as community resources; therefore, at the individual and community levels, blue spaces are often associated with a strong 'sense of place'. Multiple definitions exist regarding sense of place; however, for the purposes of this review, the concept will focus on the significance and emotional value that individuals ascribe to specific locations [26,27]. A significant body of research, spanning several decades, has considered the importance of people–place relations [26–28]. Within blue space research, there have been increasing reports of a strong sense of place associated with different inland blue space environments [29,30]. Sense of place is a significant cultural factor, allowing for improved health and wellbeing outcomes at the individual level [31]. Local waterbodies can instil a sense of place for individuals and communities through personal experiences or day-to-day observations, helping form a community identity [6,32]. Since sense of place is a key factor influencing an individual's environmental attitude towards and preference for specific environmental types, it can, in turn, influence visit frequency of inland blue space types [33,34].

The process of 'place-making' and developing an attachment to an environment involves two separate entities, the people or users of the environment and the environment itself [35]. Therefore, changes in individuals' personal circumstances alongside alterations to the landscape itself can impact human–environment interactions and place-based relations [36]. Social context is an additional underlying influence that shapes the place-making process. This is because socio-economic factors can significantly affect how an individual perceives, uses, and experiences an environment [37]. Since place-making processes are socially constructed and rely on dynamic relationships between humans and the environment, it is recognised that sense of place can evolve significantly across different timescales depending on environmental change and socio-cultural practices [38–40]. A wide range of personal and socio-economic factors such as length of residence, age, and education have been associated with levels of place attachment [38,41,42]. Additionally, numerous environmental factors including urban change, wildlife interactions and environmental threats can also influence attachment to an environment [43–45].

To advance blue space research, there is a need to focus on how spatial and temporal changes can alter place-based relations and consequently affect the health outcomes associated with blue space exposure. Short-term temporal changes in blue spaces, such as fluctuations in water quality, have been extensively studied in relation to environmental health [46–48]. However, further research is required to reach a better understanding of the potential for larger-scale and longer-term temporal factors to influence place attachment and impact the relationship between blue spaces and health. Additionally, the potential interplay between contextual and temporal factors on inland blue space exposure outcomes should not be underestimated. This critical review will evaluate three broad categories of temporal changes that can affect sense of place and consequently influence blue space exposure outcomes: (i) environmental changes; (ii) broader societal changes affecting environmental perceptions; and (iii) changes in personal circumstances for blue space users (Figure 1). Alongside this, the overarching influence of two key contextual factors, relating to the rural–urban dichotomy and differing cultural practices, will also be considered.

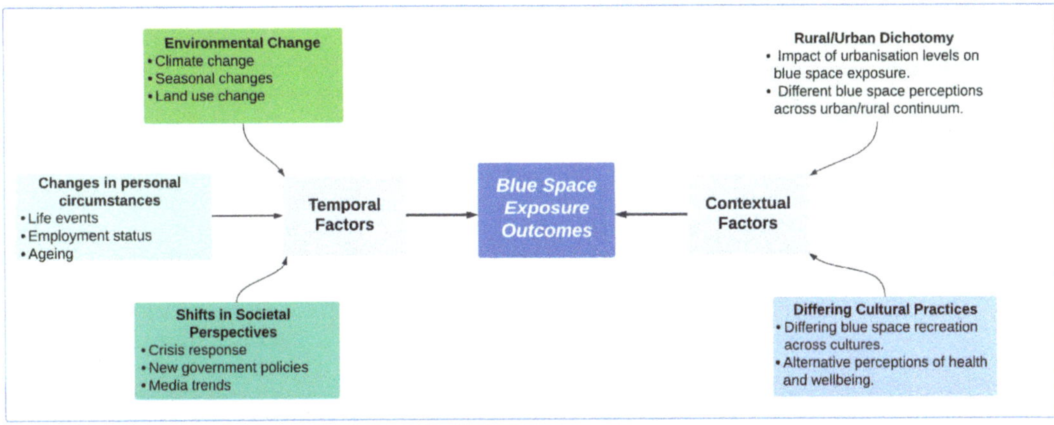

Figure 1. Examples of key temporal and contextual factors that can interrupt the relationship between blue space exposure and health outcomes.

2.1. Environmental Temporal Changes

The dynamic nature of water in both space and time means that blue space environments are continually changing. Changes in water characteristics can be observed across multiple timescales, from minutes (e.g., responses to rainfall runoff) to months (e.g., river base flow vs. spate) to years (e.g., drought year vs. flood year). The ambience of blue space environments can significantly alter as a function of time of day, wildlife presence, and shifts in visitor numbers [49]. This can, in turn, influence user groups, with research conducted at the river Rhine identifying that young adults were more likely to visit the river during evenings at the weekend whereas families and older adults used the riverside space more frequently during the daytime [50].

Over the course of a year, the nature of inland waterbodies will be significantly altered due to both changes in the surrounding scenery and vegetation as well as alterations to the flow of water. This may influence visitor behaviour at inland blue spaces and consequently influence exposure outcomes. Across the seasons, the frequency of visits to inland blue spaces can vary alongside the visit purpose and aesthetic preferences for blue spaces [51–53]; however, research findings remain inconclusive, indicating the significant complexity of seasonal effects on outdoor recreation and exposure outcomes. In line with seasonal changes, exposure outcomes can also vary because of weather conditions. Unfavourable conditions, such as high wind speeds and extended periods of drought, have been linked with blue space users avoiding these areas or reducing the length of time they spend near inland waters [32,54]. During calm conditions, exposure to a waterbody is likely to promote relaxation and stress relief [50] whereas in extreme weather conditions, such as excessive rainfall or prolonged periods of drought, blue space environments could cause severe anxiety and depression for nearby residents [55–57].

Over the next few decades, climate change has the potential to further alter inland blue spaces by detrimentally affecting hydrological regimes and negatively impacting biodiversity [58]. As noted by the United Nations, "Water is the primary medium through which we will feel the effects of climate change" [59]; therefore, blue space research should pay particular attention to this. The impact of climate change and associated environmental impact mitigation strategies can significantly affect communities by leading to the loss or alteration of areas deemed to be of cultural or historical significance [60]. For instance, the introduction of hard engineering strategies to prevent flooding can notably change the appearance of rivers. Due to a sense of place and attachment to local community sites, it is often the case that affected communities cannot be adequately compensated for their significant cultural loss resulting from environmental changes [61,62]. This is particularly pertinent for indigenous communities who have a strong connection to their land [63]. The Australian concept of 'solastalgia' is often used to refer to this pain experienced by communities when environmental changes alter their homelands [64]. Vulnerable communities are also more likely to face inequalities with regards to accessing clean drinking water and safe blue space environments [65]. This issue will continue to be exacerbated because of climate change leading to increased water shortages in future [66]. Without the introduction of sustainable management strategies, the significant interaction between climate change, sense of place, and community identity has the potential to have a considerable negative impact over time on the potential for blue space environments to promote population health and wellbeing [67].

Land-use changes affecting blue spaces and their nearby surroundings can have a significant effect on residents and blue space visitors. In relation to the urban regeneration of blue space features, the associated land-use change can have a positive impact on residents by encouraging a sense of pride in their community and creating a therapeutic experience [17,68,69]. However, whilst regeneration projects may have an overall positive effect on a community, the relationship that individuals have with a location is subjective, and so, there will inherently be a degree of variation in how people react to land-use change. Longer-term residents who have memories attached to an area and a greater knowledge of its past may feel more strongly about environmental change [70,71].

An unintended consequence that can arise from land-use changes, particularly in the case of urban regeneration projects, is gentrification. The term 'green gentrification' has recently emerged to explain the process by which green and blue space development projects can improve a neighbourhood and lead to increased house prices and, consequently, the relocation of lower-income residents [72]. The process of green gentrification is intricate and dependent on several factors including location and existing infrastructure [73,74]. Therefore, not all regeneration projects will lead to gentrification. However, when the process does occur, it can have a long-term detrimental impact on the health status of residents, particularly among groups who are already marginalised, often leaving individuals feeling like they do not belong in the new regenerated community [71,75,76]. Additional research is needed to better understand how policies and practices can facilitate the community-led, sustainable regeneration of green and blue space facilities in deprived areas without leading to green gentrification and the further establishment of environmental inequalities.

Comparative research to understand how temporal changes in environmental conditions could alter the health outcomes associated with exposure to a range of different inland blue space typologies is now warranted. Several studies have assessed the impact of specific types of blue space environments on health and wellbeing outcomes [29,32,77,78]. In comparison, few studies have researched the potential for variation in exposure outcomes across different typologies of waterbodies. Initial research in this area has indicated that exposure to rivers and canals is associated with greater mental wellbeing than exposure to lakes [79]. However, this research only focused on three broad inland blue space types—lakes, rivers, and canals—and given the diversity of blue space environments, further opportunities exist to quantify the impact of a broader range of inland waterways, including wetlands, ponds, streams, and waterfalls [79]. In recognition of the overarching influence of environmental temporal changes on the appearance and ambience of blue space environments, research should also assess whether alterations to inland waterways and their waterside environments (both natural as well as management-related) can further magnify the variation in potential exposure outcomes from visiting different types of blue spaces.

Given the significant influence of environmental factors on blue space interactions, as well as the confounding effect of other variables including sense of place and cultural perceptions, it cannot be assumed that all inland blue space visits lead to positive exposure outcomes. Initial research has documented the more complex relationship between blue space and health, highlighting the potential for blue space to create a sense of isolation and cause frustration for user groups [80–82]. However, so far, this area of research has focused on coastal blue space, and so, further insight is needed to capture a broader range of experiences at inland blue spaces. Additionally, most findings relating to negative blue space experiences are derived from specific case studies; therefore, larger samples are required to create generalisable results.

Environmental changes affecting inland blue spaces, such as seasonal variation in weather, land-use change, and climate change, are, to a certain extent, interlinked. It would therefore be worthwhile to assess the potential cumulative effect of these environmental changes on the relationship between blue spaces and health through the application of versatile mixed methods research approaches. Earlier research focusing on aesthetic preference for blue space environments largely relied on the use of cross-sectional photo-based preference studies conducted in laboratory settings [83,84]. Building on this, a mixed methods research approach using large-scale dataset analysis combined with a national survey and focus groups identified the significant range of factors that may influence aesthetic preferences for blue space users, including biodiversity levels, perceived busyness of an environment, and surrounding green-space quality [85]. However, the application of additional novel research approaches is required to consolidate these findings and account for subjective differences across populations. In situ methodologies may be particularly beneficial for capturing individuals' environmental perspectives of freshwater environments without the influence of recall bias.

2.2. *Changing Perceptions at the Societal Level*

Whilst several studies have considered the benefits of blue space exposure at the wider community and population level [24,86,87], considerably fewer have considered how populations collectively regard blue space areas and the nature by which these perceptions change over time. Public perception is a crucial factor that can alter people's willingness to adopt certain behaviours, and so, perceptions of inland blue spaces could have a major impact on environmental usage.

A pertinent example of how societal changes can impact the relationship between blue spaces and health is the range of lifestyle shifts brought about due to the onset of the COVID-19 pandemic. Research across twenty European countries discovered that at the beginning of the pandemic, there was a significant increase in the number of online searches for topics relating to nature and the environment, which suggests an increased awareness or appreciation of nature at the population level [88]. During the initial phases of the pandemic, blue spaces were associated with providing stress relief and mental wellbeing at the individual level [89,90]. However, for many, access to blue spaces was restricted during the pandemic due to public health interventions; this, in turn, affected overall wellbeing levels and altered the therapeutic nature of these environments for individuals [91,92]. As the impact of the SARS-CoV-2 virus, which causes COVID-19, continues to manifest, this will make a lasting change in society's perception of the importance of accessing natural environments and could consequently alter the health outcomes associated with blue space exposure. To enhance environmental and public health policies, the impact of the COVID-19 pandemic on the relationship between blue space and health across time provides a global exemplar for further investigation.

The introduction of new government policies can lead to a shift in how populations regard inland blue spaces and other natural environments as health-promoting resources. One noticeable blue space policy trend across Europe is the increasing establishment of designated marine and inland bathing areas [93,94]. Bathing waters are distinct from other water environments because the microbial quality of water in these areas is regularly monitored and action is taken to ensure that the quality is within pre-defined pollution limits [95]. As the environment of these blue space areas has been enhanced through the introduction of stringent policies on pollution, it is likely that bathing areas may positively impact the relationship between blue spaces and health. Furthermore, healthy and vibrant bathing waters provide social spaces, sources of wellbeing for recreational users, and key sources of jobs and revenue for local economies. A link has been determined between improved water quality and an increased frequency of visits and improved attitudes towards a blue space area [96–98]. In coastal environments, sites with lower water quality have also been linked with lower ratings of perceived restorative potential [99]. However, in relation to bathing areas, a focus on water quality alone fails to account for the value attributed to bathing-water environments by the 'hidden majority' who rarely use the sea for immersive activities. For example, waterside environments promote social interactions and wider cultural ecosystem services [100,101].

Accounting for the wider value of coastal and inland bathing waters is critical to effectively managing and promoting, more generally, the role of bathing-water environments as important socio-economic resources. Considering the range of uses and the significant number of interlinked benefits associated with bathing areas, it would be valuable to conduct a range of comparative research studies to determine whether health outcomes and other cultural ecosystem services (CESs) provided by inland bathing sites vary significantly from the benefits provided by marine bathing sites. The CESs associated with environments are continually evolving and can be affected by environmental and societal changes; therefore, the potential for the perceived benefits of inland and coastal bathing sites to vary over time should also be considered [102]. This comparative and temporal insight will help underpin a robust evidence base for the creation of tailored environmental policies.

Social media offers the potential to track real-time changes in societal perceptions and can be used as an effective tool to aid in environmental planning processes. A growing body of research has assessed the CESs (such as aesthetic enjoyment and health and wellbeing outcomes) provided by a range of environments through the application of social media Big Data analytics [103–106]. The information gained from social-media-derived environmental research studies has proven to be highly valuable and can both aid in the evaluation of infrastructure and interventions as well as help inform land-use policy decision-making processes [107–109]. Currently, however, only a small number of CES studies have considered inland blue spaces in detail, and fewer still have used social media analytics to assess the CESs of inland blue spaces; therefore, there is scope to build on this. There are limitations to social media data. For instance, the datasets can include much noise due to fake or spam social media accounts, and establishing the generalisability of the findings can be challenging [110]. However, this form of data offers a host of opportunities for advancing the blue space research field in terms of enabling the opinions of the local community and blue space users to be quickly assessed and consequently helping to establish sustainable management strategies. If Big Data analytics were to be integrated with qualitative exploratory methods, this could provide rich, valuable insight into blue space perceptions and the factors influencing blue space usage at the societal level.

2.3. *Changes in Personal Circumstances*

Within a population, individuals will have faced significant and unique life events that can change how they view blue spaces at a personal level. This aligns with Conradson's concept of 'Therapeutic Landscape Experiences', whereby he suggests that the relationship between an environment and an individual is highly subjective and so different visitors of environments can experience vastly different health outcomes [111]. Furthermore, changes in personal circumstances can alter the way in which individuals value natural environments, and so, could significantly affect the health benefits obtained from blue space exposure [112]. Most research exploring the subjective relationship between the environment and health has focused broadly on natural environments, rather than exclusively on blue spaces, and so, more specific investigation of how different life events can alter perceptions of blue spaces is warranted. Combining in-depth personal information with broader population-level data trends could help inform more effective management of blue spaces and ensure that these health-promoting resources are accessible to all.

As people age, their relationship with blue spaces can alter, leading to different age groups attributing different benefits to visiting blue space areas [23]. Some studies have identified that for children and young adults, an association exists between coastal and inland blue space exposure and mental health and wellbeing [113–116]. Research involving older generations has also highlighted the fundamental importance of being close to blue spaces for maintaining quality of life and wellbeing [7,8,80,117]. To advance blue space research, it would be worthwhile utilising cohort study designs to follow how the ageing process can directly impact the relationship between blue spaces and health. Cohort research studies are longitudinal and, typically, more resource-intensive and time-consuming than other research methods [118]. Despite these limitations, conducting longitudinal blue space research would be advantageous to effectively correlate available studies on blue space exposure and help create a stronger evidence base than can be achieved through the application of multiple cross-sectional studies. Having more information on the effect of nearby environments on different age-groups can help inform both environmental and public health policies to encourage healthy ageing across a population.

Coupled with ageing, an evolving employment status throughout an individual's life course could lead to changes in the health benefits gained from accessing blue space areas. The effect of the neighbourhood environment on health and wellbeing has been identified as having a greater effect on those who spend a larger proportion of their time at home or within their neighbourhood [119]. As a result of this, those who are retired or work from home may be more impacted by nearby inland blue space compared to those who work away from home. For those who are employed, a modifying factor is the type of commute taken by an individual; commutes that pass through natural environments are often positively regarded by individuals and linked with improved levels of mental health [120,121]. Those who are in higher income brackets typically have better access to high-quality blue space environments [122–125]; therefore, employment status is likely to add further complexity to the relationship between blue space exposure and health outcomes. Whilst specific research into the effects of employment on inland blue space access and exposure outcomes has not yet been conducted, the insights outlined from relevant research fields indicate that this is an important knowledge gap to address.

Research has identified the potential for sociodemographic factors to influence inland blue space usage. However, to date, most research has been cross-sectional, and so, the reasoning behind sociodemographic trends in blue space access cannot be established. Developing a greater understanding of sociodemographic influences is vital to effectively address existing environmental injustices relating to inland blue space access. Given the potential for lifestyle factors to change across time, the application of more innovative and longitudinal qualitative methods would help derive richer data and gain greater insight into temporal influences on blue space experiences. In line with this, diary keeping is one data collection method that could provide rich contextual knowledge relating to how changes in personal circumstances affect blue space experiences and exposure outcomes. Qualitative diaries can provide an effective means of unobtrusively assessing participants' emotions in real-life situations [126]. Research diaries are also useful for documenting short-lived events such as interactions with nature as these types of fleeting-events often do not have an overarching influence on people's lives and so could easily be forgotten [127].

Diary methods have previously proved effective in documenting the personal and wider-spread effects of experiencing flooding in the city of Hull [128]. This diary project had a national impact with findings subsequently used to inform water management policies. More recently, diaries have been utilised to assess the restorative outcomes associated with inland blue spaces for Scottish adults [129]. Conducting further diary-based research studies to track how personal and environmental changes affect blue space exposure outcomes clearly has potential to provide detailed personal insight into blue space exposure. The combination of diary methods with follow-up interviews has the potential to provide rich ethnographic-style insight into inland blue space experiences.

3. The Influence of Spatial and Contextual Factors

The relationship between blue space exposure and health is highly contextual, and so, alongside researching the temporal changes associated with environmental and socio-economic factors, it is also important to consider the broad location trends that could have an overarching influence on exposure outcomes.

3.1. The Urban–Rural Dichotomy

The degree to which the area surrounding the blue space environment is urbanised could have an impact on blue space exposure outcomes. There has been continued debate as to whether the binary classification of 'rural' and 'urban' is still relevant and useful [130]. Industrialisation and globalisation are continually blurring the boundaries between rural and urban areas [131,132]; therefore, these two categories may no longer be distinct from one another. Additionally, there is no universal definition of what constitutes an urban or rural area, and so, this can lead to uncertainties and variations when comparing the health statuses of different populations. Despite the limitations of the urban–rural dichotomy, environmental exposure outcomes are heavily context-dependent, with research continuing to identify that individuals from rural and urban settlements have significantly different place-based narratives [133–135]. As such, it remains worthwhile to consider whether different levels of urbanisation can impact blue space exposure outcomes. There is also scope for investigating whether temporal factors affect individuals from contrasting backgrounds, such as rural and urban residents, in different ways [112].

Limited research has contrasted the impact of rural and urban blue space exposure [23,29,113]. Findings from such studies have been inconsistent with further in-depth research required to provide a greater understanding of the magnitude of the urban–rural effect and the reasoning behind any variations in health outcomes arising from blue space exposure. The current inconsistencies in findings may relate to the different blue space classifications included in the research, varying ages of the sample groups, and the context-specific nature of blue space exposure outcomes. This variability suggests that longer and larger coordinated studies across a greater urban–rural spectrum may yield stronger evidence to underpin our understanding of how urbanisation can influence any variations in health outcomes from blue space exposure.

Currently, most blue space research has focused on better understanding exposure outcomes for urban blue space [7,136–138]. Further research into rural blue space is required to ensure that freshwater management policies can be tailored towards the needs of both urban and rural residents. This research strategy would align with the policy of 'rural proofing' that has been adopted across the European Union to ensure that policies and resources are appropriately adapted to meet the varied needs of rural communities [139].

A greater focus on more dynamic research approaches in relation to the urban–rural continuum is required. Future research should appreciate mobility patterns, with individuals travelling from urban to rural areas and vice versa, irrespective of administrative boundaries, to access blue space and natural environments [140,141]. It is possible that quantitative research studies that consider health outcomes strictly in relation to residential proximity to natural environments may underestimate people's willingness to travel on trips and outings in order to connect with nature [140]. Therefore, a range of nuanced and complementary research approaches is required to understand motivations for accessing blue space environments across the urban–rural continuum. In addition, future research should carefully consider the way in which urban and rural areas are categorised. Relying solely upon administrative boundaries to define urban areas can lead to a significant overestimation of green- and blue space accessibility levels since administrative boundaries often include rural areas that are on the outskirts of urban settlements [142]. Considering this, sensitive methods are required to better define land-use areas and inform resource management strategies; the application of urban footprints to more accurately quantify the extents of cities could be an alternative approach.

Creating a versatile evidence base for policy decision making across the urban–rural continuum will ensure that diverse needs of communities can be accounted for. Thus, multidisciplinary research detailing blue space community case studies in rural and urban areas coupled with more broad-scale comparisons of rural and urban exposure outcomes would be advantageous for achieving this (Figure 2). Participatory photovoice methods that involve communities and individuals taking photographs to represent issues of importance in their everyday lives constitute one such technique that has previously proven effective in providing detailed insight into specific communities' blue space management concerns [143]. However, whilst participatory research can provide a greater understanding of issues relating to social and environmental justice, this type of research is typically limited in terms of scale. Therefore, the combination of contextualised participatory research findings with large-scale population-based studies of high statistical power would be an effective means of developing a robust blue space evidence base to inform the implementation of freshwater management strategies. An alternative option could be to utilise in-depth case-study data in combination with Big Data analytics to help verify rural and urban blue space trends.

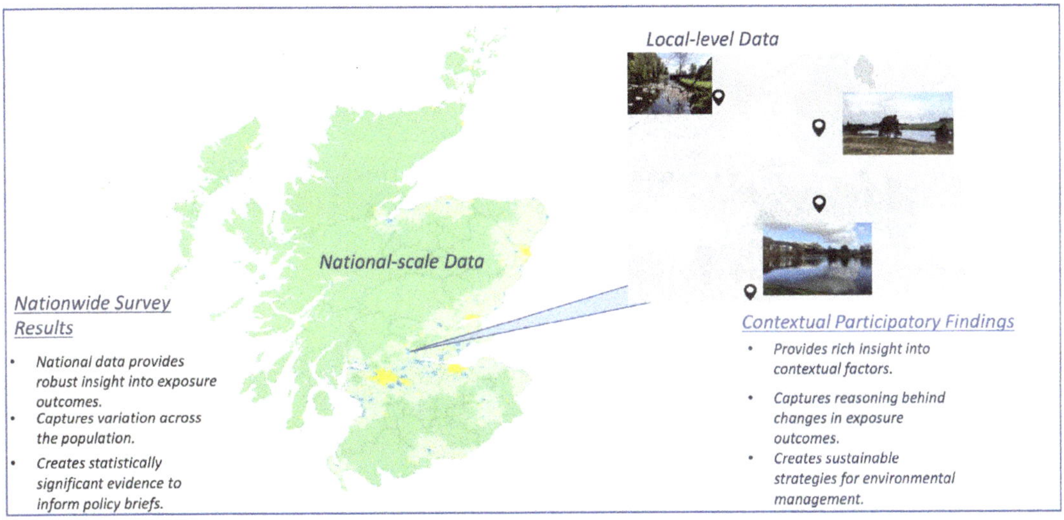

Figure 2. Exemplar of the potential for a mixed methods research approach at different spatial scales to provide detailed insight into a blue space research topic [144,145]. The colour gradients on the map represent rural and urban areas of Scotland.

3.2. Differing Cultural Practices and Blue Space Perceptions

Blue space research is increasingly challenging the presence of dominant Western perspectives on wellbeing outcomes [146–148]. Furthermore, there is an increased awareness that due to differing leisure practices and perceptions of water across cultures, the generalisability of research findings should be carefully considered [147]. Participating in surfing or canoeing, for example, can have significantly different meaning for Maori populations as opposed to non-indigenous groups who may not have the same traditional narratives or spiritual connection to the water [149,150]. However, despite this growing recognition regarding the importance of context in determining leisure practices and health and wellbeing values, to date, the majority of blue space research studies have been conducted in the Global North [151]. There is, therefore, a need to conduct research across a range of different cultural contexts. Furthermore, to gain insight into lived experiences across cultures, it is important that blue space research approaches continue to incorporate participatory and co-designed research methods that are sensitive to differing perspectives of health and recreation.

Health, happiness, and wellbeing are all subjective concepts that can vary between individuals and communities depending on social context [152,153]. Given this subjectivity, the perceived wellbeing and health benefits gained from blue space exposure could vary significantly across different countries and populations. This further justifies the benefits of conducting cross-cultural blue space research. Due to the complex nature of health-related concepts, it would also be worthwhile to adopt a multidimensional approach to measuring health and wellbeing outcomes. Currently, most health-related research on blue spaces has adopted a (semi)quantitative stance by focusing on health and wellbeing outcomes through the application of questionnaires and health-related datasets. This has provided a good overview of the potential of blue space environments to aid health outcomes. However, the current evidence base could be significantly strengthened through the application of alternative methods such as novel mixed methods research approaches. Mixed methods approaches are increasingly popular in health-based research due to their potential to provide multiple perspectives on complex problems [154,155]. In terms of blue space research, combining quantitative and qualitative methods, such as the application of robust nationally representative health-based datasets with detailed qualitative interviews, would significantly advance the research field by providing an enhanced understanding of the variables affecting freshwater experiences and health-related exposure outcomes at the individual and national levels.

4. An Enhanced Blue Space Analytical Framework

This review has highlighted the broad variety of spatial and temporal factors that influence inland blue space user experiences and identified several research priorities (Figure 3). Alongside the need to address existing research gaps within the blue space evidence base, to effectively advance the translation of blue space research into public health policy and practice, an enhanced blue space analytical framework is required. Future research methods must be flexible and responsive to capture the dynamic nature of human–environment interactions. Mixed methods research approaches offer the potential to record the spatial and temporal dynamics of blue space interactions across multiple timescales, allowing for rich insight into user experiences (Figure 4).

Figure 3. Three of the key research priorities identified in the narrative review, presented alongside potential research methods for addressing the priorities and a feasibility assessment.

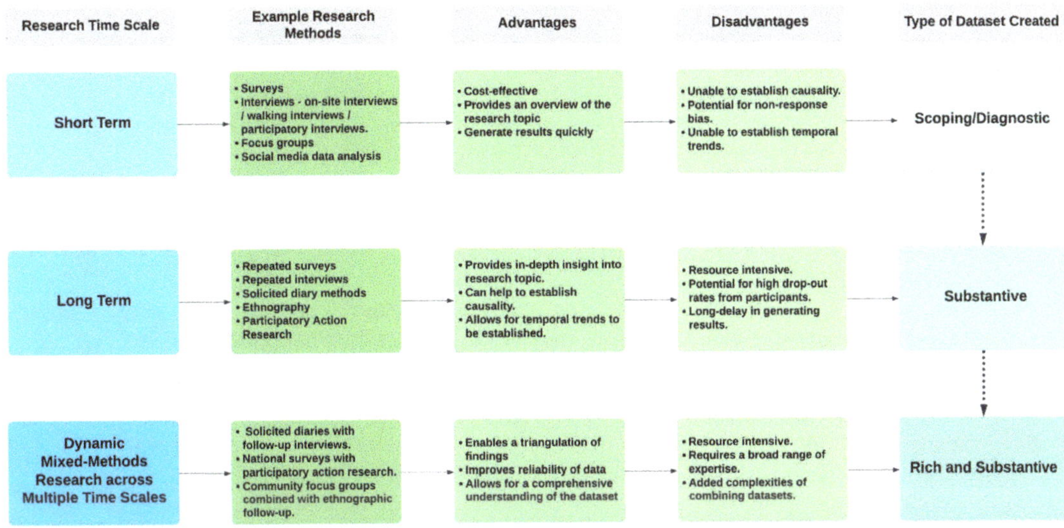

Figure 4. An outline of the advantages and disadvantages associated with conducting research across different timescales. The arrows link the factors associated with different research time scales whilst the dotted arrows highlight the increasing richness of the associated dataset.

A significant number of green-space studies have comprehensively assessed human–environment interactions using mixed methods research designs [156–159]. The mixed methods approaches adopted by these studies had clear benefits, facilitating a comprehensive exploration of the factors that can predetermine green- or public-space usage and providing a detailed understanding of environmental experiences [158,159]. Further mixed methods research in the blue space research field will help consolidate our current understanding of the complex, overarching impact that environmental factors can have on exposure outcomes. The application of a range of different methods enables multiple perspectives and insights into a topic area to be recorded, creating powerful insights into environmental interactions. For instance, the combination of qualitative walking interviews with geospatial data can help create contextualised knowledge and provide a better understanding of how individuals relate to different environments [160]. In the field of blue space research, this merging of methods may, in turn, provide a greater understanding of the wide range of factors that influence blue space exposure outcomes. A mixed methods approach can also help achieve the 'complementarity' of research findings, whereby the findings gained from one method can be used to further understand or enhance the evidence gained from another method [161]. This overlapping of research findings can create a more nuanced understanding of blue space exposure.

Alongside the application of mixed methods research projects, conducting further collaborative and cross-disciplinary research could help facilitate the development of evidence-based policies. Typically, cross-disciplinary research involves engagement from a range of stakeholders. This engagement is beneficial to the research process as it helps to ensure that the research outcomes remain relevant and applicable for aiding the target audience and facilitating a wider dissemination of research findings [162,163]. The BlueHealth project is an example of how a multidisciplinary research approach can help build a better understanding of the value, impact, and public-health potential of blue spaces across different populations [96,164,165]. The focus of the BlueHealth project was, however, largely on oceans and coastal environments, so the opportunity remains to advance inland blue space research through the application of innovative multidisciplinary research approaches. Inland blue space research aligns closely with public health, sociology, human geography, and environmental science research fields. This alignment with other disciplines offers a host of opportunities for close research collaborations to provide more powerful analytical methods, both helping strengthen research findings as well as increasing public awareness.

It would be particularly beneficial if future blue space research approaches utilised longitudinal data collection strategies. Whilst cross-sectional research projects are quicker and cheaper to conduct, these short-term data collection methods can only provide a relative snapshot into environmental interactions and often rely on participants reflecting on and considering their blue space experiences from memory. Longitudinal blue space research would provide a better understanding of whether, and to what extent, temporal changes can subsequently impact blue space exposure outcomes for individuals whilst minimising the risk of recall bias. A range of different methods could be used to carry out longitudinal research, including the application of wearable research technology over a prolonged period of time, ethnographic observation, qualitative and quantitative research diaries, and videography. Incorporating longitudinal research alongside short-term research methods such as interviews and surveys could provide novel insight into blue space exposure outcomes. The creation of geospatial longitudinal datasets, through the application of mixed methods research approaches, would help develop an increased understanding of the dynamic spatial and temporal processes that impact human–environment interactions [166]. This would allow for a triangulation of research methods to capture multiple perspectives on blue space temporal dynamics and help create a more comprehensive understanding of the research topic [167].

Future research should also seek to include creative practice into mixed methods research designs. Blue spaces, particularly rivers, have commonly been used as inspirations for creative practice research in the fields of education and psychology [168]. Community-led projects have also focused on blue space environments to encourage engagement from residents. The 'Our Living Rivers and Glens' project is one such example of this, whereby professional musicians were informed by the words, sounds, videos, and images collected from local residents to create music representing environmental experiences during the COVID-19 pandemic [169]. Whilst some blue space studies have used novel arts-based methods to gain in-depth insight into community perspectives, the application of creative practice remains underutilised [170,171]. In light of the success of previous projects, there is scope to develop creative methods further within the blue space academic research field to create a rich understanding of environmental exposure outcomes.

5. Conclusions

Blue space research is increasingly helping uncover the wide range of complex and interlinked factors that can affect blue space exposure outcomes. Whilst an increasing number of research studies demonstrate the importance of blue spaces for population health and wellbeing, further research is required to ascertain the overall influence and impact of different temporal and spatial variables on blue space exposure outcomes in order to better inform environmental policy. Arguably one of the most effective means of facilitating the translation of blue space research into policy and practice is to continue to develop transdisciplinary research projects and interventions. This would involve the collaboration of different stakeholders, including policy makers, blue space users, and healthcare practitioners, as well as involvement from scientific researchers. Alongside this collaborative research, further focused research is necessary to address the key gaps in the blue space evidence base, which could prevent the development of evidence-based policies. One such gap is a lack of longitudinal research. This type of long-term prospective research is necessary to help establish causality and draw definitive links between blue space exposure and health and wellbeing outcomes. Future studies should also move beyond traditional research approaches and seek to capture embodied knowledge through a combination of sensitive quantitative and qualitative methodologies such as solicited diaries, wearable technology, and videography. The application of a range of methodologies alongside the effective integration of research findings will help increase the breadth of blue space research and enable a wider range of blue space exposure variables to be considered.

Whilst an increasing portfolio of research findings indicates the potential of blue space environments to be utilised as public health resources, further investigation is needed to fully understand the range of factors that can affect blue space experiences and modify the potential of these natural environments to improve population health and wellbeing.

Author Contributions: Conceptualisation, M.J.G., D.M.O., J.D., P.B. and C.B.; methodology, writing—original draft preparation, M.J.G.; writing—review and editing, M.J.G., D.M.O., J.D., P.B. and C.B.; supervision, D.M.O., J.D., P.B. and C.B.; funding acquisition, M.J.G., D.M.O., J.D., P.B. and C.B. All authors have read and agreed to the published version of the manuscript.

Funding: This research was funded by the Natural Environment Research Council as part of the IAPETUS Doctoral Training Programme, with additional CASE support funding provided by the Scottish Environment Protection Agency.

Data Availability Statement: No new data were created or analysed in this study. Data sharing is not applicable to this article.

Conflicts of Interest: The authors declare no conflict of interest.

References

1. Pradinaud, C.; Northey, S.; Amor, B.; Bare, J.; Benini, L.; Berger, M.; Boulay, A.-M.; Junqua, G.; Lathuillière, M.J.; Margni, M.; et al. Defining Freshwater as a Natural Resource: A Framework Linking Water Use to the Area of Protection Natural Resources. *Int. J. Life Cycle Assess.* **2019**, *24*, 960–974. [CrossRef] [PubMed]
2. Bayart, J.-B.; Bulle, C.; Deschênes, L.; Margni, M.; Pfister, S.; Vince, F.; Koehler, A. A Framework for Assessing Off-Stream Freshwater Use in LCA. *Int. J. Life Cycle Assess.* **2010**, *15*, 439–453. [CrossRef]
3. Żywiec, J.; Szpak, D.; Piegdoń, I.; Boryczko, K.; Pietrucha-Urbanik, K.; Tchórzewska-Cieślak, B.; Rak, J. An Approach to Assess the Water Resources Reliability and Its Management. *Resources* **2023**, *12*, 4. [CrossRef]
4. Hartig, J.H. Rewilding the Detroit, Michigan, USA–Windsor, Ontario, Canada Metropolitan Area. *Resources* **2023**, *12*, 117. [CrossRef]
5. Adeeyo, A.O.; Ndlovu, S.S.; Ngwagwe, L.M.; Mudau, M.; Alabi, M.A.; Edokpayi, J.N. Wetland Resources in South Africa: Threats and Metadata Study. *Resources* **2022**, *11*, 54. [CrossRef]
6. Volker, S.; Kistemann, T. "I'm Always Entirely Happy When I'm Here!" Urban Blue Enhancing Human Health and Well-Being in Cologne and Dusseldorf, Germany. *Soc. Sci. Med.* **2013**, *78*, 113–124. [CrossRef]
7. Garrett, J.K.; White, M.P.; Huang, J.; Ng, S.; Hui, Z.; Leung, C.; Tse, L.A.; Fung, F.; Elliott, L.R.; Depledge, M.H.; et al. Urban Blue Space and Health and Wellbeing in Hong Kong: Results from a Survey of Older Adults. *Health Place* **2019**, *55*, 100–110. [CrossRef]
8. Finlay, J.; Franke, T.; McKay, H.; Sims-Gould, J. Therapeutic Landscapes and Wellbeing in Later Life: Impacts of Blue and Green Spaces for Older Adults. *Health Place* **2015**, *34*, 97–106. [CrossRef]
9. McCartan, C.; Davidson, G.; Bradley, L.; Greer, K.; Knifton, L.; Mulholland, A.; Webb, P.; White, C. 'Lifts Your Spirits, Lifts Your Mind': A Co-Produced Mixed-Methods Exploration of the Benefits of Green and Blue Spaces for Mental Wellbeing. *Health Expect.* **2023**, *26*, 1679–1691. [CrossRef]
10. Jarvis, I.; Koehoorn, M.; Gergel, S.E.; van den Bosch, M. Different Types of Urban Natural Environments Influence Various Dimensions of Self-Reported Health. *Environ. Res.* **2020**, *186*, 109614. [CrossRef]
11. Haeffner, M.; Jackson-Smith, D.; Buchert, M.; Risley, J. Landscape and Urban Planning. *Landsc. Urban Plan.* **2017**, *167*, 136–146. [CrossRef]
12. Frantzeskaki, N. Seven Lessons for Planning Nature-Based Solutions in Cities. *Environ. Sci. Policy* **2019**, *93*, 101–111. [CrossRef]
13. Sarabi, S.; Han, Q.; Romme, A.G.L.; de Vries, B.; Valkenburg, R.; den Ouden, E. Uptake and Implementation of Nature-Based Solutions: An Analysis of Barriers Using Interpretive Structural Modeling. *J. Environ. Manag.* **2020**, *270*, 110749. [CrossRef]
14. Maller, C.; Townsend, M.; Pryor, A.; Brown, P.; St Leger, L. Healthy Nature Healthy People: 'Contact with Nature' as an Upstream Health Promotion Intervention for Populations. *Health Promot. Int.* **2006**, *21*, 45–54. [CrossRef] [PubMed]
15. Shanahan, D.F.; Astell-Burt, T.; Barber, E.A.; Brymer, E.; Cox, D.T.C.; Dean, J.; Depledge, M.; Fuller, R.A.; Hartig, T.; Irvine, K.N.; et al. Nature-Based Interventions for Improving Health and Wellbeing: The Purpose, the People and the Outcomes. *Sports* **2019**, *7*, 141. [CrossRef]
16. Richardson, M.; Corral-Verdugo, V.; Summers, J.K.; Vivian, D.N. Ecotherapy—A Forgotten Ecosystem Service: A Review. *Front. Psychol.* **2018**, *9*, 1389. [CrossRef]
17. Brückner, A.; Falkenberg, T.; Heinzel, C.; Kistemann, T. The Regeneration of Urban Blue Spaces: A Public Health Intervention? Reviewing the Evidence. *Front. Public Health* **2022**, *9*, 782101. [CrossRef]
18. Elliott, L.R.; White, M.P.; Grellier, J.; Garrett, J.K.; Cirach, M.; Wheeler, B.W.; Bratman, G.N.; van den Bosch, M.A.; Ojala, A.; Roiko, A.; et al. Research Note: Residential Distance and Recreational Visits to Coastal and Inland Blue Spaces in Eighteen Countries. *Landsc. Urban Plan.* **2020**, *198*, 103800. [CrossRef]

19. Garrett, J.K.; White, M.P.; Elliott, L.R.; Grellier, J.; Bell, S.; Bratman, G.N.; Economou, T.; Gascon, M.; Lõhmus, M.; Nieuwenhuijsen, M.; et al. Applying an Ecosystem Services Framework on Nature and Mental Health to Recreational Blue Space Visits across 18 Countries. *Sci. Rep.* **2023**, *13*, 2209. [CrossRef]
20. Próchniak, P.; Próchniak, A. Adventure Recreation in Blue Spaces and the Wellbeing of Young Polish Adults. *Int. J. Environ. Res. Public Health* **2023**, *20*, 4472. [CrossRef]
21. Pasanen, T.P.; White, M.P.; Wheeler, B.W.; Garrett, J.K.; Elliott, L.R. Neighbourhood Blue Space, Health and Wellbeing: The Mediating Role of Different Types of Physical Activity. *Environ. Int.* **2019**, *131*, 105016. [CrossRef] [PubMed]
22. Elliott, L.R.; White, M.P.; Grellier, J.; Rees, S.E.; Waters, R.D.; Fleming, L.E. Recreational Visits to Marine and Coastal Environments in England: Where, What, Who, Why, and When? *Mar. Policy* **2018**, *97*, 305–314. [CrossRef]
23. de Bell, S.; Graham, H.; Jarvis, S.; White, P. The Importance of Nature in Mediating Social and Psychological Benefits Associated with Visits to Freshwater Blue Space. *Landsc. Urban Plan.* **2017**, *167*, 118–127. [CrossRef]
24. Poulsen, M.N.; Nordberg, C.M.; Fiedler, A.; DeWalle, J.; Mercer, D.; Schwartz, B.S. Factors Associated with Visiting Freshwater Blue Space: The Role of Restoration and Relations with Mental Health and Well-Being. *Landsc. Urban Plan.* **2022**, *217*, 104282. [CrossRef]
25. Pitt, H. Muddying the Waters: What Urban Waterways Reveal about Bluespaces and Wellbeing. *Geoforum* **2018**, *92*, 161–170. [CrossRef]
26. Devine-Wright, P. Think Global, Act Local? The Relevance of Place Attachments and Place Identities in a Climate Changed World. *Glob. Environ. Change* **2013**, *23*, 61–69. [CrossRef]
27. Erfani, G. Reconceptualising Sense of Place: Towards a Conceptual Framework for Investigating Individual-Community-Place Interrelationships. *J. Plan. Lit.* **2022**, *37*, 452–466. [CrossRef]
28. Cesário, V.; Acedo, A.; Nunes, N.; Nisi, V. People-Place Interactions: From Pictures and Stories to Places and Sense of Place. In *[ ] With Design: Reinventing Design Modes*; IASDR, 2021; Bruyns, G., Wei, H., Eds.; Springer: Singapore, 2022. [CrossRef]
29. Julian, J.P.; Daly, G.S.; Weaver, R.C. University Students' Social Demand of a Blue Space and the Influence of Life Experiences. *Sustainability* **2018**, *10*, 3178. [CrossRef]
30. Plieninger, T.; av Rana, H.Á.; Fagerholm, N.; Ellingsgaard, G.F.; Magnussen, E.; Raymond, C.M.; Olafsson, A.S.; Verbrugge, L.N.H. Identifying and Assessing the Potential for Conflict between Landscape Values and Development Preferences on the Faroe Islands. *Glob. Environ. Change* **2018**, *52*, 162–180. [CrossRef]
31. Escalera-Reyes, J. Place Attachment, Feeling of Belonging and Collective Identity in Socio-Ecological Systems: Study Case of Pegalajar (Andalusia-Spain). *Sustainability* **2020**, *12*, 3388. [CrossRef]
32. Vaeztavakoli, A.; Lak, A.; Yigitcanlar, T. Blue and Green Spaces as Therapeutic Landscapes: Health Effects of Urban Water Canal Areas of Isfahan. *Sustainability* **2018**, *10*, 4010. [CrossRef]
33. Wilkie, S.; Clements, H. Further Exploration of Environment Preference and Environment Type Congruence on Restoration and Perceived Restoration Potential. *Landsc. Urban Plan.* **2018**, *170*, 314–319. [CrossRef]
34. Wilkie, S.; Trotter, H. Pro-Environmental Attitudes, pro-Environmental Behaviours and Nature-Relatedness: Differences Based on Place Preference. *Eur. Rev. Appl. Psychol.* **2022**, *72*, 100705. [CrossRef]
35. Bonaiuto, M.; Bonnes, M.; Devine-Wright, P.; Raymond, C.M.; Kyttä, M.; Stedman, R. Sense of Place, Fast and Slow: The Potential Contributions of Affordance Theory to Sense of Place. *Front. Psychol.* **2017**, *8*, 1674. [CrossRef]
36. Bailey, E. A Life Course Approach to the Pluralisation of Sense of Place: Understanding the Social Acceptance of Low-Carbon Energy Developments. In *Changing Senses of Place: Navigating Global Challenges*; Di Masso, A., Raymond, C.M., Williams, D.R., Manzo, L.C., von Wirth, T., Eds.; Cambridge University Press: Cambridge, UK, 2021; pp. 156–168, ISBN 9781108477260.
37. Kyle, G.; Chick, G. The Social Construction of a Sense of Place. *Leis. Sci.* **2007**, *29*, 209–225. [CrossRef]
38. Lewicka, M. Ways to Make People Active: The Role of Place Attachment, Cultural Capital, and Neighborhood Ties. *J. Environ. Psychol.* **2005**, *25*, 381–395. [CrossRef]
39. Brown, G.; Raymond, C.M.; Corcoran, J. Mapping and Measuring Place Attachment. *Appl. Geogr.* **2015**, *57*, 42–53. [CrossRef]
40. Stokowski, P.A. Languages of Place and Discourses of Power: Constructing New Senses of Place. *J. Leis. Res.* **2002**, *34*, 368–382. [CrossRef]
41. Jansen, S.J.T. Place Attachment, Distress, Risk Perception and Coping in a Case of Earthquakes in the Netherlands. *J. Hous. Built Environ.* **2020**, *35*, 407–427. [CrossRef]
42. Kaltenborn, B.P. Recreation Homes in Natural Settings: Factors Affecting Place Attachment. *Nor. Geogr. Tidsskr.* **1997**, *51*, 187–198. [CrossRef]
43. Folmer, A.; Haartsen, T.; Huigen, P.P.P. The Role of Wildlife in Emotional Attachment to a Nature-Based Tourism Destination. *J. Ecotourism* **2013**, *12*, 131–145. [CrossRef]
44. Anton, C.E.; Lawrence, C. Home Is Where the Heart Is: The Effect of Place of Residence on Place Attachment and Community Participation. *J. Environ. Psychol.* **2014**, *40*, 451–461. [CrossRef]
45. von Wirth, T.; Grêt-Regamey, A.; Moser, C.; Stauffacher, M. Exploring the Influence of Perceived Urban Change on Residents' Place Attachment. *J. Environ. Psychol.* **2016**, *46*, 67–82. [CrossRef]
46. Saha, A.; Ramya, V.L.; Jesna, P.K.; Mol, S.S.; Panikkar, P.; Vijaykumar, M.E.; Sarkar, U.K.; Das, B.K. Evaluation of Spatio-Temporal Changes in Surface Water Quality and Their Suitability for Designated Uses, Mettur Reservoir, India. *Nat. Resour. Res.* **2021**, *30*, 1367–1394. [CrossRef]

47. Pulicharla, R.; Proulx, F.; Behmel, S.; Serodes, J.B.; Rodriguez, M.J. Occurrence and Seasonality of Raw and Drinking Water Contaminants of Emerging Interest in Five Water Facilities. *Sci. Total Environ.* **2021**, *751*, 11. [CrossRef] [PubMed]
48. Tan, X.; Liu, Y.; Burford, M.A.; Zhang, Q.F. The Performance of Diatom Indices in Assessing Temporal Changes in Water Quality in a Large Lowland River Ecosystem. *River Res. Appl.* **2021**, *37*, 423–432. [CrossRef]
49. Game, A.; Metcalfe, A. 'My Corner of the World': Bachelard and Bondi Beach. *Emot. Space Soc.* **2011**, *4*, 42–50. [CrossRef]
50. Volker, S.; Kistemann, T. The Impact of Blue Space on Human Health and Well-Being—Salutogenetic Health Effects of Inland Surface Waters: A Review. *Int. J. Hyg. Environ. Health* **2011**, *214*, 449–460. [CrossRef]
51. White, M.P.; Elliott, L.R.; Grellier, J.; Economou, T.; Bell, S.; Bratman, G.N.; Cirach, M.; Gascon, M.; Lima, M.L.; Lohmus, M.; et al. Associations between Green/Blue Spaces and Mental Health across 18 Countries. *Sci. Rep.* **2021**, *11*, 8903. [CrossRef]
52. Zhou, Y.; Yang, L.; Yu, J.; Guo, S. Do Seasons Matter? Exploring the Dynamic Link between Blue-Green Space and Mental Restoration. *Urban For. Urban Green.* **2022**, *73*, 127612. [CrossRef]
53. Ma, F. Seasonal Effects on Blue-Green Space Preferences: Examining Spatial Configuration and Residents' Perspectives. *Environ. Res. Commun.* **2023**, *5*, 65009. [CrossRef]
54. Elliott, L.R.; White, M.P.; Sarran, C.; Grellier, J.; Garrett, J.K.; Scoccimarro, E.; Smalley, A.J.; Fleming, L.E. The Effects of Meteorological Conditions and Daylight on Nature-Based Recreational Physical Activity in England. *Urban For. Urban Green.* **2019**, *42*, 39–50. [CrossRef]
55. Mason, V.; Andrews, H.; Upton, D. The Psychological Impact of Exposure to Floods. *Psychol. Health Med.* **2010**, *15*, 61–73. [CrossRef] [PubMed]
56. Vins, H.; Bell, J.; Saha, S.; Hess, J.J. The Mental Health Outcomes of Drought: A Systematic Review and Causal Process Diagram. *Int. J. Environ. Res. Public Health* **2015**, *12*, 13251–13275. [CrossRef] [PubMed]
57. Yusa, A.; Berry, P.; Cheng, J.; Ogden, N.; Bonsal, B.; Stewart, R.; Waldick, R. Climate Change, Drought and Human Health in Canada. *Int. J. Environ. Res. Public Health* **2015**, *12*, 8359–8412. [CrossRef]
58. Hassan, W.H. Climate Change Impact on Groundwater Recharge of Umm Er Radhuma Unconfined Aquifer Western Desert, Iraq. *Int. J. Hydrol. Sci. Technol.* **2020**, *10*, 392–412. [CrossRef]
59. United Nations. UN-Water Side Event: Climate-Resilient Water Management Approaches. Available online: https://www.un.org/en/desa/un-water-side-event-climate-resilient-water-management-approaches#:~:text=There%20are%20significant%20co%2Dbenefits,wetlands%20and%20increasing%20energy%20efficiency (accessed on 25 September 2023).
60. Adger, W.N.; Barnett, J.; Brown, K.; Marshall, N.; O'Brien, K. Cultural Dimensions of Climate Change Impacts and Adaptation. *Nat. Clim. Chang.* **2013**, *3*, 112–117. [CrossRef]
61. Garrett, B. Drowned Memories: The Submerged Places of the Winnemem Wintu. *Archaeologies* **2010**, *6*, 346–371. [CrossRef]
62. Kirsch, S. Lost Worlds: Environmental Disaster, "Culture Loss," and the Law. *Curr. Anthropol.* **2001**, *42*, 167–198. [CrossRef]
63. Dew, A.; Barton, R.; Gilroy, J.; Ryall, L.; Lincoln, M.; Jensen, H.; Flood, V.; Taylor, K.; McCrae, K. Importance of Land, Family and Culture for a Good Life: Remote Aboriginal People with Disability and Carers. *Aust. J. Soc. Issues* **2020**, *55*, 418–438. [CrossRef]
64. Albrecht, G.; Sartore, G.-M.; Connor, L.; Higginbotham, N.; Freeman, S.; Kelly, B.; Stain, H.; Tonna, A.; Pollard, G. Solastalgia: The Distress Caused by Environmental Change. *Australas. Psychiatry* **2007**, *15*, S95–S98. [CrossRef] [PubMed]
65. Patel, A.I.; Schmidt, L.A.; Lee, P.R. Water Access in the United States: Health Disparities Abound and Solutions Are Urgently Needed. *Am. J. Public Health* **2017**, *107*, 1354. [CrossRef]
66. Abedin, M.A.; Collins, A.E.; Habiba, U.; Shaw, R. Climate Change, Water Scarcity, and Health Adaptation in Southwestern Coastal Bangladesh. *Int. J. Disaster Risk Sci.* **2019**, *10*, 28–42. [CrossRef]
67. Cunsolo Willox, A.; Harper, S.L.; Ford, J.D.; Landman, K.; Houle, K.; Edge, V.L. "From This Place and of This Place:" Climate Change, Sense of Place, and Health in Nunatsiavut, Canada. *Soc. Sci. Med.* **2012**, *75*, 538–547. [CrossRef]
68. Satariano, B. Recreating a Therapeutic Blue Urban Space through the Architectural Restoration of the Triton Fountain in Valletta, Malta. *Cities Health* **2022**, *6*, 1094–1105. [CrossRef]
69. Wang, R.; Browning, M.H.E.M.; Kee, F.; Hunter, R.F. Exploring Mechanistic Pathways Linking Urban Green and Blue Space to Mental Wellbeing before and after Urban Regeneration of a Greenway: Evidence from the Connswater Community Greenway, Belfast, UK. *Landsc. Urban Plan.* **2023**, *235*, 104739. [CrossRef]
70. Lewis, C. 'Regenerating Community'? Urban Change and Narratives of the Past. *Sociol. Rev.* **2016**, *64*, 912–928. [CrossRef]
71. Torres, S. "For a Younger Crowd": Place, Belonging, and Exclusion among Older Adults Facing Neighborhood Change. *Qual. Sociol.* **2020**, *43*, 1–20. [CrossRef]
72. Ali, L.; Haase, A.; Heiland, S. Gentrification through Green Regeneration? Analyzing the Interaction between Inner-City Green Space Development and Neighborhood Change in the Context of Regrowth: The Case of Lene-Voigt-Park in Leipzig, Eastern Germany. *Land* **2020**, *9*, 24. [CrossRef]
73. Rigolon, A.; Németh, J. "We're Not in the Business of Housing:" Environmental Gentrification and the Nonprofitization of Green Infrastructure Projects. *Cities* **2018**, *81*, 71–80. [CrossRef]
74. Rigolon, A.; Németh, J. What Shapes Uneven Access to Urban Amenities? Thick Injustice and the Legacy of Racial Discrimination in Denver's Parks. *J. Plan. Educ. Res.* **2021**, *41*, 312–325. [CrossRef]
75. Jelks, N.O.; Jennings, V.; Rigolon, A. Green Gentrification and Health: A Scoping Review. *Int. J. Environ. Res. Public Health* **2021**, *18*, 907. [CrossRef] [PubMed]

76. Kabisch, N.; Haase, D.; Annerstedt van den Bosch, M. Adding Natural Areas to Social Indicators of Intra-Urban Health Inequalities among Children: A Case Study from Berlin, Germany. *Int. J. Environ. Res. Public Health* **2016**, *13*, 783. [CrossRef] [PubMed]
77. Natarajan, S.K.; Hagare, D.; Maheshwari, B. Understanding Socio-Economic Benefits of Stormwater Management System through Urban Lakes in Western Sydney, Australia. *Ecohydrol. Hydrobiol.* **2018**, *18*, 412–419. [CrossRef]
78. Tieges, Z.; McGregor, D.; Georgiou, M.; Smith, N.; Saunders, J.; Millar, R.; Morison, G.; Chastin, S. The Impact of Regeneration and Climate Adaptations of Urban Green-Blue Assets on All-Cause Mortality: A 17-Year Longitudinal Study. *Int. J. Environ. Res. Public Health* **2020**, *17*, 4577. [CrossRef] [PubMed]
79. McDougall, C.W.; Hanley, N.; Quilliam, R.S.; Oliver, D.M. Blue Space Exposure, Health and Well-Being: Does Freshwater Type Matter? *Landsc. Urban Plan.* **2022**, *224*, 104446. [CrossRef]
80. Coleman, T.; Kearns, R. The Role of Bluespaces in Experiencing Place, Aging and Wellbeing: Insights from Waiheke Island, New Zealand. *Health Place* **2015**, *35*, 206–217. [CrossRef]
81. Jewkes, Y.; Moran, D.; Turner, J. Just Add Water: Prisons, Therapeutic Landscapes and Healthy Blue Space. *Criminol. Crim. Justice* **2020**, *20*, 381–398. [CrossRef]
82. Fox, N.; Marshall, J.; Dankel, D.J. Ocean Literacy and Surfing: Understanding How Interactions in Coastal Ecosystems Inform Blue Space User's Awareness of the Ocean. *Int. J. Environ. Res. Public Health* **2021**, *18*, 5819. [CrossRef]
83. White, M.; Smith, A.; Humphryes, K.; Pahl, S.; Snelling, D.; Depledge, M. Blue Space: The Importance of Water for Preference, Affect, and Restorativeness Ratings of Natural and Built Scenes. *J. Environ. Psychol.* **2010**, *30*, 482–493. [CrossRef]
84. White, M.P.; Cracknell, D.; Corcoran, A.; Jenkinson, G.; Depledge, M.H. Do Preferences for Waterscapes Persist in Inclement Weather and Extend to Sub-Aquatic Scenes? *Landsc. Res.* **2014**, *39*, 339–358. [CrossRef]
85. Grace, M.J.; Dickie, J.; Bartie, P.J.; Brown, C.; Oliver, D.M. How Do Weather Conditions and Environmental Characteristics Influence Aesthetic Preferences of Freshwater Environments? *Sci. Total Environ.* **2023**, *903*, 166283. [CrossRef] [PubMed]
86. Mishra, H.S.; Bell, S.; Roberts, B.; White, M.P. Theory-based design for promoting positive behaviours in an urban blue space: Pre-and-post observations of a community co-created intervention in Plymouth, United Kingdom. *Landsc. Urban Plan.* **2023**, *233*, 104708. [CrossRef]
87. Nieuwenhuijsen, M.J.; Gascon, M.; Martinez, D.; Ponjoan, A.; Blanch, J.; Garcia-Gil, M.D.M.; Ramos, R.; Foraster, M.; Mueller, N.; Espinosa, A.; et al. Air Pollution, Noise, Blue Space, and Green Space and Premature Mortality in Barcelona: A Mega Cohort. *Int. J. Environ. Res. Public Health* **2018**, *15*, 2405. [CrossRef]
88. Rousseau, S.; Deschacht, N. Public Awareness of Nature and the Environment During the COVID-19 Crisis. *Environ. Resour. Econ.* **2020**, *76*, 1149–1159. [CrossRef]
89. Jo, T.; Sato, M.; Minamoto, T.; Ushimaru, A. Valuing the Cultural Services from Urban Blue-Space Ecosystems in Japanese Megacities during the COVID-19 Pandemic. *People Nat.* **2022**, *4*, 1176–1189. [CrossRef]
90. Earl, J.; Gormally-Sutton, A.; Ilic, S.; James, M.R. 'Best Day since the Bad Germs Came': Exploring Changing Experiences in and the Value of Coastal Blue Space during the COVID-19 Pandemic, a Fylde Coast Case Study. *Coast. Stud. Soc.* **2022**, *1*, 97–119. [CrossRef]
91. Jellard, S.; Bell, S.L. A Fragmented Sense of Home: Reconfiguring Therapeutic Coastal Encounters in COVID-19 Times. *Emot. Space Soc.* **2021**, *40*, 100818. [CrossRef]
92. Usher, L.E. "Without [Surfing], It's Kind of like Groundhog Day, but the . . . Twilight Zone Version": A Phenomenological Study of Surfers' Experiences of the COVID-19 Pandemic. *Leis. Loisir* **2023**, *47*, 1–37. [CrossRef]
93. Wuijts, S.; Friederichs, L.; Hin, J.A.; Schets, F.M.; Van Rijswick, H.F.M.W.; Driessen, P.P.J. Governance Conditions to Overcome the Challenges of Realizing Safe Urban Bathing Water Sites. *Int. J. Water Resour. Dev.* **2020**, *38*, 554–578. [CrossRef]
94. European Environment Agency. European Bathing Water Quality in 2019. Available online: https://www.eea.europa.eu/publications/european-bathing-water-quality-in-2019/european-bathing-water-quality-in-2019 (accessed on 25 September 2023).
95. European Union. *Directive 2006/7/EC of the European Parliament and of the Council of 15 February 2006 Concerning the Management of Bathing Water Quality and Repealing Directive 76/160/EEC*; Official Journal of the European Union: Brussels, Belgium, 2006; Volume 4.
96. Borger, T.; Campbell, D.; White, M.P.; Elliott, L.R.; Fleming, L.E.; Garrett, J.K.; Hattam, C.; Hynes, S.; Lankia, T.; Taylor, T. The Value of Blue-Space Recreation and Perceived Water Quality across Europe: A Contingent Behaviour Study. *Sci. Total Environ.* **2021**, *771*, 145597. [CrossRef] [PubMed]
97. Pouso, S.; Uyarra, M.C.; Borja, Á. The Recovery of Estuarine Quality and the Perceived Increase of Cultural Ecosystem Services by Beach Users: A Case Study from Northern Spain. *J. Environ. Manag.* **2018**, *212*, 450–461. [CrossRef] [PubMed]
98. Bonetti, F.; Corsi, S.; Orsi, L.; De Noni, I. Canals vs. Streams: To What Extent Do Water Quality and Proximity Affect Real Estate Values? A Hedonic Approach Analysis. *Water* **2016**, *8*, 577. [CrossRef]
99. Hipp, J.A.; Ogunseitan, O.A. Effect of Environmental Conditions on Perceived Psychological Restorativeness of Coastal Parks. *J. Environ. Psychol.* **2011**, *31*, 421–429. [CrossRef]
100. Ghermandi, A.; Nunes, P.; Portela, R.; Rao, N.; Teelucksingh, S.S. Recreational, Cultural, and Aesthetic Services from Estuarine and Coastal Ecosystems. In *Treatise on Estuarine and Coastal Science, Vol 12: Ecological Economics of Estuaries and Coasts*; Wolanski, E., McLusky, D., Eds.; Elsevier Academic Press, Inc.: San Diego, CA, USA, 2011; pp. 217–237, ISBN 978-0-08-087885-0.

101. Georgiou, S.; Bateman, I.J. Revision of the EU Bathing Water Directive: Economic Costs and Benefits. *Mar. Pollut. Bull.* **2005**, *50*, 430–438. [CrossRef] [PubMed]
102. Small, N.; Munday, M.; Durance, I. The Challenge of Valuing Ecosystem Services That Have No Material Benefits. *Glob. Environ. Change* **2017**, *44*, 57–67. [CrossRef]
103. Ghermandi, A.; Camacho-Valdez, V.; Trejo-Espinosa, H. Social Media-Based Analysis of Cultural Ecosystem Services and Heritage Tourism in a Coastal Region of Mexico. *Tour. Manag.* **2020**, *77*, 104002. [CrossRef]
104. Retka, J.; Jepson, P.; Ladle, R.J.; Malhado, A.C.M.; Vieira, F.A.S.; Normande, I.C.; Souza, C.N.; Bragagnolo, C.; Correia, R.A. Assessing Cultural Ecosystem Services of a Large Marine Protected Area through Social Media Photographs. *Ocean Coast. Manag.* **2019**, *176*, 40–48. [CrossRef]
105. Richards, D.R.; Friess, D.A. A Rapid Indicator of Cultural Ecosystem Service Usage at a Fine Spatial Scale: Content Analysis of Social Media Photographs. *Ecol. Indic.* **2015**, *53*, 187–195. [CrossRef]
106. Zhang, H.; Huang, R.; Zhang, Y.; Buhalis, D. Cultural Ecosystem Services Evaluation Using Geolocated Social Media Data: A Review. *Tour. Geogr.* **2022**, *24*, 646–668. [CrossRef]
107. Oteros-Rozas, E.; Martín-López, B.; Fagerholm, N.; Bieling, C.; Plieninger, T. Using Social Media Photos to Explore the Relation between Cultural Ecosystem Services and Landscape Features across Five European Sites. *Ecol. Indic.* **2018**, *94*, 74–86. [CrossRef]
108. Barry, S.J. Using Social Media to Discover Public Values, Interests, and Perceptions about Cattle Grazing on Park Lands. *Environ. Manag.* **2014**, *53*, 454–464. [CrossRef] [PubMed]
109. Guerrero, P.; Møller, M.S.; Olafsson, A.S.; Snizek, B. Revealing Cultural Ecosystem Services through Instagram Images: The Potential of Social Media Volunteered Geographic Information for Urban Green Infrastructure Planning and Governance. *Urban Plan.* **2016**, *1*, 2. [CrossRef]
110. Qiu, L.; Chan, S.H.M.; Chan, D. Big Data in Social and Psychological Science: Theoretical and Methodological Issues. *J. Comput. Soc. Sci.* **2018**, *1*, 59–66. [CrossRef]
111. Conradson, D. Landscape, Care and the Relational Self: Therapeutic Encounters in Rural England. *Health Place* **2005**, *11*, 337–348. [CrossRef]
112. Bell, S.L.; Phoenix, C.; Lovell, R.; Wheeler, B.W. Using GPS and Geo-Narratives: A Methodological Approach for Understanding and Situating Everyday Green Space Encounters. *Area* **2015**, *47*, 88–96. [CrossRef]
113. Huynh, Q.; Craig, W.; Janssen, I.; Pickett, W. Exposure to Public Natural Space as a Protective Factor for Emotional Well-Being among Young People in Canada. *BMC Public Health* **2013**, *13*, 407. [CrossRef]
114. Dzhambov, A.; Hartig, T.; Markevych, I.; Tilov, B.; Dimitrova, D. Urban Residential Greenspace and Mental Health in Youth: Different Approaches to Testing Multiple Pathways Yield Different Conclusions. *Environ. Res.* **2018**, *160*, 47–59. [CrossRef]
115. Oostenbach, L.H.; Noall, J.; Lamb, K.E.; Pearson, A.L.; Mavoa, S.; Thornton, L.E. Associations between Coastal Proximity and Children's Mental Health in Australia. *Geogr. Res.* **2023**, *61*, 248–258. [CrossRef]
116. George, P.; Murray, K.; Christian, H. Young Children's Use of Blue Spaces and the Impact on Their Health, Development and Environmental Awareness: A Qualitative Study from Parents' Perspectives. *Health Promot. J. Aust.* 2023; *in press*. [CrossRef]
117. Costello, L.; McDermott, M.L.; Patel, P.; Dare, J. 'A Lot Better than Medicine'—Self-Organised Ocean Swimming Groups as Facilitators for Healthy Ageing. *Health Place* **2019**, *60*, 102212. [CrossRef]
118. Thadhani, R.; Tonelli, M. Cohort Studies: Marching Forward. *Clin. J. Am. Soc. Nephrol.* **2006**, *1*, 1117–1123. [CrossRef] [PubMed]
119. Roberts, H.; van Lissa, C.; Helbich, M. Perceived Neighbourhood Characteristics and Depressive Symptoms: Potential Mediators and the Moderating Role of Employment Status. *Soc. Sci. Med.* **2021**, *268*, 113533. [CrossRef]
120. Zijlema, W.L.; Avila-Palencia, I.; Triguero-Mas, M.; Gidlow, C.; Maas, J.; Kruize, H.; Andrusaityte, S.; Grazuleviciene, R.; Nieuwenhuijsen, M.J. Active Commuting through Natural Environments Is Associated with Better Mental Health: Results from the PHENOTYPE Project. *Environ. Int.* **2018**, *121*, 721–727. [CrossRef] [PubMed]
121. Marquart, H.; Stark, K.; Jarass, J. How Are Air Pollution and Noise Perceived En Route? Investigating Cyclists' and Pedestrians' Personal Exposure, Wellbeing and Practices during Commute. *J. Transp. Health* **2022**, *24*, 101325. [CrossRef]
122. Sander, H.A.; Zhao, C. Urban Green and Blue: Who Values What and Where? *Land Use Policy* **2015**, *42*, 194–209. [CrossRef]
123. Thornhill, I.; Hill, M.J.; Castro-Castellon, A.; Gurung, H.; Hobbs, S.; Pineda-Vazquez, M.; Gómez-Osorio, M.T.; Hernández-Avilés, J.S.; Novo, P.; Mesa-Jurado, A.; et al. Blue-Space Availability, Environmental Quality and Amenity Use across Contrasting Socioeconomic Contexts. *Appl. Geogr.* **2022**, *144*, 102716. [CrossRef]
124. Miró, A.; Hall, J.; Rae, M.; O'Brien, D. Links between Ecological and Human Wealth in Drainage Ponds in a Fast-Expanding City, and Proposals for Design and Management. *Landsc. Urban Plan.* **2018**, *180*, 93–102. [CrossRef]
125. Viinikka, A.; Paloniemi, R.; Assmuth, T. Mapping the Distributive Environmental Justice of Urban Waters. *Fenn. Int. J. Geogr.* **2018**, *196*, 9–23. [CrossRef]
126. Williamson, I.; Leeming, D.; Lyttle, S.; Johnson, S. Evaluating the Audio-Diary Method in Qualitative Research. *Qual. Res. J.* **2015**, *15*, 20–34. [CrossRef]
127. Hyers, L.L. *Diary Methods: Understanding Qualitative Methods*; Oxford University Press: Oxford, UK, 2018; ISBN 978-0190256692.
128. Whittle, R.; Medd, W.; Deeming, H.; Kashefi, E.; Mort, M.; Walker, G.; Watson, N. After the Rain—Learning the Lessons from Flood Recovery in Hull. Final Project Report for "Flood, Vulnerability and Urban Resilience: A Real-Time Study of Local Recovery Following the Floods of June 2007 in Hull". 2010. Available online: https://nrl.northumbria.ac.uk/id/eprint/12165/ (accessed on 1 October 2023).

129. Grace, M.; Dickie, J.; Brown, C.; Bartie, P.; Oliver, D.M. Using Solicited Research Diaries to Assess the Restorative Potential of Exposure to Inland Blue Space across Time. *Landsc. Urban Plan.* **2024**, *241*, 104904. [CrossRef]
130. Francis, R.A.; Lorimer, J.; Raco, M. What Is Special about Urban Ecologies? *Trans. Inst. Br. Geogr.* **2013**, *38*, 682–684. [CrossRef]
131. Urso, G. Metropolisation and the Challenge of Rural-Urban Dichotomies. *Urban Geogr.* **2020**, *42*, 37–57. [CrossRef]
132. Merrifield, A. The Right to the City and Beyond. *City* **2011**, *15*, 473–481. [CrossRef]
133. Bell, M.M. The Fruit of Difference: The Rural-Urban Continuum as a System of Identity1. *Rural Sociol.* **1992**, *57*, 65–82. [CrossRef]
134. Neal, S.; Walters, S. Strangers Asking Strange Questions? A Methodological Narrative of Researching Belonging and Identity in English Rural Communities. *J. Rural Stud.* **2006**, *22*, 177–189. [CrossRef]
135. Belanche, D.; Casaló, L.V.; Rubio, M.Á. Local Place Identity: A Comparison between Residents of Rural and Urban Communities. *J. Rural Stud.* **2021**, *82*, 242–252. [CrossRef]
136. Smith, N.; Georgiou, M.; King, A.C.; Tieges, Z.; Chastin, S. Factors Influencing Usage of Urban Blue Spaces: A Systems-Based Approach to Identify Leverage Points. *Health Place* **2022**, *73*, 102735. [CrossRef]
137. Fisher, J.C.; Irvine, K.N.; Bicknell, J.E.; Hayes, W.M.; Fernandes, D.; Mistry, J.; Davies, Z.G. Perceived Biodiversity, Sound, Naturalness and Safety Enhance the Restorative Quality and Wellbeing Benefits of Green and Blue Space in a Neotropical City. *Sci. Total Environ.* **2021**, *755*, 13. [CrossRef]
138. Duke, L.; Soulsbury, C.D. Human-Wildlife Interaction Networks at Urban Blue Spaces. *Area* **2021**, *53*, 122–133. [CrossRef]
139. Sherry, E.; Shortall, S. Methodological Fallacies and Perceptions of Rural Disparity: How Rural Proofing Addresses Real versus Abstract Needs. *J. Rural Stud.* **2019**, *68*, 336–343. [CrossRef]
140. Duvernoy, I.; Gambino, M. Where to Find Nature? Connecting with Nature in Urban and Non-Urban Areas in Southwestern France. *GeoJournal* **2021**, *87*, 2771–2786. [CrossRef] [PubMed]
141. Wolff, M.; Scheuer, S.; Haase, D. Looking beyond Boundaries: Revisiting the Rural-Urban Interface of Green Space Accessibility in Europe. *Ecol. Indic.* **2020**, *113*, 106245. [CrossRef]
142. Fletcher, D.H.; Likongwe, P.J.; Chiotha, S.S.; Nduwayezu, G.; Mallick, D.; Uddin Md, N.; Rahman, A.; Golovátina-Mora, P.; Lotero, L.; Bricker, S.; et al. Using Demand Mapping to Assess the Benefits of Urban Green and Blue Space in Cities from Four Continents. *Sci. Total Environ.* **2021**, *785*, 147238. [CrossRef] [PubMed]
143. Jelks, N.O.; Smith-Perry, D.J.; Fuller, C.H.; Stauber, C. Participatory Research in Northwest Atlanta's Proctor Creek Watershed: Using Photovoice to Explore Environmental Health Risks at the Water's Edge. *Health Place* **2020**, *66*, 102444. [CrossRef] [PubMed]
144. Scottish Government Urban Rural Classification 2016—Gov.Scot. Available online: https://www.gov.scot/publications/scottish-government-urban-rural-classification-2016/pages/6/ (accessed on 4 September 2023).
145. Layer: Regional Boundaries—Forestry and Land Scotland (ID:0). Available online: https://services6.arcgis.com/JF5BD4Y14a3LnPcJ/arcgis/rest/services/Regional_Boundaries_Forestry_and_Land_Scotland/FeatureServer/0 (accessed on 4 September 2023).
146. Olive, R.; Wheaton, B. Understanding Blue Spaces: Sport, Bodies, Wellbeing, and the Sea. *J. Sport Soc. Issues* **2020**, *45*, 3–19. [CrossRef]
147. Wheaton, B.; Waiti, J.; Cosgriff, M.; Burrows, L. Coastal Blue Space and Wellbeing Research: Looking beyond Western Tides. *Leis. Stud.* **2020**, *39*, 83–95. [CrossRef]
148. Gould, R.K.; Adams, A.; Vivanco, L. Looking into the Dragons of Cultural Ecosystem Services. *Ecosyst. People* **2020**, *16*, 257–272. [CrossRef]
149. Liu, L. Paddling Through Bluespaces: Understanding Waka Ama as a Post-Sport Through Indigenous Māori Perspectives. *J. Sport Soc. Issues* **2020**, *45*, 138–160. [CrossRef]
150. Wheaton, B.; Waiti, J.T.; Olive, R.; Kearns, R. Coastal Communities, Leisure and Wellbeing: Advancing a Trans-Disciplinary Agenda for Understanding Ocean-Human Relationships in Aotearoa New Zealand. *Int. J. Environ. Res. Public Health* **2021**, *18*, 450. [CrossRef]
151. McDougall, C.W.; Quilliam, R.S.; Hanley, N.; Oliver, D.M. Freshwater Blue Space and Population Health: An Emerging Research Agenda. *Sci. Total Environ.* **2020**, *737*, 140196. [CrossRef]
152. Joshanloo, M. Eastern Conceptualizations of Happiness: Fundamental Differences with Western Views. *J. Happiness Stud.* **2014**, *15*, 475–493. [CrossRef]
153. Exenberger, S.; Banzer, R.; Christy, J.; Höfer, S.; Juen, B. Eastern and Western Children's Voices on Their Well-Being. *Child. Indic. Res.* **2019**, *12*, 747–768. [CrossRef]
154. Tariq, S.; Woodman, J. Using Mixed Methods in Health Research. *JRSM Short. Rep.* **2013**, *4*, 2042533313479197. [CrossRef] [PubMed]
155. Regnault, A.; Willgoss, T.; Barbic, S.; On behalf of the International Society for Quality of Life Research Mixed Methods Special Interest Group. Towards the Use of Mixed Methods Inquiry as Best Practice in Health Outcomes Research. *J. Patient Rep. Outcomes* **2018**, *2*, 19. [CrossRef]
156. Fisher, D.; Blackstock, K.; Irvine, K. "It's on the 'nice to Have' Pile": Potential Principles to Improve the Implementation of Socially Inclusive Green Infrastructure. *Ambio* **2021**, *50*, 1574–1586. [CrossRef]
157. Nagata, M.; Liehr, P. Urban Children's Well-Being Factors and Qualities of Being and Doing in Natural Space: Nature Immersion. *J. Holist. Nurs.* **2020**, *39*, 174–184. [CrossRef]

158. Toews, B.; Wagenfeld, A.; Stevens, J.; Shoemaker, C. Feeling at Home in Nature: A Mixed Method Study of the Impact of Visitor Activities and Preferences in a Prison Visiting Room Garden. *J. Offender Rehabil.* **2020**, *59*, 223–246. [CrossRef]
159. Bojorquez, I.; Romo-Aguilar, M.D.L.; Ojeda-Revah, L.; Tena, F.; Lara-Valencia, F.; García, H.; Díaz, R.; Aranda, P. Public Spaces and Physical Activity in Adults: Insights from a Mixed methods Study. *Cad. Saúde Pública* **2021**, *37*, 1. [CrossRef]
160. Jones, P.; Evans, J. The Spatial Transcript: Analysing Mobilities through Qualitative GIS. *Area* **2012**, *44*, 92–99. [CrossRef]
161. Greene, J.C.; Caracelli, V.J.; Graham, W.F. Toward a Conceptual Framework for Mixed method Evaluation Designs. *Educ. Eval. Policy Anal.* **1989**, *11*, 255–274. [CrossRef]
162. Bracken, L.J.; Bulkeley, H.A.; Whitman, G. Transdisciplinary Research: Understanding the Stakeholder Perspective. *J. Environ. Plan. Manag.* **2015**, *58*, 1291–1308. [CrossRef]
163. Thompson, M.A.; Owen, S.; Lindsay, J.M.; Leonard, G.S.; Cronin, S.J. Scientist and Stakeholder Perspectives of Transdisciplinary Research: Early Attitudes, Expectations, and Tensions. *Environ. Sci. Policy* **2017**, *74*, 30–39. [CrossRef]
164. Mishra, H.S.; Bell, S.; Grellier, J.; White, M.P. Testing the Reliability and Effectiveness of a New Tool for Assessing Urban Blue Spaces: The BlueHealth Environmental Assessment Tool (BEAT). *Health Place* **2021**, *68*, 102526. [CrossRef] [PubMed]
165. Vert, C.; Gascon, M.; Ranzani, O.; Márquez, S.; Triguero-Mas, M.; Carrasco-Turigas, G.; Arjona, L.; Koch, S.; Llopis, M.; Donaire-Gonzalez, D.; et al. Physical and Mental Health Effects of Repeated Short Walks in a Blue Space Environment: A Randomised Crossover Study. *Environ. Res.* **2020**, *188*, 109812. [CrossRef]
166. Desjardins, M.R.; Murray, E.T.; Baranyi, G.; Hobbs, M.; Curtis, S. Improving Longitudinal Research in Geospatial Health: An Agenda. *Health Place* **2023**, *80*, 102994. [CrossRef]
167. Patton, M.Q. Enhancing the Quality and Credibility of Qualitative Analysis. *Health Serv. Res.* **1999**, *34*, 1189.
168. Stevenson, K.; Girak, S. Creative River Journeys: Using Reflective Practice within a Practice-Led Research Context. 2013. Available online: https://ro.ecu.edu.au/ecuworks2013/270/ (accessed on 1 October 2023).
169. The Dee and Don Ceilidh Collective Our Living Rivers and Glens. Available online: https://deedonceilidhcollective.com/our-living-rivers-and-glens (accessed on 25 September 2023).
170. McEwen, L.; Gorell Barnes, L.; Phillips, K.; Biggs, I. Reweaving Urban Water-Community Relations: Creative, Participatory River "Daylighting" and Local Hydrocitizenship. *Trans. Inst. Br. Geogr.* **2020**, *45*, 779–801. [CrossRef]
171. Scott-Bottoms, S.; Roe, M. Who Is a Hydrocitizen? The Use of Dialogic Arts Methods as a Research Tool with Water Professionals in West Yorkshire, UK. *Local Environ.* **2020**, *25*, 273–289. [CrossRef]

MDPI AG
Grosspeteranlage 5
4052 Basel
Switzerland
Tel.: +41 61 683 77 34

MDPI Books Editorial Office
E-mail: books@mdpi.com
www.mdpi.com/books

www.ingramcontent.com/pod-product-compliance
Lightning Source LLC
Chambersburg PA
CBHW041939260926
51850CB00018B/298
* 9 7 8 3 7 2 5 8 5 4 7 7 6 *